Biochemical Plant Pathology

Biochemical Plant Pathology

Biochemical Plant Pathology

Edited by

J. A. Callow
Department of Plant Biology
University of Birmingham, UK

A Wiley–Interscience Publication

JOHN WILEY & SONS
Chichester · New York · Brisbane · Toronto · Singapore

Library of Congress Cataloging in Publication Data:

Main entry under title:
Biochemical plant pathology.
   A Wiley–Interscience publication.
   Includes index
   1. Plant diseases.   2. Botanical chemistry.
3. Microbiological chemistry.   I. Callow, J. A.
SB731.B56   1983   581.2′192   82-19963

ISBN 0 471 90092 3

British Library Cataloguing in Publication Data:

Biochemical plant pathology.
   1. Botanical chemistry—Congresses
   I. Callow, J. A.
   581.19   QK861

ISBN 0 471 90092 3

Typeset by Preface Ltd, Salisbury Wiltshire
Printed by BAS Printers Limited, Over Wallop, Stockbridge, Hampshire

Contributors

BRENT, K. J. *Long Ashton Research Station, University of Bristol, Long Ashton, Bristol BS18 9AF, UK*

BUCHANAN, R. B. *Division of Molecular Plant Biology, University of California, Berkeley, CA 94720, USA*

CALLOW, J. A. *Department of Plant Biology, University of Birmingham, Birmingham B15 2TD, UK*

CLARKE, D. D. *Department of Botany, The University, Glasgow G12 8QQ, UK*

COFFEY, M. D. *Department of Plant Pathology, University of California, Riverside, CA 92521, USA*

COOPER, R. M. *School of Biological Sciences, University of Bath, Claverton Down, Bath, Avon, UK*

DRUMMOND, M. *ARC Unit of Nitrogen Fixation, University of Sussex, Falmer, Brighton BN1 9RQ, Sussex, UK*

DURBIN, R. D. *Plant Disease Resistance Research Unit, ARS, USDA, and Department of Plant Pathology, University of Wisconsin—Madison, Madison, WI 53706, USA*

ELSTNER, E. F. *Institut für Botanik und Mikrobiologie, Biochemisches Labor, Technische Universität, Arcisstrasse 21, 8000 München 2, West Germany*

GAY, J. L. *Department of Pure and Applied Biology, Imperial College of Science and Technology, Prince Consort Road, London SW7 2BB, UK*

GOODMAN, R. N. *Department of Plant Pathology, University of Missouri, Columbia, MO 65211, USA*

HUTCHESON, S. W. *Division of Molecular Plant Biology, University of California, Berkeley, CA 94720, USA*

KOLATTUKUDY, P. E.   *Institute of Biological Chemistry and Biochemistry/ Biophysics Program, Washington State University, Pullman, WA 99164, USA*

KÖLLER, W.   *Institute of Biological Chemistry and Biochemistry/ Biophysics Program, Washington State University, Pullman, WA 99164, USA*

LEGRAND, M.   *Laboratoire de Virologie, Institut de Biologie Moléculaire et Cellulaire, 15 rue Descartes, 67000 Strasbourg, France*

MANNERS, J. M.   *Department of Pure and Applied Biology, Imperial College of Science and Technology, Prince Consort Road, London SW7 2BB, UK*

MANSFIELD, J. W.   *Biological Sciences Department, Wye College (University of London), Ashford, Kent, UK*

MAZZUCCHI, U.   *Istituto di Patologia Vegetale, Università di Bologna, 40126 Bologna, Italy*

NOVACKY, A.   *Department of Plant Pathology, University of Missouri, Columbia, MO 65211, USA*

PAXTON, J. D.   *Department of Plant Pathology, University of Illinois, 1102 S. Goodwin Avenue, N-519 Turner Hall, Urbana, IL 61801, USA*

RIDE, J. P.   *Department of Microbiology, University of Birmingham, P.O. Box 363, Birmingham B15 2TT, UK*

STASKAWICZ, B. J.   *International Plant Research Institute, San Carlos, CA 94070, USA*

VAN LOON, L. C.   *Department of Plant Physiology, Agricultural University, 6703 BD Wageningen, The Netherlands*

YOSHIKAWA, M.   *Laboratory of Plant Pathology, Faculty of Agriculture, Kyoto Prefectural University, Kyoto 606, Japan*

Contents

IV EFFECTS ON HOST METABOLISM

V EPILOGUE

Preface

The study of biochemical interactions between plants and pathogenic microorganisms is rapidly expanding. This surge of interest is partly due to developments in techniques and experimental approaches which facilitate exploration, at the molecular level, of the principles which govern the intimate relationship between two, otherwise independent, organisms. If we have a more complete understanding of natural mechanisms of resistance and pathogenesis, then we might be able to control resistance in crops. Yet, plant breeding for resistance is presently carried out against a background in which the function of not one single gene for resistance is understood at the molecular level, and we do not yet have a molecular description of the key events in any plant disease comparable to that of certain animal diseases. We need to know for example, what the primary products of resistance genes are, what processes they control in the plant, how they interact with products of the pathogen, and where they are localized in the cell. Such a detailed appreciation may be far off but the recent advances in molecular genetics now permit, at least in theory, the isolation, characterization and cloning of resistance genes, and characterization of their products. In addition, there is a growing realization that the development of *in vitro* technologies and somatic cell genetics, based in physiology and biochemistry, offer radical new opportunities for creating disease resistant plants.

Many undergraduate schemes in biology and agricultural science include advanced courses in plant pathology, increasingly with some physiological and biochemical component. This book is therefore aimed at providing a biochemical perspective to such courses at final year degree level, but in addition, should be of considerable help to those commencing postgraduate work. Plant pathology is also a fertile field for talented individuals from other disciplines, and there is a particular need to attract scientists with specific training in biochemistry and genetics. For those students taking advanced courses in such disciplines, the study of the diseased plant offers a number of fascinating insights into fundamental problems such as metabolic regulation, the nature of surface recognition systems, gene expression, mode of action of toxins, compartmentation and assimilate partition, to name but a few. It is hoped therefore that the book may stimulate the University lecturer to spare

ix

some thought to the diseased plant when designing courses in plant biochemistry or plant genetics.

Concerning the format of the book, international authorities were asked to prepare critical, stimulating accounts that reflect the currently exciting prospects in the field of biochemical pathology, but which do not ignore basic principles. It was not the intention merely to provide a collection of exhaustively referenced research reviews. The section on Case Studies was included to provide the biochemical reader with an insight into the broader aspects of the biology of a few selected systems that for various reasons are of special significance. Thus, we probably know more about the general pathology, genetics and physiology of potato blight than any other system. The interaction between *Phytophthora megasperma* f.sp. *glycinea* and soybean is proving especially significant in studies on the biochemistry of resistance. The choice of a disease caused by a biotrophic fungal parasite was difficult but the genetics of flax rust resistance are extremely well understood, considerable progress is being made in the molecular genetics of flax, and in addition, there have been detailed studies on the ultrastructure and physiology of the host–parasite interaction. *Agrobacterium tumefaciens* (crown gall bacterium) is a unique pathogen, both in its interactions with plants, and its immense potential for genetic engineering. Finally, *Erwinia amylovora* (fireblight of apples and pears) provides an example of a contrasting necrotrophic bacterium. It may be objected that viruses have not been included as a specific chapter. This was a difficult decision balanced between the desire to be comprehensive and the pressures on volume size. In fact, a number of chapters do include important aspects of virus biochemistry, most notably Chapter 18, which deals in large part with virus replication, Chapter 17 which considers changes in phenylpropanoid metabolism induced by both viruses and fungi, and Chapter 15 which examines some of the consequences of viral infection on primary metabolism.

Section II is devoted to the processes of infection and pathogenesis, including chapters on the biochemistry of cuticle and cell wall degradation by enzymes, fungal and bacterial toxins and their molecular action, and a detailed consideration of the host–parasite interface and its role in regulating nutrient transfer.

Section III is concerned with specificity and resistance and starts with a detailed exposé of the strategies of modern molecular genetics and their application to the solution of fundamental problems in biochemical plant pathology, a field which will surely burgeon in the near future. Another very active field is concerned with the processes of host–pathogen recognition and resistance triggering. Plant defence is considered both in terms of the proposed mechanisms by which the growth of the parasite is arrested, through wall synthesis, phytoalexin accumulation, immobilization, etc., and in terms of the recognition and triggering mechanisms that may be responsible for the invocation of resistance, or in the case of mutualistic bacteria, the acceptance of the symbiont.

Section IV examines the effects of infection on host photosynthetic and respiratory metabolism, secondary metabolism, RNA transcription and translation (including viruses), the structure and activity of membranes in the diseased plant, and hormonal metabolism, although much of the material in this section is also relevant to Section III.

To return now to the theme of the first section of this preface, much debate concerns the relevance of pure, basic pathology of the physiological and biochemical type, to the applied problems of disease control, and in Section V, the concluding chapter of the book attempts to assess the value of biochemical pathology to both present and future methods of disease control.

J. A. Callow

Dept. of Plant Biology
University of Birmingham

Sept. 1982

I Case Studies

Biochemical Plant Pathology
Edited by J. A. Callow
© 1983 John Wiley & Sons Ltd

1

Potato Late Blight: A Case Study

D. D. CLARKE

Department of Botany, The University, Glasgow G12 8QQ, UK

1. INTRODUCTION

If any crop disease can be said to have spawned the science of plant pathology it is potato blight. From the first outbreaks in North America in 1835, followed by the outbreaks in Europe from 1845 onwards, particularly in Ireland where the consequences in human suffering were so devastating, the disease has excited widespread public concern and interest and has tended to dominate the centre of the stage in plant pathology. The feature of the disease which attracts so much attention is its dramatic nature. From the first visible signs of blight within the crop until the whole of the foliage is killed may be a period of no more than 10–15 days. The reader is referred to Large (1940) for an excellent historical account of the disease.

A great deal of research has been carried out into potato blight and the results have been directly responsible for, or helped to establish, many of the basic concepts of plant pathology. Thus the observations and experiments by the Rev. M. J. Berkeley immediately after the first outbreaks in Britain in 1845, led him to the firm conviction that the associated fungus was the cause not the consequence of the disease. This view pre-dated Pasteur's germ theory of disease by almost 25 years. The work of Müller and other potato breeders in the 1930s and 1940s, leading to Black's (1952) studies of the relationships between the different physiologic races of *P. infestans* and the different resistance genes in the potato, gave early support for Flor's (1956) gene-for-gene theory of host–parasite interaction. The distinction between the two types of resistance in potato to *P. infestans*, on the one hand the resistance which operates against some races but not others (race-specific resistance) and on the other the resistance which operates equally against all races (race-non-specific resistance), led Van der Plank (1963) to his concepts of vertical and horizontal resistance. Müller's work on the interactions between the different races and particular clones culminated in his phytoalexin theory of resistance (Müller and Börger, 1940). Studies with potato blight have been taken up in many laboratories and it is currently being used to probe various aspects of the biochemical basis of the host–parasite interaction, including the nature of specificity and its molecular determinants.

2. THE CAUSAL ORGANISM AND ITS VARIABILITY

The causal organism of potato blight, *Phytophthora infestans*, is classified within the Pythiaceae, in the order Peronosporales, the order to which all the downy mildews belong. It has a fairly restricted host range within the Solanaceae. Its two main crop hosts are the potato (*Solanum tuberosum*) and tomato (*Lycopersicon esculentum*), but it can also attack a number of wild species within the family and some of these wild hosts are closely enough related to the cultivated hosts to be used as sources of resistance in breeding programmes. These breeding programmes can be very useful sources of different resistance types for use in research into the biochemical nature of many aspects of resistance. It was as a direct result of the transfer of resistance from *S. demissum* to the cultivated potato that it was demonstrated that *P. infestans* exists as a series of distinct physiological races, each of which can attack some host varieties but not others. The resistance of the host varieties which differentiate the race structure has been shown to be controlled by a series of dominant genes (R-genes), each determining resistance to one group of races but not to others. *P. infestans* is not amenable to genetic analysis and so the complementary study of the inheritance of virulence/avirulence has not been carried out, but the detailed pattern of interactions between host genotypes and physiologic races demonstrated by Black (1952), is entirely consistent with a gene-for-gene interaction. In this chapter fungal races which attack host genotypes, with or without R-genes, are termed compatible races and the interaction is referred to as a compatible interaction, while races which are unable to attack host genotypes which possess particular R-genes are termed incompatible races and the interaction is referred to as an incompatible interaction. The reader is referred to Walker (1960) for a brief historical account of variability in *P. infestans* and host susceptibility.

3. THE HOST–PARASITE RELATIONSHIP

This chapter is restricted to those events which occur after the inoculum has come into contact with the host. These events are treated in three sections. The first concerns growth on the surface after initial contact up to the stage of penetration, the second deals with reactions after penetration by an incompatible race, and the third with reactions after penetration by a compatible race.

3.1. Contact, pre-penetration growth, and penetration

P. infestans can penetrate and colonize all parts of susceptible varieties except their roots (Fehrmann and Dimond, 1967). Infections are usually established from sporangia which arrive passively on the plant surface by air or water movement. After arrival the sporangia may be re-distributed by rain water, as indeed are the sporangia that are later produced on the surface by established infections. The pattern of re-distribution is largely determined by the growth form of the aerial parts, particularly of the leaves, since the orientation of the leaves can markedly affect the direction of water flow (Lapwood and Hide, 1971). Thus spores tend to be washed into the canopy and down the stem into the soil at the ridge from plants with erect leaves while a drooping tile-like arrangement of leaves leads to the spores being washed off the leaves at their tips and down into the soil in the furrows. Since tuber infections develop from inoculum washed off the leaves, the degree of tuber infection will be in part determined by the way the tubers are borne within the ridge in relation to the way the inoculum is washed into the soil. Differences between cultivars in the patterns of infections on different organs may thus reflect differences in plant growth-form rather than differences in tissue susceptibilities. This is clearly an important consideration when selecting cultivars for biochemical studies of resistance.

The surface layers of the plant constitute the first barriers to infection and their importance in restricting penetration has been shown by many studies. Thus the leaves of most cultivars are more readily penetrated through the lower surface than the upper surface. Umaerus (1969) found, using a standard inoculum, that both the number of infections and the time taken for their establishment varied between cultivars. An intact periderm on the tuber provides a total barrier to penetration and infections are usually only established through buds, lenticels, or wounds which breach the periderm (Lapwood and Hide, 1971). Some cultivars have very susceptible lenticels while others are more readily infected through buds.

Several stages of fungal growth occur prior to penetration and all of these stages may be subject to modification by features of the host surface. Sporangia usually germinate to produce zoospores which themselves, after encystment,

germinate to produce a germ tube. The tip of the germ tube differenti-
ates to form an appressorium, which becomes attached to the plant surface.
An infection peg develops from the undersurface of the appressorium and
penetrates directly into the lumen of the host cell beneath. Sporangia may
also germinate directly to produce a germ tube.

There is no evidence as yet to suggest that the nature of the host surface
may affect the germination of sporangia or zoospores, although it is a field of
study which has received scant attention. In one study, zoospores were found
to germinate just as readily on Pimpernel, a cultivar with a high degree of
resistance to penetration, as on the cultivar Bintje, which is readily penetrated
(Lapwood, 1968). On the other hand, there is evidence for surface effects on
germ tube growth and appressorial formation. Thus short germ tubes were
formed on Bintje before appressorial formation while on Pimpernel the germ
tubes were long and many did not form appressoria and penetrate. On the
cultivar Cobler, zoospore germ tubes readily formed appressoria but those
of sporangia did not (Pristou and Gallegly, 1954), suggesting that some of the
formative effects may depend on the interaction of particular features of the
germ tube and host surfaces. This is a field of study worthy of much more
attention. Studies with the scanning electron microscope might prove most
revealing.

The particular features of the host involved in these varietal effects have
not been investigated. It is only recently, through the work of Kolattukudy
(1981; see also Chapter 6), that the basic chemistry of the cuticular layers of
the stems and leaves of plants, and particularly of the suberized layers of the
tuber surface, has become known, but the work has not advanced far enough
for any of the subtle varietal differences to be established which would be
expected to be involved in interactions of the type described above.

Little is known about the mechanisms of penetration by the infection peg.
The process has been observed with the electron microscope (Shimony and
Friend, 1975, 1977), but many questions still remain unanswered. Thus it is
not clear if it is purely a mechanical process or whether enzymic degradation
of components of the cuticle or suberin and underlying cell wall assists in any
way. There is evidence to show that *P. infestans* can produce cell wall-
degrading enzymes (Knee and Friend, 1970) but nothing to show that they
are produced at such an early stage of infection.

3.2. Host reactions after penetration

The reactions of host cells immediately after penetration are basically of
two types (Pristou and Gallegly, 1954; Ferris, 1955). The first is the dramatic
hypersensitive reaction which occurs in response to penetration by an incompatible race, and the second is the delayed reaction which occurs in response
to penetration by a compatible race. The first type of reaction is clearly a
resistance reaction while the second is associated with susceptibility.

Attempts to determine the significant features of each type of reaction have involved the use of both cytological techniques and biochemical analysis. The cytological techniques have been used to identify the component stages of the reactions and their timing in relation to changes in the growth and development of the fungus, while biochemical analyses have been used to characterize the underlying metabolic processes. Many workers have used a comparative approach in their studies. One experimental procedure involves the study of the reactions of two clones to inoculation by the same race, the race chosen being compatible with one clone and incompatible with the other. The main drawback of this approach is that the resistance genes in the incompatible clone will be in a different genetic background from the susceptible alleles in the compatible clone, and so some of the differences in their reactions may result from these different genetic backgrounds. Another procedure, which suffers from the same drawback, involves inoculating one clone with two different races, one compatible and the other incompatible. These drawbacks would be overcome if isogenic or near-isogenic lines of the host or parasite were used, as appropriate, but such lines have not yet been developed.

The reactions have been studied in leaf, stem, and tuber tissue and of these three the leaf probably provides the closest approximation to the natural situation because, apart from the inevitable use of a much larger than normal inoculum, their tissues have been challenged through an intact surface. Studies with stem and tuber tissue usually involve the use of blocks of tissue with the inoculum being applied to the cut surface, a procedure which adds a wound response to the interaction. The interpretation of differences in the interactions between different race/clone combinations can therefore be problematical because the reactions of each combination will not only include reactions due to the specific resistance or susceptibility genes, but also reactions resulting from the different genetic backgrounds as they modify responses to wounding or other forms of stress. It is also likely that reactions will occur that are coincident with or even the result of resistant or susceptible reactions, but which are not themselves involved in the outcome of the host–parasite interaction.

3.2.1. Host reactions to incompatible races—the hypersensitive reaction

Müller (see Müller, 1959, for a review) first drew attention to the importance of the hypersensitive reaction as a mechanism of resistance to incompatible isolates, but much of our understanding of the component processes is based on the careful cytological studies of Tomiyama and co-workers (Tomiyama, 1966, 1968, 1971; Kitazawa and Tomiyama, 1969; Tomiyama et al., 1979). They have shown that although both compatible and incompatible clones react to the fungus before penetration, such reactions being evident as changes in the permeability of the cells immediately beneath the germinating spores, significant differences between the two do not occur until after penetration when the infection peg tip contacts the host cell membrane (Nishimura and Tomiyama, 1978). Within a few minutes of contact in an incompatible interaction, a series of degenerative changes are set in train within the host cell which rapidly culminate in its death but, in contrast, cells penetrated by compatible isolates remain alive for many days (Tomiyama, 1971). Thus specific recognition occurs when the infection peg comes into contact with the host cell membrane. Current evidence suggests that both incompatible and compatible isolates are recognized, but that the compatible isolate actively prevents the hypersensitive reaction (Doke et al., 1979).

The earliest of the changes in incompatible cells after contact is the rapid increase in permeability (Nishimura and Tomiyama, 1978). This probably allows the phenols and their oxidative enzymes, the phenolases, which are normally retained in separate compartments within healthy cells, to react together to form the brown lignin-like polymers which are deposited within the cell and the cell wall. Cell death can occur within 10 min of contact and most cells are dead within 60 min, although the time taken by cells to die varies between tissues (Tomiyama, 1971). It is generally longest in freshly cut tuber tissue, with the time decreasing the longer the wound surface is allowed to heal before inoculation (Tomiyama et al., 1979). The processes which occur during wound healing and which enable the tissue to respond more quickly to the fungus are inhibited by protein synthesis inhibitors such as blasticidin S, and by metabolic inhibitors such as sodium azide or 2,4-dinitrophenol, and so they appear to involve enzyme synthesis and be energy requiring (Tomiyama et al., 1979). Once the tissues have been inoculated, however, regardless of the stage of wound healing, the rate of cell death is not affected by blasticidin S although it is delayed by the metabolic inhibitors. Thus, while bearing in mind that the interpretation of inhibitor experiments is rarely unequivocal (see also Chapter 17), it would appear that the resistance reaction itself does not involve enzyme synthesis but may still be energy requiring. The latter requirement is supported by the observation that the effects of the metabolic inhibitors can be overcome by the application of ATP. Tomiyama et al. (1958) have also shown that resistance is not the result of changes which occur solely in the penetrated cell but that it also requires the support of a zone of cells up to ten deep surrounding the penetrated cell.

Hyphal growth continues after cell death for a period of up to 7 h (Tomiyama, 1967). This growth may be confined to the initially penetrated cell, with the hypha growing up to the wall and along it but not through it. If the hypha penetrates the wall and grows out into the intercellular space, then the surrounding cells react hypersensitively so that it is walled in by necrotic tissue. During this period the hypha remains relatively unbranched, in contrast to the branched mycelium that develops in compatible tissue.

These observations indicate that hypersensitive resistance is a multi-component process (Clarke, 1972). Firstly, recognition is involved and occurs when the infection peg contacts the host cell membrane. Secondly, resistance to hyphal penetration develops within the wall of the penetrated cell and may also develop in the walls of the surrounding cells. A third component is involved in preventing the hypha from establishing a normal pattern of branching and a fourth is responsible for the eventual cessation of hyphal growth.

Since the first significant reactions occur only after the infection peg has contacted the cell membrane, much attention has been concentrated on the identification of the fungal components which are recognized by the host. Extracts from zoospores prior to encystment are without activity but dead hyphae and crude hyphal wall preparations are very effective at eliciting hypersensitive necrosis. Thus the hyphal wall appears to be the source of the elicitors but the major wall components, the carbohydrates, appear to be relatively inactive and current evidence indicates that lipids or lipid contain-ing molecules are the major elicitors. (Lisker and Kuć, 1977). Bostock *et al.* (1981) found that the most active elicitors were two fatty acids, eicosapen-taenoic acid and arachidonic acid, but other lipids are also present whose eliciting activity appears to depend on an interaction with other components of the wall (Kurantz and Zacharius, 1981). Garas and Kuć (1981) recently showed that elicitors from some preparations were precipitated without loss of activity by potato lectins. Since potato lectins have no affinity for *N*-acetyl-D-glucosamine, it would appear that these elicitors contained a moiety which was not lipid. The non-lipid moiety does not appear to constitute a major part of the active site of the elicitor, however, otherwise it would have lost activity on precipitation. Thus, a variety of compounds, probably mainly lipid or lipid-containing, are capable of eliciting the resistance reactions but, without exception, all preparations are as active on tissues of clones which are compatible with the race from which they were prepared as they are on tissues of incompatible clones and so, clearly, specific recognition does not occur.

Although some of the browning reactions which occur in the hypersensitive reaction are the result of the oxidation and polymerization of pre-existing phenols, such as chlorogenic acid and tyrosine, there is clear evidence that *de novo* synthesis of oxygenated cinnamic acids with their subsequent conversion to lignin-like compounds also occurs (Friend, 1976; see also Chapter 17 for a general discussion). In studies using discs of tuber tissue, it was found that this synthesis was accompanied by increases in the activity of certain key enzymes of the conversions such as phenylalanine ammonia lyase and caffeic acid *O*-methyltransferase. Although the former enzyme showed some increase in the compatible interaction its increase was most marked in the incompatible interaction, while the latter enzyme did not increase at all in the compatible interaction (Friend, 1976). Smith and Rubery (1981) showed that the increase in the activity of the lyase in the compatible reaction was probably a reaction to wounding, but unfortunately they did not extend their study to include an incompatible reaction. However, there can be no question that increased lignification is a component of the hypersensitive response and not simply an enhanced wounding reaction, since lignin-like compounds accumulate in leaf tissues which have been challenged through an intact surface (Friend, 1976).

The accumulation of lignin-like compounds in the lesion could inhibit hyphal growth by a number of mechanisms (Friend, 1976; Ride, 1978; see also Chapter 11). The coating of cellwall carbohydrates by these materials could make them inaccessible to enzymic attack and thus would cut the fungus off from an available source of carbon as well as inhibit wall penetration. If the coating is extensive enough it could toughen the cell wall so that mechanical penetration by the hyphae is prevented. The materials could denature extracellular enzymes produced by the fungus and if they accumulated in the hyphal wall, particularly at the tip, could affect its plasticity and so inhibit tip growth.

The eventual cessation of hyphal growth in the hypersensitive lesion has also been attributed to the accumulation of substances called phytoalexins. In fact, Müller and Börger (1940) developed the phytoalexin theory as the result of observations and experiments on the hypersensitive reaction of potato tissue to *P. infestans*, but it was not until 1968, 8 years after the first ever phytoalexin was characterized, the isoflavanoid pisatin from pea, that the first of the potato phytoalexins was isolated (Tomiyama *et al.*, 1968). The potato phytoalexins are terpenoids and since the discovery of the first, rishitin, a group of about 18 similar compounds have been shown to be formed in incompatible interactions (Kuć *et al.*, 1976). However, the evidence to suggest that they play a major role in hypersensitive resistance to *P. infestans* is largely circumstantial, and there is good evidence to suggest that their role is of secondary importance. Thus marked accumulation occurs only after cell death at the time when hyphal growth is ceasing. They do not appear to accumulate in leaf tissues undergoing a hypersensitive reaction and the greatest accumulation in sprout tissue is in response to infection by compatible races (Kuć, 1972). In any case, *P. infestans* is a biotrophic parasite and so hypersensitive cell death should be a sufficient mechanism in itself to prevent tissue colonization, since it would prevent the fungus establishing a parasitic relationship with living host tissue. The primary role of the phytoalexins in these situations may be as a defence mechanism of the living cells surrounding the hypersensitive lesion, to prevent colonization by secondary invaders which they might otherwise do after becoming established in the necrotic tissue (Van der Plank, 1975).

3.2.2. *Host reactions to compatible races*

Resistance to incompatible races appears to be a multi-component process. It is also non-specific in the sense that when induced, the reaction products will inhibit the growth of both compatible and incompatible races (Müller and Börger, 1940). Thus a compatible race will be able to invade tissue only if it can avoid inducing or can inhibit the induction of each of the components of the resistance reaction. Since all of the living cells of a potato plant appear to have the capacity to react hypersensitively to incompatible races, then compatible races must continue to avoid and continue to inhibit the resistance reactions during the whole of the time they are growing within their host.

The infection peg of a compatible isolate enters the lumen of a cell and invaginates the protoplast (Pristou and Gallegly, 1954). However, the protoplast does not react in a dramatic way to contact with the hypha but remains alive for 2 or more days. Tomiyama (1966) showed that such cells are unable to react hypersensitively to subsequent challenge by an incompatible race, provided that the challenge occurs at least 15 h after contact with the compatible race. Thus the compatible race appears to be able to inhibit the hypersensitive reaction. Varns and Kuć (1972) showed that compatible races also inhibit the accumulation of the terpenoid phytoalexins, rishitin and phytuberin, but not the accumulation of all terpenoid compounds.

Substances have been isolated from zoospores and from hyphae which inhibit both hypersensitive cell death and sesquiterpenoid accumulation. These substances, termed suppressors, are also released into the medium by germinating zoospores and they have been partially characterized as low molecular weight anionic and neutral glucans each containing between 17 and 23 glucose residues. The glucose residues in both types of glucan are mainly β-1,3-linked but some β-1,6 linkages are also present, indicating that they are branched-chain compounds, while the anionic forms also have one or two residues esterified with a phosphoryl monoester (Doke et al., 1979). Marcan et al. (1979) tested a number of disaccharides and methyl glycosides for their ability to inhibit the hypersensitive reaction and found that only those containing β-1,3-linked glucose were active, indicating that the suppressive activity of the glucans depends on the glucose residues linked in this way. There is some indication that the suppressors act by preventing the binding of the non-specific elicitor molecules to receptor sites on the host-cell membrane, possibly by binding to the sites themselves. Thus, microsomal preparations from tuber cells bind elicitor molecules, but this binding is more inhibited by glucans from a compatible race than by glucans from an incompatible race (Doke et al., 1979). This observation also indicates that the glucans have specific activity and support for such specificity is provided by the work of Doke and Tomiyama (1980), who found that suppressors from a number of different races inhibited elicitor-induced necrosis in protoplasts from compatible tissues, to a greater extent than they did in protoplasts from incompatible tissues. However, the inhibitory activity in some of the compatible combinations was low even when high concentrations of suppressor were used and the molecular basis of the specific activity remains to be established.

Clearly then, there is evidence to indicate that the suppression of the resistance reaction is the specific event but none to indicate that the elicitors of resistance have specific activity. However, it should be pointed out that such evidence contradicts that obtained from genetic studies on a number of host–parasite systems since these studies clearly indicate that the induction of resistance is the specific event. Perhaps the specific activities, or lack of them, exhibited by suppressors and elicitors produced by the fungus in culture are different from those produced by the fungus in its host's tissue. Certainly a knowledge of the factors and mechanisms involved in these interactions is of fundamental importance for an understanding of the host–parasite interaction and further progress in this area is awaited with interest.

Very soon after penetration the infection peg develops lateral branches which grow through the cell wall into the intercellular spaces, and by continued growth a branched mycelium develops to colonise the tissues (Kitazawa and Tomiyama, 1969). The development of branches by the infection peg is probably a very significant event in the establishment of the parasitic relationship. The inoculum, whether a sporangium or a zoospore, is endowed with a limited amount of food reserve to sustain its growth until it can obtain nutrients from its host, and the most effective way of maximizing the chance of infection would be to concentrate all food reserves into one penetration rather than to dissipate them over several attempts. Thus natural selection would be expected to favour mechanisms which prevent or minimize branching prior to penetration. The removal of this inhibition would then be essential for the establishment of an infection after penetration. Some evidence that this may be the case has come from studies in culture (Clarke, 1966). It is a common observation that when zoospores are thinly seeded over a wide range of media they readily germinate to form long germ tubes bearing numerous branch initials. However, very few of these initials develop to form branch hyphae, and those which do, stop growth after a few days when sporeling growth ceases altogether. The addition of certain natural products, including various grades of apple pectin or extracts of rye, to a minimal medium, will induce a large proportion of the sporelings to continue growth. The continued growth appears to occur as the result of the removal of an inhibition on the development of the branch initials. Thus continued growth in the host may occur as the result of the induction of branch formation by some feature of the host, possibly associated with the cell wall.

Although the host's cells remain alive for some time after contact with the hyphae, as the hyphae grow through the tissues the cells do eventually die. The time taken for cell death to occur can vary considerably, being only a few days in leaves and in the wound surfaces of tuber slices, but may be several weeks in the inner tissues of the tuber. For given tissues it can also vary between cultivars and such differences probably reflect differences in the tolerance of the cells to the fungus or its metabolites. Thus tissues may be able to resist disease as well as the growth of the fungus. The two processes are not the same although they are sometimes confused.

Cell death is preceded by changes in a number of metabolic pathways including those involved in the accumulation of phenols and terpenoid compounds. One of the earliest reactions in tuber tissue, after inoculation with either a compatible or an incompatible race, is the accumulation of amides of certain cinnamic acids (Clarke, 1982). These amides first accumulate within the cell, but unlike other phenols which after synthesis remain within the cell, they then pass out into the cell wall to which they become bound. Their accumulation is evident as an intense blue fluorescence in ultraviolet light. In an incompatible reaction this fluorescence disappears just before hypersensitive cell death, but it persists for several days in a compatible interaction. It is not yet clear how these amides relate to the accumulation of the lignin-like materials in the hypersensitive reaction although their synthesis will involve common intermediates.

Much of the work with the phenols and with the terpenoid compounds has involved the use of tissue slices and it is obvious that the wound reactions can influence the end products of pathogen-stimulated metabolism. Thus the coumarin glycosides scopolin and aesculin, the accumulation of which is such a feature of blight lesions in whole tubers of most clones (Clarke, 1973), do not accumulate in infected tuber slices unless they are above a certain thickness. Fluorescence microscopy has shown that the glycosides do not accumulate in all cells, but in groups of cells scattered throughout the tissues. Since they are retained within the vacuole, hyphae can grow through the tissues without coming into contact with them. Certain terpenoids also accumulate within the tissues but it is not yet possible to determine precisely where this occurs (Varns and Kuć, 1972; Price et al., 1976). Despite some of the problems involved in the interpretation of these results, it is clear that infection by compatible isolates elicits the accumulation of phenolic and terpenoid compounds just as incompatible races do, and that these accumulations begin soon after penetration. It has been suggested that compatible isolates may prevent the accumulation of lignin-like compounds and the terpenoid phytoalexins associated with the hypersensitive reaction by channelling the metabolism of the precursors of these compounds into the synthesis of others which then accumulate at sites within the tissues where they have little effect on hyphal growth (Clarke, 1972; Kuć et al., 1976).

Although a branched mycelium develops within all tissues of the leaf, stem, and tuber, the extent of colonization can vary considerably between them. Cortical tissues are more extensively colonized than medullary tissues, while vascular tissues are often devoid of hyphae or only sparsely colonized (Kassim, 1976). None of the tissues are totally colonized, and even in the more susceptible clones there are many intercellular spaces in the middle of colonized tissues which are devoid of hyphae, indicating the operation of a mechanism which prevents rampant growth. The presence of a periderm around tuber tissue seems to play some role in inhibiting hyphal growth since hyphae grow faster in tuber slices without a periderm than they do through slices which have developed one, or in whole tubers (Rose, unpublished work). Whether the periderm operates by forming a barrier which retains volatile inhibitors within the tissues or through some other mechanism is not known.

As is typical of most biotrophs, haustoria are developed which penetrate through the cell wall and invaginate the protoplast. The haustoria vary in form from short knobs to long branched or unbranched filiform structures (Blackwell, 1953). They are not formed in all cells in contact with hyphae, and in fact the majority of such cells are not penetrated, but of those which are, a high proportion contain several haustoria (Kassim, 1976). Little work has been done to compare the extent of colonization of different tissues in different clones, although the fact that sporulation occurs more readily on some clones than on others Lapwood (1961) suggests that differences are likely to exist. This whole field requires much more work since factors which affect the rate and extent of tissue colonization, the development of haustoria, and the production of sporangia are likely to be of great significance in the determination of race-non-specific resistance. Both nutritional and inhibitory factors are likely to be involved.

Studies in culture have shown that *P. infestans* has very simple nutritional requirements (French, 1953). It requires a source of carbon and energy which can be met by glucose, sucrose, or a number of other sugars, and it has an absolute requirement for thiamine. It also generally grows better when nitrogen is supplied in organic than in inorganic form. Whilst it would be dangerous to assume that its nutritional requirements in the host are the same as in culture, it is unlikely that they are substantially different. It is known that the host's cell walls are degraded in infected tissue, probably due to the activity of fungal enzymes such as β-1,4-galactosidase (Knee and Friend, 1970) and thus could serve as a source of carbon. However, another potential carbon source, the starch within the cell, does not appear to be utilized. Other substances within the protoplast may become available in the intercellular spaces as the result of the increased permeability which occurs on infection (Nishimura and Tomiyama, 1978), but there is some indication that these changes are selective so that the protoplast does not become equally leaky for all substances (Clarke, 1982). If the host can regulate the amounts and types of nutrients available to the fungus then it would have a mechanism for limiting fungal growth. The selective leakage of either pre-formed or post-formed inhibitors, such as phenols, steroid alkaloids, or terpenoids, would provide another mechanism for limiting fungal growth. The outer cortical layers in the tuber, just under the periderm, are very resistant to hyphal growth (Rose, unpublished work) and these tissues contain high concentrations of alkaloids and phenols (Allen and Kuć, 1968), while the vascular tissue, which is also resistant (Kassim, 1976), contains high concentrations of phenols (Hughes and Swain, 1962). However, in neither case has the resistance been shown to be causally related to the presence of the metabolites, and there are other possible explanations.

This chapter has described some of the interactions which current evidence suggests are important determinants of the host–parasite interaction. Resistance in the host appears to be a multi-component system, involving factors which operate both before and after penetration, and each of these factors must be avoided or overcome in one way or another by a compatible race before an infection can become established.

REFERENCES

Allen, E. H., and Kuć, J. (1968). α-Solanine and α-chaconine as fungitoxic compounds in extracts of Irish potato tubers, *Phytopathology*, **58**, 776–781.

Black, W. (1952). Inheritance of resistance to blight (*Phytophthora infestans*) in potatoes: inter-relationships of genes and strains, *Proc. R. Soc. Edinburgh*, **65**, 36–51.

Blackwell, E. M. (1953). Haustoria of *Phytophthora infestans* and some other species, *Trans. Br. Mycol. Soc.*, **36**, 138–158.

Bostock, R. M., Kuć, J. A., and Laine, R. A. (1981). Eicosapentaenoic and arachidonic acids from *Phytophthora infestans* elicit fungitoxic sesquiterpenes in potatoes, *Science*, **212**, 67–69.

Clarke, D. D. (1966). Factors affecting the development of single zoospore colonies of *Phytophthora infestans*, *Trans. Br. Mycol. Soc.*, **49**, 177–184.

Clarke, D. D. (1972). The resistance of potato tissue to the hyphal growth of fungal pathogens, *Proc. R. Soc. London, Ser. B.*, **181**, 303–317.

Clarke, D. D. (1973). The accumulation of scopolin in potato tissue in response to infection, *Physiol. Plant Pathol.*, **3**, 347–358.

Clarke, D. D. (1982). The accumulation of cinnamic acid in the cell walls of potato tissue as an early response to fungal attack, in *Active Defence Mechanisms in Plants* (Ed. R. K. S. Wood), Plenum Press, New York, pp. 321–322.

Doke, N., Garas, N. A., and Kuć, J. (1979). Partial characterization and aspects of the mode of action of the hypersensitivity-inhibiting factor (HIF) isolated from *Phytophthora infestans*, *Physiol. Plant Pathol.*, **15**, 127–140.

Doke, N., and Tomiyama, K. (1980). Suppression of the hypersensitive response of potato tuber protoplasts to hyphal wall components by water soluble glucans isolated from *Phytophthora infestans*, *Physiol. Plant Pathol.*, **16**, 177–186.

Fehrman, H., and Dimond, A. E. (1967). Peroxidase activity and *Phytophthora* resistance in different organs of the potato plant, *Phytopathology*, **57**, 69–72.

Ferris, V. R. (1955). Histological studies of pathogen–suscept relationships between *Phytophthora infestans* and derivative of *Solanum demissum*, *Phytopathology*, **45**, 546–552.

Flor, H. H. (1956). The complementary genic systems in flax and flax rust, *Adv. Genet.*, **8**, 29–54.

French, A. M. (1953). Physiologic differences between two physiologic races of *Phytophthora infestans*, *Phytopathology*, **43**, 513–516.

Friend, J. (1976). Lignification in infected tissue, in *Biochemical Aspects of Plant–Parasite Relationships*, Phytochemical Society Symposium Series No. 13, (Eds.J. Friend and D. R. Threlfall), Academic Press, London, pp. 291–303.

Garas, N. A., and Kuć, J. (1981), Potato lectin lyases zoospores of *Phytophthora infestans* and precipitates elicitors of terpenoid accumulation preduced by the fungus, *Physiol. Plant Pathol.*, **18**, 227–237.

Hughes, J. C. and Swain, T. (1962). After-cooking blackening in potatoes. II. Core experiments, *J. Sci. Food Agric.*, **13**, 229–236.

Kassim, M. Y. A. (1976). Potato tissue resistance to the growth of *Phytophtora infestans* (Mont.) de Bary, *PhD Thesis*, Glasgow University.

Kitazawa, K., and Tomiyama, K. (1969). Microscopic observations of infection of potato cells by compatible and incompatible races of *Phytophthora infestans*, *Phytopathol. Z.*, **66**, 317–324.

Knee, M., and Friend, J. (1970). Some properties of galactanase secreted by *Phytophthora infestans* (Mont.) de Bary, *J. Gen. Microbial.*, **60**, 23–30'

Kolattukudy, P. E. (1981). Structure, biosynthesis and biodegradation of cutin and suberin, *Annu. Rev. Plant Physiol.*, **32**, 539–567.

Kuć, J. (1957). A biochemical study of the resistance of potato tuber tissue to attack by various fungi, *Phytopathology*, **47**, 676–680.

Kuć, J. (1972). Phytoalexins, *Annu. Rev. Phytopathol.*, **10**, 207–232.

Kuć, J., Currier, W. W., and Shih, M. J. (1976). Terpenoid phytoalexins, in *Biochemical Aspects of Plant–Parasite Relationships*, Phytochemical Society Symposium Series No. 13 (Eds. J. Friend and D. R. Threlfall), Academic Press, London, pp. 225–237.

Kurantz, M. J. and Zacharius, R. M. (1981). Hypersensitive response in potato tuber: elicitation by combination of non-eliciting components from *Phytophthora infestans* *Physiol. Plant Pathol.*, **18**, 67–77.

Lapwood, D. H. (1961). Potato haulm resistance to *Phytophthora infestans*. II. Lesion production and sporulation, *Ann. Appl. Biol.*, **49**, 316–330.

Lapwood, D. H. (1968). Observations on the infection of potato leaves by *Phytophthora infestans*, *Trans. Br. Mycol. Soc.*, **51**, 233–240.

Lapwood, D. H., and Hide, G. A. (1971). Potato, in *Diseases of Crop Plants* (Ed. J. H. Western), Macmillan, London, pp. 89–122.

Large, E. C. (1940). *The Advance of the Fungi*, Jonathan Cape, London.

Lisker, N., and Kuć, J. (1977) Elicitor of terpenoid accumulation in potato tuber slices, *Phytopathology*, **67**, 1356–1359.

Marcan, H., Jarvis, M. C., and Friend, J. (1979). Effect of methyl glycosides and oligosaccharides on cell death and browing of potato tuber discs induces by mycelial components of *Phytophthora infestans, Physiol. Plant Pathol.*, **14**, 1–9.

Müller, K. O. (1959). Hypersensitivity, in *Plant Pathology*, vol. 1 (Eds. J. C. Horsfall and A. E. Dimond), Academic Press, New York, pp. 469–519.

Müller, K. O., and Börger, H. (1940). Experimentelle Undertsuchngen über die Phytophthora-Resistanz der Kartoffel; Zugleich ein Beitrag zum Problem der 'Erworbenen' Resistenz im Pflanzenreich, *Arb. Biol. Reichsants. Land-Forstwirtsch., Berlin-Dahlem*, **23**, 189–231.

Nishimura, N., and Tomiyama, K. (1978). Effect of infection by *Phytophthora infestans* on the uptake of ^{3}H-leucine, ^{32}P and ^{86}Rb by potato tuber tissue, *Ann. Phytopathol. Soc. Jpn.*, **44**, 159–166.

Price, K. R., Howard, B., and Coxon, D. T. (1976). Stress metabolite production in potato tubers infected by *Phytophthora infestans*, *Fusarium avenaceum* and *Phoma exigua, Physiol. Plant Pathol.*, **9**, 189–197.

Pristou, R., and Gallegly, M. E. (1954). Leaf penetration by *Phytophthora infestans, Phytopathology*, **44**, 81–86.

Ride, J. P. (1978). The role of cell wall alterations in resistance to fungi, *Ann. Appl. Biol.*, **89**, 304–306.

Shimony, C., and Friend, J. (1975). Ultrastructure of the interaction between *Phytophthora infestans* and leaves of two cultivars of potato (*Solanum tuberosum* L.) Orion and Majestic, *New Phytol.*, **74**, 59–65.

Shimony, C., and Friend, J. (1977). The ultrastructure of the interaction between *Phytophthora infestans* (Mont.) de Bary and tuber discs of potato (*Solanum tuberosum* L.) cv. King Edward, *Physiol. Plant Pathol*, **11**, 243–249.

Smith, B. G., and Rubery, P. H. (1981). The effects of infection by *Phytophthora infestans* on the control of phenyl propanoid metabolism in wounded potato tissue, *Planta*, **151**, 535–540.

Tomiyama, K. (1966). Double infection by an incompatible race of *Phytophthora infestans* of a potato cell which has previously been infected by a compatible race, *Ann. Phytopathol. Soc. Jpn.*, **32**, 181–185.

Tomiyama, K. (1967). Further observation on the time requirement for hypersensitive cell death of potatoes infected by *Phytophthora infestans* and its relation to metabolic activity, *Phytopathol. Z.*, **58**, 367–378.

Tomiyama, K. (1971). Cytological and biochemical studies of the hypersensitive reaction of potato cells to *Phytophthora infestans*, in *Morphological and Biochemical Events in Plant–Parasite Interaction* (Eds. S. Akai and S. Ouchi), Phytopathology Society of Japan, Tokyo, pp. 387–399.

Tomiyama, K., Doke, N., Nozue, M., and Ishiguri, Y. (1979). The hypersensitive response of resistant plants, in *Recognition and Specificity in Plant–Parasite Interactions* (Eds. J. M. Daly and I. Uritani). Japan Scientific Societies Press, Tokyo, and University Park Press, Baltimore, pp. 69–84.

Tomiyama, K., Takakuwa, M., and Takase, N. (1958). The metabolic activity in healthy tissue neighbouring the infected cells in relation to resistance to *Phytophthora* in potatoes, *Phytopathol. Z.,* **31**, 237–250.

Tomiyama, K., Sakuma, T., Ishizaka, N., Sato, N., Katsui, N., Takasugi, M., and Masamune, T. (1968). A new antifungal substance isolated from resistant potato tuber tissue infected by pathogens, *Phytopathology*, **58**, 115–116.

Umaerus, V. (1969). Studies on field resistance to *Phytophthora infestans*. 1. The infection efficiency of zoospores of *P. infestans* as influenced by the host genotype. *Z. Pflanzenzüchts.*, **61**, 29–45.

Van der Plank, J. E. (1963). *Plant Diseases: Epidemics and Control*, Academic Press, New York.

Van der Plank, J. E. (1975). *Principles of Plant Infection*, Academic Press, New York.

Varns, J., and Kuć, J. (1972). Suppression of the resistance responses as an active mechanism for susceptibility in the potato–*Phytophthora infestans* interaction, in *Phytotoxins in Plant Diseases* (Eds. R. K. S. Wood, A. Ballio, and A. Graniti), Academic Press, New York, pp. 465–468.

Walker, J. C. (1960). *Plant Pathology*, 3rd ed., McGraw-Hill, New York.

Biochemical Plant Pathology
Edited by J. A. Callow
© 1983 John Wiley & Sons Ltd

2

Phytophthora Root and Stem Rot of Soybean: A Case Study

JACK D. PAXTON

Department of Plant Pathology, University of Illinois, 1102 S. Goodwin Avenue, N-519 Turner Hall, Urbana, IL 61801, USA

1. INTRODUCTION

The interaction between *Phytophthora megasperma* var. *sojae* and soybean plants is proving to be of fundamental importance in studies on the biochemistry of host–parasite interaction and specific resistance. Part of the reason for progress in understanding this disease is that a good model system exists. Fortunately, single dominant genes for resistance to *Phytophthora* root rot have been found and incorporated by plant breeders into commercially important cultivars, giving near-isogenic lines of popular cultivars of this very important annual crop (Bernard *et al.*, 1957; Mueller *et al.*, 1977; Athow *et al.*, 1980). The disease causes significant losses and at present the only economically practical means of control is to breed for genetic resistance to the pathogen. The disease is also amenable to study as soybean plants generally respond quickly to inoculation with *P. megasperma* var. *sojae* and either the fungus or the plant dies within 48 h. For these reasons, several research groups have intensively studied aspects of this disease.

The causal organism of this disease has been known as *Phytophthora cactorum* (Skotland, 1955), *P. sojae* (Kaufmann and Gerdemann, 1958), *P. megasperma* var. *sojae* (Hildebrand, 1959), and *P. megasperma* f. sp. *glycinea* (Kuan and Erwin, 1980). Here it will be referred to as *P. megasperma* var. *sojae*, since this name is most common in the literature and its host specificity is far from clear (Hamm and Hansen, 1981).

2. THE DISEASE

Phytophthora rot of soybeans is most noticeable when it causes a classic damping-off disease. In this phase plants are killed either before they emerge, ('pre-emergence damping-off'), or shortly after they emerge, ('post-emergence damping-off') (Figure 1). This phase often occurs in low spots in fields, especially after prolonged flooding in cool weather. If the plants survive this phase of the disease, they may then survive for a long time with a slowly progressing canker that completely blackens the stem for several centimetres above the ground. Cankers may also develop on the stem where mud containing the fungus has been splashed by driving rain and wind or the plant has been bent to touch the wet soil. These cankers may eventually kill the plant or they can stop and remain quiescent.

While the above-ground symptoms are obvious, this disease can take a large toll on the plant's root system. This below-ground attack can lead to an early death, but often results in a debilitating disease which may kill the plant only under drought stress or more often may simply result in reduced yields as

Resistant   Susceptible

Figure 1. Growth of soybean plants in vermiculite infested with *Phytophthora megasperma* var. *sojae*

the plant diverts photosynthate from pod filling to replacing roots that have been rotted off.

3. THE ENVIRONMENT

The environment has a large impact on this disease and some of the possible biochemical reasons for this will be discussed later. Suffice it to say at this point that the disease is much more severe with flooding of the soil. Flooding causes conditions favourable for the production of sporangia by the fungus and subsequent release of zoospores (Eye *et al.*, 1978). Zoospores are capable of swimming short distances in water and certainly can be carried considerable distances by drainage water. This allows the pathogen to increase its infective propagules several-fold and spread these propagules over a much wider area.

Low soil temperatures such as occur in the spring, favour the development of this disease. The soybean crop grows very slowly below 15 °C and is intolerant of standing water, which not only favours the pathogen but also causes anaerobic conditions around the roots, further interfering with their normal functioning. High temperatures with water stress also appear to exacerbate the disease, but this is probably only due to a lack of ability of the plant, with a reduced root system, to respond to water stress.

Other environmental factors such as herbicides (Duncan and Paxton, 1981), nematodes (Adeniji *et al.*, 1975), symbionts (Chou and Schmitthenner, 1974; Ross, 1972), and soil temperature, moisture, porosity, and bulk density (Kittle and Gray, 1979) also influence this disease.

4. THE PATHOGEN

Over sixteen races of *P. megasperma* var. *sojae* have now been described (Morgan and Hartwig, 1965; Haas and Buzzell, 1976; Laviolette *et al.*, 1979). 'Pandora's box' has been opened and the number of races of this variable pathogen is limited only by the number of cultivars used to identify them. These races may pre-exist in the field, or may arise as genes for resistance are deployed and create selection pressure on the pathogen for mutants which can attack plants containing these genes.

The zoospore appears to be the most important infective propagule in the soil. Zoospores swim toward roots of resistant and susceptible soybeans and even non-hosts in an apparently non-discriminatory manner (Ho and Hickman, 1967). It is probable that this attraction is due to nutrients diffusing from the roots.

Once the zoospore has arrived at the root surface it quickly encysts and withdraws its flagella. The greatest number of zoospores encyst along the root elongation region just behind the root apex and over wounds on the root.

Within 30–60 min the cysts start to germinate, producing a germ tube on the side closest to the root. This germ tube then penetrates the root and establishes an initial infection point (Ho and Hickman, 1967).

At this point rather rapid biochemical processes are set in motion which, within a few hours, determine whether the plant or the pathogen will die. It is apparent that DNA-dependent RNA synthesis is required in the first 4 h after infection for resistance to be expressed (Yoshikawa *et al.*, 1977). Following this stage, and for 6 h after infection, RNA translation is required for expression of resistance (Yoshikawa *et al.*, 1977, 1978a). *De novo* RNA and protein synthesis are required for resistance, demonstrating that resistance is an active phenomenon.

The second symptom seen in inoculated susceptible plants is a darkening of the stem around the area of inoculation 24–48 h after inoculation. This is probably due to the disruption of cell membranes leading to oxidation and polymerization of phenolic compounds.

By 48 h after inoculation of the susceptible plant the hypocotyl of a young seedling collapses (Figure 2). This may be due to the complete loss of turgor in the affected cells. Although enzyme attack on host cell walls cannot be precluded, attempts to find such enzymes have not been successful. In stems of older plants such collapse does not occur because the cell walls are lignified and therefore resistant to attack.

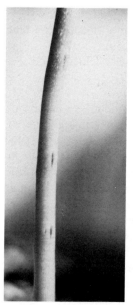  

Check inoculation   Susceptible   Resistant

Figure 2. Response of soybean hypocotyls to zoospores of *Phytophthora megasperma* var. *sojae* 48 h after inoculation

The resistant plant responds to inoculation with *P. megasperma* var. *sojae* quickly. A darkened, sometimes red lesion occurs within 24 h at the point of inoculation. This lesion is limited to a few cell layers around the point of inoculation and leads to the localized response seen in Figure 2. It is important to note that these lesions are discrete. The response in this case is quick and essentially complete within 24 h.

As in most biological systems, exceptions exist to the above descriptions. An intermediate response grading from resistance to a slowly enlarging brown lesion to susceptibility can occur between some combinations of host and pathogen genotypes and environmental conditions.

5. ELICITORS AND SPECIFICITY FACTORS

P. megasperma var. *sojae* produces compounds, both in culture and in the plant, that the soybean recognizes and to which it responds (Frank and Paxton, 1971). This response may take several forms but has most often been measured as the accumulation of the phytoalexin glyceollin. There may be several compounds released by the pathogen that can be 'recognized' by the soybean plant. These compounds have been described as 'inducers' or 'elicitors' of phytoalexin production. They have been characterized as glycoproteins, proteins, and carbohydrates (Frank and Paxton, 1971; Keen, 1975; Wade and Albersheim, 1979; Keen and Legrand, 1980).

A glucan elicitor from mycelial cell walls has been characterized as a β-1,3-linked glucan with β-1,6-branching (Ayers *et al.*, 1976a,b; Ebel *et al.*, 1976). Since laminarin and paramylon, both linear β-1,3-glucans, have little or no elicitor activity the branching must be important for elicitor activity. The native elicitor is active at concentrations as low as 10^{-9} M, and about 1 ng can be detected by its elicitor activity on a soybean cotyledon (Ayers *et al.*, 1976a). This elicitor has been degraded to an active piece of only 7–9 glucose units but the linkages in this active fragment of native elicitor remain to be determined (Valent and Albersheim, 1977).

These glucan elicitors are not race-specific and are also active in stimulating phytoalexin production in plants other than soybean (Cline *et al.*, 1978). Another elicitor has been isolated from cultures of *P. megasperma* var. *sojae* and it shows race specificity in stimulating phytoalexin production. This elicitor appears to be a glycoprotein but it has not been fully characterized (Keen and Legrand, 1980). The protein portion may confer specificity while the carbohydrate portion carries the elicitor activity. An extracellular invertase produced by this pathogen specifically inhibits accumulation of glyceollin in soybean cotyledon wounds treated with the glucan elicitor (Ziegler and Pontzen, 1982). This may also play a role in host specificity.

Other specificity factors from the fungus have also been found (Wade and Albersheim, 1979). These glycoproteins do not elicit glyceollin accumulation but do protect the plant from infection by *P. megasperma* var. *sojae*. This protection works only if the plant is resistant to the race of the pathogen from which the glycoprotein was isolated. The mechanism for this protection is not understood.

It has been demonstrated that the plant cell contains enzymes that are capable of attacking β-1,3-glucans and may be responsible for release of elicitors from the fungus cell walls (Yoshikawa et al., 1981). These plant enzymes could play an obvious role in 'alerting' the plant to the presence of the pathogen.

Although ethylene production is stimulated in this response it does not appear to have a role in phytoalexin production (Paradies et al., 1980; see also Chapter 19).

Another area of interest is the nature of the receptors for the elicitors on or in the plant cell. They may be lectins but their nature remains to be determined. A further discussion of elicitors, specificity factors, and receptors is presented in Chapter 13.

6. PHYTOALEXINS

The soybean plant responds to several stresses by producing phytoalexins (Klarman and Gerdemann, 1963; Yoshikawa et al., 1978b). These antibiotic compounds are inhibitory to pathogens and have an important role in plant disease resistance. They are either produced de novo or their production is greatly stimulated by pathogens and elicitors. Although there has been debate as to whether abiotic elicitors affect the plant differently from biotic elicitors (Yoshikawa, 1978; see also Chapters 13 and 17), it is at present thought that all elicitors generally stimulate phytoalexin accumulation by stimulating synthesis or activity of the requisite enzymes leading to phytoalexin synthesis rather than by blocking the degradation of phytoalexins (Moesta and Grisebach, 1980).

It appears that all parts of the soybean plant (root, stem, cotyledons, and true leaves) are capable of producing phytoalexins although the specific phytoalexin, and amount, can vary with the plant part and the conditions to which it is subjected (Bridge and Klarman, 1973; Keen and Paxton, 1975; Yoshikawa et al., 1979; Lazarovits et al., 1981; Moesta and Grisebach, 1981; Weinstein et al., 1981). The environment also has a strong influence on phytoalexin accumulation. Temperature, salinity, light, and oxygen tension all affect glyceollin accumulation in patterns that approximate to what would be expected from field observations of disease susceptibility (e.g. under anaerobic conditions the soybean produces less phytoalexin and field-flooding encourages disease) (Murch and Paxton, 1980a,b). A very interesting observation is that heating soybean plants at 40 °C for 2 h will *reversibly* inhibit both phytoalexin accumulation and disease resistance.

Several phytoalexins have now been identified from soybean plants (Lyne

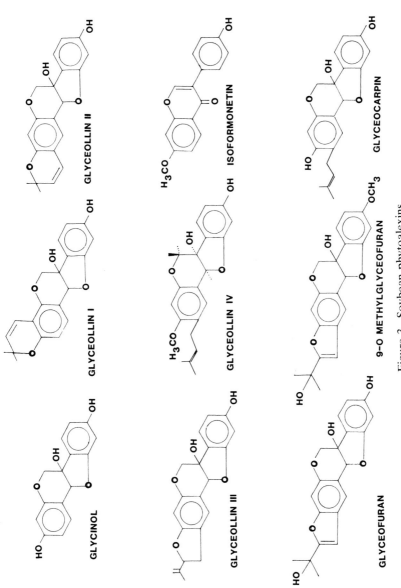

Figure 3. Soybean phytoalexins

et al., 1976; Lyne and Mulheirn, 1978; Ingham *et al.*, 1981). The most prominent are the glyceollin isomers listed in Figure 3. Unfortunately, glyceollin was initially misidentified and the incorrect name hydroxyphaseollin also occurs in the literature. The other phytoalexins that are produced by soybeans are also listed in Figure 3 and others undoubtedly will be found and identified. The strategy of producing several phytoalexins rather than just one is sound, based on the fact that resistance to several different compounds at one time is much harder for the potential pathogen to achieve.

Good progress is being made at present in the elucidation of the biosynthetic pathways for glyceollin synthesis. It appears that phenylalanine ammonia lyase is an early and key enzyme in glyceollin synthesis (Zaehringer *et al.*, 1977, 1978, 1981). However, Partridge and Keen (1977) would debate this point. Another enzyme, dimethylallylpyrophosphate:trihydroxypterocarpan–dimethylallyltransferase, attaches the C_5 isoprenoid group to the isoflavan ring (Zaehringer *et al.*, 1979). The complete biosynthetic pathway for any soybean phytoalexin, however, remains to be elucidated. Both the plant and the pathogen can degrade glyceollin. This may be important for pathogenesis.

7: DISEASE CONTROL

While the only present control of *Phytophthora* rot is through genetically resistant cultivars, these fundamental biochemical studies suggest some novel methods of disease control (Vaartaja *et al.*, 1979). Phytoalexins may be sprayed directly on to the plant to prevent the growth of pathogens. This approach has been used with other phytoalexins and other plants (Ward *et al.*, 1975 and see Chapter 20). However, phytoalexins at present would be very expensive to use and tend to damage the plant as well as the pathogen!

Another approach that suggests itself is the use of phytoalexin elicitors. These materials appear to be very stable and undoubtedly are safe in the environment. Theoretically they would elicit phytoalexin production only where the pathogen was attempting to enter the plant as this would carry them into cells. Work is underway to see if these or other compounds from *P. megasperma* var. *sojae* can indeed be used to protect the plant (Wade and Albersheim, 1979).

An equally exciting prospect is the use of compounds that have been called sensitizers. These materials can sensitize the plant to respond to the pathogen vigorously, when it normally would not. The sensitizer has no activity against the fungus or in the plant but, when administered to the plant, sensitizes it to respond to the pathogen in a resistant fashion (e.g. Wade and Albersheim, 1979). This could be expressed in increased phytoalexin production after inoculation of the sensitized plant in comparison with the unsensitized plant.

Rhizobacteria also have been shown to protect the soybean plant against *P. megasperma* var. *sojae*. Rhizoplane bacteria isolated from healthy roots of field plants, when applied to soybean seed, have greatly increased soybean yields in *Phytophthora*-infested soil (Paxton, unpublished work). This may establish a root environment favourable to the host but unfavourable to pathogens. Cross-protection with non-pathogens is also possible (Svoboda and Paxton, 1972).

Finally, cultural practices have and will be used to control this disease. These include draining fields so that water does not stand around the plant roots, creating conditions favourable for infection of the plant by *Phytophthora*. Later planting, when soil temperatures have risen, could also reduce this disease, as suggested by environmental studies.

REFERENCES

Adeniji, M. O., Edwards, D. I., Sinclair, J. B., and Malek, R. B. (1975). Interrelationship of *Heterodera glycines* and *Phytophthora megasperma* var. *sojae* in soybeans, *Phytopathology*, **65**, 722–725.

Athow, K. L., Laviolette, F. A., Mueller, E. H., and Wilcox, J. R. (1980). A new major gene for resistance to *Phytophthora megasperma* var. *sojae* in soybean, *Phytopathology*, **70**, 977–980.

Ayers, A. R., Ebel, J., Finelli, F., Berger, N., and Albershein, P. (1976). Host pathogen interactions. 9. Quantitative assays of elicitor activity and characterization of the elicitor present in the extracellular medium of cultures of *Phytophthora megasperma* var. *sojae, Plant Physiol.*, **57**, 751–759.

Ayers, A. R., Ebel, J., Valent, B., and Albersheim, P. (1976). Host pathogen interactions. 10. Fractionation and biological activity of an elicitor isolated from the mycelial walls of *Phytophthora megasperma* var. *sojae, Plant Physiol.*, **57**, 760–765.

Bernard, R. L., Smith, P. E., Kaufmann, M. J., and Schmitthenner, A. F. (1957). Inheritance of resistance to *Phytophthora* root and stem rot in soybean, *Agron. J.*, **49**, 391.

Bridge, M. A., and Klarman, W. L. (1973). Soybean phytoalexin hydroxyphaseollin induced by UV irradiation, *Phytopathology*, **63**, 606–609.

Chou, L. G., and Schmitthenner, A. F. (1974). Effect of *Rhizobium japonicum* and *Endogone mosseae* on soybean root rot caused by *Pythium ultimum* and *Phytophthora megasperma* var. *sojae, Plant Dis. Rep.*, **58**, 221–225.

Cline, K., Wade, M., and Albersheim, P. (1978). Host pathogen interactions. 15. Fungal glucans which elicit phytoalexin accumulation in soybean also elicit the accumulation of phytoalexins in other plants, *Plant Physiol.*, **62**, 918–921.

Duncan, D. R., and Paxton, J. D. (1981). Trifluralin enhancement of *Phytophthora* root rot of soybean, *Plant Dis.*, **65**, 435–436.

Ebel, J., Ayers, A. R., and Albersheim, P. (1976). Host pathogen interactions. 12. Response of suspension cultured soybean cells to the elicitor isolated from *Phytophthora megasperma* var. *sojae* a fungal pathogen of soybeans, *Plant Physiol.*, **57**, 775–779.

Eye, L. L., Sneh, B., and J. L. Lockwood, (1978). Factors affecting zoospore production by *Phytophthora megasperma* var. *sojae, Phytopathology*, **68**, 1766–1768.

Frank, J. A., and Paxton, J. D., (1971). An inducer of soybean phytoalexin and its role in the resistance of soybean to *Phytophthora* rot, *Phytopathology*, **61**, 954–958.

Haas, J. H., and Buzzell, R. I. (1976). New races 5 and 6 of *Phytophthora megasperma* var. *sojae* and differential reactions of soybean cultivars for races 1 to 6, *Phytopathology*, **66**, 1361–1362.

Hamm, P. B., and Hansen, E. M. (1981). Host specificity of *Phytophthora megasperma* from Douglas fir, soybean and alfalfa, *Phytopathology*, **71**, 65–68.

Hildebrand, A. A. (1959). A root and stalk rot of soybeans caused by *Phytophthora megasperma* Drechsler. var. *sojae* var. nov., *Can. J. Bot.*, **37**, 927–957.

Ho, H. H., and Hickman, C. J. (1967). Factors governing zoospore responses of *Phytophthora megasperma* var. *sojae* to plant roots, *Can. J. Bot.*, **45**, 1983–1994.

Ingham, J. L., Keen, N. T., Mulheirn, L. J., and Lyne, R. L. (1981). Inducibly formed isoflavonoids from leaves of soybean, *Phytochemistry*, **20**, 795–798.

Kaufmann, M. J., and Gerdemann, J. W. (1958). Root and stem rot of soybean caused by *Phytophthora sojae* n. sp., *Phytopathology*, **48**, 201–208.

Keen, N. T. (1975). Specific elicitors of plant phytoalexin production: determinants of race specificity in pathogens?, *Science*, **187**, 74–75.

Keen, N. T., and Legrand, M. (1980). Surface glycoproteins: evidence that they may function as the race specific phytoalexin elicitors of *Phytophthora megasperma* f. sp. *glycinea*, *Physiol. Plant Pathol.*, **17**, 175–192.

Keen, N. T., and Paxton, J. D. (1975). Coordinate production of hydroxyphaseollin and the yellow-fluorescent compound PA_k in soybeans resistant to *Phytophthora megasperma* var. *sojae*, *Phytopathology*, **65**, 635–637.

Kittle, D. R., and Gray, L. E. (1979). The influence of soil temperature, moisture, porosity and bulk density on the pathogenicity of *Phytophthora megasperma* var. *sojae*, *Plant Dis. Rep.*, **63**, 231–234.

Klarman, W. L., and Gerdemann, J. W. (1963). Resistance of soybean to three *Phytophthora* species due to production of a phytoalexin, *Phytopathology*, **53**, 1317–1320.

Kuan, T. L., and Erwin, D. C. (1980). Formae speciales differentiation of *Phytophthora megasperma* isolates from soybean and alfalfa, *Phytopathology*, **70**, 333–338.

Laviolette, F. A., Athow, K. L., Mueller, E. H., and Wilcox, J. R. (1979). Inheritance of resistance in soybeans to physiologic races 5, 6, 7, 8, and 9 of *Phytophthora megasperma* var. *sojae*, *Phytopathology*, **69**, 270–271.

Lazarovits, G., Stossel, R., and Ward, E. W. B. (1981). Age related changes in specificity and glyceollin production in the hypocotyl reaction of soybeans to *Phytophthora megasperma* var. *sojae*, *Phytopathology*, **71**, 94–97.

Lyne, R. L., and Mulheirn, L. J. (1978). Minor pterocarpinoids of soybean, *Tetrahedron Lett.*, 3127–3128.

Lyne, R. L., Mulheirn, L. J., and Leworthy, D. P. (1976). New pterocarpanoid phytoalexins of soybean, *J. Chem. Soc., Chem. Commun.*, 497–498.

Moesta, P., and Grisebach, H. (1980). Effects of biotic and abiotic elicitors on phytoalexin metabolism in soybean, *Nature (London)*, **286**, 710–711.

Moesta, P., and Grisebach, H. (1981). Investigation of the mechanism of glyceollin accumulation in soybean infected by *Phytophthora megasperma* f. sp. *glycinea*, *Arch. Biochem. Biophys.*, **212**, 462–467.

Morgan. F. L., and Hartwig, E. E. (1965). Physiologic specialization in *Phytophthora megasperma* var. *sojae*, *Phytopathology*, **55**, 1277–1279.

Mueller, E. H., Athow, K. L., and Laviolette, F. A. (1977). Inheritance of resistance to physiologic races 1, 2, 3, and 4 of *Phytophthora megasperma* var. *sojae* in 4 soybean cultivars, *Proc. Am. Phytopathol. Soc.*, **4**, 103.

Murch, R. S. and Paxton, J. D. (1980a) Temperature and glyceollin accumulation in *Phytophthora* resistant soybean, *Phytopathol. Z.*, **97**, 282–285.

Murch, R. S., and Paxton, J. D. (1980b). Rhizosphere anaerobiosis and glyceollin accumulation in soybean, *Phytopathol. Z.*, **96**, 91–94.

Paradies, I., Konze, J. R., Elstner, E. F., and Paxton, J. (1980). Ethylene indicator but not inducer of phytoalexin synthesis in soybean, *Plant Physiol.*, **66**, 1106–1109.

Partridge, J. E., and Keen, N.T. (1977). Soybean phytoalexins rates of synthesis are not regulated by activation of initial enzymes in flavonoid biosynthesis, *Phytopathology*, **67**, 50–55.

Ross, J. P. (1972). Influence of *Endogone* mycorrhiza on *Phytophthora* rot of soybean, *Phytopathology*, **62**, 896–897.

Skotland, C. B. (1955) A *Phytophthora* damping-off disease of soybean, *Plant Dis. Rep.*, **39**, 628–683.

Svoboda, W. E., and Paxton, J. D. (1972). Phytoalexin production in locally cross protected Harosoy and Harosoy 63 soybeans, *Phytopathology*, **62**, 1457–1460.

Vaartaja, O., Pitblado, R. E., Buzzell, R. I., and Crawford, L. G. (1979). Chemical and biological control of *Phytophthora megasperma* var. *sojae* root and stalk rot of soybean, *Can. J. Plant Sci.*, **59**, 307–312.

Valent, B. S., and Albersheim, P. (1977). Role of elicitors of phytoalexin accumulation in plant disease resistance, in *Host Plant Resistance to Pests*, (Ed. P. A. Hedin), *Am. Chem. Soc. Sympo. Ser.* No. 62, 27–34.

Wade, M., and Albersheim, P. (1979). Race specific molecules that protect soybeans from *Phytophthora megasperma* var. *sojae*, *Proc. Natl. Acad. Sci. USA*, **76**, 4433–4437.

Ward, E. W. B., Unwin, C. H., and Stoessl, A. (1975). Experimental control of late blight of tomatoes with capsidiol, the phytoalexin from peppers, *Phytopathology*, **65**, 168–169.

Weinstein, L. I., Hahn, M. G., and Albersheim, P. (1981). Host pathogen interactions. 18. Isolation and biological activity of glycinol, a pterocarpan phytoalexin synthesized by soybeans, *Plant Physiol.*, **68**, 358–363.

Yoshikawa, M. (1978). Diverse modes of action of biotic and abiotic phytoalexin elicitors, *Nature (London)*, **275**, 546–547.

Yoshikawa, M., Masago, H., and Keen, N. T. (1977). Activated synthesis of polyriboadenylic acid containing messenger RNA in soybean hypocotyls inoculated with *Phytophthora megaspema* var. *sojae, Physiol. Plant Pathol.*, **10**, 125–138.

Yoshikawa, M., Yamauchi, K., and Masago, H. (1978a). *De-novo* messenger RNA and protein synthesis are required for phytoalexin mediated disease resistance in soybean hypocotyls, *Plant Physiol.*, **61**, 314–317.

Yoshikawa, M., Yamauchi, K., and Masago, H. (1978b). Glyceollin: its role in restricting fungal growth in resistant soybean hypocotyls infected with *Phytophthora megasperma* var. *sojae, Physiol. Plant Pathol.*, **12**, 73–82.

Yoshikawa, M., Yamauchi, K., and Masago, H. (1979). Biosynthesis and biodegradation of glyceollin by soybean hypocotyls infected with *Phytophthora megasperma* var. *sojae, Physiol. Plant Pathol.*, **14**, 157–169.

Yoshikawa, M., Matama, M. and Masago, H. (1981). Release of a soluble phytoalexin elicitor from mycelial walls of *Phytophthora megasperma* var. *sojae* by soybean tissues, *Plant Physiol.*, **67**, 1032–1035.

Zaehringer, U., Ebel, J., Kreuzaler, F., and Grisebach, H. (1977). Biosynthesis of the elicitor induced phytoalexin glyceollin in soybean, *Hoppe-Seyler's Z. Physiol. Chem.*, **358**, 1303–1304.

Zaehringer, U., Ebel, J., and Grisebach, H. (1978). Induction of phytoalexin synthesis in soybean elicitor induced increase in enzyme activities of flavonoid biosynthesis and incorporation of mevalonate into glyceollin, *Arch. Biochem. Biophys.*, **188**, 450–455.

Zaehringer, U., Ebel, J., Mulheirn, L. J., Lyne, R. L., and Grisebach, H. (1979). Induction of phytoalexin synthesis in soybean dimethylallylpyrophospate : trihydroxypterocarpan dimethylallyl transferase from elicitor induced cotyledons, *Fed. Eur. Biochem. Soc. Lett.*, **101**, 90–92.

Zaehringer, U., Schaller, E., and Grisebach, H. (1981). Induction of phytoalexin synthesis in soybean. Structure and reactions of naturally occurring and enzymatically prepared prenylated pterocarpans from elicitor-treated cotyledons and cell cultures of soybean, *Z. Naturforsch., Teil C*, **36**, 234–241.

Ziegler, E., and Pontzen, R. (1982). Specific inhibition of glucan-elicited glyceollin accumulation in soybeans by an extracellular mannan-glycoprotein of *Phytophthora megasperma* f. sp. *glycinea, Physiol. Plant Pathol.*, **20**, 321–331.

Further Reading

Athow, K. L. (1973). Fungal Diseases in *Soybeans: Improvement, Production, and Uses* (Ed. B. E. Caldwell), Am. Soc. Agron., Madison, WI, pp. 459–489.

Cruickshank, I. A. M. (1963). Phytoalexins, *Annu. Rev. Phytopathol.*, **1**, 351–374.

Keen, N. T. (1981). Evaluation of the role of phytoalexins, in *Plant Disease Control: Resistance and Susceptibility* (Eds. R. Staples and G. Toenniessen), Wiley, New York, pp. 166–177.

Keen, N. T., and Bruegger, B. (1977). Phytoalexins and chemicals that elicit their production in plants, in *Host Plant Resistance to Pests* (Ed. P. A. Hedin), *Am. Chem. Soc. Symp. Ser.* No. 62, 1–26.

Paxton, J. D. (1975). Phytoalexins, phenolics and other antibiotics in roots resistant to soil-borne fungi, in *Biology and Control of Soil-Borne Plant Pathogens* (Ed. G. W. Bruehl), Am. Phytopathol. Soc., St. Paul, MN, pp. 185–192.

Biochemical Plant Pathology
Edited by J. A. Callow
© 1983 John Wiley & Sons Ltd

3

Flax Rust: A Case Study

MICHAEL D. COFFEY

Department of Plant Pathology, University of California, Riverside, CA 92521, USA

1. INTRODUCTION

Flax rust, *Melampsora lini* (Ehrenb.) Lév., is the cause of a serious disease of cultivated flax, *Linum usitatissimum* L., in the major producing regions of the world including Australia, New Zealand, Egypt, and the USA. The life cycle of flax rust is termed macrocyclic since it consists of all five possible spore stages, namely basidiospore, pycniospore, aeciospore, urediospore and teliospore. It is also referred to as autoecious since all spore stages occur on one host, as opposed to heteroecious where more than one host is involved. The urediospore stage is a repeating stage and the one involved in spread of the disease from plant to plant.

Traditionally, rust fungi have been regarded as obligate parasites incapable of growth on culture media; however, in recent times this concept has been challenged. Several rust species, including *M. lini*, have been successfully cultured on simple defined media containing a sugar such as sucrose or glucose and amino acids including a sulphur amino acid (Scott and Maclean, 1969; Coffey, 1975). Nevertheless, there is no evidence for a saprophytic mode of existence for these fungi in nature, and in the ecological sense they can still be regarded as obligate parasites.

Rust fungi such as *M. lini* possess a complete dependence on their living hosts for nutrition and completion of their life cycle. This biotrophic dependence on the host is usually regarded as the ultimate outcome of evolution away from necrotrophism where the pathogen lives saprophytically on dead host tissue. The evolution of biotrophism undoubtedly involves the development of sophisticated physiological mechanisms which allow the metabolisms of host and parasite to operate in concert. The nature of biotrophism, however, still eludes definition in even the barest physiological terms.

An important underlying principle of biotrophism is the requirement for either the passive avoidance or active suppression of a host resistance response. A large number of biotypes of the pathogen exist, each biotype or 'physiological race' having a unique set of avirulent genes which act in a highly specific fashion with yet another set of resistance genes in the host (Flor, 1971).

2. LIFE CYCLE OF THE DISEASE

The pathogen *M. lini* belongs to the order Uredinales, class Basidiomycetes. The basidiospore has a haploid nucleus and is borne on a structure termed a sterigma, one of four which develop from a four-celled metabasidium. The basidiospore germinates to produce a small appressorium and penetrates the host leaf directly (Littlefield and Heath, 1979).

A pycnium develops on the upper surface of the leaf from 4 to 6 days following basidiospore infection. The pycnia of *M. lini* are somewhat flattened structures, positioned so that the host stomata comprise definite openings or ostioles. They produce the pycniospores or spermatia which constitute the male gametes of the rust fungi. The pycniospores are released into a matrix of honeydew in the pycnial cavity, exude through the ostiole, and are transferred by insects to other pycnia. Pycnia can be of two mating types, + or −, and a successful dikaryon is established only where opposite types interact. The process of plasmogamy is believed to involve the production of a hyphal branch by the flexuous hypha and a small papilla by the pycniospore. The two structures subsequently fuse to establish a binucleate dikaryon (Buller, 1950).

An alternative mechanism has been proposed for formation of the dikaryon in *M. lini* where the pycniospores germinate to produce infection hyphae, which grow into the host tissues and subsequently anastomose with similar hyphae of the opposite mating type (Allen, 1934). Good evidence substantiating either mechanism of dikaryotization is lacking at present.

Following formation of the dikaryon, aecia are formed which produce aeciospores. The aeciospores are a non-repeating spore stage, which infect the host to produce uredia. Urediospores are oval in shape with a spiny surface and, since they are a repeating spore stage, they are very important in the spread of the pathogen. Infections caused by urediospores result in the development of further uredia or the formation of the next stage in the life-cycle, telia.

Telia in *Melampsora* spp. comprise a sub-epidermal layer of single-celled, columnar teliospores. Teliospores do not infect the host, but germinate to form a structure known as a metabasidium, which lies on the plant surface. The two nuclei of the dikaryon fuse in the metabasidium, meiosis takes place and four haploid nuclei are produced. The metabasidium becomes separated by septa into four cells. Each cell develops a conical sterigma which bears a haploid basidiospore (Littlefield and Heath, 1979). Following its release the badidiospore can germinate, and infect a new flax host.

3. GENETICS OF THE HOST–PARASITE INTERACTION

The work of H. H. Flor in North Dakota established the existence of genetic complementarity between a series of genes for avirulence in the pathogen and a similar series of genes for resistance in the host (Flor, 1971). This gene-for-gene concept serves to explain the phenotypic expression of resistance or susceptibility in terms of the interaction of specific genes in the host and parasite. Using a classical Mendelian approach, Flor examined the segregation patterns of the F2 generations of both host and pathogen and found that resistance in the host and avirulence in the pathogen were always inherited as dominant traits. An important consequence of the concept is that the phenotypic expression of susceptibility results whenever either the host genome contains a double recessive for susceptibility (rr), or the pathogen genome has a double recessive for virulence (aa). Resistance is expressed only when both the host and pathogen have genomes containing the dominant genes for resistance (R) and avirulence (A), either in a homozygous or heterozygous condition. This is illustrated in Figure 1.

To date, 29 genes for resistance have been identified, occurring as closely linked or allelic series at 5 loci in flax (Figure 2). The question of whether the genes at a single locus are truly allelic or are closely linked has been questioned (e.g. Shepherd and Mayo, 1972). Flor (1965) produced evidence that crossing over in the progenies of M paired with either M^1, M^3, or M^4 ranged from 0.17 to 0.39%. Interestingly, no carryover occurred when M^1, M^3, or M^4 were paired. No crossing over was detected in over 30 000 progeny of L heterozygotes. He deduced that the genes at the M locus were closely linked whereas those at the L locus were allelic. Although at first sight the distinction between these two possibilities may seem academic, there is one important

	KK	Kk	kk
$A_k A_k$	R	R	S
$A_k a_k$	R	R	S
$a_k a_k$	S	S	S

Figure 1. Quadratic check of the interaction of the host and pathogen genomes and their phenotypic expression, either susceptible (S) or resistant (R). The genomes in question are KK : Kk : kk in the host and the corresponding avirulence/virulence genes $A_k A_k$:$A_k a_k$:$a_k a_k$ in the pathogen

practical consideration. With close-linkage the possibility exists of incorporating more than five resistant genes in flax.

The evolutionary significance of close-linkage and allelism is unclear, although some healthy speculation on their possible origins has been made by Shepherd and Mayo (1972). They proposed that starting with a single resistance gene, rare unequal crossing over could have resulted in the production of serial duplications of the gene, and led to a divergence in gene function. The pattern of close-linkage as seen at the M locus could have evolved in this fashion. With allelism typified by the L locus the most likely origin would

Loci	No. of genes
K	1
L	13
M	7
N	3
P	5

Figure 2. Number of resistance genes in flax (data taken from Lawrence et al., 1981)

have been through mutation at different sites within the same gene. The implications of such an origin of allelism and close-linkage are that the products of different genes at a single locus would be variants of the original gene product. In the case of the L locus this would imply the existence of at least 13 variant products.

The data available on the inheritance of pathogenicity in flax rust is less complete than that of host resistance, although 24 genes have been identified, and in some cases close-linkage has been demonstrated between specific avirulence genes: A_L7 to A_L5 and A_L6, A_L10 to A_L3 and A_L4. In several instances not one but two allelic gene pairs in the pathogen have been shown to be involved in pathogenicity. The additional pair of genes has been termed an 'inhibitor' gene pair (I/i) because in its presence the avirulence gene pair (A/a) is not expressed phenotypically (Lawrence et al., 1981). The known inhibitor genes are all dominant in character and have been designated I_{M1}, I_{L1}, I_{L7}, and I_{L10}, being associated with the M^1, L^1, L^7, and L^{10} resistant genes, respectively, in the host. Since both the avirulence and 'inhibitor' genes are dominant it follows that the only combinations where avirulence is expressed phenotypically are genotypes ii Aa and ii AA. With the other genotypes II Aa, Ii Aa, II AA, and Ii AA, II aa, Ii aa, and ii aa the phenotypic expression is one of virulence.

An important conclusion from a consideration of the genetics of the interaction of flax and flax rust is that there is high specificity between individual resistance genes and individual avirulence genes in host and pathogen. In contrast, susceptibility and virulence appear to represent the phenotypic expression of a lack of any specific interaction of the two genomes (see also Chapter 13).

4. MORPHOLOGICAL AND STRUCTURAL FEATURES OF THE HOST–PARASITE INTERACTION

4.1. The compatible state

There are two parasitic phases in the life cycle where the fungus establishes a biotrophic relationship with its host. In the haploid or monokaryotic phase the basidiospores infect the host and parasitize it, intercellular hyphae growing and branching in the intercellular spaces of the host. In addition, the host cells are penetrated by hyphal-like 'intracellular' structures, but are separated from the host cytoplasm by the host plasma membrane. These fungal organs have been termed monokaryotic or M-haustoria (Littlefield and Heath, 1979). In the dikaryotic phase of the cycle, aeciospores or urediospores give rise to a parasitic colony comprising intercellular hyphae and characteristic intracellular structures with a narrow specialised neck region and an expanded roughly spherical body. These latter fungal structures have been termed dikaryotic or D-haustoria (Littlefield and Heath, 1979).

Monokaryotic or M-haustoria are generally filamentous in shape and may also be branched. Visually M-haustoria lack any specialization of their cell walls, in contrast to the situation to be described with D-haustoria. M-haustoria generally possess a septum which often forms close to the site of penetration, and is identical with the septal structures present in intercellular hyphae. The septal wall tapers towards a central pore which is plugged with a pulley wheel-shaped structure and associated with this is a group of small microbodies usually situated at the periphery of a small zone of organelle-free cytoplasm (Coffey, et al., 1972). This relatively complex arrangement of the septum serves to separate the nuclei and cytoplasmic organelles of each cell, while presumably still allowing interchange of soluble metabolites.

The M-haustoria are surrounded by a membrane continuous with that of the host plasmalemma. This extrahaustorial membrane is separated from the haustorial wall by a space which has been termed the extrahaustorial matrix (Bushnell, 1972). The matrix of M-haustoria is generally regarded as a real entity and not an artifact of fixation since it contains some stainable material, although its origin and chemical nature are undetermined. Overall the M-haustorium shows little of the fine-structural specialization characteristic of the D-haustorium.

The D-haustorium is not merely an intracellular extension of the intercellular hyphal network. Rather it represents a highly specialized morphological adaptation. Such haustoria possess a narrow neck region, the most distinctive feature of it being an extremely electron-dense band midway along its length (Figure 3). In addition, the host membrane surrounding the haustorium and fungal plasmalemma are tightly associated with the haustorial wall in the neck region. Apparently, this neckband region presents an effective barrier to the apoplastic transport of solutes (Littlefield and Heath, 1979). This situation would favour direct uptake of host solutes into the haustorial body, since it would severely restrict any passive apoplastic flow out of the host cell via the fungal cell wall. A neckband is also associated with the haustoria of several powdery mildew species and seems to function similarly (Bushnell and Gay, 1978). By contrast, no structural feature that would prevent apoplastic flow has been characterized as a component of the M-haustorium of rust fungi.

The membrane surrounding the D-haustorium is in direct continuity with the host plasmalemma (Figure 3; Littlefield and Bracker, 1970; Coffey et al., 1972). Whether the membrane around the haustorium is functionally similar to the host plasmalemma is less clear. The extrahaustorial membrane does not stain intensely with a phosphotungstic acid/chromic acid procedure which selectively stains the plasmalemma of host and fungus (Littlefield and Bracker, 1972). However, this does not necessarily indicate a fundamentally

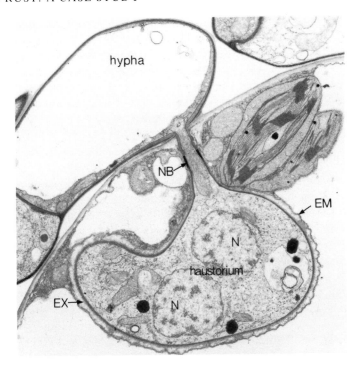

Figure 3. A D-haustorium with the two nuclei (N) visible within a host cell. The neck region of the fungal haustorium is distinguished by the presence of a dark-staining neckband (NB). Surrounding the haustorium is the extra-haustorial membrane (EM), which is separated from the body of this fungal organ by a zone termed the extrahaustorial matrix (EX). The infection is 5 days old and in a moderately incompatible host possessing the KK genome. Note the presence of patches of fibrillar material which have accumulated in the matrix. ×8 000. (From Coffey, 1975; reproduced by permission of Cambridge University Press)

different function for this membrane, but may merely reflect its lack of involvement in cell wall synthesis. The staining technique could be picking up carbohydrate or glycoprotein components present on the plasma membrane surface and destined for the cell wall, rather than actually staining specific membrane components such as phospholipid and protein. It is the inhibition of cell wall synthesis by the host plasmalemma when present around the haustorium that deserves more critical attention. The matrix between the host membrane and the haustorial body wall has an electron-lucent appearance (Figure 3) and it is generally assumed to be a fluid in consistency (Coffey, 1975; Littlefield and Heath, 1979). There is also the possibility that this electron-lucent extrahaustorial matrix is an artifact caused by the shrinkage, due to loss of turgor, of the haustorial wall away from the plasmalemma during fixation for electron microscopy (Allen et al., 1979).

A feature associated with both the M- and D-haustoria is the appearance of profiles of rough endoplasmic reticulum (RER) in the host cytoplasm. An equally consistent feature is the absence of any obvious activity of the Golgi apparatus. The significance of this in physiological terms is unclear, although there is the possibility that the host is induced to synthesize new materials within the RER and these may be destined for the haustorium, perhaps being transported across the extrahaustorial membrane.

4.2. The incompatible state

Several cytological studies with flax rust have been made using various iso-genic lines of the host containing different major genes for resistance (Little-field and Aronson, 1969; Littlefield, 1973). Criteria such as the number of haustoria and the area of necrotic host tissue produced were used to evaluate the degree of resistance expressed by particular host genotypes (Littlefield, 1973). Macroscopically, symptoms on the genotypes LL, MM, NN, and PP were almost indistinguishable, consisting at the most of minute chlorotic flecks. Microscopically, differences were apparent with the MM and PP reactions giving much more host necrosis than the LL and NN types, where necrosis was more rapid and more restricted in nature (Littlefield, 1973). Additionally, small differences were noted between MM and PP in the timing of the onset of necrosis, with host cell death being observed 18–24 h sooner in PP than in MM. An intermediate type of resistance, involving the KK genotype where small- to medium-sized uredia were reproduced, behaved differently (Littlefield and Aronson, 1969; Littlefield, 1973). At 7 days after inoculation the fungus had produced almost as many haustoria at a single infection site as in the susceptible interaction (Littlefield and Aronson, 1969). Some host cell necrosis was observed in KK 36 h after infection and, eventually, the total area of necrosis often became greater than in the more resistant interactions, although the 'severity of necrosis' was stated to be 'much less' (Littlefield, 1973). In an ultrastructural study of the same interaction it was determined that indeed some host cells had become necrotic and collapsed, constituting about 13% of those infected at 4 days after infection (Coffey, 1976). More significantly, the band of dark-staining host cells which developed centripetally in association with the expanding fungal colony were not necrotic as visualized in the electron microscope (Coffey, 1976). These cells frequently contained large, dense osmiophilic bodies. Conceivably these bodies contained material which was responsible for the dark-staining charac-teristics seen in the light microscope. Another feature correlated with resis-tance was the appearance of glycogen-like granules in place of starch in the chloroplasts of infected and non-infected cells of the KK interaction (Coffey, 1976). These ultrastructural changes indicated a profound alteration in metabolism in the incompatible host, but did not demonstrate the presence of any marked necrosis.

In all of the incompatible interactions studied, the development of the first haustorium takes place and there is no measurable difference in its final size or its morphology (Littlefield, 1973). In the KK interaction more subtle ultrastructural changes are detectable in the haustorial apparatus several days after inoculation. The normally electron-lucent extrahaustorial matrix becomes partially filled with moderately electron-dense material (Figure 3). The nature and origin of this material is unknown, although the fact that it is continuous with a distal region of the haustorial neck wall may indicate that it is fungal in origin (Coffey, 1976). Similar ultrastructural studies of the other incompatible genotypes should prove illuminating.

5. BIOCHEMICAL AND PHYSIOLOGICAL FEATURES OF THE HOST–PARASITE INTERACTION

5.1. The compatible state

The nature of biotrophism as exemplified by pathogens such as flax rust has received much attention. A dependence on the host for highly specific metabolites such as nucleoproteins was long proposed as a basis for such parasitism, although the ability to culture these fungi axenically detracts from this possibility. The possibility remains that at the most fundamental level there are transcriptional changes in the patterns of RNA synthesis and/or its translation into protein and these may determine the outcome of disease development (Chakravorty and Shaw, 1977a). A study of the *in vivo* ^{32}P-labelling pattern of RNA in flax infected with *M. lini* revealed a 2–3-fold increase in ^{32}P incorporation 48 h after infection. Further, an NaCl-soluble fraction of the newly synthesized RNA species had base ratio characteristics atypical of either the host or pathogen prior to inoculation. A hypothesis has been proposed that involves derepression of host genes, and intercistronic complementation of oligomeric enzymes in host and pathogen, leading to the production of unique enzymes such as RNA polymerases and RNases, critical for the establishment of a compatible interaction (Chakravorty and Shaw, 1977b; see also Chapter 18). An incompatible interaction would arise when genetic changes in either the host or pathogen lead to a change intercistronic complementation such that the function of the oligomeric enzyme is altered resulting in a partial or complete malfunction. A tacit assumption of the above hypothesis seems to be that successful biotrophic parasitism and the avoidance of resistance are represented by the same cellular and molecular events.

A critical consequence of the establishment of the various molecular and metabolic conditions which regulate biotrophism is the establishment of a successful nutritional relationship in which the parasite acts as a powerful sink for host photosynthetic assimilates (Lewis, 1976). Conceivably, the increasing demand of the parasite for nutrients as it develops in the host tissue in itself leads to the characteristic changes in the host, including alterations in translocation patterns, accumulation of starch and sucrose, and enhancement of invertase levels (Clancy and Coffey, 1980). Apart from providing hexoses for uptake and utilization by the pathogen, increased invertase activity probably stimulates starch synthesis. The hexoses not immediately taken up by the pathogen can undergo glycolysis to triose phosphates which are capable of transportation across the chloroplast membranes where they can be converted to starch (Lewis, 1976). The effectiveness of the rust fungus in creating a sink for metabolites is illustrated by the fact that at 11 days after infection 45% of the labelled metabolites were in the form of fungal polyols only 30 min after supplying $^{14}CO_2$ to the host (Clancy and Coffey, 1980).

5.2. The incompatible state

In genetic terms the expression of incompatibility has been regarded as a single phenomenon, which in reality it is not, as can be seen by a cytological examination of individual interactions (Littlefield, 1973). Nevertheless, the genetic approach affords one fair conclusion, namely that the specificity of the host–parasite interaction involves avirulence in the pathogen and resistance in the host. In the absence of either there is no specific interaction. Physiologically the implication is that the products of host resistance genes and pathogen avirulence genes interact with each other in a specific fashion. Theories have been advanced to explain the molecular basis of the host–parasite interaction (e.g. Chakravorty and Shaw, 1977a,b). Experimental data have been obtained on the susceptible interaction and relate to the possible derepression of host genes and the production of specific RNA polymerases (Chakravorty and Shaw, 1977a). These could reflect changes in the host–parasite interaction necessary for the establishment of a successful 'source–sink' relationship characteristic of flax rust disease. Any malfunction in this presumably complex series of metabolic interactions could lead to a breakdown in the compatability of the interaction. Alternatively, there is the possibility that resistance operates at an entirely different level of metabolism, unrelated to the establishment of a 'source–sink' relationship.

Evidence has been obtained that the flax host produces phytoalexin-like compounds which may directly interfere with the development of *M. lini* (Keen and Littlefield, 1979). These compounds were determined to be the phenylpropanoids coniferyl alcohol and coniferyl aldehyde. They were produced in minimal amounts in compatible interactions, and in incompatible interactions there was a positive correlation between their rate of production and the restriction of growth of the pathogen. For instance, in a comparison of different genes of the L locus, the highly resistant LL genotype caused the accumulation of approximately 5 μg of coniferyl aldehyde per gram fresh weight by day 2 after inoculation. The L^8L^8 genotype, which allowed more development of the pathogen, did not produce similar amounts of the aldehyde until 3 days after inoculation. In the moderately incompatible L^7L^7 genotype, the amount produced was significantly less than either LL or L^8L^8 at 4 days after inoculation. In *in vitro* tests inhibition of urediospore germination occurred at 140 μg ml^{-1} with coniferyl alcohol and 40 μg ml^{-1} with coniferyl aldehyde (Keen and Littlefield, 1979). The antifungal properties of these phenylpropanoids and the good correlation which exists between the timing and amount of their production following infection indicates that they probably have a role in restricting fungal development *in vivo*. Coniferyl alcohol is a precursor of lignin, although no evidence was obtained that lignin biosynthesis took place in incompatible interactions. The *modus operandi* of these compounds remains unclear, since more information needs to be acquired on how, and at what concentrations, they exert their effects *in vivo*.

In the moderately incompatible interactions, such as those involving the genotypes L^7L^7 and KK, the production of phenylpropanoids is considerably delayed, and the size of the 'disease lesion' is large compared with the more incompatible types (Keen and Littlefield, 1979). Another mechanism of resistance may operate here, possibly one involving a restriction of the transfer of nutrients from host to parasite. In a study of the KK genotype, the presence of infection sites failed to influence the import or export of host assimilates even at advanced stages of disease, whereas in the susceptible interaction there was an increased import into infection sites (Clancy and Coffey, 1980). Acid invertase levels did not increase in KK, whereas in the susceptible interaction levels increased from day 4 after inoculation coincident with the onset of sporulation. In the susceptible interaction fungal polyols, both hexitols and pentitols, became increasingly labelled with ^{14}C, derived from ^{14}CO$_2$ fed to the host as the disease progressed. In contrast, even at 11 days after inoculation only 9% of the ethanol-soluble ^{14}C extracted from infected KK tissue 30 min after exposure to ^{14}CO$_2$ was in the form of fungal polyols (Clancy and Coffey, 1980). It is probable that a critical biomass of fungus is required to establish an effective 'source–sink' relationship between host and parasite (Holligan, *et al.*, 1974). It is unclear what causes the failure to achieve this biomass. Is it due to the lack of activation of acid invertase, for instance? Or is it due to the production of coniferyl alcohol and aldehyde, which retards fungal development sufficiently to prevent achievement of this biomass? An important role of invertase is in providing hexoses for uptake and utilization by the pathogen. It also facilitates the 'sugar-feeding' of infected host tissues, since hexoses are readily broken down via glycolysis to triose phosphates, which are then available to the chloroplast for starch synthesis (Lewis, 1976). Since flax rust mycelium grown *in vitro* is capable of utilizing sucrose (Coffey, 1975) and produces an active acid invertase (Clancy, 1979), perhaps the surprising fact is that *in vivo* this activity is suppressed until the onset of sporulation (Clancy and Coffey, 1980).

In summary, at its most fundamental level changes in transcription patterns of RNA synthesis may determine the outcome of disease development (Chakravorty and Shaw, 1977a,b) and these changes could be the products of the resistance and avirulence genes (Flor, 1971). In more general terms the expression of high incompatibility is seen as involving the production of fungitoxic phenylpropanoids by the host which restrict further development of the parasite (Keen and Littlefield, 1979). In moderately incompatible interactions failure to establish an efficient metabolic sink for transfer of host nutrients may be a factor (Clancy and Coffey, 1980).

6. FUTURE PROSPECTS

The future for research in this field is challenging and exciting. We have a genetic model *par excellence*, upon which to base molecular biological studies of the type pioneered by Chakravorty and Shaw (1977a,b). The most critical need, and undoubtedly the most elusive, will be identification of the initial biomolecular interactions which take place between host and parasite. Electron microscopy has revealed that a likely site of such events will be intracellular at the interface between the host protoplast and the fungal haustorium. A combination of biochemical, cytochemical and genetic approaches will be needed in order to reveal the physiological basis of biotrophism and incompatibility in the interaction between flax rust and its host.

REFERENCES

Allen, R. F. (1934). A cytological study of heterothallism in flax rust, *J. Agric. Rs.*, **49**, 765–791.

Allen, F. H. E., Coffey, M. D., and Heath, M. C. (1979). Plasmolysis of rusted flax: a fine-structural study of the host-pathogen interface, *Can. J. Bot*, **57**, 1528–1533.

Buller, A. H. R. (1950). *Researches on Fungi*, Vol. VII, University of Toronto Press, Toronto.

Bushnell, W. R. (1972). Physiology of fungal haustoria, *Annu. Rev. Phytopathol.*, **10**, 151–176.

Bushell, W. R., and Gay, J. (1978). Accumulation of solutes in relation to the structure and function of haustoria in powdery mildews, in *The Powdery Mildews* (Ed. D. M. Spencer), Academic Press, London, New York and San Francisco, pp. 183–235.

Chakravorty, A. K., and Shaw, M. (1977a). The role of RNA in host–parasite specificity, *Annu. Rev. Phytopathol.*, **15**, 135–151.

Chakravorty, A. K., and Shaw, M. (1977b). A possible molecular basis for obligate host–pathogen interactions, *Biol. Rev.*, **52**, 147–179.

Clancy, F. G. (1979). Comparative carbohydrate physiology of susceptible and resistant flax in response to inoculation with the flax rust, *Doctoral Thesis*, Trinity College, Dublin.

Clancy, F. G., and Coffey, M. D. (1980). Patterns of translocation, changes in invertase activity, and polyol formation in susceptible and resistant flax infected with rust fungus *Melampsora lini*, *Physiol. Plant Pathol.*, **17**, 41–52.

Coffey, M. D. (1975). Obligate parasites of higher plants, particularly rust fungi, *Symp. Soc. Exp. Biol.*, **29**, 297–323.

Coffey, M. D. (1976). Flax rust resistance involving the K gene: an ultrastructural survey, *Can. J. Bot.*, **54**, 1443–1457.

Coffey, M. D., Palevitz, B. A., and Allen, P. J. (1972). The fine structure of two rust fungi, *Puccinia helianthi* and *Melampsora lini, Can. J. Bot.*, **50**, 231–240.

Flor, H. H. (1965). Tests for allelism of rust-resistance genes in flax, *Crop Sci.*, **5**, 415–418.

Flor, H. H. (1971). Current status of the gene-for-gene concept, *Annu. Rev. Phytopathol.*, **9**, 275–296.

Holligan, P. M., Chen, C., McGee, E. E. M., and Lewis, D. H. (1974). Carbohydrate metabolism in healthy and rusted leaves of coltsfoot, *New Phytol.*, **73**, 881–888.

Keen, N. T., and Littlefield, L. J. (1979). The possible association of phytoalexins with resistance gene expression in flax to *Melampsora lini, Physiol. Plant Pathol.*, **14**, 265–280.

Lawrence, G. J., Mayo, G. M. E., and Shepherd, K. W. (1981). Interactions between genes controlling pathogenicity in the flax rust fungus, *Phytopathology*, **71**, 12–19.

Lewis, D. H. (1976). Interchange of metabolites in biotrophic symbioses between angiosperms and fungi, in *Perspectives in Experimental Biology, Vol. 2, Botany* (Ed. N. Sutherland), Pergamon Press, Oxford and New York, pp. 207–219.

Littlefield, L. J. (1973). Histological evidence for diverse mechanisms of resistance to flax rust, *Melampsora lini* (Ehrenb.) Lev., *Physiol. Plant Pathol.*, **3**, 241–247.

Littlefield, L. J., and Aronson, S. J. (1969). Histological studies of *Melampsora lini* resistance in flax, *Can. J. Bot.*, **47**, 1713–1717.

Littlefield, L. J., and Bracker, C. E. (1970). Continuity of host plasma membrane around haustoria of *Melampsora lini, Mycologia*, **62**, 609–614.

Littlefield, L. J., and Bracker, C. E. (1972). Ultrastructural specialization at the host–pathogen interface in rust infected flax, *Protoplasma*, **74**, 271–305.

Littlefield, L. J., and Heath, M. C. (1979). *Ultrastructure of Rust Fungi*, Academic Press, New York, San Francisco and London.

Scott, K. J., and Maclean, D. J. (1969). Culturing of rust fungi. *Annu. Rev. Phytopathol.*, **7**, 123–146.

Shepherd, K. W., and Mayo, G. M. E. (1972). Genes conferring specific plant disease resistance, *Science*, **175**, 375–380.

Further Reading

Alexopoulos, C. J., and Mims, C. W. (1979). Class Basidiomycete Subclass Teliomycetidae Rusts, Smuts, Basidiomycetous yeasts, in *Introductory Mycology*, 3rd ed., Wiley, New York, pp 449–521.

Chakravorty, A. K., and Shaw, M. (1977). The role of RNA in host-parasite specificity, *Annu. Rev. Phytopathol.*, **15**, 135–151.

Coffey, M. D. (1975). Obligate parasites of higher plants, particularly rust fungi, *Symp. Soc. Exp. Biol.*, **29**, 297–323.

Littlefield, L. J., and Heath, M. C. (1979). *Ultrastructure of Rust Fungi*, Academic Press, New York, San Francisco and London.

Biochemical Plant Pathology
Edited by J. A. Callow
© 1983 John Wiley & Sons Ltd

4

Fireblight—A Case Study

ROBERT N. GOODMAN

Department of Plant Pathology, University of Missouri, Columbia, MO 65211, USA

1. INTRODUCTION

The history of fireblight from its earliest report in 1793 to 1979 was compiled by van der Zwet and Keil (1979) and published by the US Department of Agriculture. This compendium on the most serious disease of rosaceous plants, including apple, pear, quince, and numerous ornamentals, caused by the Gram-negative bacterium *Erwinia amylovora* (Burrill) (Winslow *et al.*, 1920), contains data from 1077 journal articles and 62 graduate student dissertations. It is clearly a labour of love by its authors and is an anthology on the subject with which every student of the pathogen should be familiar.

The pathogen attacks 21 genera containing 39 species in the family Rosaceae. Most severely affected genera and perhaps most important economically are *Malus, Pyrus, Cotoneaster, Crataegus, Cydonia, Pyracantha,* and *Sorbus*. Within these genera there are species and varieties that vary widely in their susceptibility to the pathogen.

It is not the purpose of this case history either to recapitulate or to summarize the broad spectrum of investigational effort and results thereof that have been catalogued by van der Zwet and Keil (1979). Rather it is the author's intention to acquaint the reader with basic information concerning the pathogen, the infection process, and the disease syndrome.

2. THE PATHOGEN

Erwinia amylovora is a Gram-negative short rod, it is peritrichously flagellated (Bryan, 1927) and in culture has an absolute requirement for nicotinic acid (Starr and Mandel, 1950). Pathogenesis is correlated with those isolates of the pathogen that are encapsulated (Bennett and Billing, 1978; Ayers *et al.*, 1979). Capsular material that is tightly adsorbed may be discerned with negative stains of skimmed milk (Vörös and Goodman, 1965) or Indian ink (Politis and Goodman, 1980). Apparently, excess capsule which is loosely held, is released into the environment of the pathogen *in vivo* (Eden-Green and Knee, 1974; Goodman *et al.*, 1974) and *in vitro* (Ayers *et al.*, 1979). The generation time of the pathogen varies with temperature, being approximately 80 min at its optimum of 28 °C *in vitro* (Hildebrand, 1954).

Although a number of mutant forms have been detected, the organism is stable in culture for long periods of time and has been kept in lyophile for periods as long as 10 years (unpublished data). This bacterium is, however, extremely osmotically sensitive in distilled water, e.g. suspensions of virulent isolates at 10^4 cells cm^{-3} reveal no live cells after 2 h (unpublished data). Verification of an isolate on the selective medium of Crosse and Goodman (1973) can be made with assurance owing to the unique cratering of the pathogen on this high sucrose medium. Further differentiation between virulent and avirulent isolates can be made as the former develop a much greater density of craters on their colony surfaces than the latter. It is also possible to differentiate virulent from avirulent isolates on triphenyltetrazolium medium (Ayers *et al.*, 1979) as the colonies of the virulent isolates have broad whitish mucoid peripheries with pink centres. Colonies of the avirulent ones, on the other hand, are smaller and are bounded by a narrow white border and have a deep reddish purple centre (Figures 1 and 2).

2.1. Inoculum source

One of the most perplexing facets concerning fireblight is the inoculum source for infections in the orchard, nursery, forest, hedgerow, and home planting

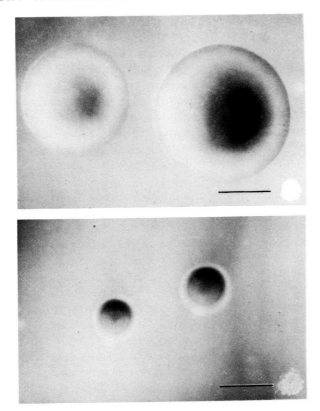

Figures 1 and 2. Bacterial colonies of *Erwinia amylovora* on triphenyltetrazolium chloride agar. Figure 1 (top): virulent encapsulated E_9 strain. Figure 2 (bottom): avirulent unencapsulated E_8 strain. Bar = 1 mm. (After Politis and Goodman, 1980)

from one growing season to the next. Reports that the 'primary' inoculum issues from hold-over cankers which is subsequently transported to suscept-ible organs such as blossoms by pollinating insects dominate the literature (Pierstorff, 1931; Rosen, 1932, 1933, 1936; Hildebrand, 1937; Keitt and Ivanoff, 1941). The overwintering canker fails to satisfy the observations of fireblight in the nursery row of year-old trees. Nor does it explain the appear-ance of blossom blight before or at the same time as cankers 'ooze' their bacterial masses. The oozing of comparatively few cankers and the appear-ance of uniformly infected acres of orchard and thousands of blossoms per tree requires a particularly efficient delivery system such as pollinating insects (Keitt and Ivanoff, 1941; Powell, 1964). However, the frequency with which bees and other insects have been observed to visit cankers reduces the univer-sality of this source–vector system as a regular occurrence where epiphytotics are concerned.

Attention is called to the frequent observation that it is difficult to make positive isolations from sites of infection in the autumn or winter after the growing season in which infection occurred. Rosen (1929, 1933, 1936) and others (Pierstorff, 1931; Ritchie and Klos, 1975) recorded positive isolations of 0.6–2.5%, depending on variety, from about 12 000 twigs. In addition, the numbers of bacteria obtained from tissue macerates (prior to incubation) were comparatively few. Hence, generally and importantly, few *E. amylovora* cells are actually retrieved from cankers in winter. The point for consideration here is that, given the paucity of holdover sites, the inoculum potential may be even further reduced because of the low numbers of bacteria that survive at the site.

The pathogen has been detected in wood of apple and pear over a range of caliper from a few millimetres to many centimetres. In general, greater success in isolating the pathogen, according to Rosen (1929) one of the most assiduous and competent early investigators, was from cankers on small branches (twigs). This was also the experience of Pierstorff (1931) and Ritchie and Klos (1975).

Rosen (1930, 1932) also stirred a significant controversy in his persistent and successful isolations of the pathogen from beehives collected throughout the summer, winter, and early spring. In addition, the pathogen was recovered from bees *per se* in early spring prior to the appearance of the disease in the orchard (Rosen, 1932).

Erwinia amylovora has been recovered from mummied fruits (Goodman, 1954) and buds (Baldwin and Goodman, 1963; Dueck and Morand, 1975; Paulin, 1981). Although the bud or leaf trace has been regarded as a logical site of 'primary' inoculum development in the cases of *Pseudomonas morsprunorum* (Crosse, 1957) in cherry and seems likely for *Xanthomonas pruni* in peaches (Feliciano and Daines, 1970) it has yet to be enthusiastically accepted for *E. amylovora*. However, it is the *opinion* of this author that bud infections are more prevalent than currently surmised. Some logic for this will be presented in the section on infection process.

3. INFECTION PROCESS

3.1. Reaching the host

The first step in the infection process is for the pathogen to reach the host. The manner in which *E. amylovora* usually, generally, or even frequently accomplishes this has yet to be established. most of our information comes from observations made after the fact or from observations of probable 'delivery systems.' For example, several researchers have reported the existence of aerial strands of polysaccharide-containing bacteria that could be carried by wind currents and/or washed from the atmosphere by rain on to blossoms. A detailed study of these aerial strands, their formation, and bacterial content was made by Keil and van der Zwet (1972). It has yet to be established that these strands are responsible for a significant number of infections under natural conditions. The studies of Bauske (1967), however, of large nursery stands becoming infected suggests this route as a possibility. The stand of year-old trees in an Iowa nursery was more uniformly infected in the centre of the field than those trees which were sheltered by windbreaks along the edges of the planting.

Bees that were infested with the pathogen and that had their flight carefully monitored clearly link this agent to widespread uniform blossom infections (Keitt and Ivanoff, 1941). Hence, given an inoculum source of an infected blossom or two, a few bees visiting several blossoms could cause a large number of infections. That number would necessarily increase geometrically if the inoculum was brought back to the hive during a major nectar and pollen collection period.

3.2. Probable portals of entry

The literature indicates that blossoms are the organs most susceptible to infection (Rosen, 1929, Hildebrand, 1937; Keitt and Ivanoff, 1941). This was substantiated for the author on seeing Williams (Barlett) and Laxton Superb pears growing side by side in an orchard on July 9, 1966, in Kent, England. On that date Laxton Superb was still developing late blossoms and was exhibiting infections in epidemic proportions. Williams, growing in alternate rows, had long since stopped blooming, most of its terminals had matured, and was completely free of infections. Williams is exceedingly susceptible in the USA where it blooms at a temperature conducive for the growth of the pathogen. In the UK, however, Williams blossoms during a period in which temperatures are too low for significant bacterial multiplication.

Since infections are frequently detected in apples and pears that are too young to bloom, other routes of penetration must be projected. There is good evidence that nectaries, hydathodes (Rosen, 1929; Keitt and Ivanoff, 1941; Goodman et al., 1974), exudative glands on leaves (Keil and van der Zwet, 1972; Huang et al., 1975) and, perhaps most importantly, wounds of leaves and stems, are all sites for pathogen ingress (Crosse et al., 1972). For example, leaf tearing or stem abrasion in a driving rain provide thousands of wounds in a given tree. In the event of accompanying hail these numbers are surely increased. Rosen (1929) and Tullis (1929) presented histological evidence of trans-stomatal penetration and infection of apple leaves by E. amylovora; however, whether such infections generally become systemic remains unclear. Experiments by Plurad et al. (1967) revealed that interveinal tissues of apple leaves were particularly recalcitrant to systemic infection by needle prick inoculation.

Discussion of establishment of infection by *E. amylovora* requires some comprehension of (a) the number of bacteria necessary to cause a systemic infection, (b) the size and fate of the inoculum at the carryover site (ICS) (overwintering site), and (c) the intrinsic sensitivity of the various portals of penetration to infection.

3.3. Number of bacteria necessary for systemic infections

By inoculating still expanding Jonathan apple leaves by clipping their apices and placing a measured dose of bacterial cells on the freshly exposed mid-rib, Crosse *et al.* (1972) found that an ID-50 was regularly achieved with 38 cells. Dilutions of the inoculum (Figure 3) suggested that 3–5% infection of Jonathan apple shoots could be achieved with one or two virulent cells. Hildebrand (1938) had previously concluded that an inoculum of a single *E. amylovora* cell is sufficient to cause systemic infection in an apple blossom.

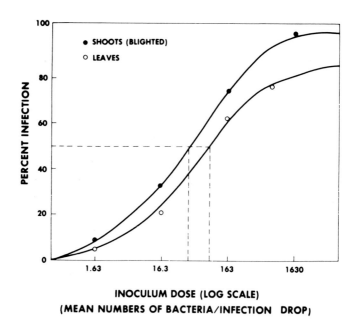

INOCULUM DOSE (LOG SCALE)
(MEAN NUMBERS OF BACTERIA/INFECTION DROP)

Figure 3. Effect of inoculum dose on the incidence of shoot and leaf blight (systemic infection) caused by *Erwinia amylovora*. Inoculum was administered as a measured droplet deposited on the exposed midrib of a Jonathan apple leaf following excision of its apical 0.5 mm. (After Crosse *et al.*, 1972)

3.4. Inoculum at carryover site, its size, and fate

Since the inoculum carryover at any site is probably comparatively few cells (based on the paucity of positive isolations and low bacterial numbers per site when made in winter), the conditions of temperature and moisture are crucial to the further development of the infection process and the establishment of the disease in the host. Although moisture during bloom seems to accentuate disease severity (van der Zwet and Keil, 1979), temperature during blossom development (from bud break to petal fall) appears to be the decisive factor determining disease intensity.

Several formulae for forecasting fireblight intensity have been proposed (Powell, 1964); however, the most recent and apparently most predictive one was devised by Billing (1980), based on fireblight incidence in different hosts over a 20-year period in southeast England. It is concerned primarily with rates of multiplication of the pathogen and hence incubation periods. Potential daily doublings of the bacterium (P.D.s) are related to the daily maximum and minimum temperatures. For maxima of 18–23 °C PD values range from 3.5–8.0, and for 24–30 °C, 9.0–12.5. Keeping in mind the data of Crosse *et al.* (1972) (Figure 3), the ID-50 of 38 cells should be rapidly reached at an ICS of one or two *E. amylovora* cells. What has not been devised is a rapid predictive measure for the number of ICSs likely to exist in a given orchard. The intensity of the developing disease would also depend on the nature of the ICS, e.g. canker *vs.* blossom bud. The latter would probably have the greater *dispersal potential*. Of importance here also would be the intrinsic susceptibility of tissues at the ICS.

3.5. Intrinsic susceptibility of the infectible site

An assessment of the voluminous literature summarized by van der Zwet and Keil (1979) suggests that the blossom is probably the most susceptible organ to the fireblight pathogen. This may be due to an insufficiently developed resistance capability, e.g. low levels of arbutin (in pear) and phloretin (in apple tissue) (Hildebrand and Schroth, 1963; Powell and Hildebrand, 1967, 1970). However, *E. amylovora* appears capable of entering the blossoms' nectarthodes, colonizing the surfaces of uncutinized stigmas, undehisced anthers, and stomata on sepals. There the parenchyma cells subsequently collapse owing to plasmolysis and necrosis develops (Hildebrand, 1937). From the early studies it is not clear how bacteria on the surfaces of organs such as the stigma or anther, or indeed the nectarthodal chamber, gain entry into xylem vessels for subsequent long-distance systemic movement (i.e. several centimetres in 24 h). These distances are greater than can be achieved by intracellular migration (Lewis and Goodman, 1965, 1966). However, from a study by Goodman and White (1981) to be described subsequently, a hypothesis is offered.

Although blossom infections are probably the most important early infections from an epidemiological point of view, movement of the pathogen from the blossom has not been as critically studied ultrahistologically as it has been from wound inoculations. An early histological study by Rosen (1929) of spray-inoculated (unwounded) blossoms suggests that in the process of petal unfolding, these fragile tissues 'rupture,' causing a wound through and into which the inoculum enters.

Systemic infection of either stem tissue or leaf petioles at increasing distances from a vigorously vegetative apex becomes proportionately difficult to effect. In the experience of the author there is no significant variation in the infectibility of blossoms or actively meristematic tissue of a susceptible apple cultivar such as the Jonathan. It would appear, then, that in the intercellular space and vessel lumina of immature fruit, blossom, and vegetative tissue, *E. amylovora* multiplies without significant restriction. Rosen (1933) concluded that primary blossom blight may originate as (1) internal extensions of the previous year's infection, (2) buds infected in the previous year, (3) infections from bacterial exudate produced the previous growing season, or (4) twig blight (which usually occurs later) caused by inoculum from overwintering cankers. It was Rosen's contention, apparently, that twig blight inoculum originated from overwintering cankers whereas blossom infections resulted from conditions 1–3 above. It is the author's opinion that the origin of blossom infections has been misunderstood and that extensions of the previous year's infections and bud infections *per se* may be regularly more important than cankers as ICSs; van der Zwet and Keil (1979) seem to have come to the same conclusion.

That the bud infection in the previous season is or may be an important ICS was suggested by Baldwin and Goodman (1963) and confirmed by Dueck and Morand (1975) and again most recently by Paulin (1981). Given this thought, it becomes difficult to attribute blossom infections to cankers if, in early spring, they both have low levels of virulent bacteria and both depend on favourable temperatures for significant numbers of P.D.s.

4. INFECTION PROCESS FROM ARTIFICIAL INOCULATIONS

4.1. Leaf inoculations

The study of Lewis and Goodman (1965) called attention to the rapidity with which streptomycin-resistant *E. amylovora* traversed long distances in apple shoots from leaf surface inoculations. They referred to this procedure as the leaf-smear method of inoculation, that apparently caused superficial injuries to hydathodes and exudative glands, exposing xylem vessels from which systemic infections developed.

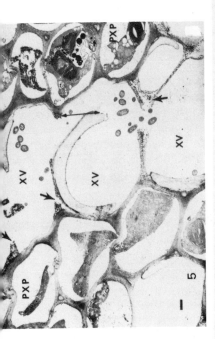

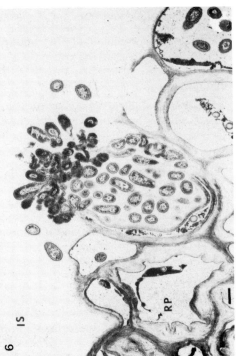

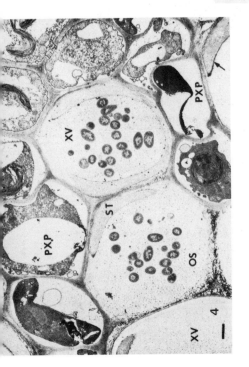

Figures 4–6. Infected apple leaf petiole tissue taken after 24 h, 2 mm from the site of inoculation with 5×10^4 cells of *Erwinia amylovora*. Figure 4: eight xylem parenchyma cells (PXP) reveal various stages of plasmolysis. Two xylem vessels (XV) contain bacteria and occluding substance (OS). Figure 5: a file of xylem vessels (XV) bordered by files of intensely plasmolysed parenchyma cells (PXP). Both middle lamella, secondary wall, and secondary wall thickenings have ruptured permitting bacteria to spill into what will become enlarged intercellular space. The rupture of XV occurs as a consequence of the plasmolysis and collapse of XP. Figure 6: a xylem vessel discharging bacteria into intercellular space (IS) created by the plasmolysis and subsequent collapse of both ray parenchyma cells (RP) and a large area of xylem parenchyma opposite the bacteria (not shown).
Bar = 1 μm. (After Goodman and White, 1981)

4.2. Leaf petiole and stem apex inoculations

Huang and Goodman (1976) and subsequently Suhayda and Goodman (1981a,b) established the xylem vessel lumen as the initial site of intense early proliferation of the fireblight pathogen. The deposition of 5×10^4 streptomycin-resistant *E. amylovora* cells on freshly delaminated leaf petioles of Jonathan apple revealed that for the first 48 h after inoculation, the bacteria grew only in the xylem lumina (exclusive of the apical 0.5 mm of the petiole, where bacteria also grew profusely in intercellular space). The same was true in young Jonathan apple shoots inoculated by needle puncture between the apex and first node. With the exception of the 0.5 mm above and below the needle puncture, bacterial proliferation was, during the initial 48 h after inoculation, exclusively in xylem lumina.

The ultrastructural study of Goodman and White (1981) confirmed the observations of Huang and Goodman (1976) and provided supporting evidence for the often reported plasmolysis of parenchyma cells and the development of lysigenous cavities filled with bacteria (Bachmann, 1913; Rosen, 1929; Hildebrand, 1937). The electron micrographs in Figures 4 and 5 suggest the following sequence of events.

Within 24 h after petiole inoculation, xylem vessels throughout the length of the petiole contain bacteria and bordering xylem parenchyma cells are in the process of plasmolysis. After 48 h many parenchyma cells are completely plasmolysed and their cell walls have begun to collapse (Figure 5). The vessel lumina contain considerably larger populations of bacteria and some of these are released from rupturing vessels into the intercellular space that has enlarged at the expense of collapsing xylem parenchyma cells (Figure 6). It is in this manner that the lysigenous cavities form and the bacteria in them proliferate luxuriously on the substrate being released from the plasmolysing parenchyma cells. It is from these sites, teeming with bacteria, that bacterial movement of *E. amylovora* proceeds in concert with the visual symptoms of wilt, tissue reddening, and necrosis. The author has hypothesized that vessel rupture, releasing bacteria, is due to the loss of xylem parenchyma turgor and collapse. A sequence such as this probably permits bacteria to enter small vessels in the area of hydathodes and nectaries (Hildebrand, 1937).

5. THE ROLE OF EXTRACELLULAR POLYSACCHARIDE (AMYLOVORIN) IN VIRULENCE

5.1. Amylovorin, its toxigenic activity and chemical and physical properties

The first inkling that amylovorin, the capsular extracellular polysaccharide (EPS) of *E. amylovora*, might be related to virulence came with Billing's association of capsule with phage sensitivity (Billing, 1960). However, the association was not clarified until Bennett and Billing (1978) and Ayers *et al.*

(1979) drew attention to the fact that bacteriophage-insensitive, EPS-deficient mutants of *E. amylovora* were avirulent.

The polysaccharide was incorrectly judged, by the author, to be a host-specific toxin (Goodman *et al.*, 1974) responsible for the wilt symptom of the disease syndrome. The EPS is toxic; however, it is non-specific (Ayers *et al.*, 1979). It does, nevertheless, induce wilt in susceptible varieties of rosaceous species more rapidly than in resistant ones (Figure 7) (Goodman *et al.*, 1978) and was given the trivial name amylovorin (Goodman *et al.*, 1974). The author initially reported that amylovorin was produced only in host tissue (e.g. immature green apple or pear fruits), since the polysaccharide's wilt-inducing activity was not detectable in culture filtrates (CF). However, this misconception was acknowledged when EPS was precipitated from CF by ethanol in the purification procedure (Ayers *et al.*, 1979).

Amylovorin was first cited as having a molecular weight of 165 000 daltons (Goodman *et al.*, 1974). Subsequently, Ayers *et al.* (1979) reported it to be *ca.* 40×10^6 daltons and our most recent calculation is *ca.* 10^8 daltons following purification on DEAE-Bio-Gel A and Bio-Gel 150. Attempts to depolymerize this massive polysaccharide were without success until Sijam (1982) reduced its size by incubating with the bacteriophage depolymerase from the phages PEal(h) (Ritchie and Klos, 1975) or Sϕ3 (Ayers *et al.*, 1979). The enzyme is located in the tail spikes of these EPS-specific phages and its presence on bacterial lawns is evident by the broad halo it produces around the plaques.

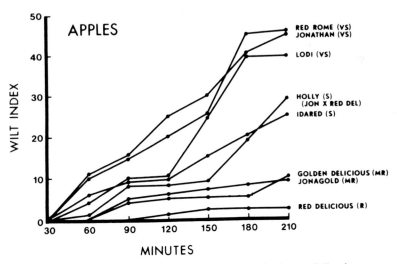

Figure 7. The rate and intensity of wilt of 5 cm long apple shoots, following exposure to 100 μg cm^{-3} of amylovorin. Sensitivities of the apple varieties to *Erwinia amylovora per se* are indicated as VS = very susceptible, S = susceptible, MR = moderately resistant, and R = resistant. (After Goodman *et al.*, 1978)

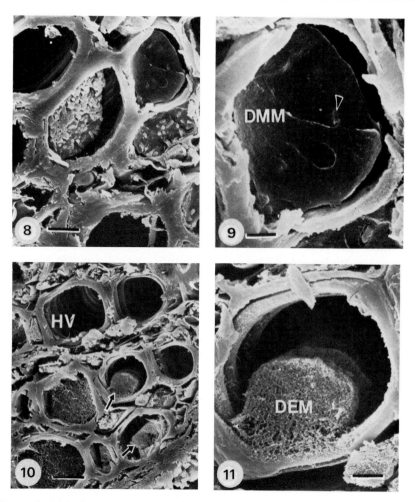

Figures 8–11. Xylem vessels infected by *Erwinia amylovora* or exposed to EPS. Figure 8: three helical vessels totally occluded by *E. amylovora* and dense matrix material (DMM), ostensibly capsule, 48 h after inoculation. Bar = 5 μm. Figure 9: higher magnification of one of the helical vessels showing the DMM. Bar = 10 μm. Figure 10: mature occluded (arrows) helical xylem vessels (HV) in shoots treated with 100 μg cm^{-3} solution EPS. Bar = 5 μm. Figure 11: higher magnification of one of the HV showing dense fibrous material (DFM), which is presumed to be partially EPS. Bar = 20 μm. (After Suhayda and Goodman, 1981b)

The polysaccharide toxin has always been associated with its wilt-inducing capability. This activity seemed recently to have been verified ultrastructurally (Figure 8–11) wherein the occlusion of vessels was observed in both apple petiole and stem tissue, either inoculated with *E. amylovora* or treated with amylovorin (Suhayda and Goodman, 1981b).

Structural analysis of amylovorin was initially performed by Albersheim (personal communication), and reported by Ayers *et al.* (1979) as having a galactan 'backbone' with β-1,3- and 1,6-linkages and side-chains of galactose, glucuronic acid, and glucose. This structure has since been modified to include one pyruvate molecule, in ketal form, per side-chain (A. Karr, personal communication). Amylovorin is clearly a molecule with a high net negative charge. In addition, there seems to be little immunologically detectable difference between amylovorin and EPS produced in axenic culture by *E. amylovora* (Sijam, 1982).

Complementing its wilt-inducing activity, Huang and Goodman (1976) and subsequently Goodman and White (1981) ascribed the oft-reported plasmolysis of xylem parenchyma to the action of amylovorin. This claim has been questioned by Sjulin and Beer (1978). The concentrations of amylovorin used to evoke wilt in Jonathan apple shoots in the author's laboratory have been generally $100 \mu g$ cm^{-3} (galactose equivalents). This concentration of amylovorin significantly increases the viscosity of the solution. A sound physiological *raison d'être* for plasmolysis could not be proposed, notwithstanding the ultrastructural evidence that was derived (Figures 4 and 5). However, the recent depolymerization of amylovorin with bacteriophage enzymes has yielded a polysaccharide that has molecular weight of *ca.* 2×10^4 daltons, which at $100 \mu g$ cm^{-3} galactose equivalents causes wilt with the same intensity as its much larger parent molecule yet does not alter the viscosity of the wilt-inducing solution appreciably (Sijam, 1982). The vessel-occluding and plasmolytic capabilities of the much smaller fragment of amylovorin are currently under investigation in the author's laboratory.

5.2. Role of amylovorin as a resistance suppressor

Huang *et al.* (1975) reported that avirulent strains of *E. amylovora* were totally agglutinated in apple petiole xylem vessels within 1–2 mm of the site of inoculum deposit (Figures 12 and 13). On the other hand, the virulent isolates quickly moved the length of the petiole. Some virulent cells passed through the length of the petiole into the vessels of the stem in less than 2 min (Suhayda and Goodman, 1981a,b).

In searching for the agglutinin, Romeiro *et al.* (1981b) isolated an agglutinating factor with a high degree of activity for the avirulent strain in apple seed, leaf, and stem tissue. The agglutinin, given the trivial name malin (Romeiro *et al.*, 1981b), has a comparative titre of 1000–2000 for avirulent EPS-deficient strains and 16–32 for virulent wild-type isolates. It was initially thought that malin was a lectin; however, in the absence of a sugar hapten capable of reversing its activity this seems less likely. Malin is a highly positively charged protein with a molecular weight of 12 500 and agglutinating activity is suppressed by less than microgram quantities of negatively charged EPS (Romeiro *et al.*, 1981a).

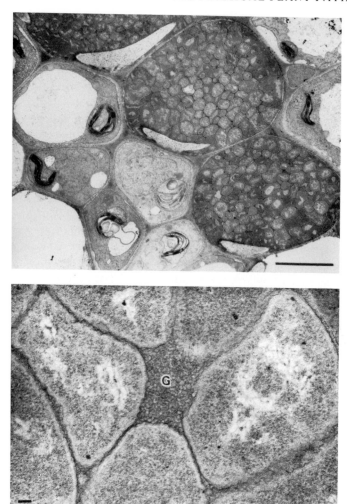

Figures 12 and 13. A file of xylem vessels completely occluded with the acapsular avirulent E_8 strain of *E. amylovora* localized *in situ*, in apple leaf petiole tissue <1 mm from the site of inoculation. Bar = 10 μm. Figure 12: A high magnification showing the bacteria and the granules (G) that localize them. Bar = 0.5 μm. (Goodman. unpublished work)

EPS (amylovorin) on the surface of virulent *E. amylovora* cells exists in two forms (Figure 14), the first a tightly bound distinguishable capsule and the second possibly an excess of the polymer that is released into the environment of the bacterium (Ayers *et al.*, 1979). A hypothesis has been formulated that suggest that malin may be the indigenous agglutin, observed by Huang *et al.* (1975), and which may be a part of the apple's disease resistance capability.

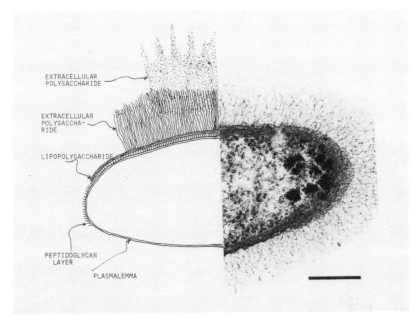

EXTRACELLULAR
POLYSACCHARIDE

EXTRACELLULAR
POLYSACCHA-
RIDE

LIPOPOLYSACCHARIDE

PEPTIDOGLYCAN
LAYER

PLASMALEMMA

Figure 14. A model relating the appearance of an ultrathin section of the outer envelope of *E. amylovora* to components that have been detected biochemically and immunologically. The fibrillar extracellular polysaccharide (EPS) is apparently anchored in the bacterial cell wall and is not easily removed. The stippled EPS is present in excess amounts and is released into the environment. Bar = 0.5 μm. (After Politis and Goodman, 1980)

The pathogen may, on the other hand, have evolved as its mechanism for virulence, a toxic agent (amylovorin), that is released into the environment and which is capable of neutralizing the host's resistance factor (malin). The validity of this hypothesis is also currently under intensive investigation. It is of further interest that the small 10^4 dalton fragment of amylovorin is capable of agglutinating malin as effectively as the parent molecule.

It is apparent from this brief account and from the more comprehensive compendium of van der Zwet and Keil (1979) that several very important questions concerning fireblight disease remain incompletely answered. Foremost among these appears to be the nature of the inoculum carryover site—ICS. Equally perplexing is the precise role of amylovorin in the infection process, i.e. is it essentially a resistance-countering molecule, acting by complexing the agglutinating factor malin (Romeiro *et al.*, 1981a,b), or does it have an elicitor-like action that is directly related to the observed plasmolysis of xylem parenchyma cells (Goodman and White, 1981; Huang and Goodman, 1976)? Ultrastructural studies from the author's laboratory suggest that pathogenesis engendered by *E. amylovora* involves host cell wall disorientation (Goodman and White, 1981). Although most published recent accounts have failed to detect cell wall-degrading enzyme activity in *E. amylovora* cultures (van der Zwet and Keil, 1979), a study by Wallis and Goodman (unpublished work) has measured significant growth of the pathogen on a substrate containing highly purified citrus pectin as a sole carbon source.

6. CONTROL OF FIREBLIGHT

The pathogen has been the limiting factor in pear production in the USA and in countries where the pathogen has become endemic (van der Zwet and Keil, 1979). Pears and other normally susceptible rosaceous species continue to be grown even where the pathogen has become established. In these instances resistant varieties, of which there are many, are planted. However, where highly desirable fruit varieties, e.g. Bartlett (Williams) pear and Jonathan, Ida Red, Yellow Transparent, and other susceptible apple varieties are grown commercially, an integrated management programme of chemical control and sanitation is essential. The sanitation feature of the programme is primarily careful pruning of diseased wood with special attention being paid to the excision of cankers. The regular remedial spray programme that has been most effective (Shaffer and Goodman, 1964) utilizes streptomycin. This antibiotic is sanctioned for use in orchards in the USA, Canada and New Zealand but not in western European countries into which the pathogen has spread since 1957 (Crosse *et al.*, 1958).

REFERENCES

Ayers, A. R., Ayers, S. B., and Goodman, R. N. (1979). Extracellular polysaccharide of *Erwinia amylovora*: a correlation with virulence, *Appl. Environ. Microbiol.*, **38**, 659–666.

Bachmann, F. M. (1913). Migration of *Bacillus amylovorus* in the host tissues, *Phytopathology*, **3**, 3–14.

Baldwin, C. H., Jr., and Goodman, R. N. (1963). Prevalence of *Erwinia amylovora* in apple buds as detected by phage typing, *Phytopathology*, **53**, 1299–1303.

Bauske, R. J. (1967). Dissemination of waterborne *Erwinia amylovora* by wind in nursery plantings, *Am. Soc. Hort. Sci. Proc.*, **91**, 795–801.

Bennett, R. A., and Billing, E. (1978). Capsulation and virulence in *Erwinia amylovora*, *Ann. Appl. Biol.*, **89**, 41–45.

Billing, E. (1960). An association between capsulation and phage sensitivity in *Erwinia amylovora*, *Nature (London)*, **186**, 819–820.

Billing, E. (1980). Fireblight in Kent, England in relation to weather (1955–1976), *Ann. Appl. Biol.*, **85**, 341–364.

Bryan, M. K. (1927). The flagella of *Bacillus amylovorus*, *Phytopathology*, **17**, 405–406.

Crosse, J. E. (1957). Bacterial canker of stone-fruits. III. Inoculum concentration and time of inoculation in relation of leaf-scar infection of cherry, *Ann. Appl. Biol.*, **45**, 19–35.

Crosse, J. E. and Goodman, R. N. (1973). A selective medium for a definitive colony characteristic of *Erwinia amlovora, Phytopathology, 63,* 1425–1426.

Crosse, J. E., Bennett, M., and Garrett, C. M. E. (1958). Fire blight of pear in England, *Nature, London,* **182,** 1530.

Crosse, J. E., Goodman, R. N., and Shaffer, W. H. (1972). Leaf damage as a predisposing factor in the infection of apple shoots by *Erwinia amylovora, Phytopathology,* **62,** 176–182.

Dueck, J., and Morand, J. B. (1975). Seasonal changes in the epiphytic population of *Erwinia amylovora* on apple and pear, *Can. J. Plant. Sci.,* **55,** 1007–1012.

Eden-Green, S. J., and Knee, M. (1974). Bacterial polysaccharide and sorbitol in fireblight exudate, *J. Gen. Microbiol.,* **81,** 509–512.

Feliciano, A., and Daines, R. H. (1970). Factors influencing ingress of *Xanthomonas pruni* through peach leaf scars and subsequent development of spring cankers, *Phytopathology,* **60,** 1720–1726.

Goodman, R. N. (1954). Apple fruits a source of overwintering fireblight inoculum, *Plant Dis. Rep.,* **38,** 414.

Goodman, R. N. (1965). *In vitro* and *in vivo* interactions between components of mixed bacterial cultures isolated from apple buds, *Phytopathology,* **55,** 217–221.

Goodman, R. N., and White, J. A. (1981). Xylem parenchyma plasmolysis and vessel wall disorientation are early signs of pathogenesis caused by *Erwinia amylovora, Phytopathology,* **71,** 844–852.

Goodman, R. N., Huang, J. S., and Huang, P. Y. (1974). Host specific phytotoxic polysaccharide from apple tissue infected by *Erwinia amylovora, Science,* **183,** 1081–1082.

Goodman, R. N., Stoffl, P. R., and Ayres, S. M. (1978). The utility of the fireblight toxin, Amylovorin, for the detection of resistance of apple, pear, and quince to *Erwinia amylovora, Acta Hort.,* **86,** 51–56.

Hildebrand, D. C., and Schroth, M. N. (1963). Relation of arbutin-hydroquinone in pear blossoms to invasion by *E. amylovora, Nature (London),* **197,** 513.

Hildebrand, E. M. (1973). The blossom-blight phase of fire blight and methods of control, *N. Y. (Cornell) Agric. Exp. Stn. Mem.,* no. 207, 40 pp.

Hildebrand, E. M. (1938). Infectivity of the fireblight organism, *Phytopathology,* **27,** 850–852.

Hildebrand, E. M. (1954). Relative stability of fireblight bacteria, *Phytopathology,* **44,** 192–197.

Huang, P. Y., and Goodman, R. N. (1976). Ultrastructural modifications in apple stems induced by *Erwinia amylovora* and the fireblight toxin, *Phytopathology,* **66,** 269–276.

Huang, P. Y., Huang, J. S., and Goodman, R. N. (1975). Resistance mechanisms of apple shoots to an avirulent strain of *Erwinia amylovora, Physiol. Plant Pathol.,* **6,** 283–287.

Keil, H. L., and van der Zwet, T. (1972). Aerial strands of *Erwinia amylovora*: structure and enhanced production by pesticide oil, *Phytopathology,* **62,** 335–361.

Keitt, G. W., and Ivanoff, S. S. (1941). Transmission of fireblight by bees and its relation to nectar concentration of apple and pear blooms, *J. Agric. Res.,* **62,** 745–753.

Lewis, S. M., and Goodman, R. N. (1965). Mode of penetration and movement of fireblight bacteria in apple leaf and stem tissue, *Phytopathology,* **55,** 719–723.

Lewis, S. M., and Goodman, R. N. (1966). The glandular trichomes, hydathodes and lenticels of Jonathan apple and their relation to infection by *Erwinia amylovora, Phytopathol. Z.,* **55,** 325–358.

Paulin, J. P. (1981). Overwintering of *Erwinia amylovora*: sources of inoculum in spring, *Acta Hort.*, **117**, 49–54.

Pierstorff, A. L. (1931). Studies on the fire-blight organism, *Bacillus amylovorus*, *N. Y. (Cornell) Agric. Exp. Stn. Mem.*, No. **136**, 53pp.

Plurad, S. B., Goodman, R. N., and Enns, W. R. (1967). Factors influencing the efficacy of *Aphis pomi* as a potential vector for *Erwinia amylovora*, *Phytopathology*, **37**, 1060–1063.

Politis, D. J., and Goodman, R. N. (1980). Fine structure of extracellular polysaccharide of *Erwinia amylovora*, *Appl. Environ. Microbiol.*, **40**, 596–607.

Powell, C. C., and Hildebrand, D. C. (1967). ß-Glucosidase content of pear blossom tissues and its relation to antibiotic activity of the tissues against *Erwinia amylovora* (Abstract), *Phytopathology*, **57**, 826.

Powell, C. C., and Hildebrand, D. C. (1970). Fire blight resistance in *Pyrus*: involvement of arbutin oxidation, *Phytopathology*, **60**, 337–340.

Powell, D. (1964). Prebloom freezing as a factor in the occurrence of the blossom blight phase of fireblight on apples, *Ill. State Hort. Soc. Trans.*, **97**, 144–148.

Ritchie, D. F., and Klos, E. J. (1975). Overwinter survival of *Erwinia amylovora* in apple and pear cankers (Abstract), *Am. Phytopathol. Soc. Proc.*, **2**, 67–68.

Romeiro, R., Karr, A. L., and Goodman, R. N. (1981a). *Erwinia amylovora* cell wall receptor for apple agglutinin, *Physiol. Plant Pathol.*, **19**, 383–390.

Romeiro, R., Karr, A. L., and Goodman, R. N. (1981b). Isolation of a factor from apple that agglutinates *Erwinia amylovora*, *Plant Physiol.*, **68**, 772–777.

Rosen, H. R. (1929). The life history of the fireblight pathogen, *Bacillus amylovorus*, as related to the means of overwintering and dissemination, *Arkansas Agric. Exp. Stn. Bull.*, No. 244, 96 pp.

Rosen, H. R. (1930). Overwintering of the fire blight pathogen, *Bacillus amylovorus*, within the beehive, *Science*, **72**, 301–302.

Rosen, H. R. (1932). Relation of fire blight to honey bees, *Mo. State Hort. Soc. Proc.*, 1931–32, 67–77.

Rosen, H. R. (1933). Further studies on the overwintering and dissemination of the fire-blight pathogen, *Arkansas Agric. Exp. Stn. Bull.*, No. 283, 102 pp.

Rosen, H. R. (1936). Mode of penetration and of progressive invasion of fire-blight bacteria into apple and pear blossoms, *Arkansas Agric. Exp. Stn. Bull.*, No. 331, 68 pp.

Shaffer, W. H., Jr., and Goodman, R. N. (1964). Compatibility of streptomycin with some fungicides and insecticides, *Plant Dis. Rep.*, **48**, 180–181.

Sijam, K. (1982). Comparative study of the properties of amylovorin and EPS produced by *Erwinia amylovora*, *PhD Thesis*, University of Missouri, 180 pp.

Sjulin, T. M., and Beer, S. V. (1978). Mechanism of wilt induction by amylovorin in rosaceous shoots and its relationship to wilting of shoots infected by *Erwinia amylovora*, *Phytopathology*, **68**, 89–94.

Starr, M. P., and Mandel, M. (1950). The nutrition of phytopathogenic bacteria. IV. Minimal nutritive requirements of the genus Erwinia, *J. Bacteriol.*, **60**, 669–672.

Suhayda, C. G., and Goodman, R. N. (1981a). Infection courts and systemic movement of ^{32}P-labeled *Erwinia amylovora* in apple petioles and stems, *Phytopathology*, **71**, 656–660.

Suhayda, C. G., and Goodman, R. N. (1981b). Early proliferation and migration of *Erwinia amylovora* and subsequent xylem occlusion, *Phytopathology*, **71**, 697–707.

Tullis, E. C. (1929). Studies on the overwintering and modes of infection of the fire blight organism, *Mich. Agric. Exp. Stn. Tech. Bull.*, No. 97, 32 pp.

van der Zwet, T., and Keil, H. L. (1979). Fire blight, a bacterial disease of rosaceous plants, *US Dep. Agric. Handb.*, No. 510, 200 pp.

Vörös, J., and Goodman, R. N. (1965). Filamentous forms of *Erwinia amylovora, Phytopathology*, **55,** 876–879.

Winslow, C. E. A., Broadhurst, J., Buchanan, R. E., Krummwiede, C., Jr., Rogers, L. A., and Smith, G. H. (1920). The families and genera of the bacteria; Erwineae, *J. Bacteriol.*, **5,** 209.

Biochemical Plant Pathology
Edited by J. A. Callow
© 1983 John Wiley & Sons Ltd

5

Crown Gall Disease: A Case Study

M. DRUMMOND

ARC Unit of Nitrogen Fixation, University of Sussex, Falmer, Brighton BN1 9RQ, UK

1. INTRODUCTION

When *Agrobacterium tumefaciens*, a gram-negative soil bacterium, is inoculated into freshly wounded plant tissue, it reprograms the cells with which it comes into contact such that they divide chaotically, giving rise to a crown gall tumour. During this transformation, part of a large plasmid carried by the bacterium is integrated into plant nuclear DNA. This has two principal physiological effects; it disrupts the metabolism of growth hormones in the tissue, bringing about oncogenesis, and it causes the synthesis of opines, plant metabolites not found in normal tissue (Nester and Kosuge, 1981). The disease is widespread, affecting most dicotyledonous angiosperms and gymnosperms.

2. OPINE METABOLISM

Opines can serve as a carbon or nitrogen source for oncogenic agrobacteria, and it has long been recognized that the opine produced is determined by the inducing bacterium (Goldmann *et al.*, 1968). This constituted the first evidence for DNA transfer from the bacterium to the plant cell, and led to the suggestion that the enzyme catabolizing a given opine in the bacterium was identical with that producing it in tumour tissue. In fact, however, two different enzymes are involved, distinguishable both genetically and by their biochemical characteristics (Klapwijk *et al.*, 1976; Petit and Tempé, 1978). The enzyme responsible for bacterial catabolism is referred to as an opine oxidase, and that in the plant is termed a synthase. Both are encoded by the Ti plasmid (Van Larebeke *et al.*, 1974; Schröder *et al.*, 1981). Four distinct groups of opines have been recognized, their type members being octopine, nopaline, agropine, and agrocinopine (Figure 1) (Menagé and Morel, 1964; Goldmann *et al.*, 1969; Guyon *et al.*, 1980; Ellis and Murphy, 1981). Each group is metabolized by a separate oxidase and synthase of loose substrate specificity. Thus octopine synthase, for example, also synthesizes histopine, lysopine, and octopinic acid. No naturally occurring Ti plasmid codes for the synthesis and breakdown of both octopine and nopaline, but octopine tumours often synthesize agropine and nopaline tumours synthesize agrocinopine. Agropine and agrocinopine metabolisms are also sometimes encoded by the same plasmid (Ellis and Murphy, 1981). A plasmid-encoded permease exists for each group of opines, (Firmin and Fenwick, 1978; Klapwijk *et al.*, 1977; Ellis and Murphy, 1981), and it has been shown that the expression of oxidases and permeases for octopine and nopaline are coordinately controlled, both enzymes being inducible by their particular substrate

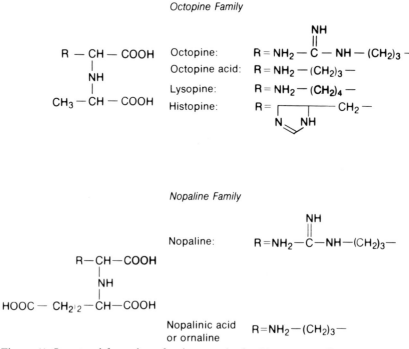

Figure 1. Structural formulae of opines synthesized by crown gall tumours

(Petit *et al.*, 1979). In certain strains the catabolism of arginine, a breakdown product of both octopine and nopaline, is a plasmid-borne trait which is also inducible by the appropriate opine (Ellis, *et al.*, 1979).

Ti plasmids promote bacterial conjugation (Kerr *et al.*, 1977) and the genes required for their transfer (tra functions) are closely associated with opine metabolism, being inducible by the characteristic opine (Petit *et al.*, 1978). This makes good teleological sense, since in the presence of its substrate the catabolic plasmid will spread rapidly through a mixed population of oncogenic and non-oncogenic bacteria. The three classes of regulatory mutant that can be isolated are most simply explained by postulating separate operons for plasmid transfer and opine catabolism, controlled by the same repressor (Klapwijk *et al.*, 1978).

Tumourigenesis is almost always accompanied by opine synthesis, but since one can isolate both mutants that induce opine synthesis without oncogenesis, and that give tumours failing to produce any opine, such synthesis cannot be necessary for tumour maintenance (Klapwijk *et al.*, 1978; Leemans *et al.*, 1982). The most plausible rationale for the presence of opines in tumour tissue is that they provide a specific ecological niche for the oncogenic bacterium (Guyon *et al.*, 1980). Since only bacteria capable of using opines as carbon and nitrogen sources can exploit the niche, the plasmid thereby generates selection pressure for itself as well as promoting its own transfer through the population. The logical link between opine production and uncontrolled cell division is that the plasmid need only transform a single cell to procure a relatively abundant supply of its opine substrate. Possibly other plasmids are also capable of introducing some of their genetic information into eucaryotic cells, but go undetected because they lack the more conspicuous oncogenic phenotype.

3. STRUCTURE AND FUNCTION OF Ti PLASMIDS

A complete description of the crown gall transformation must explain how plasmid DNA enters the plant cell, how it becomes integrated into the plant DNA, and how it disturbs phytohormone metabolism. Wounding, an absolute prerequisite for transformation (Rack, 1954) might be necessary to expose bacterial attachment sites on the plant cell surface, which is consistent with the finding that plant protoplasts can be transformed by *Agrobacterium* only when limited regeneration of the cell wall has occurred (Márton *et al.*, 1979).

The formal resemblance of the plasmid transfer phase of oncogenesis to bacterial conjugation may be reflected at the molecular level. Both transformation and conjugation are temperature-sensitive, with critical temperatures around 30 °C, suggesting that one or more of the oncogenic (*onc*) and transfer functions could be identical (Hooykaas *et al.*, 1979; Tempé *et al.*, 1977). However, mapping of the *tra* and *onc* genes in both octopine and nopaline plasmids has so far failed to reveal any overlap of these functions (Klapwijk *et al.*, 1978).

Ti plasmids vary in size from 150 to 230 kilobases (kb) and presumably share at least some of the genetic information involved in oncogenesis. However, DNA solution hybridization shows that sequence homology between the various groups can be as low as 6%, sufficient nevertheless to represent several *onc* genes scattered throughout the plasmid (Thomashow *et al.*, 1981). Some *onc* genes must be involved in recognition of the plant cell surfaces by the bacterium. Divergence of such information could explain the plasmid-borne variations in host range which have been observed (Loper and Kado, 1979). The segment of plasmid DNA transferred to the plant is designated T-DNA, part of which is highly conserved on most but not all Ti plasmids (Depicker *et al.*, 1978; Chilton *et al.*, 1978; Thomashow *et al.*, 1981).

Transposon mutagenesis of octopine and nopaline plasmids reveals *onc* genes lying outside the T-DNA in sequences conserved between these two groups of plasmid (Ooms *et al.*, 1980; Holsters *et al.*, 1980). Clearly such genetic information cannot be involved in maintaining the transformed state, but must be involved in some phase in the process of transferring T-DNA to the plant, for example recognition of the plant cell surface.

T-DNA is extensive, being about 23 kb in nopaline tumours and up to 15 kb in octopine tumours (Lemmers *et al.*, 1980; Chilton *et al.*, 1977). It is usually present in multiple copies in each plant cell. Southern blot hybridization of crown gall DNA using cloned Ti plasmid fragments as probes indicates that nopaline T-DNA is co-linear with Ti plasmid sequences internal to the T-DNA (Lemmers *et al.*, 1980). In certain octopine tumour lines, on the other hand, T-DNA occurs in two segments, T_L-DNA and T_R-DNA, the short section of Ti plasmid DNA separating them being absent from the plant cell (Thomashow *et al.*, 1980).

By allowing the detection of restriction fragments containing both plant and plasmid sequences, Southern blotting experiments also demonstrate integration of T-DNA to the plant genome, and further, by separating crown gall DNA into nuclear and organellar fractions, it was possible to demonstrate that the site of integration was in the nucleus (Willmitzer *et al.*, 1980). In nopaline tumours, Southern blotting reveals single restriction fragments homologous to sequences from both the right and left ends of the T-DNA, indicating that it is present in tandem repeats. The presence of T-DNA as an independent replicon is excluded by the absence from undigested crown gall DNA of a discrete entity homologous to Ti plasmid sequences (Lemmers *et al.*, 1980). The borders of the T-DNA of nopaline tumours are rather precisely defined and are constant from tumour to tumour. In octopine tumours the right-hand border seems to be variable, the T_R-DNA sometimes being absent or sometimes present at high copy number, whereas the T_L-DNA is always present at low copy number. The length of the T-DNA seems to be independent of the host plant species (De Beuckeleer *et al.*, 1981). These differences may reflect some dissimilarity in the mechanism of integration of T-DNA in octopine and nopaline tumours.

From one nopaline tumour a restriction fragment containing the junction between plant and plasmid DNA has been cloned. When this fragment is used as a probe in a Southern blotting experiment with untransformed tobacco DNA, the pattern of bands hybridizing suggests that the T-DNA is covalently joined to a repeated DNA element of the plant genome (Yadav *et al.*, 1980). The variety of fragments containing both plant and plasmid sequences indicates that the T-DNA must be integrated at different sites in different transformation events (Lemmers *et al.*, 1980).

DNA sequence determination of one nopaline plasmid shows that the T-region is flanked by short direct repeats which are not integrated into the plant genome, but probably provide recognition sites for the integration event. This suggests that the mechanism of transposition differs from that of conventional bacterial transposons which are usually flanked by longer inverted repeats that move with the transposed sequence. Furthermore, saturation mutagenesis of the T-DNA indicates that none of the genetic information it carries is necessary for transposition, unlike those transposons which have been shown to encode their own transposase (Garfinkel *et al.*, 1981). Integration of the plasmid RP4 into octopine T-DNA, however, leads to lack of tumourigenicity, which probably reflects an upper limit to the size of T-DNA that can be transposed (Depicker *et al.*, 1978). When a transposon such as Tn7 is introduced into nopaline plasmid T-DNA, the segment transferred to the plant genome is correspondingly larger (Hernalsteens *et al.*, 1980).

Genetic analysis of the early events in oncogenesis has hitherto been hampered by the difficulty of detecting discrete steps in the transformation process and the laborious screening for mutants. Thus we do not yet have answers to such crucial questions as whether the entire Ti plasmid enters the plant cell and whether it dissociates to form a smaller intermediate before integration. However, *in vitro* transformation of regenerating plant protoplasts is now possible, and as higher efficiencies of transformation are obtained, this model system will undoubtedly prove a powerful experimental tool to approach such problems (Márton *et al.*, 1979). Transformation with isolated Ti plasmid DNA is also possible in this system, which incidentally indicates that the bacterium itself is not essential for oncogenesis (Krens *et al.*, 1982).

4. HORMONES

The physiological basis of transformation seems to be a disruption of hormone levels in the plant cell. Cytokinins, which promote cell division, are abundant in crown gall tissue, as are auxins, a class of plant hormones which stimulate cell enlargement (Jablonski and Skoog, 1954; Skoog and Armstrong, 1970; see also Chapter 19). Untransformed tissue cultured *in vitro* does not contain physiologically effective levels of either hormone. Furthermore, plant tumours caused by Black's wound tumour virus as well as those occurring spontaneously in interspecific hybrids of *Nicotiana* also contain abnormally high levels of these hormones (Kehr and Smith, 1954; Black, 1972). Such hormone autotrophy might thus be a sufficient condition for plant neoplasia in general.

There is extensive homology between the T_L-DNA of octopine tumours and a segment of the T-DNA of nopaline tumours which seems to play an essential role in oncogensis (Depicker *et al.*, 1978). Mutagenesis of this region results in tumours of altered morphology or growth rate (Ooms *et al.*, 1981). This suggests that these sequences somehow modulate hormone metabolism in the plant, since it has been shown that the morphology of callus tissue cultured *in vitro* is determined by the relative quantities of auxin and cytokinin in the medium (Skoog and Miller, 1957). T-DNA is extensively transcribed, nopaline tumours containing about thirteen transcripts and octopine tumours at least seven (Figure 2). Transcription is carried out by host RNA polymerase II from promoters within the T-DNA which appear to have typically eucaryotic structure (Willmitzer *et al.*, 1981b, 1982). The transcripts are polyadenylated and present in very low concentrations, constituting no more than 0.0005–0.001% of the total polyadenylated population in octopine tumours (Willmitzer *et al.*, 1981a). The pattern of transcription varies bet-

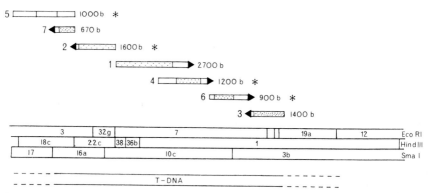

Figure 2. The pattern of transcription of octopine T-DNA in crown gall tissue as determined by Northern blots. Transcripts are numbered in order of size (number of bases including poly-A tails). The boxes indicate the region from which each transcript is derived, the stippled area its probable exact position and strength of transcription, and the arrows the direction of transcription. Transcripts which probably have functional counterparts in nopaline crown gall tissue are marked with an asterisk. Redrawn from Willmitzer *et al.* (1982)

ween actively growing callus and cells in the stationary phase, indicating some degree of transcriptional control. Weak transcription of T-DNA also occurs in the bacterium (Gelvin et al., 1981).

Some, if not all, of these transcripts are translated in crown gall tissue. The only protein product that has been characterized beyond a determination of molecular weight is octopine synthase (McPherson et al., 1980; Schröder et al., 1981).

5. REDIFFERENTIATION OF CROWN GALL TUMOURS

Crown gall tumours vary considerably in their morphology, some producing recognizable stems or roots, while others remain a chaotic mixture of cell types. Tumour morphology is influenced by position of the overgrowth on the plant, and by the species of the host as well as by plasmid-borne genes (Hooykaas et al., 1977). Nopaline tumours typically organize stem primordia on their surfaces, in which case they are referred to as teratomata. Agrobacterium rhizogenes carries a large plasmid of a different incompatibility group from the Ti plasmids. Like Agrobacterium tumefaciens, it inserts plasmid sequences into the host plant genome, thereby causing a proliferation of adventitious roots from the wound site (Chilton et al., 1982). It is relatively easy to cause such organized neoplasms cultured in vitro to redifferentiate into whole plants. Use of cloned teratoma tissue removes the possibility of selection of untransformed cells. The redifferentiated plant contains T-DNA, synthesizes opines and reverts to hormone autotrophy when emplanted into tissue culture, indicating that this redifferentiation has a purely epigenetic basis (Yang et al., 1980; Chilton et al., 1982).

Redifferentiated plants may flower and set viable seed. In some cases, plants grown from this seed do not synthesize any opine and require a hormone supplement in tissue culture. Since haploid anther tissue from the parent shares these characteristics it appears that this reversion occurs during meiosis. Such reversion is accompanied by loss of most of the T-DNA sequences and hence is termed genotypic reversion (Yang et al., 1980). In other cases it has been shown that the F1 progeny retain the capacity to synthesize opines (Wullems et al., 1981). The opine synthesis marker is transmitted through both pollen and eggs as a single dominant factor with Mendelian segregation ratios typical for monohybrid crosses (Otten et al., 1981).

6. GENETIC ENGINEERING

Ti plasmids may be used as vectors to insert a given DNA sequence into the genome of any dicotyledenous plant. If plant protoplast systems with very high transformation frequencies are used the oncogenic phenotype is no longer necessary to select transformed cells, simple screening for opine synthesis being sufficient. Thus 'disarmed' Ti plasmids which do not perturb phytohormone levels may be used, making redifferentiation of morphologically normal plants a simple matter.

At present, *Agrobacterium* cannot be used as a vector for monocotyledenous plants since these are resistant to the crown gall disease, a very serious limitation, as this excludes all the cereal crops. There is evidence that some difference in composition of the monocotyledon cell wall prevents binding of *Agrobacterium*, in which case the difficulty might possibly be circumvented by transformation of protoplasts with Ti plasmid DNA (Lippincott and Lippincott, 1978). If the incompatibility lies at the level of intracellular mechanisms, say a block in transposition, it may be insurmountable. However, there remains the ironic possibility that on infection, plasmid DNA is integrated to the monocotyledon genome but fails to disrupt the normal morphogenetic controls of the host. There are formidable difficulties involved in introducing any procaryotic gene into a crop plant and having it expressed approximately. For example, simply inserting *nif* (nitrogen fixation) genes into the potato genome would be useless, since even if they were processed correctly, their expression must be subjected to eucaryotic control and some system devised to protect the nitrogenase from oxygen poisoning. Because the molecular basis of many desirable plant traits is obscure, making available the genetic resources of the entire biosphere has not revolutionized plant breeding. The most immediate practical applications of the technique may involve restoring to their original genome plant DNA sequences which have been modified *in vitro*, say to increase the lysine contents of a storage protein.

Because the Ti plasmids are so large, they are extensively cleaved by the available restriction enzymes and hence are not amenable to the simple *in vitro* techniques used for cloning into smaller vectors. This difficulty has been elegantly circumvented by a method involving recombination *in vivo* from a smaller intermediate cloning vector. This approach has been used to insert the gene for yeast alcohol dehydrogenase into tobacco and the structural gene for the bean storage protein phaeolin into the sunflower genome (Matzke and Chilton, 1981; Kemp, unpublished work). First a T-DNA segment is cloned into a broad-host-range vector in *E. coli*. Then the gene of interest is inserted into a restriction site in the T-DNA sequence along with a selectable drug marker. This chimaera is then transformed into *Agrobacterium*, where recombination can occur between the T-DNA sequences flanking the gene of interest and the resident Ti plasmid. These rare recombinants may be identified by displacing the intermediate vector with a plasmid from the same incompatibility group and selecting for retention of the drug marker accompanying the gene of interest, which now lies within the T-DNA of a functional Ti plasmid. Plant cells transformed with Ti plasmids engineered in this way have been shown to transcribe the novel genetic material, albeit weakly, but the protein products are not detectable. It is probable that in the near future *in vitro* manipulation of the inserted sequences will be used to obtain more efficient translation in the plant cell. In this way studies with Ti plasmids may be expected to contribute substantially to our understanding of gene expression in the nuclear genome of higher plants.

REFERENCES

Black, D. M. (1972). Plant tumours of viral origin, *Prog. Exp. Tumor Res.,* **15,** 110–137.

Chilton, M. D., Drummond, M. H., Merlo, D. J., Sciaky, D., Montoya, A. L., Gordon, M. P., and Nester, E. W. (1977). Stable incorporation of plasmid DNA into higher plant cells: the molecular basis of crown gall tumorigenesis, *Cell,* **11,** 263–271.

Chilton, M. D., Drummond, M. H., Merlo, D. J., and Sciaky, D. (1978). Highly conserved DNA of Ti plasmids overlaps T-DNA maintained in plant tumors, *Nature (London),* **275,** 147–149.

Chilton, M. D., Tepfer, D. A., Petit, A., David, C., Casse-Delbart, F., and Tempé, J. (1982). *Agrobacterium rhizogenes* inserts T-DNA into the genome of the host plant root cells, *Nature (London),* **295,** 432–434.

De Beuckeleer, M., Lemmers, M., De Vos, G., Willmitzer, L., van Montagu, M., and Schell, J. (1981). Further insight on the T-DNA of octopine crown gall, *Mol. Gen. Genet.,* **183,** 283–288.

Depicker, A., Van Montagu, M., and Schell, J. (1978). Homologous DNA sequences in different Ti plasmids are essential for oncogenicity, *Nature (London),* **275,** 150–153.

Ellis, J. G., and Murphy, J. (1981). Four new opines from crown gall tumors: their detection and properties, *Mol. Gen. Genet.,* **181,** 36.

Ellis, J. G., Kerr, A., Tempé, J., and Petit, A. (1979). Arginine catabolism—new function of both octopine and nopaline Ti plasmids of *Agrobacterium, Mol. Gen. Genet.,* **173,** 263–270.

Engler, G., Depicker, A., Maenhaut, R., Villarroel, R., Van Montagu, M., and Schell, J. (1981). Physical mapping of DNA base sequence homologies between an octopine and a nopaline Ti plasmid of *Agrobacterium tumefaciens, J. Mol. Biol.,* **152,** 183–208.

Firmin, J. K., and Fenwick, R. G. (1978). Agropine: a major new plasmid-determined metabolite in crown gall tumors. *Nature (London),* **276,** 842–844.

Garfinkel, D. J., Simpson, R. B., Ream, L. S., White, F. F., Gordon, M. P., and Nester, E. W. (1981). Genetic analysis of crown gall, fine structure map of the T-DNA by site-directed mutagenesis, *Cell,* **27,** 143–153.

Gelvin, S. B., Gordon, M. P., Nester, E. W., and Aronson, A. (1981. Transcription of the *Agrobacterium* Ti plasmid in the bacterium and in crown gall tumors, *Plasmid,* **6,** 17–29.

Goldmann, A., Tempé, J., and Morel, G. (1968). Quelques particularités de diverses souches d'*Agrobacterium tumefaciens, C. R. Soc. Biol.,* **162,** 630–631.

Goldmann, A., Thomas, D. W., and Morel, G. (1969). Sur la structure de la nopaline metabolite anormal de certaines tumeurs de crown-gall, *C. R. Acad. Sci.,* **268,** 852–854.

Guyon, P., Chilton, M. D., Petit, A., and Tempé, J. (1980) Agropine in null-type crown gall tumors: Evidence for generality of the opine concept, *Proc. Natl. Acad. Sci. USA,* **77,** 2693–2697.

Hernalsteens, J. P., van Vliet, F., de Beuckeleer, M., Depicker, A., Engler, G., Lemmers, M., Holsters, M., van Montagu, M., and Schell, J. (1980). The *Agrobacterium tumefaciens* Ti plasmid as a host vector system for introducing foreign DNA into plant cells, *Nature (London),* **287,** 654–656.

Holsters, M., Silva, B., van Vliet, F., Genetello, G., De Block, M., Dhaese, P., Depicker, A., Inzé, D., Engler, G., Villarroel, R., van Montagu, M., and Schell, J. (1980). The functional organisation of the nopaline *A. tumefaciens* plasmid pTiC58, *Plasmid*, **3**, 212–230.

Hooykaas, P. J. J., Klapwijk, P. M., Nuti, M. P., Schilperoort, R. A., and Rörsch, A. (1977). Transfer of the *Agrobacterium* Ti plasmid to avirulent *Agrobacteria* and to *Rhizobium ex planta*, *J. Gen. Microbiol.*, **98**, 477–484.

Hooykaas, P. J. J., Roobol, C., and Schilperoort, R. A. (1979). Regulation of the transfer of Ti plasmids of *Agrobacterium tumefaciens*, *J. Gen. Microbiol.*, **110**, 99–109.

Jablonski, J. R., and Skoog, F. (1954). Cell enlargement and cell division in excised tobacco pith tissue, *Physiol. Plant*, **7**, 16–24.

Kehre, A. E., and Smith, H. H. (1954). Genetic tumours in *Nicotiana* hybrids, *Brookhaven Symp. Biol.*, **6**, 55–76.

Kerr, A., Manigault, P., and Tempé, J. (1977). Transfer of virulence *in vivo* and *in vitro* in *Agrobacterium*, *Nature (London)*, **265**, 560–561.

Klapwijk, P. M., Hooykaas, P. J. J., Kester, H. C. M., Schilperoort, R. A., and Rörsch, A. (1976). Isolation and characterisation of *Agrobacterium tumefaciens* mutants affected in the utilisation of octopine, octopinic acid and lysopine, *J. Gen. Microbiol.*, **96**, 155–163.

Klapwijk, P. M., Oudshoorn, M., and Schilperoort, R. A. (1977). Inducible permease involved in the uptake of octopine, lysopine and octopinic acid by *Agrobacterium tumefaciens* strains carrying virulence-associated plasmids, *J. Gen. Microbiol.*, **102**, 1–11.

Klapwijk, P. M., Scheulderman, T., and Schilperoort, R. A. (1978). Coordinated regulation of octopine degradation and conjugative transfer of Ti plasmids in *Agrobacterium tumefaciens*: evidence for a common regulatory gene and separate operons, *J. Bacteriol.*, **136**, 775–785.

Krens, F. A., Molendijk, L., Wullems, G. J., and Schilperoort, R. A. (1982). *In vitro* transformation of plant protoplasts with Ti plasmid DNA, *Nature (London)*, **296**, 72–74.

Leemans, J., Deblaere, R., Willmitzer, L., De Greve, H., Hernalsteens, J.-P., van Montagu, M., and Schell, J. (1982). Genetic identification of functions of T_L-DNA transcripts in octopine crown galls, *EMBO J.*, **1**, 147–152.

Lemmers, M., De Beuckeleer, M., Holsters, M., Zambryski, P., Depicker, A., Hernalsteens, J.-P., van Montagu, M., and Schell, J. (1980). Internal organisation, boundaries and integration of Ti plasmid DNA in nopaline crown gall tumors, *J. Mol. Biol.*, **144**, 353–376.

Lippincott, J. A., and Lippincott, B. B. (1978). Cell walls of crown gall tumors and embryonic plant tissues lack *Agrobacterium* adherence sites, *Science*, **199**, 1075–1078.

Loper, J. E., and Kado, C. I. (1979). Host range conferred by the virulence specifying plasmid of *Agrobacterium tumefaciens*, *J. Bacteriol.*, **139**, 591–596.

Márton, L., Wullems, G. J., Molendijk, L., and Schilperoort, R. A. (1979). *In vitro* transformation of cultured cells from *Nicotiana tabacum* by *Agrobacterium tumefaciens*, *Nature (London)*, **277**, 129–131.

Matzke, A. J. M., and Chilton, M.-D. (1981). Site-specific insertion of genes into the T-DNA of the *Agrobacterium* Ti plasmid: and approach to the genetic engineering of higher plant, *J. Mol. Appl. Genet.*, **1**, 39–49.

McPherson, J. C., Nester, E. W., and Gordon, M. P. (1980). Proteins encoded by *Agrobacterium tumefaciens* Ti plasmid DNA in crown gall tumors, *Proc. Natl. Acad. Sci. USA*, **77**, 2666–2670.

Ménage, A., and Morel, G. (1964). Sur la présence d'octopine dans les tissus de crowngall, *C. R. Acad, Sci.*, **259**, 4795–4796.

Nester, E. W., and Kosuge, T. (1981). Plasmids specifying plant hyperplasias. *Annu. Rev Microbiol.*, **35**, 531–565.

Ooms, G., Klapwick, P. M., Poulis, J. A., and Schilperoort, R. A. (1980). Characterisation of Tn904 insertions in octopine Ti plasmid mutants of *Agrobacterium tumefaciens, J. Bacteriol.*, **144**, 82–91.

Ooms, G., Hooykaas, P. J. J., Moolenaar, G., and Schilperoort, R. A. (1981). Crown gall plant tumors of abnormal morphology induced by *Agrobacterium tumefaciens* carrying mutated octopine Ti plasmids: analysis of T-DNA functions. *Gene*, **14**, 33–50.

Otten, L., De Greve, H., Hernalsteens, J.-P., van Montagu, M., Schieder, O., Straub, J., and Schell, J. (1981). Mendelian transmission of genes introduced into plants by the Ti plasmids of *Agrobacterium tumefaciens, Mol. Gen. Genet.*, **183**, 209–213.

Petit, A., and Tempé, J. (1978). Isolation of *Agrobacterium* Ti plasmid regulatory mutants, *Mol. Gen. Genet.*, **167**, 147–155.

Petit, A., Tempé, J., Kerr, A., Holsters, M., van Montagu, M., and Schell, J. (1978). Substrate induction of conjugative activity of *Agrobacterium tumefaciens* Ti plasmids, *Nature (London)*, **271**, 570–571.

Petit, A., Dessaux, Y., and Tempé, J. (1979). *Proc. Fourth Int. Conf. Plant. Pathol. Bacteria (Angers, France).*

Rack, K. (1954). The wound response in crown gall tumor induction, *Phyopathol. Z.*, **21**, 1.

Schröder, J., Hillebrand, A., Klipp, W., and Pühler, A. (1981). Expression of plant tumor-specific proteins in minicells of *Escherichia coli*: a fusion protein of lysopine dehydrogenase with chloramphenicol acetyltransferase, *Nucleic Acids Res.*, **9**, 5187–5202.

Skoog, F., and Armstrong, D. J. (1970). Cytokinins, *Annu. Rev. Plant Physiol.*, **21**, 359–384.

Skoog, F., and Miller, C. O. (1957). Chemical regulation of growth and organ formation in plant tissues cultured *in vitro, Symp. Soc. Exp. Biol.*, **11**, 118–130.

Tempé, J., Petit, A., Holsters, M., van Montagu, M., and Schell, J. (1977). Thermosensitive step associated with transfer of the Ti plasmid during conjugation: possible relation to transformation in crown gall, *Proc. Natl. Acad. Sci. USA*, **74**, 2848–2849.

Thomashow, M. F., Nutter, R., Montoya, A. L., Gordon, M. P., and Nester, E. W. (1980). Integration and organisation of Ti plasmid sequences in crown gall tumors, *Cell*, **19**, 729–739.

Thomashow, M. F., Knauf, G., and Nester, E. W. (1981). Relationship between the limited and wide host range octopine Ti plasmids of *A. tumefaciens, J. Bacteriol.*, **146**, 484–493.

Van Larebeke, N., Engler, G., Holsters, M., van den Elsacker, S., Zaenen, I., Schilperoort, R. A., and Schell, J. (1974). Large plasmid in *Agrobacterium tumefaciens* essential for crown-gall inducing ability, *Nature (London)*, **252**, 169–170.

Willmitzer, L., De Beuckeleer, M., Lemmers, M., van Montagu, M., and Schell, J. (1980). DNA from Ti plasmid present in nucleus and absent from plastids of crown gall plant cells, *Nature (London)*, **267**, 359–361.

Willmitzer, L., Otten, L., Simons, G., Schmalenbach, W., Schröder, J., Schröder, G., van Montagu, M., de Vos, G., and Schell, J. (1981a). Nuclear and polysomal transcripts of T-DNA in octopine crown gall suspension and callus cultures, *Mol. Gen. Genet.*, **182**, 255–262.

Willmitzer, L., Schmalenbach, W., and Schell, J. (1981b). Transcription of T-DNA in octopine and nopaline crown gall tumors is inhibited by low concetrations of α-amanitin, *Nucleic Acids Res.*, **9**, 4801–4812.

Willmitzer, L., Simons, G., and Schell, J. (1982). The T_L-DNA in octopine crown gall tumors codes for seven well-defined poyladenylated transcripts, *EMBO J.*, **1**, 139–146.

Wullems, G. J., Molenijk, L., Ooms, G., and Schilperoort, R. A. (1981). Retention of tumor markers in Fl progeny plants from *in vitro* induced octopine and nopaline tumor tissues, *Cell*, **24**, 719–727.

Yadav, N. S., Postle, K., Saiki, R. K., Thomashow, M. F., and Chilton, M.-D. (1980). T-DNA of a crown gall teratoma is covalently joined to host plant DNA, *Nature (London)*, **287**, 456–461.

Yang, F. M., Montoya, A. L., Merlo, D. J., Drummond, M. H., Chilton, M.-D., Nester, E. W., and Gordon, M. P. (1980). Foreign DNA sequences in crown gall teratomata and their fate during the loss of tumorous traits, *Mol. Gen. Genet.*, **177**, 707–714.

Further Reading

Kahl, G., and Schell, J. (Eds) (1981). The Molecular Biology of Plants, Academic Press, New York.

Schell, J., Van Montagu, M., Holsters, M., Hernalsteens, J.-P., Lemmens, J., De Greve, H., Shaw, C., Zambryski, P., Willmitzer, L., Otten, L., Schröder, G., and Schröder, J. (1982), in *The Molecular Biology of Plant Development* (Eds H. Smith and D. Grierson), Blackwell Scientific, Oxford, pp. 498–514.

II INFECTION AND PATHOGENESIS

Biochemical Plant Pathology
Edited by J. A. Callow
© 1983 John Wiley & Sons Ltd

6

Fungal Penetration of the First Line Defensive Barriers of Plants[*]

P. E. KOLATTUKUDY and WOLFRAM KÖLLER

*Institute of Biological Chemistry and Biochemistry/Biophysics Program,
Washington State University, Pullman, WA 99164, USA*

1. INTRODUCTION

Fungi must penetrate the natural defensive barriers of plants before infection can be established. In intact aerial organs of plants the cuticle constitutes the

[*]This work was supported in part by grants from the National Science Foundation, PCM-8007908, and from the Washington Tree Fruit Research Commission.

first barrier whereas in periderms suberized cell walls assume this role. Whether the penetration of the first barrier is achieved merely by the physical force of growth of an infection structure or by an enzymic digestion of the cuticle has been debated for a long time (van den Ende and Linskens, 1974). This chapter deals with the nature of the defensive barriers and the mode of penetration of this barrier by fungi. Entry of pathogens into plants via wounds will not be covered.

2. NATURE OF THE DEFENSIVE BARRIERS

2.1. Cuticle

Plant cuticle consists of the insoluble polymer cutin, embedded in a complex mixture of hydrophobic materials collectively called wax. Waxy materials are also found in crystalline form on the surface of plants (Jeffree *et al.*, 1976). Extensive electron microscopic studies revealed the great diversity of crystalline structures on plant surfaces (Martin and Juniper, 1970). Transmission electron microscopy shows that the cuticle usually has an amorphous appearance with some reticular structures; the outer regions of the cuticle of some plants show lamellar structure (Figure 1).

The cuticular wax, which can be readily extracted with organic solvents, contains many classes of relatively non-polar $C_{20}-C_{32}$ aliphatic compounds, principally hydrocarbons, wax esters, primary alcohols, and fatty acids (Martin and Juniper, 1970; Kolattukudy, 1980a). The most common major hydrocarbons are $n\text{-}C_{29}$ and $n\text{-}C_{31}$ whereas the most common primary alcohols in the wax esters and in free form are $n\text{-}C_{26}$ and $n\text{-}C_{28}$. Many other less common components are also found.

Cutin, the structural component of cuticle, can be isolated by disruption of the underlying pectin layer with ammonium oxalate–oxalic acid or pectinase. The adhering carbohydrates are removed by cellulase and pectinase treatment and the waxes by extensive extraction with organic solvents. The resulting insoluble polymer can be depolymerized by hydrolysis with alcoholic KOH, transesterification with methanol and a catalyst such as BF_3 or $NaOCH_3$, or by hydrogenolysis with $LiAlH_4$ in tetrahydrofuran. The cutin monomers can be readily analysed by chromatographic techniques, especially GC–MS (Walton and Kolattukudy, 1972). Such analyses of cutin isolated from many species of plants show that this polymer is a polyester composed of C_{16} and C_{18} families of hydroxy and epoxy fatty acids (Figure 2). The major component of the former is a dihydroxypalmitic acid with one hydroxyl group at the ω-C position and the other at C-10, C-9, C-8 or C-7. The major components of the C_{18} family are ω-hydroxyoleic acid,

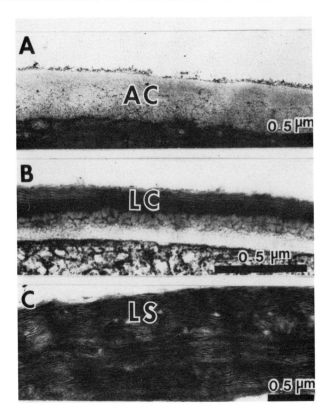

Figure 1. Electron micrographs showing amorphous cuticle (A, leaf of *Frankenia grandifolia*), lamellar cuticle (B, developing leaf of *Phormium tenax*), and lamellar suberin (C, wall of a cell in the chalazal region of the inner seed coat of *Citrus paradisi*). (Courtesy of Dr. D. W. Thomson, Dr. A. B. Wardrop, and R. W. Davis). AC, amorphous cuticle; LC, lamellar cuticle; LS, lamellar suberin. (From Kolattukudy *et al.*, 1981. Reproduced by permission of Springer-Verlag)

ω-hydroxy-9,10-epoxystearic acid, 9,10,18-trihydroxystearic acid and their analogues containing one additional double bond at C-12 (Kolattukudy, 1980b, 1981). Cutin on most plant organs consists of a combination of these monomers although the cutin of rapidly expanding organs of plants is chiefly composed of the C_{16} family of monomers. The monomers are linked together mainly via primary alcohol ester linkages although cross-links involving secondary alcohols are also present (Deas and Holloway, 1977; Kolattukudy, 1977). The model shown in Figure 2 is consistent with the currently available data, but the structural details of the inter-monomer linkages remain to be elucidated.

CUTIN

Major Monomers

C₁₆- FAMILY

CH₃(CH₂)₁₄COOH

CH₂(CH₂)₁₄COOH
|
OH

CH₂(CH₂)ₓCH(CH₂)ᵧCOOH
| |
OH   OH

(y = 8, 7, 6, or 5   x + y = 13)

C₁₈- FAMILY

CH₃(CH₂)₇CH= CH(CH₂)₇COOH

CH₂(CH₂)₇CH= CH(CH₂)₇COOH
|
OH

CH₂(CH₂)₇CH-CH(CH₂)₇COOH
| \ /
OH   O

CH₂(CH₂)₇CH-CH(CH₂)₇COOH
| | |
OH   OH OH

Polymer

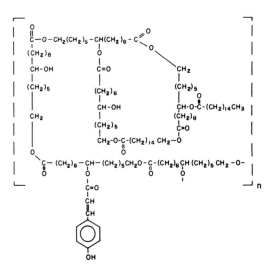

Figure 2. Structure of the cutin monomers and a model of the polymer containing only the C_{16} family of acids

2.2. Suberized cell walls

Suberization of the walls of a few layers of cells at the surface of wounds appears to be the general response of plants to wounding (Kolattukudy, 1980a). Suberization is also used by plants to erect barriers to diffusion in many other anatomical regions. Examples include endodermis (Casparian bands), bundle sheaths of grasses, the seed coat cells in the area of the attachment of vascular bundles, endodermis, epidermis, and hypodermis of roots, pigment strands of cereal grains, sheaths around idioblasts containing calcium oxalate crystals, and basal wall of trichomes and oil glands (Kolattukudy, 1980a; Kolattukudy *et al.*, 1981b). Although the chemical nature of some of these barrier layers remains unknown the lamellar appearance typical of suberized layers suggests the presence of suberin. The alternating light and dark layers in the electron micrographs of suberized layers (Figure 1) probably represent waxes and polymeric materials, respectively (Figure 3). Thus the polymeric material, defined as suberin, is associated with waxes just as cutin is associated with cuticular wax. Ultrastructural studies indicate that suberized layers are deposited on the cell wall just outside the plasma membrane and as suberization proceeds the cellular content degenerates, leaving only cytoplasmic remnants in the lumen of highly suberized cells. In such cells the entire cell wall shows a lamellar appearance, indicating that suberization has progressed throughout the wall (Figure 1).

The chemical composition of suberized layers is not well understood. However, from a limited number of recent studies the following general picture

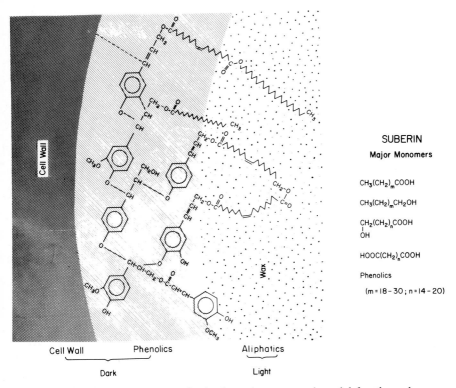

Figure 3. The major monomers of suberin and a proposed model for the polymer. Dark and light refer to the appearance in the electron micrographs

has emerged. The waxes associated with suberin contain components similar to those found in cuticular waxes (Kolattukudy *et al.*, 1981b). However, the chain length range of the aliphatic components of suberin-associated waxes appears to be shorter than those of cuticular waxes. The hydrocarbon fraction of suberin-associated waxes contains a relatively large proportion of even-chain components and, unlike the cuticular hydrocarbons, one component usually does not predominate. Further, a high content of free fatty acids appears to be a unique feature of suberin-associated waxes.

On the basis of the limited amount of information available, the following tentative conclusions can be drawn about the structure and composition of suberin (Figure 3). An aromatic matrix similar to lignin is attached to the cell wall and aliphatic components are attached to this matrix. The major aliphatic components are ω-hydroxy acids, dicarboxylic acids, and *n*-fatty acids and alcohols with 20–30 carbon atoms, with minor amounts of the more polar acids similar to those found in cutin. These are attached to the polymer by ester linkages. The cleavage methods so far tested reveal only esterified aliphatics and therefore the occurrence of other aliphatic components attached via linkages such as ether bonds which are resistant to such cleavage techniques remains a possibility. Structural considerations strongly suggest that the aliphatic components cannot, by themselves, form extensive polymers. These aliphatic components might serve to cross-link aromatic components and the limited amounts of aliphatic domains also probably provide hydrophobicity for association with the waxes. Thus deposition of the aromatics and covalently attached aliphatics provides a means for providing a hydrophobic environment on the hydrophilic cell wall and with the deposition of the waxes the suberized wall becomes an excellent barrier to diffusion.

3. GERMINATION AND DIFFERENTIATION OF FUNGAL SPORES

The germination of fungal spores is the first and most crucial event in the propagation of the species. This transition from dormancy to active growth is accompanied by profound structural and biochemical changes. Early events in germination include the mobilization of stored carbon sources like lipids, carbohydrates, or proteins, a process which makes spores of many fungi nutri-tionally independent. The requirement for additional exogenous nutrients, however, might vary even within one taxonomic group (Blakeman, 1980). Germination is necessarily accompanied by the biosynthesis of nucleic acids, proteins, membrane systems, and the cell wall. Most biochemical studies on spore germination, however, are concerned with saprophytic fungi with a low degree of plant pathogenicity, and a more comprehensive treatment of this subject is beyond the scope of this chapter.

If fungal spores germinate in an unfavourable environment, they are unlikely to survive. Many plant pathogenic fungi depend entirely on the host cells for a successful life cycle. Thus, it is crucial for spores of these specialized species to remain dormant until a host plant is near. The specific interactions between spores of the pathogen and a corresponding host plant leading to spore germination and differentiation of infection structures is treated only briefly in this chapter. This subject has been recently reviewed in greater detail elsewhere (Allen, 1976; Macko, 1981; Trione, 1981).

Dormancy, the state of metabolic inactivity, is maintained by natural endogenous germination inhibitors detected in a wide range of fungal spores. These inhibitors prevent germination until they leach out during water uptake and hydration of the spore. Most research has focused on rust fungi, and the inhibitors were shown to be a group of similar *cis*-isomers of cinnamic acid derivatives. The high biological activity of these compounds, with inhibitory concentrations ranging from 10^{-8} to 10^{-10} M, might explain the phenomenon of self-inhibition often observed in dense spore suspensions. The chemical nature of these endogenous inhibitors seems to vary among taxonomic groups. Quiesone, a β-ionone derivative, for example, was identified as the active compound in *Peronospora tabacina*, a pathogen causing downy mildew. This class of germination inhibitors, perplexingly enough, was able to stimulate the germination of bean rust uredospores. The endogenous germination inhibitor found in rust fungi is thought to prevent the digestion of the germ pore plug, a prerequisite for germ tube growth (Hess *et al.*, 1975). Our knowledge of the biochemical mode of action of these inhibitors, however, is rudimentary and different mechanisms might be operative in different spore types.

Chemical stimulants released by the host plant may counteract the endogenous germination inhibitors and thus initiate germination. These compounds may play a crucial role in an early stage of plant–pathogen interaction. Our current understanding of these signals, which allow a pathogen to recognize the host, is limited. It has been suggested that a variety of volatile carbonyl compounds derived from lipid peroxidation might be responsible for the regulation of spore germination in soil (Harmann *et al.*, 1980). The infection of aerial parts of a plant often requires, in addition to germination, the formation of highly specialized infection structures such as appressoria. In some cases mechanical stimulants presented by micro-structures of the plant surface might regulate the differentiation of infection structures (Staples and Hoch, 1982), while in other cases chemical signals are involved. The endogenous signal which induces appressorium formation in *Puccina graminis* f. sp. *tritici in vitro* was identified as acrolein (Macko *et al.*, 1978). On the other hand, formation of infection structures by the same fungus was reported to be induced by *cis*-hex-3-en-1-ol in the presence of a phenol fraction, both of which were detected in the host plant (Grambow and Grambow, 1978).

Plant surfaces often contain fungistatic or fungitoxic compounds. Components isolated from cuticular waxes of a number of plants were reported to inhibit spore germination of pathogenic fungi (Blakeman and Atkinson, 1976, 1981). Only a few of these compounds have been chemically identified (e.g. Cruickshank *et al.*, 1977), and convincing evidence for their function in nature is lacking. Both the chemical nature and the biochemical mode of action of these wax components, which probably constitute part of a pre-formed defense mechanism of a plant against fungal attack, need further investigation.

4. PENETRATION OF THE CUTICLE

4.1. Indirect evidence for the involvement of cutinase in penetration

Direct invasion of fungal pathogens through intact surfaces of the host plant has been observed in numerous host–parasite interactions. The question of whether the penetration of the cuticle occurs merely by physical force or by enzymatic degradation of this protective layer has been a controversial subject for almost a century. On the basis of ultrastructural examinations of the penetration area it was suggested that enzymatic hydrolysis of the cuticle was involved in cuticular penetration (Aist, 1976; Dodman, 1979; Verhoeff, 1980). Since the polymer cutin constitutes the main physical barrier in the cuticle, It was proposed that the penetrating fungi secreted a cutin-degrading enzyme, cutinase. Early research on enzymatic cutin degradation using saprophytes and various pathogenic fungi grown with cutin as the sole carbon source has been summarized by van den Ende and Linskens (1974). The conclusions drawn from these studies were based on the results of indirect assay methods applied to crude extracts. More recently a direct enzyme assay for cutinase was developed using radioactive cutin as substrate and with this method cutinase activity was found in the culture filtrates of all phytopathogenic fungi investigated thus far (Kolattukudy, 1981). The presence of cutinase in the extracellular fluids of cultured fungi was consistent with the hypothesis that cuticular penetration of infection structures involves enzymatic degradation of cutin.

4.2. Isolation of cutinase from pathogenic fungi

Cutinase was first isolated in homogeneous form from the extracellular fluid of *Fusarium solani pisi* grown on cutin as a sole source of carbon (Purdy and Kolattukudy, 1975a,b). Using basically the same procedures originally developed for *F. solani pisi* and recently summarized (Kolattukudy *et al.*, 1981a), cutinase has been purified from *F. roseum culmorum*, *F. roseum sambucinum*, *Ulocladium consortiale*, *Colletotrichum gloeosporioides*, *Phytophcactorum*, *Botrytis cinerea*, *Helminthosporium sativum*, and *Strep-*

tomyces scabies (Soliday and Kolattukudy, 1976; Lin and Kolattukudy, 1980a; Dickman *et al.*, 1982). Removal of phenolics from the dark-coloured extracellular fluids is an essential step in the purification of the enzyme and is accomplished by chromatography on QAE-Sephadex. Highly purified enzyme can usually be obtained by additional chromatography on SP-Sephadex, but in some cases further purification by hydrophobic chromatography using octyl-Sepharose with octyl glycoside as the eluent is necessary.

4.3. Properties of cutinase

4.3.1. Catalytic properties

All of the fungal cutinases which have been purified to homogeneity show similar catalytic properties. All hydrolyse cutin at maximal rates at pH 9–10. Both the generation of oligomers and further hydrolysis of the oligomers to monomers are catalysed by cutinase. This enzyme does show specificity for primary alcohol esters. Fungal cutinases also hydrolyse *p*-nitrophenyl esters of short-chain fatty acids and thus such model substrates can be used in a

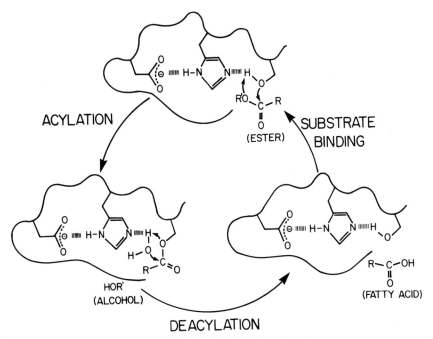

Figure 4. Mechanism of action of cutinase

convenient spectrophotometric assay. As the chain length of the acyl moiety increases, the rate of hydrolysis decreases. However, the chain length dependence of the rate and specific activity for cutin hydrolysis can vary, depending on the source of the enzyme.

Fungal cutinases are severely inhibited by reagents directed towards 'active serine.' Experiments with radioactive diisopropyl fluorophosphate showed that this reagent inhibited the enzyme by covalent modification of one active serine per molecule of the enzyme. Enzymes which catalyse hydrolysis using an active serine would be expected to have an interacting catalytic triad: serine hydroxyl group, a carboxyl group, and an imidazole group of histidine as in serine proteases (Kraut, 1977). As expected from the presence of such a catalytic triad, selective chemical modification of the carboxyl group by carbodiimide inactivated cutinase and a single carboxyl group in the enzyme was thus shown to be involved in catalysis. Selective chemical modification of the imidazole group of histidine by treatment with diethyl pyrocarbonate also inhibited the enzyme. Modification of one histidine per molecule of the enzyme was required for complete inactivation of the enzyme. Thus presence of the catalytic triad in this enzyme has been established (Köller and Kolattukudy, 1982) (Figure 4).

4.3.2. Molecular and structural properties

The enzymes from all fungal sources so far examined are similar in size ($M_r \approx 25\,000$) and amino acid composition. Most fungal cutinases contain a single disulphide bridge and no free SH group. They also contain one methionine, one histidine, and one tryptophan per molecule. All of them are glycoproteins containing 3.5–6% of carbohydrate (Table 1). The carbohydrates are attached to all of these cutinases, except those from S. scabies and H. sativum, by O-glycosidic linkages and most of the carbohydrate residues are monosaccharides (Lin and Kolattukudy, 1980b). The hydroxyl groups of the following amino acid residues were found to be involved in the O-glycosidic linkages: serine, threonine, β-hydroxyphenylalanine and β-hydroxytyrosine in the enzyme from F. solani pisi, threonine and β-hydroxyphenylalanine in U. consortiale, serine and β-hydroxyphenylalanine in F. roseum culmorum, and only serine in F. roseum sambucinum. β-Hydroxyphenylalanine and β-hydroxytyrosine, the novel amino acids heretofore not found in any other protein, do not appear to have any speical functional significance in this enzyme, as they are not even found in cutinase from all fungal sources. The N-terminal amino group of the fungal cutinases is attached to glucuronic acid by an amide linkage, and this is a novel N-terminal modification. The close similarity in properties of the enzyme for all of the fungal sources so far examined leaves in doubt the function of the regions of the protein where carbohydrates are attached. It is possible that these struc-

Table 1. Properties of fungal cutinase

| Source of cutinase | Host | Properties | | |
		Molecular weight	Polypeptide[a] composition	Carbohydrate content (%)
Fusarium solani pisi[b]	Pea	22 000	Single[b]	4–5
Fusarium roseum culmorum	Wheat	24 300	13 000 and 10 500	6.0
Fusarium roseum sambucinum	Squash	24 800	13 800 and 12 700	6.0
Ulocladium consortiale	Cucumber	25 100	13 000 and 10 000	4.0
Streptomyces scabies	Potato	26 000	Multiple	4.5
Helminthosporium sativum	Barley	26 300	Single	3.5
Phytophthora cactorum	Apple	21 000	Single	—
Colletotrichum gloeosporioides	Papaya	26 000	Single	16.0

[a] When two polypeptides are found they are linked by a disulphide bridge and are most likely generated by a proteolytic nick.
[b] Four cutinases have been purified from two different field isolates. One enzyme (cutinase II) contained a proteolytic nick.

tural features provide stability to the enzyme in the extracellular environment.

Cutinase from *F. solani pisi* is stable to proteolytic enzymes such as trypsin, chymotrypsin, proteinase K, elastase or clostripain. However, conformational changes brought about by the presence of sodium dodecylsulphate make it susceptible to proteolysis (Köller and Kolattukudy, 1982). The conformational tightness suggested by this observation is also reflected in the observation that both histidine and the carboxyl group of the catalytic triad are not accessible to chemical modification in the absence of sodium dodecylsulphate. Further, the only disulphide bond present in this protein reacts readily with dithioerythritol only in the presence of sodium dodecylsulphate and reduction of this disulphide causes irreversible inactivation of the enzyme. The conformational tightness suggested by all of these observations probably provides the stability required for functioning in the extracellular environment.

4.4. Biosynthesis of cutinase

How cutinase is induced by the presence of insoluble polymeric cutin in the growth medium is an intriguing question. The following observations suggest that the real inducer of cutinase synthesis is the small amounts of cutin hydrolysate generated by the small amounts of cutinase which are probably excreted under starvation conditions. In cultures of *F. solani pisi* grown on glucose, cutinase was induced by very small amounts (80 μg cm^{-3}) of exogenous cutin hydrolysate when glucose was depleted (Lin and Kolattukudy, 1978). Under these conditions cutinase was the major (*ca.* 70%) protein in the extracellular medium. From such induced cultures poly A-containing mRNA was isolated and translated *in vitro* in rabbit reticulocyte lysate and wheat germ cell-free preparations (Flurkey and Kolattukudy, 1981). Analysis of the total products of *in vitro* translation showed that mRNA from the induced culture gave rise to one protein which was not generated by the mRNA from the non-induced culture. Immunoprecipitation of the translation products with antibodies prepared against cutinase resulted in the recovery of a 25.5 kDa peptide which was 2.1 kDa larger than the peptide in the mature extracellular enzyme. Therefore, it is concluded that cutinase is synthesized as a precursor slightly larger than the mature peptide and this precursor is processed by proteolysis and glycosylation before excretion.

The post-translational events must include introduction of the glucuronic acid moiety at the *N*-terminal amino group which might occur concurrently with the proteolytic processing of the precursor. However, the mechanism of introduction of the glucuronamide at the *N*-terminus is not understood. Introduction of the hydroxyl group at the β-carbon of phenylalanine and tyrosine probably involves a hydroxylation reaction similar to that involved in the synthesis of hydroxyproline and hydroxylysine (Cardinale and Udenfriend, 1974).

Introduction of the *O*-glycosidically linked mannose was shown to be catalysed by microsomal membranes from *F. solani pisi* which was induced to synthesize cutinase by cutin hydrolysate (Soliday and Kolattukudy, 1979). This glycosylation involved dolicholphosphorylmannose and the endogenous dolichol phosphate from the microsomal preparation was shown to be a $C_{90}-C_{105}$ mixture. The mechanism of introduction of the other *O*-glycosidically attached monosaccharides remains to be elucidated.

4.5. Direct proof for the involvement of cutinase in fungal penetration

To prove the involvement of cutinase in penetration of a pathogen into its host, secretion of cutinase by the penetrating fungus must be conclusively shown and conclusive evidence that cutinase is required for infection must be provided. Both of these have been accomplished for infection of *Pisum sativum* by *F. solani pisi* under controlled experimental conditions.

Ferritin-conjugated antibodies prepared against cutinase were used to determine whether the fungus penetrating into intact pea stem secretes cutinase. Scanning electron microscopy showed that the spores placed in a suspension on pea stem germinated and penetrated into the stem in 18 h.

Treatment of the penetrating area with ferritin-labelled anticutinase followed by electron microscopy demonstrated the presence of cutinase–antibody complex, showing that the penetrating fungus secreted cutinase during penetration of its host (Shaykh *et al.*, 1977) (Figure 5).

If the secreted cutinase is necessary for penetration, specific inhibition of cutinase should prevent penetration and therefore infection. Rabbit antiserum prepared against cutinase prevented infection whereas control serum had no effect. Chemicals known to be inhibitors of cutinase such as diisopropyl fluorophosphate also prevented infection, although such chemicals did not appear to affect the growth of the organism. Further, the inhibitors and the antiserum prevented infection only when intact cuticular barrier was present and not when the cuticle–cell wall barrier was mechanically breached (Maiti and Kolattukudy, 1979; Köller *et al.*, 1981). More recently a similar approach was used to demonstrate that cutinase is essential in the infection of papaya fruits by *Colletotrichum gloeosporioides* (Dickman *et al.*, 1982). Infection of fruits with intact cuticle by this organism was also prevented by inhibition of cutinase either by chemical inhibitors or by specific antibodies and no such protection was observed when the cuticle–cell wall barrier was mechanically breached prior to placing the spore suspension on the fruit. Although it is difficult to generalize the essential role of cutinase in fungal infection of plants,the two examples where such a role has been proved raise the strong possibility that cutinase plays an essential role in infection of plants by those fungi which possess the ability to infect plants with an intact cuticular barrier.

Recent results raised the possibility that the level of cutinase in the germinating spores can determine the degree of virulence. Highly sensitive radioassays have shown that germinating spores of *F. solani pisi* isolate T-8 secrete a cutinase which is extremely similar to, if not identical with, that secreted during saprophytic growth on cutin (Köller *et al.*, 1981). Bioassays using pea stems with or without intact cuticle–cell wall barriers showed that *F. solani pisi* isolate T-8 was highly virulent in both assays whereas isolate T-30, which was highly virulent on wounded stems, was almost avirulent on the intact surface (Köller *et al.*, 1981). Measurement of cutinase released by the germinating spores from the two isolates showed that the avirulence of T-30 on intact surface was caused by lack of cutinase. This low virulence could be enhanced to the level expressed on wounded stems by supplementation of the T-30 spores with exogenous enzymes only when cutinase was included (Figure 6).

A pathogenic fungus has to breach two major physical barriers before it can infect a plant organ with intact surface. After breaching the first barrier with cutinase the fungus encounters the cell wall. Therefore, successful penetration requires cutinase and cell wall-degrading enzymes which might be collectively termed a set of penetration enzymes. Although the number and nature of the cell wall-degrading enzymes released by the germinating spores remain

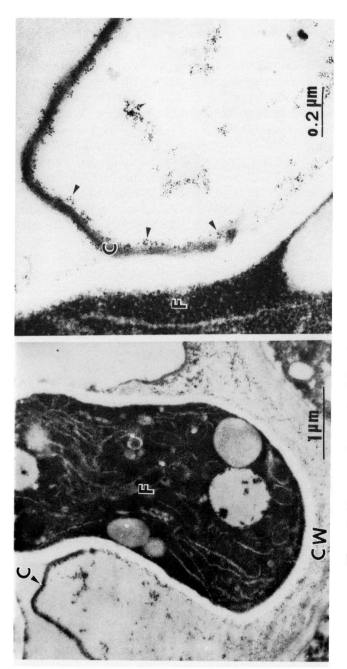

Figure 5. Electron micrograph of *F. solani pisi* penetrating into the host plant *P. sativum* (left) and a magnified region of the cuticle (right). The dark spots (arrows) represent ferritin-conjugated antibody prepared against cutinase. C, cuticle; CW, cell wall; F, *Fusarium*. (Reproduced by permission of the American Society of Plant Physiology)

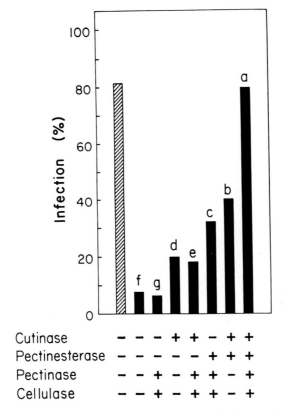

Figure 6. Infection efficiency of *F. solani pisi* (isolate T-30) conidia after enzyme supplementation. A conidial suspension was mixed with (+) or without (−) indicated enzymes. The mixtures were used to inoculate intact (■) or wounded (▨) surfaces of epicotyl segments of *P. sativum*. [Reproduced with permission from Köller *et al.* (1981) Copyright: Academic Press Inc. (London) Ltd.]

obscure, some recent results strongly suggest that cellulase, pectinase, and pectin methylesterase are essential for successful penetration and infection of pea stems with intact cell wall barrier by *F. solani pisi*. Isolate T-30, which was avirulent because of its inability to penetrate the barrier, became virulent when the spores were supplemented with exogenous cutinase, pectinase, cellulase, and pectin methylesterase (Köller *et al.*, 1981). All four enzymes were required to enhance the virulence of T-30 on intact stems to the level expressed on wounded stems (Figure 6). Thus it appeared that T-30 lacked all of the penetration enzymes. The detection of pectinase in germinating spores of *Colletotrichum orbiculare* (Porter, 1969) and *Botrytis cinerea* (Verhoeff and Liem, 1978) is consistent with this hypothesis. It is possible that the level of penetration enzymes as a group in the spores is a factor which can determine the degree of virulence of pathogenic strains and that the production of these enzymes might be under coordinate control. Cell wall-degrading enzymes are considered in Chapter 7 in more detail.

5. INHIBITORS OF CUTINASE AND THEIR POTENTIAL AS ANTIPENETRANTS

The finding that specific inhibition of cutinase can prevent infection by some fungi under controlled conditions in bioassays in the laboratory raises the possibility of developing cutinase-directed chemicals as antipenetrants to protect plants. The presence of the catalytic triad involving an active serine provides one avenue for designing effective inhibitors against cutinase. Thus organic phosphates and phosphorothioates, a potent class of pesticides (Eto, 1974), are inhibitors of cutinase (Table 2). Among them the aromatics such as O,O-dimethyl-O-(3,5,6-trichloro-2-pyridyl) phosphate (Fospirate) appear to be the most potent. Even a small structural alteration makes a substantial difference to the inhibitory activity of these chemicals. Thus, on replacement

Table 2. Cutinase inhibitors as antipenetrants

Structure	Name[a]	Current use	I_{50} (μM)[b]	Prevention of infection at I_{50} (%)[c]
	Fospirate	Insecticide	13 (4 ppm)	67
	Hinosan	Fungicide	19 (6 ppm)	88
	Benomyl	Fungicide	10 (3 ppm)	69

[a] Trade name.
[b] I_{50} is concentration for 50% cutinase inhibition after incubation for 60 min.
[c] Determined with segments of pea hypocotyls inoculated with a spore suspension of *F. solani pisi* containing the inhibitor at I_{50} concentration.

of the methyl groups by ethyl the potency of inhibition increased 400-fold, as indicated by the lowering of the I_{50} value by 400-fold. These findings raise the possibility of being able to maximize the inhibitory effect of such chemicals by structural alterations. In the pea stem bioassay with intact surface these chemicals were effective in preventing infection by *F. solani pisi*, and their effectiveness reflected, to a certain degree, their cutinase inhibitory activity.

Hinosan, a phosphorodithioate currently used as a fungicide against rice blast disease (Eto, 1974), is a powerful cutinase inhibitor and a potent protectant against infection under *in vitro* conditions (Table 2). The same was true for Benomyl, a systemic fungicide with a broad action spectrum (Erwin, 1973). Carbamate derivatives such as Benomyl are known to inhibit serine hydrolases and are in wide use as insecticides (Kuhr and Dorough, 1976). Our finding that these two widely used fungicides are cutinase inhibitors opens the possibility that their protective action is due not only to their fungitoxic properties, but also to their activity as antipenetrants.

6. PENETRATION OF SUBERIZED LAYERS

Suberized periderm layers protect underground parts of plants. Ultrastructural studies have shown that fungi do breach such barriers (Peek *et al.*, 1972). This process is probably aided by enzymes but they have not been studied. In this case, the penetration enzymes would have to deal with cell wall carbohydrate polymers, the phenolic matrix and the aliphatic components. Cutinase from *F. solani pisi* which catalyses hydrolysis of ester bonds can release the aliphatic components from suberin. Several pathogenic fungi grow on suberin from potato tuber periderm as the sole source of carbon. The medium from *F. solani pisi* and *Streptomyces scabies* grown on suberin contained enzymes which catalysed the hydrolysis of ester bonds contained in a cutin polymer (Kolattukudy, 1978). Using radioactive cutin as substrate, such hydrolytic enzymes were purified by methods similar to those used for cutinase. These fractionation procedures revealed the presence of several proteins which hydrolysed cutin and model substrates. At least one of these hydrolases from *F. solani pisi* grown on suberin cross-reacted with rabbit antibodies prepared against cutinase purified from cultures grown on cutin. Thus, growth on cutin and suberin results in induction of similar ester-hydrolysing enzymes. Although breaching of the cutin barrier would involve hydrolysis of ester bonds only, breaking of a suberin barrier would require disruption of the phenolic matrix and the cell wall to which suberin is attached. To determine the nature of the enzymes involved in the disruption of suberin a detailed study of the extracellular enzymes in cultures of *F. solani pisi* isolate T-8 grown on suberin was initiated. In an attempt to detect enzymes which degrade the phenolic matrix, potato wound periderm biosynthetically labelled with [³H]cinnamate was used as substrate. To test for enzymes which cleave the aliphatic components characteristic of suberin (ω-hydroxyoleic acid and the corresponding dicarboxylic acid), potato wound periderm biosynthetically labelled with [^{3}H]oleic acid was used as substrate. The extracellular fluid from cultures grown on suberin catalysed the release of label from both of these insoluble labelled polymers. Fractionation of the culture filtrate showed that the release of the label from the two substrates was catalyzed by different proteins (C. Allan, C. Soliday and P. E. Kolattukudy, unpublished work). These enzymes have not been purified and characterized and their role in the penetration of suberized walls remains to be established.

7. CONCLUSION

There are many observations which suggest that cuticular components play a role in the interaction between the plant and pathogenic fungi. However, the nature of the stimulators and inhibitors of spore germination and differentiation of infection structures has not been elucidated in most cases. Further, whether such compounds play a meaningful role in regulating the infection capability of pathogens in the field is not known. Another possibility which has not been explored is that ligands such as fungitoxic phenolics covalently attached to the cuticular polymer might be released by the fungal enzymes and thus protect the plant. During the past decade the nature of the physical barriers in the first line defense of plants against pathogens has become clearer, although much remains unknown. Even the chemical composition of suberized layers and in particular the nature of the phenolic domains remain obscure. The long-standing controversy concerning the mode of penetration of the cuticle by fungi has been settled to some degree by providing direct proof that at least some fungi use cutinase to gain access through the cuticle. However, such direct proof is currently available for only two pathogens and therefore it is not known whether this generalization is widely applicable. The enzymology of suberin degradation is extremely poorly understood and therefore it is not yet possible to test directly whether such enzymes are involved in penetration of pathogens into plant organs protected by suberin. Since the defensive barriers of most plants show similar structural features it is unlikely that the fungal enzymes which degrade them would show much species specificity for hydrolysis of the polymer. Therefore, it is highly unlikely that hydrolytic enzymes would play a significant role in determining host–pathogen specificity. Nevertheless, the polymer-hydrolysing enzymes are required for establishing infection and therefore agents directed against such enzymes can be used as antipenetrants to protect plants. Since penetration of the first line defensive barrier is the first step in infection, exploration of this process and the role of cuticular components on the regulation of development of pathogens is likely to reveal mechanisms of natural defense of plants. An understanding of such mechanisms could provide novel avenues for protecting plants.

REFERENCES

Aist, J. R. (1976). Cytology of penetration and infection-fungi, in *Encyclopedia of Plant Physiology, New Series, Vol. 4, Physiological Plant Pathology* (Eds R. Heitefuss and P. H. Williams), Springer-Verlag, New York, pp. 197–221.

Allen, P. J. (1976). Spore germination and its regulation, in *Encyclopedia of Plant Physiology, New Series, Vol. 4, Physiological Plant Pathology* (Eds R. Heitefuss and P. H. Williams), Springer Verlag, New York, pp. 51–85.

Blakeman, J. P. (1980). Behavior of conidia on aerial plant surfaces, in *The Biology of Botrytis* (Eds J. R. Coley-Smith, K. Verhoeff, and W. R. Jarvis), Academic Press, New York, pp. 115–151.

Blakeman, J. P., and Atkinson, P. (1976). Evidence for a spore germination inhibitor co-extracted with wax from leaves, in *Microbiology of Aerial Plant Surfaces* (Eds C. H. Dickinson and T. F. Preece), Academic Press, New York, pp. 441–449.

Blakeman, J. P., and Atkinson, P. (1981). Antimicrobial substances associated with the aerial surfaces of plants, in *Microbial Ecology of the Phylloplane* (Ed. J. P. Blakeman), Academic Press, New York, pp. 245–263.

Cardinale, G. J., and Udenfriend, S. (1974). Prolyl hydroxylase, *Adv. Enzymol.*, **41**, 245–300.

Cruickshank, I. A. M., Perrin, D. R., and Mandryk, M. (1977). Fungitoxicity of duvatriendediols associated with the cuticular wax of tobacco leaves, *Phytopathol, Z.*, **90**, 243–249.

Deas, A. H. B., and Holloway, P. J. (1977). The intermolecular structure of some plant cutins, in *Lipids and Lipid Polymers in Higher Plants* (Eds M. Tevini and H. K. Lichtenthaler), Springer-Verlag, New York, pp. 293–299.

Dickman, M. B., Patil, S. S., and Kolattukudy, P. E. (1982). Purification and characterization of an extracellular cutinolytic enzyme from *Colletotrichum gloeosporioides* on *Caria papaya, Physiol. Plant Pathol.*, **20**, 333–347.

Dodman, R. L. (1979). How the defenses are breached, in *Plant Disease—An Advanced Treatise, Vol. 4, How Pathogens Induce Disease* (Eds J. G. Horsfall and E. B. Cowling), Academic Press, New York, pp. 135–153.

Erwin, D. C. (1973). Systemic fungicides: disease control, translocation, and mode of action, *Annu. Rev. Phytopathol.*, **11**, 389–422.

Eto, M. (1974). *Organophosphorus Pesticides: Organic and Biological Chemistry*, CRC Press, Cleveland.

Flurkey, W. H. and Kolattukudy, P. E. (1981). *In vitro* translation of cutinase mRNA: evidence for a precursor form of an extracellular fungal enzyme, *Arch. Biochem. Biophys.*, **212**, 154–161.

Grambow, H. J., and Grambow, G. E. (1978). The involvement of epicuticular and cell wall phenols of the host plant in the *in vitro* development of *Puccinia graminis* f. sp. *tritici, Z. Pflanzenphysiol.*, **90**, 1–9.

Jeffree, C. E., Baker, E. A., and Holloway, P. J. (1976). Origins of the fine structure of plant epicuticular waxes, in *Microbiology of Aerial Plant Surfaces* (Eds C. H. Dickinson and T. F. Preece), Academic Press, New York, pp. 119–158.

Harman, G. E., Mattick, L. R., Nash, G., and Nedrow, B. L. (1980). Stimulation of fungal spore germination and inhibition of sporulation in fungal vegetative thalli by fatty acids and their volatile peroxidation products, *Can. J. Bot.*, **58**, 1541–1547.

Hess, S. L., Allen, P. J., Nelson, D., and Lester, H. (1975). Mode of action of methyl-*cis*-ferulate, the self inhibitor of stem rust uredospore germination, *Physiol. Plant Pathol.*, **5**, 107–112.

Kolattukudy, P. E. (1977). Lipid polymers and associated phenols, their chemistry, biosynthesis, and role in pathogenesis, in *Recent Advances in Phytochemistry, Vol. 11, The Structure, Biosynthesis, and Degradation of Wood* (Eds F. A. Loewus and V. C. Runeckles), Plenum Press, New York, pp. 185–246.

Kolattukudy, P. E. (1978). Chemistry and biochemistry of the aliphatic components of suberin, in *Biochemistry of Wounded Plant Tissues* (Ed. G. Kahl), Walter de Gruyter, New York, pp. 43–84.

Kolattukudy, P. E. (1980a). Cutin, suberin and waxes, in *The Biochemistry of Plants*, Vol. 4 (Eds. P. K. Stumpf and E. E. Conn), Academic Press, New York, pp. 571–645.

Kolattukudy, P. E. (1980b). Biopolyester membranes of plants: cutin and suberin, *Science*, **208**, 990–1000.

Kolattukudy, P. E. (1981). Structure, biosynthesis, and biodegradation of cutin and suberin, *Annu. Rev. Plant Physiol.*, **32**, 539–567.

Kolattukudy, P. E., Purdy, R. E., and Maiti, I. B. (1981a). Cutinases from fungi and pollen, *Methods Enzymol.*, **71**, 652–664.

Kolattukudy, P. E., Espelie, K. E., and Soliday, C. L. (1981b). Hydrophobic layers attached to cell walls, cutin, suberin and associated waxes, in *Encyclopedia of Plant Physiology, New Series, Vol. 13B, Plant Carbohydrates* (Eds F. A. Loewus and W. Tanner), Springer-Verlag, New York, pp. 225–254.

Köller, W. and Kolattukudy, P. E. (1982). Submitted for publication.

Köller, W., Allan, C. R., and Kolattukudy, P. E. (1981). Role of cutinase and cell wall degrading enzymes in infection of *Pisum sativum* by *Fusarium solani* f. sp. *pisi*, *Physiol. Plant Pathol.*, **20**, 47–60.

Kraut, J. (1977). Serine proteases: structure and mechanism of catalysis, *Annu. Rev. Biochem.*, **46**, 331–358.

Kuhr, R. J., and Dorough, H. W. (1976). *Carbamate Insecticides: Chemistry, Biochemistry, and Toxicology*, CRC Press, Cleveland.

Lin, T. S., and Kolattukudy, P. E. (1978). Induction of a polyester hydrolase (cutinase) by low levels of cutin monomers in *Fusarium solani* f. sp. *pisi*, *J. Bacteriol.*, **133**, 942–951.

Lin, T. S., and Kolattukudy, P. E. (1980a). Isolation and characterization of a cuticular polyester (cutin) hydrolysing enzyme from phytopathogenic fungi, *Physiol. Plant Pathol.*, **17**, 1–15.

Lin, T. S., and Kolattukudy, P. E. (1980b). Structural studies on cutinase, a glycoprotein containing novel amino acids and glucuronic acid amide at the *N*-terminus, *Eur. J. Biochem.*, **106**, 341–351.

Macko, V. (1981). Inhibitors and stimulants of spore germination and infection structure formation in fungi, in *The Fungal Spore: Morphogenetic Controls* (eds G. Turian and H. R. Hohl), Academic Press, New York, pp. 565–584.

Macko, V., Renwich, J. A. A., and Rissler, J. F. (1978). Acrolein induces differentiation of infection structures in the wheat stem rust fungus, *Science*, **199**, 442–443.

Maiti, I. B., and Kolattukudy, P. E. (1979). Prevention of fungal infection of plants by specific inhibition of cutinase, *Science*, **205**, 507–508.

Martin, J. T., and Juniper, B. E. (1970). *The Cuticles of Plants*, St. Martin's Press, New York.

Peek, V. R. D., Liese, W. and Parameswaran, N. (1972). Infektion und Abbau der Wurzelrinde von Fichte durch *Fomes annosus, Eur. J. For. Pathol.*, **2**, 104–115.

Porter, F. M. (1969). Protease, cellulase, and differential localization of endo- and exopolygalacturonase in conidia and conidial matrix of *Colletotrichum orbiculare, Phytopathology*, **59**, 1209–1213.

Purdy, R. E., and Kolattukudy, P. E. (1975a). Hydrolysis of plant cuticle by plant pathogens. Purification, amino acid composition, and molecular weight of two isoenzymes of cutinase and a nonspecific esterase from *Fusarium solani* f. *pisi, Biochemistry*, **14**, 2824–2831.

Purdy, R. E., and Kolattukudy, P. E. (1975b). Hydrolysis of plant cuticle by plant pathogens. Properties of cutinase I, cutinase II, and a nonspecific esterase isolated from *Fusarium solani pisi, Biochemistry*, **14**, 2832–2840.

Shaykh, M., Soliday, C., and Kolattukudy, P. E. (1977). Proof for the production of cutinase by *Fusarium solani* f. *pisi* during penetration into its host, *Pisum sativum, Plant Physiol.*, **60**, 170–172.

Soliday, C. L., and Kolattukudy, P. E. (1976). Isolation and characterization of a cutinase from *Fusarium solani culmorum* and its immunological comparison with cutinases from *F. solani pisi, Arch. Biochem. Biophys.*, **176**, 334–343.

Soliday, C. L., and Kolattukudy, P. E. (1979). Introduction of *O*-glycosidically linked mannose into proteins via mannosyl phosphoryl dolichol by microsomes from *Fusarium solani* f. *pisi, Arch. Biochem. Biophys.*, **197**, 367–378 (1979).

Staples, R. C., and Hoch, H. C. (1982). A possible role for microtubules and microfilaments in the induction of nuclear division in bean rust urediospore germlings, *Exp. Mycol.*, **6**, 293–302.

Staples, R. C., and Macko, V. (1980). Formation of infection structures as a recognition response in fungi, *Exp. Mycol.*, **4**, 2–16.

Trione, E. J. (1981). Natural regulators of fungal development, in *Plant Disease Control* (Eds R. C. Staples and G. H. Toenniessen), Wiley, New York, pp. 85–102.

van den Ende, G., and Linskens, H. F. (1974). Cutinolytic enzymes in relation to pathogenesis, *Annu. Rev Phytopathol.*, **12**, 247–258.

Verhoeff, K. (1980). The infection process and host-pathogen interaction, in *The Biology of Botrytis* (Eds J. R. Coley-Smith, K. Verhoeff and W. R. Jarvis), Academic Press, New York, pp. 153–180.

Verhoeff, K., and Liem, J. I. (1978). Presence of endopolygalacturonase in conidia of *Botrytis cinerea* before and during germination, *Phytopathol. Z.*, **91**, 110–115.

Walton, T. J., and Kolattukudy, P. E. (1972). Determination of structures of cutin monomers by a novel depolymerization procedure and combined gas chromatography and mass spectrometry, *Biochemistry*, **11**, 1885–1897.

Further Reading

Burnett, J. H., and Trinci, A. P. J. (1979). *Fungal Walls and Hyphal Growth*, Cambridge University Press, Cambridge.

Farkas, V. (1979). Biosynthesis of cell walls of fungi, *Microbiol. Rev.*, **43**, 117–144.

Maxwell, D. P., Armentrout, V. N., and Graves, L. B. (1977). Microbodies in plant pathogenic fungi. *Annu. Rev. Plant Pathol.*, **15**, 119–134.ß

Smith, J. E., and Berry, D. R. (Eds) (1978). *The Filamentous Fungi*, Vol. 3, Edward Arnold, London.

Staples, R. C., and Yaniv, Z. (1976). Protein and nucleic acid metabolism during germination, in *Encyclopedia of Plant Physiology, New Series, Vol. 4, Physiological Plant Pathology* (Eds R. Heitefuss and P. H. Williams), Springer-Verlag, New York, pp. 86–103.

Weber, D. J., and Hess, W. M., (1976). *The Fungal Spore-Form and Function*, Wiley, New York.

Wynn, W. K., and Staples, R. C. (1981). Tropism of fungi in host recognition, in *Plant Disease Control* (Eds R. C. Staples and G. H. Toenniessen). Wiley, New York, pp. 45–69.

Biochemical Plant Pathology
Edited by J. A. Callow
© 1983 John Wiley & Sons Ltd

7

The Mechanisms and Significance of Enzymic Degradation of Host Cell Walls by Parasites

RICHARD M. COOPER

School of Biological Sciences, University of Bath, Claverton Down, Bath, Avon, UK

1. INTRODUCTION

During penetration and colonization microbial plant pathogens must repeatedly encounter and penetrate cell walls of their hosts. Many but not all parasites can produce a wide range of cell wall-degrading enzymes (CWDE)

Figure 1. Penetration of host cell walls. (a) Extensive degradation and swelling of epidermal cell wall (W) during penetration of a broad bean leaf by *Botrytis cinerea*. Cuticle (C), germ tube (G), infection hyphae (H). ×4000. (From McKeen, 1974. Reproduced by permission of the American Phytopathological Society). (b) Maceration of potato tuber parenchyma by *Erwinia carotovora*. Bacteria (B) remain confined to the middle lamella region, degradation of which results in separation of adjacent cells and disruption of host cytoplasm. Swollen primary walls (W). ×14 400. (From Fox *et al.*, 1972. Reproduced by permission of the European Association for Potato Research). (c) Penetration of a flax mesophyll cell (M) by *Melampsora lini*. Degradation of the host wall (W) is not apparent but there is no indication of mechanical disruption. Haustorial mother cell (HMC). ×11 300 (From Coffey *et al.*, 1972. Reproduced by permission of the National Research Council of Canada). (d) Penetration of lettuce epidermal wall (W) by *Bremia lactucae*. The non-median section through a penetration peg (P) reveals its fluid nature extending between the apparently digested ends (↑) of host wall microfibrils. Appressorium (A). ×21 800 [Reproduced with permission from Ingram *et al.* (1976). Copyright: Academic Press Inc. (London) Ltd.] (e) Hyphae (H) of *Monilinia fructigena* growing in intercellular spaces of pear fruit. Wall degradation is restricted to the immediate vicinity of the pathogen. ×5000 (From Byrde and Willets, 1977. Reproduced by permission of Pergamon Press Ltd.). (f) Penetration of a thickened wall (EW) of wheat endodermis by *Gaeumannomyces graminis* (F). Localized degradation is evident, even of the layers impregnated with suberin (↑). ×6100 (From Manners and Myers, 1975. Reproduced by permission of the Society for Experimental Biology). (g) Early stage of penetration by *Verticillium albo-atrum* of a paired-pit between contiguous xylem vessels (Ve) in tomato stem. Compare the thin primary wall of the pit membrane (P) with the adjacent vessel wall comprising the massive overlying secondary wall (S). ×8900 (From Cooper, 1981)

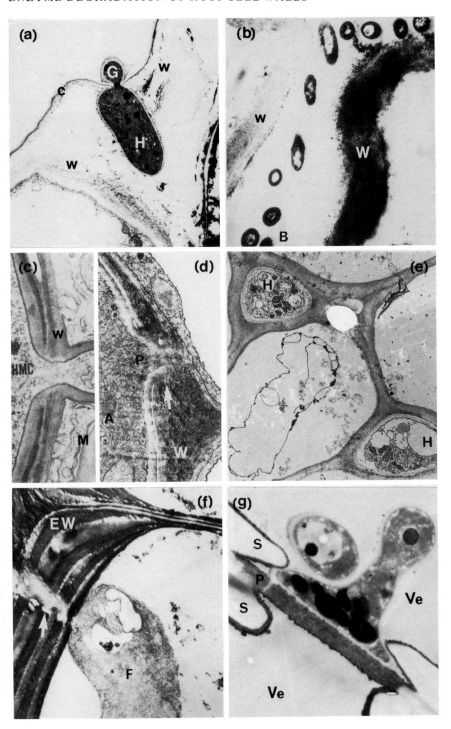

corresponding to the diverse polymers in plant cell walls. The contact between extracellular CWDE of pathogens and host cell walls is often one of the first molecular interactions between host and parasite, the outcome of which can modify the type or balance of the relationship.

This account will concentrate on the degradation of polysaccharides in primary walls, as this is probably of greatest significance in pathogenesis. Plant cell walls exist as a potential barrier to invasion but many pathogens penetrate them readily, and some have come to rely upon them as a means of intercellular growth thereby avoiding resistance reactions elicited by contact with host protoplasts. Ultrastructural studies have revealed the various ways that pathogens deal with host walls (Figure 1) (Cooper, 1981). Growth *through* walls can appear largely mechanical as hyphae constrict markedly during penetration, e.g. *Colletotrichum lindemuthianum* in bean, but often destruction of wall polymers is apparent as a localized zone of degradation, as around hyphae of *Monilinia fructigena* in pear. Penetration of the relatively massive secondary walls is routine for wood decay fungi but infrequent with plant pathogens, although *Gaeumannomyces graminis* can degrade and broach endodermal and xylem walls and vascular wilt fungi traverse xylem walls via the fragile primary walls of pit membranes (Figure 1f,g).

Intercellular growth characterizes many biotrophic fungi such as rusts, and smuts; presumably this must involve alteration to wall structure but the changes are so subtle as to leave in doubt the participation of CWDE (Figure 1c,d). In complete contrast is the extensive destruction of host walls, and consequently tissues, mediated by extracellular CWDE of 'soft rot' fungi and bacteria (Figure 1a,b). Understandably, it is this group that has provided most information on the role of CWDE in pathogenesis, and some of their enzymes have proved useful in recent analysis of the chemical structure of plant cell walls.

2. STRUCTURE OF PLANT CELL WALLS

Primary walls are laid down by undifferentiated cells that are still growing and consist mainly of polysaccharides with lesser amounts of glycoprotein; secondary walls are derived from these after completion of cell elongation and involve deposition of new polysaccharides and in specialized cells suberin (e.g. endodermis) or lignin (e.g. xylem) are also incorporated.

Cell walls may be regarded as a two-phase system with a hydrated, continuous matrix containing an organized phase of cellulose fibrils. Although frequently misconceived as inert structures, walls undergo substantial physicochemical changes during growth (Lamport, 1970) and contain various molecules of potential physiological activity (Section 5.2).

Wall polysaccharides have traditionally been divided into pectic substances, hemicelluloses, and cellulose based on relative solubilities in different extractants. However, the lack of specificity of each extractant for glycosidic bonds leads to considerable overlapping and renders analysis of the fragments complex. Recently, improved approaches such as the use of cultured plant cells, controlled solubilization of walls with pure CWDE, gas chromatography, and methylation analysis have enabled considerable advances to be made, as reviewed by Albersheim (1976) and Darvill et al., (1980).

The nine principal monosaccharides of plant cell walls consist of three hexoses [D-glucose (Glu), D-galactose (Gal) and D-mannose (Man)], two pentoses [L-arabinose (Ara) and D-xylose (Xyl)], two uronic acids [D-glucuronic acid (Glu A)) and D-galacturonic acid (Gal A)], and two deoxyhexoses [L-rhamnose (Rha) and L-fucose (Fuc)]. They exist in the pyranose form, except for arabinose , which usually occurs as a five-membered ring (furanose). Wall polysaccharides are almost invariably based on more than one type of monomer. The tentative structures of the main wall polysaccharides are shown in Figure 2, and the monosaccharide compositions of walls of a dicotyledonous and a monocotyledonous plant in Figure 5.

2.1. Pectic polysaccharides

Pectic polymers predominate in the middle lamella region, which is often regarded as an intercellular cement. They are based on long chains [e.g. degree of polymerization (DP) ca. 2000] of α-1,4-linked Gal A interspersed with 1,2-linked Rha, known as rhamnogalacturonan. Rha may occur frequently but long stretches of Gal A do exist, e.g. DP > 25 in sycamore walls (Darvill et al., 1980); Rha interrupts the tendency for ordered chain conformation as it leads to a kinking of the chain, as found in other network polysaccharides (Aspinall, 1973). The carboxyls of Gal A may be methylated (Figure 3) or cross-linked by calcium; this forms bridges between rhamnogalacturonan chains and partly explains their characteristic property of gel formation, which has profound influence on wall structure and amenability to CWDE (Pilnik and Voragen, 1970). To this backbone, arabinan and galactan sidechains are linked via the C_4 of Rha and possibly C_3 of Gal A residues. The complexity of one polymer is shown by the existence of at least six linkages between Gal and Rha (McNeil et al., 1982); this has led to the suggestion that some wall polymers may fulfil more than just a structural role (Section 5.2).

Arabinans are highly branched polymers of predominantly α-1,3- and α-1,5-linked Ara, but other linkages may be involved; they can occur as mono- or disaccharide side-chains, but arabinans of DP ca. 100 have been isolated. Pectic galactans are mainly linear β-1,4-Gal chains of DP < 50, but 6-linked galactosyl residues may be present. Arabinogalactans based on a β-1,4-Gal chain with Ara side-chains occur in walls of apple fruit and in soybean cotyledons, whereas those with a 3- and 6-linked galactan backbone are more characteristic of conifer wood (Timell, 1965), although they may represent a

Rhamnogalacturonan   (16%)

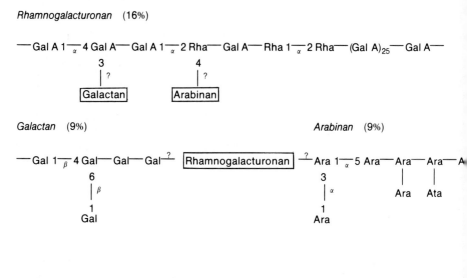

Galactan   (9%)   *Arabinan*   (9%)

Xyloglucan   (19%)

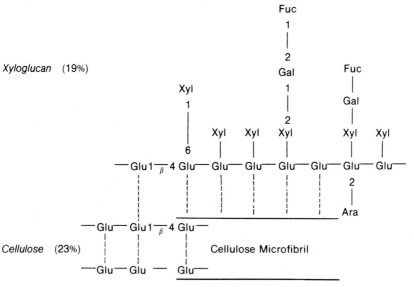

Cellulose   (23%)

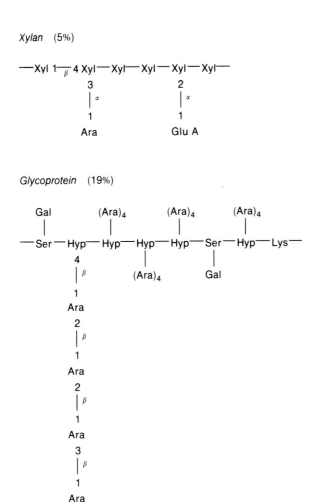

Figure 2. Possible structures and interconnections of some major polymers in plant cell walls. Hyp, hydroxyproline; Ser, serine; Lys, lysine; - - - - -, hydrogen bonding; other abbreviations as in text. Percentages represent the composition of sycamore primary walls. (Adapted from Albersheim, 1976; Aspinall, 1973; Darvill *et al.*, 1980; Northcote, 1972; and Whistler and Richards, 1970)

minor component of the primary walls of dicotyledons. Recently a highly complex branched rhamnogalacturonan was isolated as 3–4% of sycamore walls which contained ten different monosaccharides including the rarely detected wall sugars 2-O-methyl-Fuc, 2-O-methyl-Xyl, and apiose (Darvill *et al.*, 1980).

Primary walls of dicotyledons contain a relatively large amount (*ca.* 35%) of pectic polysaccharides but these compounds constitute only 8–9% of primary walls of monocotyledons and may be of different composition.

2.2. Hemicelluloses

The predominant hemicellulose in primary walls of dicotyledons is xyloglucan; this consists of a cellulose-like backbone of β-1,4-Glu, with terminal branches of 1,6-Xyl and terminal Fuc. The molecule consists of *ca.* 50 residues and is capable of hydrogen bonding to cellulose fibrils; this property has considerable implications in wall structure and may allow a redefinition of the term hemicellulose to include wall polysaccharides able to bind non-covalently to cellulose (Bauer *et al.*, 1973).

Xylans with a β-1,4 backbone are major constituents of *secondary* walls of dicotyledons (Whistler and Richards, 1970). The most frequent side-chains are α-1,2-linked Glu A (usually 4-O-methyl-Glu A) and α-1,3-linked Ara, and about half of the Xyl residues are acetylated at C_3 and C_2 (Northcote, 1972). In contrast, arabinoxylans form the main component (>40%) of monocotyledonous *primary* walls although those from secondary walls have been most extensively studied (Whistler and Richards, 1970). Xyloglucan may be found in small amounts in monocotyledons as well as non-cellulosic 1,4- and 1,3-linked β-glucans (Labavitch and Ray, 1978).

Other hemicelluloses have been described mainly from secondary walls of woods and include linear β-1,4-linked glucomannans and galactoglucomannans with 1,6-Gal side-branches.

2.3. Cellulose

Cellulose is the major structural component of plant cell walls, visible by electron microscopy as microfibrils of *ca.* 10 nm diameter. In secondary walls these are of indeterminate length (at least several micrometres) and are laid down as parallel lamellae. Microfibrils are often of much smaller cross-section and randomly orientated in primary walls and may be absent from the middle lamella. Each fibril is composed of *ca.* 40 extremely long (DP 8000–12 000) linear chains of β-1,4-Glu stabilized by inter- and intra-molecular hydrogen bonding. The degree of organization is such that fibres contain crystalline regions or micelles (Preston, 1974).

Cellulose makes up 20–30% of primary walls but often constitutes over 40% of secondary walls as in hardwoods (Timell, 1965).

2.4. Wall protein

Primary walls of dicotyledons contain between 5 and 10% of a protein exceptionally rich in hydroxyproline (20%) and certain other amino acids characteristic of the structural proteins of animals (Lamport, 1970). This analogy and the tenacity of wall protein to extraction suggest a structural role in plant cell walls. Ara is linked to hydroxyproline hydroxyls mainly as a β-linked tetraarabinoside, and single Gal units are attached to serine. The extent of glycosylation and length of side-chains are both less in the analogous protein of walls of monocotyledons (Darvill *et al.*, 1980).

2.5. Lignin

Lignin is laid down during secondary thickening, in place of water, as a three-dimensional aromatic polymer throughout the walls of certain specialized cells, e.g. hardwoods contain *ca.* 18–25% lignin (Northcote, 1972). Lignin can form covalent linkages with polysaccharides and confers upon them resistance to CWDE; its structure, formation, and role in resistance are considered by Ride (Chapter 11).

2.6. Interconnections between wall polymers

Research into wall structure and degradation was stimulated by a model of the primary wall of dicotyledons proposed by Albersheim's group (Keegstra *et al.*, 1973), in which it was suggested that polymers from different regions of the wall are interlinked as one macromolecule. The xyloglucan was conceived as a monolayer hydrogen bonded to cellulose fibrils *and* covalently linked to pectic polysaccharides, which would serve to unite the amorphous and fibrillar phases of the wall. Also, it was apparent that the glycoprotein is covalently linked via arabinogalactan to the pectic fraction.

A more recent picture (Darvill *et al.*, 1980) reveals the infrequency of the xyloglucan–pectic linkage and lack of evidence for glycoprotein–pectic connection, but still stresses the intimate association between these polymers. Thus, degradation of rhamnogalacturonan is a prerequisite to extraction of xyloglucan (by urea or endo-glucanase) and to release of glycoprotein by protease (Albersheim, 1976). Several hydroxyproline-rich glycoproteins are lectins which would (if wall bound and of appropriate specificity) participate in the non-covalent cross-linking of wall polysaccharides. The protein may even cross-link to itself, as suggested by the wall-shaped protein network which remains after chemical removal of wall polysaccharides. Also, the wall structures of primitive protists such as *Chlamydomonas* consist mainly of a non-covalent lattice of a similar glycoprotein (Lamport, 1980). Some of the wall rhamnogalacturonan is not linked to other wall polymers but is bound by Ca^{2+}, as revealed by its partial release with chelating agents or following methylation of carboxyl groups (Knee, 1978).

Monocotyledonous primary walls, although chemically dissimilar, may be arranged on similar architectural plans to those of dicotyledons, but there could be a greater reliance on the binding ability of arabinoxylan to cellulose and even to itself (Darvill *et al.*, 1980).

Obviously an understanding of wall structure is a prerequisite to evaluating the significance of a CWDE in parasitism, but advances in the two fields are likely to be interdependent.

3. TYPES OF CELL WALL-DEGRADING ENZYMES

The ability to degrade plant cell wall polysaccharides is widespread amongst microorganisms and is not unique to plant pathogens. The enzymes are often low molecular weight, stable, extracellular glycoproteins and one 'activity' may be represented by multiple forms or isoenzymes which differ in charge, size, regulation, stability to host factors, and ability to degrade cell walls. The complexity of one organism is well illustrated for CWDE of *M. fructigena* (Byrde and Willetts, 1977). The properties and types of enzymes involved will be briefly considered but have been reviewed in detail elsewhere (Bateman and Basham, 1976; Dekker and Richards, 1976; Reese, 1977; Rombouts and Pilnik, 1980); synthesis was described by Byrde and Archer (1977).

3.1. Degradation of pectic polysaccharides

The rhamnogalacturonan chain is degraded by hydrolytic polygalacturonases, or by lyases which form an unsaturated bond between C_4 and C_5 of Gal A at the non-reducing ends of cleaved chains (Figure 3). These may be subdivided into forms that attack internal regions of chains at random (*endo-*) or terminally (*exo-*), and on the basis of their specificity for polymers with esterified (methoxy) or non-esterified carboxyls groups which are known as pectin and polygalacturonic acid, respectively; the corresponding enzymes are named polymethylgalacturonase (PMG), polygalacturonase (PG), pectin lyase (PL), and polygalacturonide lyase (PGL). The methoxy groups of pectin are removed by pectin methylesterase (PME) (Figure 3), which can therefore act in concert with PGs or PGLs. In *Clostridium multifermentans* PME and PGL exist as a bound complex which facilitates a rapid alternation of de-esterification and depolymerization (Sheiman *et al.*, 1976). Lyases have a partial or complete requirement for Ca^{2+} and usually alkaline pH optima (8–10), whereas PGs are most active at low pH (4–5) and may be inhibited by Ca^{2+} (Cooper *et al.*, 1978). In general, fungi produce PG and/or PL and bacteria secrete PGL.

Endo-enzymes cause a 50% decrease in substrate viscosity after cleavage of only *ca.* 1% of the glycosidic bonds, as distinct from *exo*-forms, which must cause 20–40% hydrolysis to effect a similar change. However, classification

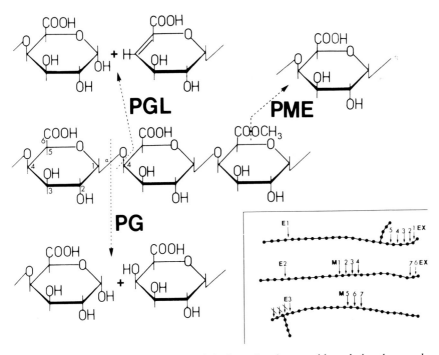

Figure 3. Cleavage of an α-1,4-linked polygalacturonide chain by *endo*-polygalacturonide lyase (PGL) and *endo*-polygalacturonase (PG), and demethoxylation by pectin methylesterase (PME). Hydrolysis by PG represents a mechanism typical of most polysaccharidases. Inset: modes of action of polysaccharidases; arrows represent the splitting of glycosidic bonds between sugar residues (●), numbers represent the sequence of attack by a single glycanase molecule with either *endo*-(E), *exo*-(EX) or 'multiple' (M) action. *Endo*-glycanases require a minimum substrate DP of usually 3–5, and most will not degrade bonds adjacent to side-chains

into simply *endo*- or *exo*- forms requires revision as some *endo*-enzymes attack chains at random then release monomers or oligomers (Figure 3). This so-called 'multiple attack' is shown by various polysaccharidases including the PG of *C. lindemuthianum* and *Verticillium albo-atrum* (Cooper *et al.*, 1978; Talmadge *et al.*, 1973) and may have implications in pathogenesis, possibly allowing wall degradation without maceration (Section 5.2) and simultaneously releasing low molecular weight residues for nutrition and enzyme induction (Section 4.2). Other pathogens may produce mixtures of *endo*- and *exo*-enzymes, e.g. *Sclerotium rolfsii* PGs, or in addition intracellular oligogalacturonidases, e.g. *Erwinia carotovora* (Stack *et al.*, 1980).

At least one pathogen (*S. rolfsii*) releases Rha from rhamnogalacturonan, presumably as a result of its α-rhamnosidase, but whether this was exposed only after degradation of internal Rha–Rha or Gal–Rha linkages is not known. Enzymes specific for mixed linkages do exist (Reese, 1977), but the significance of such enzymes in wall degradation remains unrealized.

The neutral polymers linked to rhamnogalacturonan are degraded by *endo-* and *exo-β*-1,4-galactanases and *β*-galactosidase (which may act synergistically), whereas breakdown of *α*-1,3- and *α*-1,5-arabinan is almost invariably by *exo*-enzymes (Bauer *et al.*, 1977; Byrde and Willetts, 1977; Cooper *et al.*, 1978; Baker *et al.*, 1979); many enzymes of this group have low pH optima (2.5–5). Degradation of the 3- and 6-linked galactan may be infrequent (Van Etten and Bateman, 1969) but is achieved by the wilt pathogen *V. albo-atrum* (Cooper and Wood, 1975).

3.2. Degradation of hemicelluloses

The xyloglucan chain is degraded by *endo-β*-1,4-glucanase into small fragments which can no longer bind to cellulose (Bauer *et al.*, 1973). The *β*-1,4-xylans and mannans are broken down by *endo*-enzymes to yield oligomers which may be further degraded by *exo*-enzymes or glycosidases (Baker *et al.*, 1977; Cooper *et al.*, 1978; Van Etten and Bateman, 1969). Many pathogens produce specific glycosidases (English *et al.*, 1971) which can release some terminal residues from polymers such as removal of Ara from xylans and Gal from galactoglucomannans.

3.3. Cellulose degradation

In cell walls degradation of cellulose is restricted by its insoluble, partly crystalline form. Conversion into glucose requires a complex of enzymes originally designated by Reese (1977) as C_1, C_x, and cellobiase. C_1 renders crystalline cellulose amenable to subsequent action by C_x, which must involve breaking of hydrogen bonds as well as glucosidic bonds. The *endo-β*-1,4-glucanase (C_x) converts the loosened chains into oligomers, especially the dimer cellobiose, which is finally degraded to glucose by cellobiase (*β*-glucosidase). The number of components and precise modes of action, particularly of C_1, are in dispute but the overall concept of a synergistic system in which the individual enzymes are relatively ineffective against native cellulose is well established for various microorganisms including the pathogen *Fusarium solani* (Wood, 1969). In mature tissue cellulose breakdown is often limited by lignin; thus it is of interest that a wood decay fungus degrades the two polymers simultaneously via the enzyme cellobiose : quinone oxidoreductase (Eriksson, 1977).

3.4. Degradation of cell walls *in vitro*

Proof that cell walls can act as substrates for polysaccharidases of pathogens is readily demonstrated by exposing walls isolated from plants to microbial enzymes from cultures or diseased tissue. Thus extensive solubilization of walls occurs with crude mixtures of enzymes from pathogens such as *Fusarium roseum, Rhizoctonia solani*, and *Botrytis alli* (Bateman *et al.*, 1969; Mullen and Bateman, 1975; Mankarios and Friend, 1980). However, with the exception of *endo*-pectic enzymes, there have been few attempts to study the efficacy of single CWDE alone, or in combination. The need for such an approach is apparent as some enzymes of high activity against model substrates may be ineffective against insoluble cell walls (Section 6.2), or may exhibit different modes of action, pH optima, or response to cofactors, as with some CWDE of *V. albo-atrum* (Cooper *et al.*, 1978).

The central role of rhamnogalacturonan in wall structure is revealed by the dramatic effects of *endo*-polygalacturonidases on cell walls. Thus purified *endo*-PGs and *endo*-PLs not only solubilize much Gal A but also cause a marked depletion of neutral sugars, which confirms their covalent linkage to rhamnogalacturonan, e.g. a PGL from *Erwinia chrysanthemi* released from tobacco walls *ca.* 65% of the Ara, Rha, and Gal but none of the hemicellulosic fraction (Xyl, Glu, and Man) (Basham and Bateman, 1975); an *endo*-PG of *C. lindemuthianum* is similarly destructive (Talmadge *et al.*, 1973). This key role of *endo*-pectic enzymes is further emphasized by claims that their action is a prerequisite to attack of enzymes degrading other wall polymers. Thus *endo*-PG treatment of sycamore walls enhanced *ca.*10-fold the subsequent degradation by *endo*-glucanase and protease, and enabled several glycosidases, an *exo*-PG, and cellulase of *C. lindemuthianum* to perform as CWDE (Bauer *et al.*, 1973; Keegstra *et al.*, 1973). Anomalously, several CWDE have since been obtained which readily act upon isolated walls without prior 'wall modification'; these include arabinanases, galactanases, xylanases, and a cellulase (Knee *et al.*, 1975; Baker *et al.*, 1977, 1979; Bauer *et al.*, 1977; Cooper *et al.*, 1978). An attempt to show synergism between CWDE of *V. albo-atrum* failed to reveal any interdependence in degradation of isolated tomato walls (Cooper *et al.*, 1978). Nevertheless, results with model substrates and even isolated walls must be interpreted with caution, and it remains likely that enzymes degrading the wall matrix (especially rhamnogalacturonan) *in vivo* would open the way for CWDE with less accessible substrates such as hemicelluloses and cellulose.

In spite of the possible structural role of wall glycoprotein, its degradation by plant pathogens has been neglected. The only pure proteases studied, from *C. lindemuthianum* and *M. fructigena*, proved ineffective in releasing protein from cell walls, even after acid stripping of Ara residues which renders wall protein susceptible to trypsin (Ries and Albersheim, 1973; Darvill *et al.*, 1980; Hislop *et al.*, 1982).

4. REGULATION OF SYNTHESIS OF CELL WALL-DEGRADING ENZYMES

Understanding the mechanisms regulating synthesis of CWDE helps in the interpretation of pathogenesis and also in establishing culture conditions suitable for selective enzyme biosynthesis at high rates, which is required as a starting point to obtain a pure CWDE or when comparing enzyme production by mutants or isolates; the subject was reviewed by Cooper (1977).

Much of our knowledge of enzyme regulation is derived from studies on prokaryotes, in particular the *lac* operon of *E. coli*; in eukaryotes the increased complexity of regulation has not provided a basis for a general model. Catabolic enzymes which attack exogenous substrates (e.g. CWDE) are, as a rule, *induced* by specific compounds structurally related to the substrate of the enzyme and are usually formed to a small extent in the absence of inducer. This *basal* synthesis results in up to five molecules of β-galactosidase per *E. coli* cell, whereas in the presence of galactoside inducers the level rises to 5000 molecules. Basal enzyme is not to be confused with *constitutive* synthesis, which occurs in some mutants in the absence of inducer at rates similar to induced wild-type strains. Basal, constitutive, and induced enzymes are identical as they are specified by the same structural gene.

Synthesis of many inducible enzymes is reduced by readily metabolized carbon sources; once known as the 'glucose effect', the term *catabolite repression* (CR) better describes this non-specific phenomenon exerted by numerous compounds. Control of the *lac* operon effected by two proteins; the repressor governs the cell's response to β-galactosides by blocking the binding of RNA polymerase to its site on the promoter region on the *E. coli* DNA where transcription is initiated. This binding is facilitated by $3',5'$-cyclic AMP (cAMP) and its receptor protein. The repressive effect of catabolites appears to be mediated by decreasing intracellular cAMP levels, as shown for an *endo*-PGL of *E. carotovora* (Hubbard *et al.*, 1978).

4.1. Regulation of synthesis *in vitro*

Most CWDE are produced in culture at high levels during growth on the appropriate polysaccharide substrate but are absent or at low activities when monosaccharides serve as the carbon source. This is interpreted (often erroneously) as induced synthesis, but repressible, constitutive enzymes would also appear in higher yields on polysaccharides as the gradual rate of polymer breakdown may not exert CR. Also, attempts to show that mono- or oligosaccharides may act as inducers fail in batch cultures as when used as sole carbon source they effect CR. Therefore, the control mechanisms can only be established when CR is avoided, by supplying wall sugars to cultures at low constant rates close to those immediately required by the microorganism; this simulates the gradual release of oligomers which results during growth on polysaccharides. With the wilt fungi *V. albo-atrum* and *F. oxysporum* this approach revealed that synthesis of each CWDE was induced rapidly and specifically by the *monomer* predominant in the substrate of the enzyme (Cooper and Wood, 1973, 1975). Thus, synthesis of PG and PL was induced by Gal A, arabinanase by Ara, xylanase by Xyl, etc. It is interesting that the dimeric repeating unit of cellulose, cellobiose, acts as inducer of cellulase rather than monomeric glucose. In view of the abundance of glucose in the free state and as various polymers, the evolution to a disaccharide inducer was a necessary step, as was also found with amylases and the corresponding dimer, maltose.

It is possible that enzymes of related function, such as pectic enzymes or those of the cellulase complex, may be under coordinate control; this is suggested by the existence of complexes of mixed types, e.g. PME/PGL (Section 3.1) and PGL/PG (Stack *et al.*, 1980) and in the wood-decaying fungus *Polyporus adjusta*, induction of a group of CWDE of different but probably related function, viz. mannanase, xylanase, and cellulase, are controlled by a single regulatory gene (Eriksson, 1977).

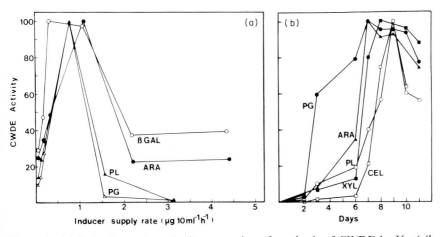

Figure 4. (a) Induction and catabolite repression of synthesis of CWDE by *Verticillium albo-atrum* in cultures growing on a low, continuous supply of monosaccharide inducers. The onset of CR coincides with detection of free sugar in the culture fluid indicating saturation of the organism's 'metabolic pool.' (b). Sequential production of CWDE by *V. albo-atrum* grown on tomato stem cell walls; enzyme levels were similar on walls from resistant and susceptible cultivars. PG, polygalacturonase; PL, pectin lyase; ARA, arabinanase; XYL, xylanase; CEL, cellulase; β-GAL, β-galactosidase. (After Cooper, 1977; Cooper and Wood, 1975) Reproduced with permission from *Physiological Plant Pathology*, **5**, 135–136 (1975), Copyright: Academic Press, Inc. (London) Ltd.

The fine balance between induction and CR is shown in Figure 4a; increasing the amounts of inducing monosaccharides supplied to cultures eventually exceeds the immediate needs of the pathogen, at which point CR operates. Syntheses of pectic enzymes are repressed by >90% at only twice the optimum supply rate for induction.

Most data for CWDE synthesis by plant pathogens can be interpreted as dual control by induction and CR but the general applicability remains to be established; especially as the few other critical studies have revealed that synthesis of some pectic enzymes and cellulases is constitutive (e.g. Hsu and Vaughn, 1969; Hulme and Stranks, 1971). In a few cases CWDE are produced constitutively and are not subject to CR; these include a PG and xylanase of *Helminthosporium maydis* and cellulase of *Pseudomonas solanacearum* (Kelman and Cowling, 1965; Bateman, 1976). The significance in pathogenesis of these different levels of control of CWDE is not known but deserves study, especially as methods are available for obtaining the appropriate mutants.

4.2. Regulation of synthesis *in vivo*

An indication of the influence of the host on synthesis of CWDE can be obtained with extracted, insoluble cell walls as sole carbon source in culture. Cell walls are highly effective inducers of CWDE and under these conditions production by several pathogenic fungi, including *C. lindemuthianum* and *V. albo-atrum*, occurs in sequence, in the order pectic enzymes, hemicellulases, and finally cellulase (Figure 4b) (English *et al.*, 1971; Cooper and Wood, 1975). The gradual appearance of different CWDE presumably reflects the physico-chemical susceptibilities of corresponding wall polymers, resulting in induction as successive substrates are released during progressive degradation of the wall; this again bears on the possible need for CWDE to act in sequence (Section 3.4).

Analogously, in several diseases pectic enzymes are the first to appear and cellulases the last, often only detectable when tissues are moribund. This sequence occurs in *Pyrenochaeta*-infected onions, leaf spots of pea, wheat roots infected with *G. graminis*, and *Verticillium* wilt of lucerne (Cooper, 1977).

Polysaccharides in native walls are insoluble and as such cannot induce CWDE. Soluble inducers are probably released after action by low levels of basal enzymes, and once initiated should become autocatalytic; presumably *exo*-acting enzymes would be more effective than *endo*-enzymes in this respect, or intracellular oligogalacturonidases may effect the final conversion to inducers (Collmer and Bateman, 1981). Thus, speed of induction may partly depend on basal synthesis and is of obvious significance in infection by necrotrophs. A comparison of *E. carotovora* with the saprophyte *Pseudomonas fluorescens* revealed that both organisms synthesized an inducible PGL but the pathogen produced more basal enzyme which enabled much faster synthesis *in vitro* (Zucker and Hankin, 1970). In inoculated potato tubers rapid synthesis of PGL by *E. carotovora* (rotting is evident after only 2–3 h) allows ingress before host suberization, which restricts saprophytic species.

In view of the sensitivity of CWDE synthesis to CR, it should be considered that if free sugars in host tissues exceed certain levels then infection could be prevented or delayed. Such analyses would apply only to pathogens which require CWDE for infection and cause protoplast disruption during invasion; intercellular pathogens would not encounter sugars of the host cytoplasm with the exception of haustorium-forming biotrophs for which CR may have particular significance (Section 4). One study of a disease involving extensive tissue damage, root rot of onion by *Pyrenochaeta terrestris*, clearly revealed that alterations in host sugar levels influenced pathogenesis and production of PG and cellulase in parallel (Horton and Keen, 1966). However, tolerance of a tomato cultivar to *P. lycopersici* was not related to sugar levels in roots, although the concentrations were sufficiently high to repress CWDE production *in vitro* for 4 days (Goodenough *et al.*, 1976).

The influence on CWDE synthesis of cell walls from different tissues, varieties, and species is considered later (Section 6).

5. ROLE OF CELL WALL-DEGRADING ENZYMES IN PATHOGENICITY

5.1. Criteria

There are numerous criteria which should be fulfilled before a CWDE can be implicated in pathogenesis. These ideals have been achieved in remarkably few cases.

5.1.1. *Ability to produce CWDE* in vitro

Synthesis of CWDE *in vitro* is no proof of involvement in disease. Nevertheless, an indication of their range and characteristics can be obtained from cultures grown on host cell walls, although properties occasionally differ from forms detected *in vivo* (Cooper, 1977). For pathogens with low potential for CWDE synthesis or with wall-bound enzymes, the ability of cells to reduce the viscosity of a polysaccharide or to grow on a polysaccharide implies production of the relevant polysaccharidase; however, lack of growth can sometimes reflect inability to metabolize the end product, e.g. *P. solanacearum* degrades soluble cellulose but cannot utilize cellobiose (Kelman and Cowling, 1965).

5.1.2. Detection of CWDE in infected tissue

Obviously enzymes should be detected in diseased tissue before, or coincident with symptoms. However, lack of activity does not discount involvement as plants contain several inhibitors of CWDE, particularly oxidized phenolics and wall-bound PG inhibitors (Byrde and Archer, 1977). For example in *Monilinia*-infected fruits PG activity was only detected after removal of inhibitor(s) by isoelectric focusing (Fielding, 1981). Also some CWDE remain bound to the wall of the parasite or may bind ionically to host walls (Section 6.2). Thus extraction media should contain a reducing agent and phenol adsorbent and be of suitable pH and high ionic strength (to desorb bound enzyme) (e.g. Cooper and Wood, 1980).

An increase in CWDE activity may not derive from the pathogen. Healthy plants possess a number of CWDE involved in processes such as germination, pollination, dissolution of xylem vessel end walls, abscission, and fruit ripening. Some of these are under the control of hormones such as ethylene and auxin, which commonly increase markedly during infection (e.g. Bal *et al.*, 1976; Pesis *et al.*, 1978). Also, wall degradation by a parasite may release previously bound host enzymes. Pectic lyases are not produced by plants, but host PME and glycosidases often contribute to the activity of extracts from diseased tissue, e.g. increased PME activity in *Fusarium* wilt of tomato derives mainly (90%) from the host (Langcake *et al.*, 1973). Low levels of *exo*-PG occur in many plants (e.g. Bartley, 1978) but *endo*-PG only reaches significant activity in some ripening fruits or sometimes along with cellulase in abscission zones (Section 5.2). Therefore, characterization of CWDE from host as well as pathogen may be required in order to establish the origin of increased activity. Invasion of diseased tissue by saprophytes presents another potential source of CWDE.

Multiplicity of forms should be considered, as assay of total activity may overshadow the contribution of one key isoenzyme (e.g. Sexton *et al.*, 1980). Also, individual CWDE should not be considered in isolation as they may act synergistically, e.g. precise localization of an *exo*-PG to abscission zones of *Phaseolus vulgaris* strongly implicates its participation, but no increase in its activity during abscission suggests it is only one of several CWDE involved in the associated wall changes (Berger and Reid, 1979).

5.1.3. Depletion of cell wall polysaccharides

Improved procedures for the analysis of cell wall polysaccharides (Jones and Albersheim, 1972) facilitate comparisons between walls from infected and healthy tisue. Depletion of specific monomers could implicate previous action of a CWDE (Figure 5), but the ability of pure enzymes such as PG and xylanase to release covalently linked sugars along with the main polymers complicates interpretation. Alternatively, the presence of free degradation products can give a qualitative reflection of prior activity of a CWDE, e.g. methanol (from PME), Gal A (PG), and Ara (arabinanase) in *Monilinia*-infected apples (Byrde and Willetts, 1977).

5.1.4. Correlation of enzyme production with pathogenicity

Comparisons of natural isolates for pathogenicity and production of CWDE usually only reveal the great variability within a species for numerous factors, many of which may influence but be unrelated to CWDE synthesis. Induction of mutants in a common genetic background is a powerful but underused approach in which current knowledge of enzyme regulation allows the choice of media to select mutants deficient, super-productive, or constitutive for several CWDE (Howell, 1976; Cooper, 1977). Obviously this relies on corre-lations based on numbers amenable to statistical analysis (Howell, 1975) as loss of pathogenicity can reflect mutations other than for CWDE produc-tion; nevertheless, maintained pathogenicity of a single mutant deficient in a specific CWDE is strong evidence against involvement of that enzyme.

For bacteria, conjugational transfer of sections of genome coding for synth-esis of a CWDE can be exploited, and has recently confirmed the role for a PGL but against PG in maceration by *E. chrysanthemi* (Chatterjee and Starr, 1977). Manipulation of plasmids for CWDE production in fungi is also being attempted.

5.1.5. Microscopic alterations in walls of infected tissue

Transmission electron microscopy can give valuable indications as to the par-ticipation of CWDE as already shown (Figure 1). Displacement of wall fibrils indicates mechanical penetration, wall-bound or rapidly inactivated enzymes would probably give localized degradation, whereas extensive wall dissolution signifies freely diffusible extracellular CWDE. Loss of the middle lamella implies action by pectic enzymes (Baker *et al.*, 1980), which may also be revealed by lost affinity for the specific stains, ferric hydroxylamine or ruthenium red. Alteration or loss of wall microfibrils resulting from cellulase activity is also apparent as reduced birefringence under polarized light.

Ultrastructural localization of specific CWDE during infection has particu-lar merit, especially in diseases not characterized by extensive tissue disinteg-ration for which other approaches are likely to fail. An example of the use of elegant immuno-cytochemical techniques in this context is shown in Chapter 6, for cutinase. Such approaches have yet to be successful for other CWDE although appropriate antisera have been raised against some wall-degrading isoenzymes of *M. fructigena* (Hislop *et al.*, 1974b). Also, cytochemical visual-ization of α-arabinosidase in *M. fructigena* has revealed activity in spherosome-like bodies in hyphae *in vitro* but not during infection (Hislop *et al.*, 1974a). Conventional histochemical localization by light microscopy can be made for glycosidases (e.g. Mace, 1973), but it is not possible to ascribe activity to a particular isoenzyme and host glycosidases often give interference.

5.1.6. Reproduction of wall changes or disease symptoms with purified enzymes

As a molecular form of Koch's postulates, reintroduction of a purified CWDE at physiological activities into the healthy host plant or appropriate tissue can be attempted in order to simulate changes in ultrastructure of walls and cytoplasm (Ben Arie *et al.*, 1979; Hislop *et al.*, 1979), tissue integrity (Basham and Bateman, 1975), or disease symptoms (Cooper and Wood, 1980). Purification of enzymes from cultures is far simpler than from diseased tissues but is only valid if the forms produced *in vitro* and *in vivo* are the same. Complications arise with multiplicity of forms of many CWDE, and the inability of some CWDE to act in the absence of other enzymes or metabolites, or without the pathogen to aid in their distribution. It is axiomatic that this approach cannot reproduce conditions during infection.

5.1.7. Inhibition of CWDE during pathogenesis

Repression of enzyme synthesis (via CR effected by increasing host sugar levels) or inhibition of activity (e.g. by phenolics) during pathogenesis has been attempted but the effects are too non-specific to allow clear interpretation. The prevention of abscission by application of antiserum specific to a cellulase isoenzyme (Sexton *et al.*, 1980) and prevention of infection by *Fusarium solani* with anti-cutinase (see Chapter 6) demonstrates the potential of this approach if applied to CWDE of parasites.

5.2. Significance of polygalacturonide degradation in pathogenesis

Extensive wall degradation is found in a wide range of diseases which include soft rots, damping-off, leaf and stem lesions, and dry rots. The major role of *endo*-polygalacturonidases is often evident from their rapid production and the concomitant depletion of Gal A from infected tissue, e.g. $\leqslant 82\%$ of pectic substances were removed from sunflower stems by *Sclerotinia sclerotiorum* and $\leqslant 90\%$ loss of Gal A was detected in young lesions on bean hypocotyls by *S. rolfsii* (Figure 5) and *R. solani* (Bateman, 1976).

Induced changes in pH or polygalacturonides in infected tissue may favour activity of the predominant CWDE of the invading pathogen. Thus in *S. sclerotiorum* lesions $\leqslant 93\%$ demethoxylation of wall pectin and a drop in pH from 6.2 to 4.5 enhance the action of its PG. Similar depletion of methoxy

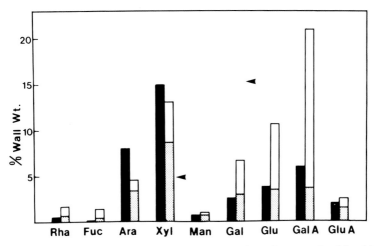

Figure 5. Monosaccharide composition of cell walls from hypocotyls of healthy (□) and *Sclerotium rolfsii*-infected (▨) 10-day-old *Phaseolus* bean plants, 3 days after inoculation; arrows show the Xyl and Gal content of younger (7-day old) seedlings. Composition of walls from a monocotyledon (■), 5-day old leaves of *Zea mays*, are included for comparison; note particularly their lower Rha and Gal A and higher Ara and Xyl content. [Adapted from Baker *et al.*, 1977; and Bateman, 1976. Reproduced from the former by permission of the American Phytopathological Society and from the latter by permission of Academic Press Inc. (London) Ltd.]

groups and *increase* in pH should favour the PGL produced in stems rotted by *F. solani* f. sp. *cucurbitae*. All of the CWDE of *S. rolfsii* have acidic optima (pH 2–4) which is created *in vivo* by copious secretion of oxalic acid; this metabolite also renders calcium pectate amenable to *endo*-PG by chelating calcium (Bateman and Beer, 1965).

Inhibition of polygalacturonidases may still allow invasion of the host but can result in different symptoms. Thus *M. fructigena* causes a firm, dark brown rot of apple fruits in which CWDE are difficult to detect and polygalacturonide is depleted by only *ca.* 15%. In contrast, infection by *Penicillium expansum* results in a soft, light brown rot with a 50–70% loss of pectic substances, by virtue of inhibition of the host's phenol oxidase system which allows the *endo*-PMG of the pathogen to remain active. In this instance limited wall degradation confers an advantage on *M. fructigena* as it can overwinter in the still intact fruit; however, inhibition of CWDE by oxidized phenolics can form the basis of host resistance, as in certain cider apple cultivars (Byrde and Willetts, 1977).

Some necrotrophs such as *B. cinerea* cause profound wall changes even prior to or during penetration (Figure 1a); *endo*-PG occurs in ungerminated conidia of this pathogen (Verhoeff and Liem, 1978) and is secreted on the surface of onion leaves before penetration by *B. squamosa* (Hancock *et al.*, 1963). Presumably this reflects the need for some pathogens to damage cells in advance of invasion, and for 52 induced mutants of *B. fabae* virulence was correlated with their ability to kill host cells, apparently as a consequence of wall degradation (Mansfield and Richardson, 1981). Wall damage frequently occurs well ahead of organisms with high pectolytic capacity, e.g. *R. solani* and *E. carotovora*, and is usually apparent as selective dissolution of the middle lamella. However, a fine balance must exist between prevention of host defences by rapid cell killing and triggering of host resistance, as endo-PGs and endo-PLs can 'elicit' phytoalexin production (e.g. West, 1981).

Extensive wall degradation is invariably accompanied by cytoplasmic disruption. Studies with homogeneous preparations of *endo*-pectic hydrolases and lyases from various fungi and bacteria have clearly established their unique role as the factors responsible for maceration and simultaneous protoplast damage—the characteristic symptoms of soft rot disease (Bateman and Basham, 1976). A study of the ultrastructural changes in apple tissue treated with a pure *endo*-PL showed many features typical of invasion by necrotrophs, such as wall degradation, retraction of the plasmalemma, dilation, and vesiculation of endoplasmic reticulum (Hislop *et al.*, 1979). Cells with conspicuously altered walls were invariably 'dead,' but many cells with apparently normal or only slightly degraded walls were often injured. This is highly significant in view of the frequency of wall degradation in various diseases, and it follows that this may be one of the most common means of cell damage during pathogenesis.

The mechanism of cell killing is generally considered to involve degradation of the wall matrix so that the weakened structure can no longer counteract the pressure exerted by the turgid protoplast, resulting in a damaged plasmalemma. However, some consider this an over-simplification as ion leakage, indicative of cell injury, occurs almost immediately following exposure of tissues to *endo*-pectic enzymes (Hislop *et al.*, 1979), which could suggest release of a toxic component from the wall. Cell walls could be regarded as a component of the cell's lysosome system as they contain many enzymes and molecules of potential physiological or damaging effects, which may be exposed or released from degraded walls, e.g. peroxidase, glucose oxidase (and indirectly hydrogen peroxide and ethylene), acid phosphatase, PME, glycosidases, lectins, PG inhibitors (Section 6.2), and wall carbohydrate fragments which may function as elicitors of phytoalexins or even as regulatory molecules (Lund and Mapson, 1970; Mussell and Strand, 1977; Cooper *et al.*, 1981; Hahn *et al.*, 1981; McNeil *et al.*, 1982). However, attempts to establish the toxicity *per se* to plant cells of these components or their products have proved negative (Bateman and Basham, 1976). Nevertheless, it is evident that numerous side-effects can result from degradation of wall rhamnogalacturonan, and in this way diverse symptoms of more complex diseases may arise, such as vascular wilts.

Polygalacturonidases have been implicated in wilt diseases as infected xylem vessels are often occluded by pectic gels arising from pit membranes, which may contribute to the characteristic water stress (Cooper and Wood, 1980); other diverse symptoms could originate indirectly from wall break-down, viz. epinasty and abscission (from increased ethylene), chlorosis (hydrogen peroxide), and vascular browning (peroxidase). The evidence is contradictory but for *Verticillium* wilts PL in lucerne and tomato and PG in cotton have been detected in xylem before appearance of symptoms; when purified and reintroduced into vessels some enzymes have reproduced symptoms such as wilting and chlorosis (Mussell and Strand, 1977; Cooper and Wood, 1980). Claims that UV-induced mutants of *V. dahliae* deficient for *either* PG *or* PL retained pathogenicity for cotton have cast some doubt on the role of these enzymes (Howell, 1976). However, for organisms such as *Verticillium* spp. which produce both highly active hydrolases and lyases, mutants simultaneously deficient for both enzymes must be obtained in order to establish their involvement.

In healthy plants solubilization of wall polygalacturonide by *endo*-PG occurs during ripening of fruits such as avocado, tomato, and peach (Gross and Wallner, 1979) and abscission (Berger and Reid, 1979). Loss of the middle lamella strikingly resembles ultrastructural wall changes induced during pathogenesis (Sexton and Hall, 1974; Ben Arie *et al.*, 1979), but the comparative lack of damage to the cytoplasm of contiguous cells reveals that the process is more carefully controlled.

5.3. Degradation of non-uronide matrix polymers during pathogenesis

The involvement of enzymes degrading non-uronide polymers is still poorly understood. They are produced in large amounts by many pathogens *in vitro*, and often detected in infected tisues in which the levels of corresponding wall sugars are depleted, e.g. there is a rapid loss of arabinan and galactan ($\leq 75\%$ and 97%, respectively) in lesions caused by *S. sclerotiorum* on sunflower, and about 60% reduction in neutral sugars of matrix polysaccharides (Ara, Gal, Man, and Xyl) in bean hypocotyls soon after colonization by *S. rolfsii* (Figure 5), as well as the full range of polysaccharidases needed for removal of these polymers (Hancock, 1967; Bateman, 1976). However, their role in pathogenesis is often relegated to one of nutrition as they do not effect maceration (Bauer *et al.*, 1977; Cooper *et al.*, 1978; Baker *et al.*, 1979). Reliance on such dramatic wall changes may well underestimate the contribution of these CWDE. Thus, α-arabinosidase occurs in apple tissue infected with *M. fructigena* and production has been correlated with pathogenicity of induced mutants (Howell, 1975). It was tentatively suggested that this enzyme might remove Ara from wall glycoprotein, but in fact the β-linkages between these Ara residues would preclude this.

Degradation of non-uronides may be significant during invasion of tissues especially rich in a neutral polymer, as in late blight of potato. The matrix polysaccharides of tubers consist of 45% galactan (cf. 9% in sycamore walls) which is rapidly depleted on infection, and galactanase is the predominant CWDE produced *in vitro* by *P. infestans*, (Friend, 1973).

Galactan degradation has been implicated in the wall changes accompanying ripening of apple fruits. Hydrolysis by a β-galactosidase may contribute to the softening as loss of Gal precedes solubilization of polygalacturonide. It is unusual that this appears to result from an *exo*-PG (Bartley, 1978). However, in general, CWDE other than *endo*-polygalacturonidases do not appear to cause detectable changes in wall integrity. Thus, in ripening tomato fruits, walls undergo a marked loss of Gal and Ara but wall loosening depends on solubilization of Gal A and Rha as revealed by a non-ripening mutant (*rin*) which lacked *endo*-PG of the parent line (Gross and Wallner, 1979). Also, maceration of potato tissue was not evident even with a fungal *endo*-galactanase capable of releasing *ca.* 56% of matrix carbohydrates from isolated walls of tubers (Bauer *et al.*, 1977).

Perhaps consideration should be given to the remarkable and rapid wall changes which must occur during plant cell elongation, to allow wall polymers to move relative to one another without loss of wall strength. It is likely that CWDE are involved, possibly those that degrade Ara and Gal side-branches (Darvill *et al.*, 1978), but bonds may be broken and reformed elsewhere by the *trans*-glycosylating ability of some polysaccharidases (Albersheim, 1976). It is conceivable that CWDE of some biotrophic pathogens act in this subtle way to allow wall penetration. In this light, demonstration of the ability of a parasite's CWDE to cause maceration appears to be a crude and unnecessary prerequisite to involvement in pathogenesis.

5.4. Cellulases in pathogenesis

As already mentioned cellulases usually appear only during later stages of infection but in a few diseases may be active during early critical stages, as in lesions caused by *R. solani* and *S. sclerotiorum* and in tobacco stems invaded by *P. solanacearum*, for which pathogenicity of isolates was clearly correlated with cellulase production (Kelman and Cowling, 1965). The insoluble nature and location of fibrils in cell walls may explain the infrequent participation of cellulases in pathogenesis (similar comments may apply to xylanases of pathogens of dicotyledons); activity is usually absent from tissues invaded by soft rot pathogens, and is not involved in ripening of most fruits in which glucan levels remain unchanged (Knee, 1978). The inevitable exceptions are to be found with a correlation between endoglucanase (C_x) and ripening of avocado (Pesis *et al.*, 1978) and C_x activity in abscission zones of many species (Sexton *et al.*, 1980) which suggests a role in wall loosening. However, *endo*-glucanase can reverse the bonding of xyloglucan to cellulase fibrils and, in view of the probable significance of this link in wall structure, C_x may be regarded in a different light than as just one of a 'cellulase complex.'

5.5. Biotrophic parasitism

The fundamental difference between necrotrophic and biotrophic parasitism is often reflected in the extent of host wall degradation. As wall breakdown usually involves cellular disruption it follows that infection by ecologically obligate parasites must involve minimal changes to host walls. This was once considered to result from mechanical penetration, but ultrastructural studies suggest highly localized action by CWDE, as microfibrils are not distorted around penetration pegs which may be highly convoluted and lacking a cell wall, e.g. *Bremia lactucae* (Figure 1) (Ingram *et al.*, 1976). Exceptionally, marked wall disruption is evident as during haustorium formation by *Peronospora spinaciae* (Cooper, 1981).

If CWDE are produced by biotrophs presumably they must be under strict regulatory control, cell-bound [and therefore probably too large to diffuse through the host wall matrix (Section 6.3)], of different mode(s) of action to CWDE of necrotrophs, or especially prone to binding or inhibition. Investigations are complicated by the fact that few biotrophs seem amenable to axenic culture. An early report of *endo*-PG is germinated urediospores of *Puccinia graminis* can be questioned on the grounds of microbial contamination, but the detection of cellulase(s) and hemicellulases in non-germinated spores appears valid (Van Sumere *et al.*, 1957).

Recent studies on urediospores of *Uromyces* spp. and sporidia of *Ustilago maydis* in culture suggest that biotrophs do produce CWDE but in very small amounts and partly in bound forms (Table 1). Particularly significant is the absence of the destructive *endo*-polygalacturonidases, minute amounts of which can affect wall structure (Baker *et al.*, 1980). The lack of cellulases may reflect the contact mainly with primary walls and also suggests that cellulose degradation is not required for wall penetration. It seems that the subtle growth of biotrophs through walls may be facilitated by enzymes which degrade neutral polymers of the wall matrix.

Biotrophy may depend on additional control as exerted by CR of CWDE synthesis resulting from the characteristic influx of photosynthates to infected areas (Lewis, 1974). Some CWDE of *U. maydis* appear to be subject to CR but those from *Uromyces* urediospores are not repressible (unpublished results). It is likely that repression of CWDE synthesis (or inhibition) does occur with some biotrophs *in vivo*; thus a few mycorrhizal fungi are capable of producing CWDE *in vitro*, but more significant are the benign relationships between plants and certain facultative parasites with high potential to produce CWDE, e.g. *V. albo-atrum*. Biotrophic and necrotrophic parasitism may be in delicate balance as suggested by pathogens such as *C. lin-*

Table 1. Comparison of types and levels of CWDE from a facultative and two obligate parasites

Pathogens/enzymes[a]	Endo-PG	Exo-PG	Endo-PL	Cellulase		Arabanase	Galactanase	Xylanase	Glycosidases		
				C_1	C_x				β-Glu	β-Gal	β-Xyl
Verticillium albo-atrum[b]	2000	0	150	+	775	61	4	62	0.3	30	0
Uromyces fabae:											
Extracellular[c]	0	2	0	0	0	2	5	5	3	6	4
Cell bound[d]	0	3	0	0	0	8	9	4	3	0	4
Ustilago maydis[b]	0	0.8	0	0.6	0	10	3	45	11	45	79

[a]Activities (as μmol min^{-1} cm^{-3}) are the maximum attained in [b]fluids from liquid cultures, [c]fluids containing germinating urediospores, and [d]sequential extractions from germinated urediospores (Harding and Cooper, unpublished work).

demuthianum and *P. infestans* which have initial phases of biotrophy before causing extensive wall breakdown and necrosis.

6. INFLUENCE OF WALL COMPOSITION AND STRUCTURE ON COMPATIBILITY

The effect of host cell walls *per se* on synthesis and activity of CWDE of parasites will be considered as this may influence initial infection; cytoplasmic constituents such as phenols probably act later and are discussed elsewhere (Byrde and Archer, 1977).

6.1. Wall polysaccharide composition

The primary walls of different cultivars and species are similar in terms of major polysaccharides and glycosidic linkages (Albersheim, 1976; Darvill *et al.*, 1980). It follows that walls isolated from cultivars with differential resistance are equally susceptible to CWDE and induce similar types and levels of CWDE *in vitro* (Cooper, 1977). This may also be reflected by the nearly identical rates of penetration of the epidermis of resistant and susceptible cultivars by fungi (Cooper *et al.*, 1981). Non-host resistance sometimes appears to be expressed at the level of cell walls and, in culture, walls isolated from host or non-host species can have a differential effect on rate and extent of synthesis of pectic enzymes by pathogens. Although this shows no consistent relationship with resistance or susceptibility, instances of low enzyme production on non-host walls or the converse on host walls, e.g. high PG synthesis by *B. fabae* on *Vicia faba* walls, may be reflections of the level of compatibility (Cooper *et al.*, 1981).

The proportions of monosaccharides can vary in cell walls from different organs of a plant (Jones and Albersheim, 1972); little is known concerning the influence of this on infection except for much higher synthesis of α-galactosidase of *C. lindemuthianum* on walls isolated from susceptible regions, such as hypocotyls, than on walls from roots which are resistant (English *et al.*, 1971).

The disparity between the wall compositions of mono- and dicotyledonous plants could possibly contribute to the resistance of one group to pathogens of the other group. Thus the differences would predict the much more effective degradation of bean walls compared with rice by an *endo*-PGL and conversely of maize walls compared with bean by a xylanase (Baker *et al.*, 1977, 1980). However, it is not known if pathogens of either group have evolved their range of CWDE accordingly, but production by *H. maydis* of a *constitutive* xylanase (Bateman, 1976) as well as much higher β-xylosidase and β-1,3-glucanase levels than by a pathogen of bean (Anderson, 1978), and the predominance of xylanase and β-xylosidase in CWDE of *U. maydis* (Table 1), could represent such an adaptation. Although pectic polymers are present in smaller amounts in walls of monocotyledons they may serve a similar structural function as in dicotyledons, for at least one monocotyledon, onion, is vulnerable to maceration and cell killing by pure *endo*-PG; also degradation of rhamnogalacturonan appears to be a key step in wall degradation by pathogens of *Allium* (Mankarios and Friend, 1980).

Extensive wall changes accompany maturation, which include conversion of pectin to calcium polygalacturonate, alterations in the proportions of monomers, e.g. loss of Gal and incorporation of Xyl (Figure 5) (Jones and Albersheim, 1972), and lignification or suberization. Disease resistance of mature tissue can sometimes be explained in terms of the resistance of its walls to CWDE, e.g. bean to *R. solani* (Bateman *et al.*, 1969), whereas fruits become more susceptible to infection with age which coincides with substantial changes in wall polysaccharides (Knee *et al.*, 1975).

Even in the absence of such marked changes in wall structure, many pathogens with highly active CWDE cause negligible damage to cell walls during invasion. There are several ways in which their CWDE might be restricted in cell walls.

6.2. Enzyme inhibition and binding in walls

Proteins ionically bound to plant cell walls have been isolated, which can completely inhibit *endo*-PG but not other CWDE (Albersheim and Anderson, 1971). The inhibitors are relatively non-specific to PGs from different races and species of pathogens and any role in resistance remains to be established. Their specificity for *endo*-PG again suggests a key role for this enzyme in wall degradation and it is feasible that the inhibitors are present to limit *plant* PG activity. They may function during events such as abscission and xylem maturation in which wall degradation is remarkably localized to avoid extensive damage to adjacent tissues.

Wall proteins have been implicated by some in the high capacity of cell walls to bind fungal PGs ionically. Claims of specific binding of PGs from parasites such as *Cladosporium cucumerinum* to walls from host but not non-host plants, and lack of binding of PGs from incompatible parasites or saprophytes, imply that restriction of PG activity is *beneficial* to pathogenesis. However, in a recent survey of six pathogens, selective binding of PG and PL was not apparent; also, the efficacy of PG inhibitors in native walls was questioned as protein-free walls bound similar amounts of enzyme activity (Cooper *et al.*, 1981). Nevertheless, under conditions of appropriate ionic strength and pH much of the soluble activity of pectic enzymes can be immobilized by cell walls (e.g. 98% of *F. oxysporum* PL by pea walls at pH 4), with the likely effect of restriction of wall degradation to the immediate vicinity of invading hyphae. Positively charged enzymes (with high isoelectric point) would be especially vulnerable to binding to negative carboxyl groups of wall rhamnogalacturonan.

6.3. Wall porosity

An alternative mechanism of restriction of activity concerns the molecular sieving properties of cell walls of microorganisms (Chang and Trevithick, 1974) and plants. The size of an enzyme can determine its location in the parasite (intercellular, wall bound, or extracellular) and activity in host walls. Thus cellulases are probably restricted in degradation of secondary wall as most fungal cellulases (mean diameter 60 Å) are larger than the wall capillaries (diameter generally <40 Å) (Figure 6a) (Preston, 1974; Cowling, 1975).

Little is known about movement of proteins within primary walls but two recent studies have suggested pore sizes of 35–52 Å (Carpita *et al.*, 1979) and 70 Å (Tepfer and Taylor, 1981); the latter estimate would allow permeation by globular proteins of <60 000 daltons. This relates to an earlier suggestion by Knee *et al.* (1975) that apple walls are permeable only to CWDE of <100 000 daltons, based on the relative abilities of various CWDE to degrade walls, e.g. a low molecular weight PG (37 000 daltons) and a PL released most of the wall uronides, a larger PG (75 000 daltons) released about half, and the high molecular weight PG of *P. infestans* (>200 000 daltons) was relatively ineffective (Figure 6b). It is noteworthy that *endo*-polygalacturonidases of most necrotrophic pathogens are small (*ca.* 30 000

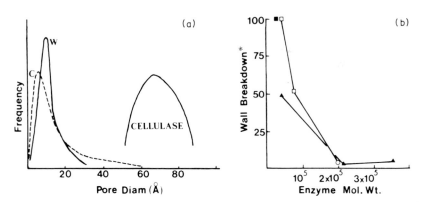

Figure 6. (a) Comparison of pore diameter of cell wall capillaries in secondary walls of cotton (C) and spruce wood (W) with diameter of fungal cellulases (after Cowling, 1975; and Reese, 1977. Reproduced from the latter by permission of Plenum Publishing Corp.). (b) Ability of CWDE of differing molecular weights to degrade apple cell walls. PG (□), PL (■), and α-arabinosidases (▲) were all from *Monilinia fructigena*, except one PG (○) from *Phytophthora infestans*. *Wall breakdown as % of available monomer in wall. (After Knee *et al.*, 1975)

daltons) and therefore potentially capable of penetration of the wall matrix. The minimal degradation during invasion by *P. infestans* may depend on the localized action of its large PG; it remains to be seen whether other facultative biotrophs employ a similar strategy.

REFERENCES

Albersheim, P. (1976). The primary cell wall, in *Plant Biochemistry* (eds J. Bonner and J. E. Varner), Academic Press, New York, pp. 225–273.

Albersheim P., and Anderson, A. J. (1971). Proteins from plant cell walls inhibit polygalacturonase secreted by plant pathogens, *Proc. Natl. Acad, Sci. USA,* **68**, 1815–1819.

Anderson, A. J. (1978). Extracellular enzymes produced by *Colletotrichum lindemuthianum* and *Helminthosporium maydis* during growth on isolated bean and corn cell walls, *Phytopathology,* **68**, 1585–1589.

Aspinall, G. O. (1973). Carbohydrate polymers of plant cell walls, in *Biogenesis of Plant Cell Wall Polysaccharides* (Ed. F. Loewus). Academic Press, New York and London, pp. 95–115.

Baker, C. J., Whalen, C. H., and Bateman, D. F. (1977). Xylanase from *Trichoderma pseudokoningii*: purification, characterization, and effects on isolated plant cell walls, *Phytopathology,* **67**, 1250–1258.

Baker, C. J., Whalen, C. H., Korman, R. Z., and Bateman, D. F. (1979). α-*L*-Arabinofuranosidase from *Sclerotinia sclerotiorum*: purification, characterization and effects on plant cell walls and tissue, *Phytopathology,* **69**, 789–793.

Baker, C. J., Aist, J. R., and Bateman, D. F. (1980). Ultrastructural and biochemical effects of *endo*-pectate lyase on cell walls from cell suspension cultures of bean and rice, *Can. J. Bot.*, **58**, 867–880.

Bal, A. K., Verma, D. P. S., Byrne, V. M., and Maclachlin, G. A. (1976). Subcellular localization of cellulases in auxin-treated pea, *J. Cell Biol.,* **69**, 97–105.

Bartley, I. M. (1978). *Exo*-polygalacturonase of apple, *Phytochemistry,* **17**, 213–216.

Basham, H. G., and Bateman, D. F. (1975). Relationship of cell death in plant tissue treated with a homogeneous endopectate lyase to cell wall degradation, *Physiol. Plant Pathol.*, **5**, 249–262.

Bateman, D. F. (1976). Plant cell wall hydrolysis by pathogens, in *Biochemical Aspects of Plant–Parasite Relationships* (Eds J. Friend and D. R. Threlfall), Academic Press, London, pp. 79–103.

Bateman, D. F., and Basham, H. G. (1976). Degradation of plant cell walls and membranes by microbial enzymes, in *Encyclopedia of Plant Physiology,* Vol. 4 (Eds R. Heitefuss and P. H. Williams), Springer-Verlag, New York, pp. 316–355.

Bateman, D. F., and Beer, S. V. (1965). Simultaneous production and synergistic action of oxalic acid and polygalacturonase during pathogenesis by *Sclerotium rolfsii, Phytopathology,* **55**, 204–211.

Bateman, D. F., Van Etten, H. D., English, P. D., Nevins, D. J., and Albersheim, P. (1969). Susceptibility to enzymic degradation of cell walls from bean plants resistant and susceptible *to Rhizoctonia solani* Kuhn, *Plant Physiol.*, **44**, 641–648.

Bauer, W. D., Talmadge, K. W., Keegstra, K., and Albersheim, P. (1973). The structure of plant cell walls. II. The hemicellulose of the walls of suspension-cultured sycamore cells, *Plant Physiol,* **51**, 174–187.

Bauer, W. D., Bateman, D. F., and Whalen, C. H. (1977). Purification of an *endo*-β-1, 4-galactanase produced by *Sclerotinia sclerotiorum*: effects on isolated plant cell walls and potato tissue, *Phytopathology,* **67**, 862–868.

Ben Arie, R., Kislev, N., and Frenkel, C. (1979). Ultrastructural changes in the cell walls of ripening apple and pear fruit, *Plant Physiol.*, **64**, 197–202.

Berger, R. K., and Reid, P. D. (1979). Role of polygalacturonase in bean leaf abscission, *Plant Physiol.*, **63**, 1133–1137.

Byrde, R. J. W., and Archer, S. A. (1977). Host inhibition or modification of extracellular enzymes of pathogens, in *Cell Wall Biochemistry Related to Specificity in Host–Plant Pathogen Interactions* (Eds B. Sólheim and J. Raa). Universitetsforlaget, Oslo, pp. 213–245.

Byrde, R. J. W., and Willetts, A. J. (1977). *The Brown Rot Fungi of Fruit: Their Biology and Control*, Pergamon Press, Oxford.

Carpita, N., Sabularse, D., Montezinos, D., and Delmer, D. P. (1979). Determination of the pore size of cell walls of living plant cells, *Science*, **205**, 1144–1147.

Chang, P. L. Y., and Trevithick, J. R. (1974). How important is secretion of *exo-*enzymes through apical cell walls, of fungi?, *Arch. Microbiol.*, **101**, 281–293.

Chatterjee, A. K., and Starr, M. P. (1977). Donor strains of the soft-rot bacterium *Erwinia chrysanthemi* and conjugational transfer of the pectolytic capacity, *J. Bacteriol.*, **132**, 862–869.

Coffey, M. D., Palevitz, B. A., and Allen, P. S. (1972). The fine structure of two rust fungi, *Puccinia helianthi* and *Melampsora lini, Can. J. Bot.*, **50**, 231–240.

Collmer, A., and Bateman, D. F. (1981). Impaired induction and self-catabolite repression of extracellular pectate lyase in *Erwinia chrysanthemi* mutants deficient in oligogalacturonide lyase, *Proc. Natl. Acad. Sci. USA*, **78**, 3920–3924.

Cooper, R. M. (1977). Regulation of synthesis of cell-degrading enzymes of plant pathogens, in *Cell Wall Biochemistry Related to Specificity in Host–Plant Pathogen Interactions* (Eds B. Solheim and J. Raa), Universitetsforlaget, Oslo, pp. 163–211.

Cooper, R. M. (1981). Pathogen-induced changes in host ultrastructure, in *Plant Disease Control: Resistance and Susceptibility* (Eds R. C. Staples and G. H. Toenniesen), Wiley, New York, pp. 105–142.

Cooper, R. M., and Wood, R. K. S. (1973). Induction of synthesis of extra-cellular cell wall-degrading enzymes in vascular wilt fungi, *Nature*, **246**, 309–311.

Cooper, R. M., and Wood, R. K. S. (1975). Regulation of synthesis of cell wall-degrading enzymes by *Verticillium albo-atrum* and *Fusarium oxysporum* f. sp. *lycopersici, Physiol. Plant, Pathol.*, **5**, 135–156.

Cooper, R. M., and Wood, R. K. S. (1980). Cell wall-degrading enzymes of vascular wilt fungi. III. Possible involvement of endopectin lyase in *Verticillium* wilt of tomato, *Physiol. Plant Pathol.*, **16**, 285–300.

Cooper, R. M., Rankin, B., and Wood, R. K. S. (1978). Cell wall-degrading enzymes of vascular wilt fungi. II. Properties and modes of action of polysaccharides of *Verticillium albo-atrum* and *Fusarium oxysporum* f. sp. *lycopersici, Physiol. Plant Pathol.*, **13**, 101–134.

Cooper, R. M., Wardman, P. A., and Skelton, J. E. M. (1981). The influence of cell walls from host and non-host plants on the production and activity of polygalacturonide-degrading enzymes from fungal pathogens, *Physiol. Plant Pathol.*, **18**, 239–255.

Cowling, E. B. (1975). Physical and chemical constraints in the hydrolysis of cellulose and lignocellulosic materials, in *Biochemical and Bioengineering Symposium 5* (Ed. C. R. Wilkie), Wiley, New York, Wiley, pp. 163–181.

Darvill, A. G., Smith, C. J., and Hall, M. A. (1978). Cell wall structure and elongation growth in *Zea mays* coleoptile tissue, *New Phytol.*, **80**, 503–516.

Darvill, A. G., McNeil, M., Albersheim, P., and Delmer, D. P. (1980). The primary walls of flowering plants, in *The Biochemistry of Plants*, Vol. 1 (Ed. N. E. Tolbert), Academic Press, New York, pp. 91–162.

Dekker, R. F. H., and Richards, G. N. (1976). Hemicellulases: their occurrence, purification, properties, and mode of action, *Adv. Carbohydr. Chem.*, **32**, 277–352.

English, P. D., Jurale, J. B., and Albersheim, P. (1971). Host parasite interactions. II. Parameters affecting polysaccharide-degrading enzyme secretion by *Colletotrichum lindemuthianum* grown in culture, *Plant Physiol.*, **47**, 1–6.

Eriksson, K.-E. E. (1977). Degradation of wood cell walls by the rot fungus *Sporotrichum pulverulentum*—enzyme mechanisms, in *Cell Wall Biochemistry Related to Specificity in Host–Plant Pathogen Interactions*. (Eds B. Solheim and J. Raa), Universitetsforlaget, Oslo, pp. 71–83.

Fielding, A. H. (1981). Natural inhibitors of fungal polygalacturonases in infected fruit tissues, *J. Gen. Microbiol.*, **123**, 377–381.

Fox, R. T. V., Manners, J. G., and Myers, A. (1972). Ultrastructure of tissue disintegration and host reactions in potato tubers infected with *Erwinia carotovora* var. *atroseptica, Potato Res.*, **15**, 130–145.

Friend, J. (1973). Resistance of potato to *Phytophthora*, in *Fungal Pathogenicity and the Plant's Response* (Eds R. J. W. Byrde and C. V. Cutting), Academic Press, London, pp. 383–399.

Goodenough, P. W., Kempton, R. J., and Maw, G. A. (1976). Studies on the root rotting fungus *Pyrenochaeta lycopersici*: extracellular enzyme secretion by the fungus on cell wall material from susceptible and tolerant plants, *Physiol. Plant Pathol.*, **8**, 243–251.

Gross, K. C., and Wallner, S. J. (1979). Degradation of cell wall polysaccharides during tomato fruit ripening, *Plant Physiol.*, **63**, 117–120.

Hahn, M. G., Darvill, A. G., and Albersheim, P. (1981). Host–pathogen interactions, XIX. The endogenous elicitor, a fragment of a plant cell wall polysaccharide that elicits phytoalexin accumulation in soybeans, *Plant Physiol.*, **68**, 1161–1169.

Hancock, J. G. (1967). Hemicellulase degradation in sunflower hypocotyls infected with *Sclerotinia sclerotiorum, Phytopathology*, **57**, 203–206.

Hancock, J. G., Millar, R. L., and Lorbeer, J. W. (1963). Role of pectolytic and cellulolytic enzymes in *Botrytis* leaf blight of onion, *Phytopathology*, **53**, 932–935.

Hislop, E. C., Barnaby, V. M., Shellis, C., and Laborda, F. (1974a). Localization of α-L-arabinofuranosidase and acid phosphatase in mycelium of *Sclerotinia fructigena, J. Gen. Microbiol.*, **81**, 79–99.

Hislop, E. C., Shellis, C., Fielding, A. H., Bourne, F. J., and Chidlow, J. W. (1974b). Antisera produced to purified extracellular pectolytic enzymes from *Sclerotinia fructigena, J. Gen. Microbiol.*, **83**, 135–143.

Hislop, E. C., Keon, J. P. R., and Fielding, A. H. (1979). Effect of pectin lyase from *Monilinia fructigena* on viability, ultrastructure and localization of acid phosphatase of cultured apple cells, *Physiol. Plant Pathol.*, **14**, 371–381.

Hislop, E. C., Paver, J. L., and Keon, J. P. R. (1982). An acid protease produced by *Monilinia fructigena in vitro* and in infected apple fruits and its possible role in pathogenesis, *J. Gen. Microbiol.*, **128**, 799–807.

Horton, J. C., and Keen, N. T. (1966). Sugar repression of *endo*-polygalacturonase and cellulase synthesis during pathogensis by *Pyrenochaeta terrestris* as a resistance mechanism in onion pink root rot, *Phytopathology*, **56**, 908–916.

Howell, C. R. (1976). Use of enzyme-deficient mutants of *Verticillium dahliae* to assess the importance of pectolytic enzymes in symptom expression of *Verticillium* wilt of cotton, *Physiol. Plant Pathol.*, **9**, 279–283.

Howell, H. E. (1975). Correlation of virulence with secretion *in vitro* of three wall-degrading enzymes in isolates of *Sclerotinia fructigena* obtained after mutagen treatment, *J. Gen. Microbiol.*, **90**, 32–40.

Hsu, G. J., and Vaughn, R. H. (1969). Production and catabolite repression of the constitutive polygalacturonic acid *trans*-eliminase of *Aeromonas liquefaciens*, *J. Bacteriol.*, **98**, 172–181.

Hubbard, J. P., Williams, J. D., Niles, R. M., and Mount, M. S. (1978). The relation between glucose repression of *endo*-polygalacturonate *trans*-eliminase and adenosine 3′5′-cyclic monophosphate levels in *Erwinia carotovora*, *Phytopathology*, **68**, 95–99.

Hulme, M. A., and Stranks, D. W. (1971). Regulation of cellulase production by *Myrothecium verrucaria* grown on non-cellulosic substrates. *J. Gen. Microbiol.*, **69**, 145–155.

Ingram, D. S., Sargent, J. A., and Tommerup, I. C. (1976). Structural aspects of infection by biotropic fungi, in *Biochemical Aspects of Plant–Parasite Relationships* (Eds J. Friend and D. R. Threlfall), Academic Press, London, pp. 43–78.

Jones, T. M., and Albersheim, P. (1972). A gas chromatographic method for determination of aldose and uronic acid constituents of plant cell wall polysaccharides, *Plant Physiol.*, **49**, 926–936.

Keegstra, K., Talmadge, K. W., Bauer, W. D., and Albersheim, P. (1973). The structure of plant cell walls. III. A model of the wall of suspension-cultured sycamore cells based on interconnections of the macromolecular components, *Plant Physiol.*, **51**, 188–197.

Kelman, A., and Cowling, E. B. (1965). Cellulase of *Pseudomonas solanacearum* in relation to pathogenesis, *Phytopathology*, **55**, 148–155.

Knee, M. (1978). Properties of polygalacturonate and cell cohesion in apple fruit cortical tissue, *Phytochemistry*, **17**, 1257–1260.

Knee, M., Fielding, A. H., Archer, S. A., and Laborda, F. (1975). Enzymic analysis of cell wall structure in apple fruit cortical tissue, *Phytochemistry*, **14**, 2213–2222.

Labavitch, J. M., and Ray, P. M. (1978). Structure of hemicellulosic polysaccharides of *Avena sativa* coleoptile cell walls, *Phytochemistry*, **17**, 933–937.

Lamport, D. T. A. (1970). Cell wall metabolism, *Annu. Rev. Plant Physiol.*, **21**, 235–269.

Lamport, D. T. A. (1980). Structure and function of plant glycoproteins, in *The Biology of Plants*, Vol. 3 (Ed. J. Preiss), Academic Press, New York, pp. 501–541.

Langcake, P., Bratt, P. M., and Drysdale, R. B. (1973). Pectin methylesterase in *Fusarium*-infected susceptible tomato plants, *Physiol. Plant Pathol.*, **3**, 101–106.

Lewis, D. H. (1974). Microorganisms and plants: the evolution of parasitism and mutualism, in *Evolution in the Microbial World* Eds M. J. Carlile and J. J. Skehel), Cambridge University Press, Cambridge, *Symp. Soc. Gen. Microbiol.*, Vol. 24, pp. 367–392.

Lund, B. M., and Mapson, L. W. (1970). Stimulation by *Erwinia carotovora* of the synthesis of ethylene in cauliflower tissue, *Biochem. J.*, **119**, 251–263.

Mace, M. E. (1973). Histochemistry of beta-glucosidase in isolines of *Zea mays* susceptible or resistant to northern corn leaf blight, *Phytopathology*, **63**, 243–245.

Mankarios, A. T., and Friend, J. (1980). Polysaccharide-degrading enzymes of *Botrytis allii* and *Sclerotium cepivorum*. Enzyme production in culture and the effect of the enzymes on isolated onion cell walls, *Physiol. Plant Pathol.*, **17**, 93–104.

Manners, J. G., and Myers, A. (1975). The effect of fungi (particularly obligate pathogens) on the physiology of higher plants, *Symp. Soc. Exp. Biol.*, **29**, 279–296.

Mansfield, J. W., and Richardson, A. (1981). The ultrastructrue of interactions between *Botrytis* species and broad bean leaves, *Physiol. Plant Pathol.*, **19**, 41–48.

McKeen, W. E. (1974). Mode of penetration of epidermal cell walls of *Vicia faba* by *Botrytis cinerea*, *Phytopathology*, **64**, 461–467.

McNeil, M., Darvill, A. G., and Albersheim, P. (1982). Structure of plant cell walls.

XII. Identification of seven differently-linked glycosyl residues attached to C-4 of the 2, 4-linked L-rhamnosyl residues of rhamnogalacturonan I, *Plant Physiol.*, **70**, 1586–1591.

Mullen, J. M., and Bateman, D. F. (1975). Enzymatic degradation of potato cell walls in potato virus X-free and potato virus X-infected potato tubers by *Fusarium roseum* 'avenaceum', *Phytopathology*, **65**, 797–802.

Mussell, H., and Strand, L. L. (1977). Pectic enzymes: involvement in pathogenesis and possible relevance to tolerance and specificity, in *Cell Wall Biochemistry Related to Specificity* in *Host–Plant Pathogen Interactions* (Eds B. Solheim and J. Raa), Universitetsforlaget, Oslo, pp. 31–70.

Northcote, D. H. (1972). Chemistry of the plant cell wall, *Annu. Rev. Plant Physiol.*, **23**, 113–132.

Pesis, E., Fuchs, Y., and Zauberman, G. (1978). Cellulase activity and fruit softening in avocado, *Plant Physiol.*, **61**, 416–419.

Pilnik, W., and Voragen, A. G. J. (1970). Pectic substances and other uronides, in *The Biochemistry of Fruits and their Products* (Ed. A. C. Hulme), Academic Press, London, pp. 53–87.

Preston, R. D. (1974). *The Physical Biology of Plant Cell Walls*, Chapman and Hall, London.

Reese, E. T. (1977). Degradation of polymeric carbohydrates by microbial enzymes, in *The Structure, Biosynthesis and Degradation of Wood* (Eds F. A. Loewus and V. C. Runeckles), Plenum Press, New York, London, *Recent Adv. Phytochem.*, Vol 11, 311–367.

Ries, S. M., and Albersheim, P. (1973). Purification of a protease secreted by *Colletotrichum lindemuthiamum*, *Phytopathology*, **63**, 625–629.

Rombouts, F. M., and Pilnik, W. (1980). Pectic enzymes, in *Microbial Enzymes and Transformation* (Ed. A. H. Rose), Academic Press, London, pp. 228–283.

Sexton, R., Durbin, M. L., Lewis, L. N., and Thomson, W. W. (1980). Use of cellulase antibodies to study leaf abscission, *Nature (London)*, **283**, 873–874.

Sexton, R., and Hall, J. L. (1974). Fine structure and cytochemistry of the abscission zone of *Phaseolus* leaves. I. Ultrastructural changes occurring during abscission, *Ann. Bot.*, **38**, 849–854.

Sheiman, M. I., Macmillan, J. D., Miller, L., and Chase, T. (1976). Coordinated action of pectinesterase and polygalacturonate lyase complex of *Clostridium multifermentans*, *Eur. J. Biochem.*, **64**, 565–572.

Stack, J. P., Mount, M. S., Berman, P. M., and Hubbard, J. P. (1980). Pectic enzyme complex from *Erwinia carotovora*: a model for degradation and assimilation of host pectic fractions, *Phytopathology*, **70**, 267–272.

Talmadge, K. W., Keegstra, K., Bauer, W. D., and Albersheim, P. (1973). The structure of plant cell walls. I. The macromolecular components of the walls of suspension-cultured sycamore cells with a detailed analysis of the pectic polysaccharides, *Plant Physiol.*, **51**, 158–173.

Tepfer, M., and Taylor, I. E. P. (1981). The permeability of plant cell walls as measured by gel filtration chromatography, *Science*, **213**, 761–763.

Timell, T. E. (1965). Wood and bark polysaccharides, in *Cellular Ultrastructure of Woody Plants* (Ed. W. A. Côté), University of Syracuse, Press, Syracuse, NY, pp. 127–156.

Van Etten, H. D., and Bateman, D. F. (1969). Enzymatic degradation of galactan, galactomannan, and xylan by *Sclerotium rolfsi*, *Phytopathology*, **59**, 968–972.

Van Sumere, C. F., Van Sumere-de Preter, C., and Ledingham, G. A. (1975). Cell wall-splitting enzymes of *Puccinia graminis* var. *tritici*, *Can. J. Microbiol.*, **3**, 761–770.

Verhoeff, K., and Liem, J. I. (1978). Presence of *endo*-polygalacturonase in conidia of *Botrytis cinerea* before and during germination, *Phytopathol. Z.*, **91**, 110–115.
West, C. A. (1981). Fungal elicitors of the phytoalexin response in higher plants, *Naturwissenschaften*, **68**, 447–457.
Whistler, R. L., and Richards, E. L. (1970). Hemicelluloses, in *The Carbohydrates,* Vol. IIA (Eds W. Pigman and D. Horton), Academic Press, New York, pp. 447–469.
Wood, T. M. (1969). The cellulase complex of *Fusarium solani.* Resolution of the enzyme complex, *Biochem. J.*, **115**, 457–464.
Zucker, M., and Hankin, L. (1970). Regulation of pectate lyase synthesis in *Pseudomonas fluorescens* and *Erwinia carotovora, J. Bacteriol.*, **104**, 13–18.

Further Reading

Albersheim, P. (1976). The primary cell wall, in *Plant Biochemistry* (Eds J. Bonner and J. E. Varner), Academic Press, New York, pp. 225–273.
Bateman, D. F. (1976). Plant cell wall hydrolysis by pathogens, in *Biochemical Aspects of Plant–Parasite Relationships* (Eds J. Friend and D. R. Threlfall), Academic Press, London, pp. 79–103.
Byrde, R. J. W., and Archer, S. A. (1977). Host inhibition or modification of extracellular enzymes of pathogens, in *Cell Wall Biochemistry Related to Specificity in Host–Plant Pathogen Interactions* (Eds B. Solheim and J. Raa), Universitetsforlaget, Oslo, pp. 213–245.
Cooper, R. M. (1977). Regulation of synthesis of cell-degrading enzymes of plant pathogens, in *Cell Wall Biochemisty Related to Specificity in Host–Plant Pathogen Interactions* (Eds B. Solheim and J. Raa), Universitetsforlaget, Oslo, pp. 163–211.

Biochemical Plant Pathology
Edited by J. A. Callow
© 1983 John Wiley & Sons Ltd

8

The Biochemistry of Fungal and Bacterial Toxins and Their Modes of Action

R. D. DURBIN

Plant Disease Resistance Research Unit, ARS, USDA, and Department of Plant Pathology, University of Wisconsin–Madison, Madison, WI 53706, USA

I. INTRODUCTION

Since the early part of this century, when plant pathology was still in its infancy, scientists have been intrigued with the mechanisms by which pathogens attack plants—and conversely how plants resist potential pathogens. During this period, the concept that the pathogen produces substances toxic to the host emerged as one of the important mechanisms of disease causation. How broadly this concept can be applied is still a matter of considerable debate, however. Some workers, most notably Gäumann (1954), have taken the stand that to be pathogenic, a pathogen must be toxigenic. Others have taken a less encompassing position, and some even feel that toxin production is of minor significance.

Several major epiphytotics in which toxins played a decisive role have been recorded. These outbreaks had an important impact on the agricultural economy of the USA, and more importantly for the long run, precipitated a re-evaluation of the strategies for the genetic control of diseases in major crop plants (Horsfall *et al.*, 1972).

The first example concerns the use of the V_b, or Victoria gene, for crown rust resistance, which by the mid-1940s had been incorporated into about 80% of the US cultivars (Coffman, 1961). With favourable weather conditions in 1945, a new seedling and leaf blight disease, caused by a species of the fungus *Helminthosporium* not previously known to occur on oats, appeared on these cultivars. The disease was so devastating that by 1946 most cultivars had to be eliminated, and new ones, which did not contain the V_b gene, introduced. Subsequent research showed that the fungus produces a toxin which is extremely potent, but only on oat cultivars possessing the V_b gene.

A similar but far more serious example arose in maize (Horsfall *et al.*, 1972; Scheffer, 1976). Here the genetic culprit conferring susceptibility to a toxin was a cytoplasmic sterility system called Texas male sterile (TMS). This system was once widely used throughout the US corn belt in double-cross hybrid seed production. However, the genetic uniformity it created allowed a new, aggressive mutant of the fungus *Helminthosporium maydis*, long known to cause a minor disease of maize, to build up quickly. Its pathogenicity was due to the production of a new toxin, trivially named T-toxin, to which the TMS system conferred susceptibility. The disease caused an estimated loss in 1970 of between $500 million and $1 billion, about 15% of the crop. In addition, the need to eliminate the TMS system from hybrid maize forced seed producers back to the expensive and labour-intensive practice of hand-pulling tassels from rows of one of the two adjacent single-cross lines to achieve cross-pollination.

Other less dramatic examples could be cited. However, their recounting does not bear directly on the fundamental question of how widespread the occurrence of toxins that play a causal role in disease production is. Certainly it is true that toxic compounds occur widely. Individual species within all major taxonomic groups of plant pathogens, except viruses, have been reported to produce them, and the number of examples is rising rapidly. However, so far, compelling evidence that toxins play a causal role has been provided in only a small number of cases.

2. DEFINITION

Various workers have defined toxins in different ways (Dimond and Waggoner, 1953; Ludwig, 1960; Wheeler and Luke, 1963; Graniti, 1972; Rudolph, 1976; Durbin and Steele, 1979; Yoder, 1980). Basically, the definitions have differed in their degree of restrictiveness and in the number and technical complexity of the criteria, both specified and implied, that must be met before a compound can be classified as a toxin. All researchers would agree that toxins are (a) metabolic products of the pathogen, (b) not enzymes, (c) injurious to the host, and (d) factors that condition either pathogenicity or virulence. Pathogenicity is defined here as the ability to cause disease. Factors responsible for pathogenicity, sometimes referred to as primary disease determinants (Yoder, 1980), are qualitative in nature and are essential for disease to occur; some toxins are classified within this category. Virulence is defined as the measure of disease severity. The factors responsible for virulence, or secondary disease determinants, are thus quantitative characters which contribute to the severity of a disease and by themselves are not essential for disease production; the majority of toxins fall within this category.

There is less agreement among researchers about whether compounds known to occur in healthy plants, such as plant-growth substances or organic acids, should be included in the definition of a toxin (even though we would be concerned only with that portion produced by the pathogen). Likewise, should high molecular weight substances be excluded, either because of their size (where does one draw the line between high and low molecular weight substances?) or because those reported to date appear to act in a physical rather than a chemical manner? Must the effect of a toxin always be expressed visually, or may ultrastructural and/or biochemical alterations qualify? If action at low concentrations is a requirement, how does one define a low concentration? Sampling procedures could become important here, for one would expect to find in diseased tissues relatively rapid changes in toxin concentration over short distances. Lastly, must toxins be produced *in planta,* or can they be produced by pathogens in the rhizosphere or phyllosphere and subsequently taken up by the host?

Arguments have been made on both sides of these questions. The point here is not to marshal the arguments for and against each, but simply to point out that the number of examples of diseases in which toxins play a crucial role will vary greatly with one's definition of a toxin. As more host–parasite systems are studied, I suspect our views on this matter will broaden, because current knowledge about toxins, most notably those produced by nematodes, mycoplasmas, spiroplasmas, and rickettsias, is very rudimentary and because past experience has shown us that the more examples we study, the broader the limits of diversity become.

Irrespective of any shortcomings current definitions may have, they have served the very useful purpose of keeping our attention focused on the diseased plant, rather than on *in vitro* situations. Indeed, it was precisely this problem that caused Dimond and Waggoner (1953), to formulate the first really modern definition of a toxin emphasizing complicity in disease causation. Since then others have extended this definition, but always with disease considerations paramount.

Regardless of the exact details of any of these definitions, it has proved to be technically very difficult to satisfy their criteria fully. This fact more than any other is why the number of widely accepted examples of toxin involvement in disease is not larger. This is not to say that the criteria are over-restrictive, but rather that proving that a 'toxin' is involved in disease causation is a complex undertaking that requires a number of kinds of evidence. The technical difficulties exist mainly because (a) toxins are biologically active at very low concentrations (usually $<10^{-6}$–10^{-8} M), (b) they generally cannot be easily or quickly isolated, especially from diseased plant material, (c) they may be chemically unstable, (d) artifacts can be introduced when purified toxins are administered to plants, and (e) it is difficult to synthesize (either biologically or chemically) toxins containing a high specific activity stable isotope or radioactive label.

3. ROLE IN PATHOGENESIS

From the viewpoint of a pathogen, the host can be thought of simply as a source of nutrients, the rate of utilization of which controls the pathogen's ability to grow, reproduce, and/or form overwintering structures. To be successful at these ventures the pathogen must break down and assimilate host materials and, at the same time, overcome whatever resistance mechanism(s) the host might possess. It is within this context that toxins function in pathogenesis. There are a number of potential activities they could have. However, as yet, we have very little definitive information as to which one(s) a particular toxin is assuming. Some which can be envisaged include (a) counteracting the initiation or maintenance of host resistance mechanisms, (b) damaging host cells which then release nutrients for pathogen growth, (c) causing the release of degradative enzymes from host organelles, (d) providing a conducive micro-environment for the pathogen (e.g. pH, water potential, or redox potential), (e) facilitating movement of the pathogen through the plant, (f) accelerating senescence of the host, and (g) inhibiting secondary invasion by other microorganisms.

Paradoxically, while still focusing their primary attention on the role toxins might play in the diseased plant, researchers have had, by necessity, to rely in large measure on *in vitro* experimental systems. This has brought up the following potential problem: can the complex properties of a dynamic, interactive host–parasite system be properly inferred from studies of component parts viewed under *in vitro* conditions? In the case of toxins one would not necessarily expect them to act independently in the host but rather in concert with other attack mechanisms in a parallel and/or sequential fashion. Consequently, by using *in vitro* systems we may be overlooking some important features of toxin interactions.

Most of the information we have on the role of toxins in disease production is focused on the early portion of the disease cycle. For instance, we know that toxins can be synthesized by microorganisms growing outside the host. An extreme example of this is 1-amino-2-nitrocyclopentanecarboxylic acid (ANCPA), which causes unusual plant growth abnormalities (Woltz, 1978). The fungus producing this toxin, *Aspergillus wentii*, lives in the soil and never invades the affected plant (Baker and Cook, 1974). It is a pathogen without being a parasite. Some pathogens cannot successfully infect living cells but require that the infection site be a necrotic lesion, provided either by the action of toxins and degradative enzymes, or by other means. After infection these microorganisms spread through the host, killing in advance by the same method.

Host-specific fungal toxins that affect only certain cultivars, like the two examples described in the Introduction, are synthesized simultaneously with spore germination. They also are required for the initial establishment of the pathogens that produce them. With *H. victoriae* and *Alternaria mali*, Tox⁻ mutants will behave like pathogenic isolates if the respective toxin is added to the inoculum; otherwise they are non-pathogenic (Scheffer, 1976; Yoder, 1972). Within 6 hours after inoculation with pathogenic isolates the cells of susceptible but not resistant cultivars begin to lose electrolytes, a response observed also with purified toxins.

Comparisons between a wild-type isolate and a Tox⁻ mutant strain of *Pseudomonas syringae* pv. *phaseolicola* which attacks beans, has shown that at the site of inoculation in susceptible tissues both types will multiply at the same rate (Gnanamanickam and Patil, 1976). However, the Tox⁻ strain will not subsequently spread beyond the primary lesion and become systemic, whereas the wild-type isolate will. There is also some evidence that the toxin is able to inhibit the formation by the host of compounds inhibitory to the bacteria, the so-called phytoalexins (Patil and Gnanamanickam, 1976; see also Chapter 12).

Evidence for a toxin acting late in the disease cycle has been obtained from studies on wheat stem rust, caused by *Puccinia graminis* f. sp. *tritici* (Silverman, 1960). A necrotic area, which normally would form around the developing urediospore pustule, can be displaced by applying an electrical potential across the developing lesion. Whether, in fact, the necrosis is caused by a toxin and, if so, what function it might play, is still unknown. Teleologically, it appears that necrosis-inducing toxins might be particularly useful just before a pathogen sporulates. At this stage there is an intense burst of metabolic activity associated with reproduction which proceeds at the expense of all other pathogen activities. Also, at this time the pathogen has become less dependent upon living host cells for nourishment and much of the material being translocated into the reproductive structures comes directly from the mycelium rather than the host. Another possibility is that the onset of necrosis might serve as a metabolic 'signal' for the initiation of sporulation! Thus, it may be significant that many fungal reproductive structures form in necrotic tissue.

Besides playing various roles in pathogenesis, toxins may also be important in the saprophytic phase of pathogens. Acting like classical antibiotics, they could minimize the competition from other microflora occupying the same microsite. Some toxins, which have the capability of reversibly binding metal ions, might also be important nutritionally.

4. CHEMICAL CONSIDERATIONS

4.1. Production

The production of toxins for biological and biochemical studies involves growing the pathogen under controlled environmental conditions (e.g. temperature and aeration) in liquid or on solid media (Shaw, 1981). These media vary from the very complex, containing ill-defined natural substances, to those containing only inorganic salts and a simple sugar or amino acid. Production occurs at different phases of the growth curve; however, it generally is closely related to a particular phase (Figure 1). After a suitable incubation period, the toxin is isolated from the medium using, in sequence, a range of chemical separation techniques. Each technique is designed to effect separation of the toxin from the other constituents in the medium by a different means (e.g. differences due to electrical charge at different pHs, molecular size, and solubility in aqueous and organic solvents). If done properly—and this is as much a matter of art as it is of science—maximum resolution of the toxin will result, preferably to a state of absolute purity. At each step in the purification scheme suitable assays, either biological or biochemical, are used to determine the quantity and location of the toxin. In most instances the final yield of toxin is in the range $1-10$ mg dm^{-3} of culture medium. Only rarely are synthetic toxin preparations available for use.

At present there is no good evidence that the production of toxins *sensu strictu* depends upon a plasmid or plasmids for either structural or regulatory gene action, although this possibility has been suggested several times (Sands *et al.*, 1978). It is, however, a well known phenomenon responsible for the production of toxic metabolites in other systems (Elwell and Shipley, 1980), so we always need to be aware of this eventuality when studying a new toxin.

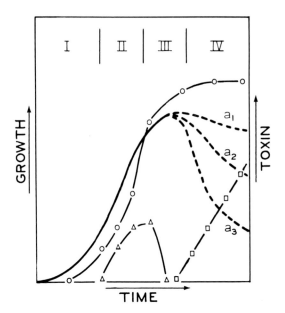

Figure 1. Schematic representation of the relationship between toxin production and *in vitro* growth of pathogens. The growth curve is commonly divided into four phases: lag (I), logarithmic or exponential (II), stationary or equilibrium (III), and autolytic or death (IV). Phases I and II also are called the trophophase and phases III and IV the idiophase. The broken portions of the growth curve indicate that after the stationary phase (III) different pathogens (a_1–a_3) can behave diversely. The three toxin production curves are for AM-toxin I ($\triangle$), a host-specific toxin from *Alternaria alternata* f. sp. *lycopersici* ($\bigcirc$) and lycomarasmin ($\square$)

However, as discussed earlier, if one subscribes to a relatively broad view of what a toxin is and includes plant growth substances, two well documented examples can be cited. They involve indole-3-acetic acid (IAA) production as directed by the plasmids of *Agrobacterium tumefaciens* and *Pseudomonas syringae* pv. *savastanoi* (Nester and Kosuge, 1981; see also Chapter 19). Production of this compound is necessary for gall formation in the diseases caused by these bacteria; strains with increased capacity for IAA synthesis incite larger galls. In at least the latter case, the extrachromosomal DNA element bears the structural genes coding for its synthesis. Thus, IAA⁻ mutants lacking the plasmid do not contain enzyme activities for the two reactions immediately preceeding IAA, and do not form galls (Comai and Kosuge, 1980). Reintroduction of the plasmid via transformation restores these characters (see also Chapter 10). It has been proposed that a similar system operates in *A. tumefaciens*; however, definitive evidence, such as has been provided for pv. *savastanoi*, is not yet at hand. Hopefully, the near future will provide additional evidence on this general topic, since the number of pathogens which have been examined even for the presence of plasmids is still relatively small.

4.2. Composition

The toxins produced by plant pathogens are predominantly of low molecular weight (*ca.* 1000 daltons), except those causing wilting, which have molecular weights between 10^4 and 2×10^6 daltons (Strobel, 1977; Mitchell, 1981). As

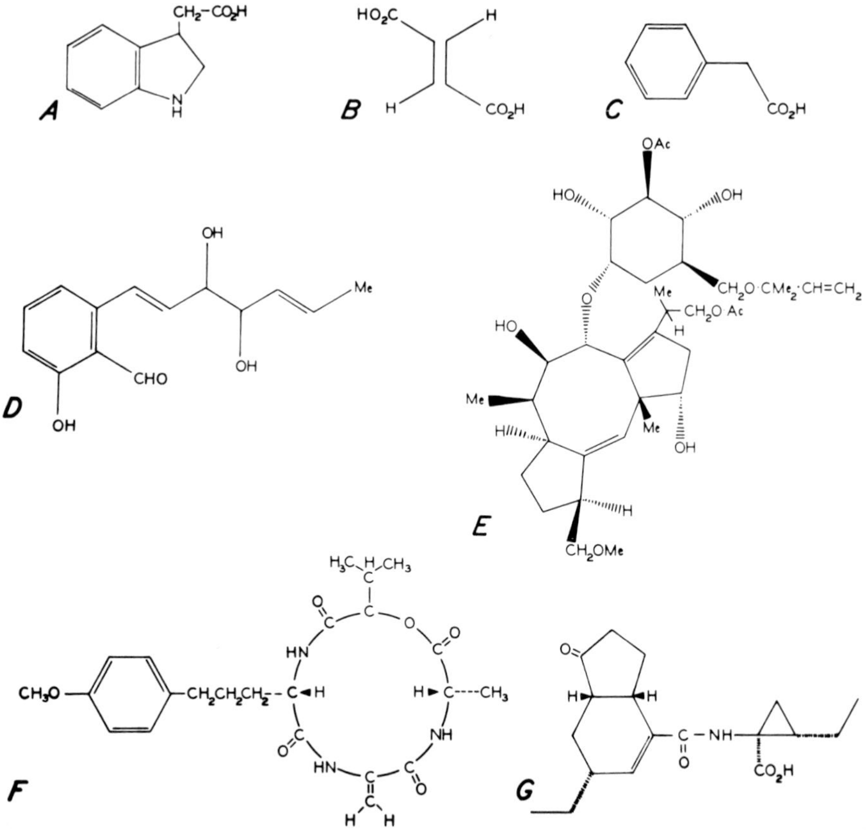

Figure 2. Examples of the chemical diversity of fungal and bacterial toxins synthesized on different biosynthetic pathways. The pathways are noted in parenthesis. (A) Indole-3-acetic acid from *Pseudomonas syringae* pv. *savastanoi* (shikimic acid); (B) *trans*-fumaric acid from *Rhizopus* spp. (TCA cycle); (C) phenylacetic acid from *Rhizoctonia solani* (fatty acid); (D) pyriculol from *Pyricularia oryzae* (acetate–polymalonate); (E) fusicoccin A from *Fusicoccum amygdali* (acetate–mevalonate); (F) AM-toxin I from *Alternaria mali* (amino acid); (G) coronatine from *Pseudomonas syringae* pv. *atropurpurea* (mixed)

mentioned earlier, there is some question of whether the latter substances, chiefly carbohydrate in nature, should be considered as toxins since they appear to act in a physical rather than chemical manner. Fungal toxins contain carbon, hydrogen, oxygen, generally nitrogen, and rarely sulphur. Bacterial toxins have the same constituents and in addition some contain phosphorus. Fungal toxins with known structures now number in excess of 150 and are produced by over 100 taxonomically different pathogens (Rudolph, 1976). In contrast, the structures of only five bacterial toxins are known in their entirety (Mitchell, 1981). Four of these are produced by different pathovars of *Pseudomonas syringae* and the fifth by *Rhizobium japonicum*. Selected examples of fungal and bacterial toxins are presented in Figure 2.

4.3. Biosynthesis and its regulation

On the basis of structural analyses and biosynthetic studies, toxins appear to be synthesized along already well known, main biochemical pathways. Fungal toxins are much more diverse in this regard, with toxins being derived from (a) the shikimic acid pathway, (b) the TCA cycle, (c) the fatty acid pathway, (d) a branch of the acetate–polymalonate pathway, (e) the acetate–mevalonate pathway, (f) various amino acid pathways branching off from the shikimic acid pathway, glycolysis, and the TCA cycle, and (g) combinations of two or more of the preceeding pathways (Figure 3). Bacterial toxins can be viewed as products of either (f) plus derivatization, or (g).

Most toxins are synthesized via soluble enzyme systems, and essentially all of the cell-free biosynthetic studies have been carried out with this kind of system. Based on what we know about their structures, a few toxins, such as cerato-ulmin, a small protein containing 128 amino acid residues, would appear to require the ribosomal protein-synthesizing system for their formation. A third group, because of their cyclic nature and composition, may be made by means of what is known as protein templates. These templates consist of specific aggregations of enzymes to which activated 'building blocks' bind in a sequence and form dictated by the aggregate's architecture. Such systems are already known to be responsible for the synthesis of a variety of cyclic, peptide antibiotics (Perlman and Bodanszky, 1979).

Essentially all toxins can be considered to be secondary metabolites, that is, products of secondary metabolism. Secondary metabolism has been defined in many ways (see also Chapter 17). For our purposes it is easiest to define it simply as anything not primary—primary being those metabolic reactions considered essential for maintaining the living state. This definition should not obscure the fact that primary and secondary metabolism are highly integrated (Figure 3). Secondary metabolites, although not required for life, are fairly common, being found in plants, animals, and microorganisms. Most of them, however, are produced by plants and, to a lesser extent, fungi.

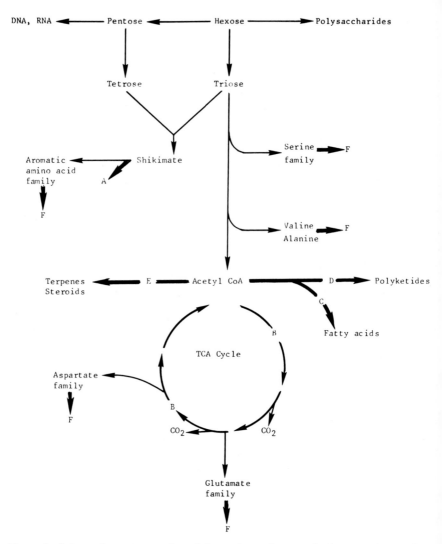

Figure 3. Schematic representation of the major pathways of primary and secondary metabolism showing their interrelationships. The pathways of secondary metabolism are indicated by heavy arrows. They are lettered corresponding to Figure 2 and the text

Microbial toxins share many general properties with other secondary metabolites. For instance, each is produced by a restricted number of microorganisms. Their production is highly dependent on nutritional and environmental conditions. They are produced at specific phases in the microorganism's life cycle. Many are the result of complex and unusual biochemical transformations. They either have no obvious function or else it is highly specialized. Lastly, they are not readily metabolized.

From a consideration of these shared properties, it seems probable that toxin production is controlled by the same regulatory processes that control the synthesis of other secondary metabolites. Unfortunately, essentially nothing is known about this matter beyond some correlative results on toxin production in culture under different nutritional conditions. However when taken together, these results do support the notion that toxin production is under metabolic control. Assuming that this is indeed the case, it would be instructive for us to consider the kinds of control mechanisms that influence the production of antibiotics, since the effects of different nutritional conditions on both antibiotic and toxin production appear to be similar (Drew and Demain, 1977). We might then expect at least the following regulatory processes to be operative: (a) *Carbon catabolic regulation*. During periods of rapid utilization of the carbon source—particularly of glucose or sucrose—either the formation of enzymes in the secondary metabolic pathways leading to toxins would be repressed, or the activity of these pathways would be inhibited. (b) *Nitrogen catabolic repression*. Excessive levels of rapidly assimilated forms of nitrogen (e.g. ammonium ion) could repress the formation of enzymes concerned with nitrogen transformations of toxin intermediates. (c) *Feedback regulation*. As toxins accumulate they would, in some instances, limit their own biosynthesis by inhibiting the activity of one or more enzymes earlier in their synthetic pathway. (d) *Feedback regulation by primary precursors*. Primary metabolites that are precursors of toxins could act similarly by inhibiting enzymes in primary pathways prior to where they branch off into secondary ones. (e) *Energy charge regulation*. High phosphate levels could reduce the availability of high-energy phosphate (i.e. ATP and ADP). This would effectively inhibit a number of key reactions in primary metabolism which, in turn, would cause a reduction in the activity of secondary pathways linked to toxin production. (f) *Induction*. The addition of certain primary metabolites (termed effectors) could induce the formation of enzymes in secondary pathways leading to toxin production. This effect would be aside from any function the effectors might have as precursors of the toxins.

The role of the host plant *vis-à-vis* metabolic regulation is another topic of great potential interest, but again it is virtually unstudied. Even the direct demonstration of a toxin *in planta* has only very rarely been reported. Here, we are substituting a host plant for the artificial culture medium. The result is a vastly more complicated situation because of the host's chemical complexity and the dynamic metabolic disequilibrium that exists between host and parasite. There is, though, some biological evidence which provides support for the idea that toxin production can be regulated by means of chemical cues from the plant. Several examples are known in which selected plant cultivars will not support toxin production even though the pathogen multiplies to a level that would have resulted in toxin production in other cultivars (Gnanamanickam and Patil, 1976; Owens and Wright, 1965). Secondly, Tox⁻ strains of *Helminthosporium sacchari* can resume toxin production if serinol (2-aminopropane-1,3-diol) is added to the culture medium (Babczinski *et al.*, 1978). Interestingly, serinol appears to be present on the surface of and within susceptible sugar cane cultivars at higher levels than in resistant cultivars, suggesting that this compound may be a natural modulator of toxin production. In a similar example, certain strains of *P. syringae* pv. *tagetis* will produce tagetitoxin only *in planta*, whereas other strains of the bacterium will also produce the toxin in synthetic media (Styer and Durbin, unpublished data).

A knowledge of the regulatory role of the host may have important practical consequences. If, for example, it can be shown experimentally that toxin production is under metabolic control, it would open up the possibility that man could manipulate conditions so as to modulate toxin synthesis. In the host plant one might be able to shut off synthesis. Conversely, conditions for toxin production in culture could be maximized for structural and biosynthetic studies.

4.4. Structure–activity

Information about the structural features of toxins responsible for their toxicity has been obtained by comparing the inhibitory effects of a toxin with those of its analogues at different concentrations (Ballio, 1981). The sources of these analogues are three-fold. First, it is not unusual for pathogens to synthesize simultaneously a mixture of closely related toxins. Some may be intermediates along the synthetic pathway, whereas others may represent multiple end products. Particularly with amino acid-containing toxins (e.g. malformins, phaseolotoxin, and tabtoxin), one amino acid may be substituted for another since the enzymes involved in synthesis do not have rigid substrate specificity. Second, by a process called semisynthesis, purified toxins have been modified, usually by chemical oxidation or reduction reactions, or enzymatic hydrolysis. The last source involves pure chemical synthesis. By varying the precursors and the synthetic scheme used, different analogues can be prepared. So far about 50 toxins have been made by this approach, all but two being fungal toxins (Rich, 1981).

The specific structural features determining toxicity are not easily summarized because they vary so much with the toxin in question. For instance, saturation of carbon—carbon double bonds generally results in a drastic reduction or elimination of toxicity. This happens with the AM-toxins, rhizobitoxine and tentoxin but not with fusaric acid and fusicoccin. An additional complicating factor is that the relative order of toxicity of a group of related compounds will often vary with the kind of assay used (e.g. seed germination, hypocotyl elongation, or enzyme activity). We can, however, make some comments about which structural features are generally important. They include the following: (a) size of side-chains, (b) type, number, and position of the functional groups, (c) stereochemistry, (d) ability to chelate metal ions, (e) unsaturated bonds, and (f) electronic nature and oxidation state of the functional groups. In some cases, altering several features may contribute to changing only one critical property of the toxin, as for example altering its hydrophobicity, or lipid solubility, which in turn can affect transport across membranes. Likewise, altering by different means a toxin's three-dimensional or conformational 'flexibility' might determine whether its reactive groups can properly align themselves on the receptor site(s) of the target.

The chemical features responsible for host specificity are not well understood because the structures of almost all of the host-specific toxins are still unknown. However, we do know that not all of these toxins are structurally similar to each other, so that possibly the chemical explanations for their host specificity will vary as well. Also, the necessity for using bioassays makes it difficult to separate the chemical features determining specificity from those determining toxicity. The only information we do have concerns two of the toxins from *Helminthosporium* spp. When victorin is subjected to mild alkali, it separates into two components: a non-toxic small peptide of unknown structure and a terpene which is weakly toxic, but without host specificity. From these results it has been postulated that the peptide portion is responsible for the host specificity of victorin. The HC-toxin from *H. carbonum* race 1 loses specificity when it is reduced (Pringle, 1973). Further studies, especially with victorin and T-toxin, should soon provide additional information.

In phaseolotoxin it has been shown that the presence of the two C-terminal amino acids, homoarginine and alanine, allows inward transport in certain bacteria via an oligopeptide permease system (Staskawicz and Panopoulos, 1980). This is an interesting example of 'illicit' transport in which the presence of these amino acids allows the toxin sufficiently to mimic normal oligopeptides that it is transported by a system that otherwise would exclude it. Thus, it appears that the function of peptide portion of phaseolotoxin is selectively to facilitate transmembrane movement.

5. SYMPTOM EXPRESSION

By themselves toxins can reproduce many of the commonly observed symptoms of disease, including chlorosis, growth abnormalities, necrosis, water soaking, and wilting. However, we should not conclude that toxins are always responsible for these symptoms since each can be induced by other virulence and pathogenicity factors. As would be expected from the multiplicity of factors that can induce the same host response, different toxins can induce the same symptom. A case in point is chlorosis: phaseolotoxin induces it by interfering with amino acid synthesis (Patil, 1974), tentoxin by inhibiting photosynthetic ATP formation (Steele *et al.*, 1976), rhizobitoxine by inhibiting amino acid and/or ethylene formation (Giovanelli *et al.*, 1971; Lieberman, 1979), and tagetitoxin by interfering with some early, light-independent step in chloroplast development (Lukens and Durbin, unpublished data). Rather commonly, pathogens produce more than one toxin. Sometimes they all contribute to the expression of one symptom while in other instances they may induce different symptoms.

The complexity and number of physio-chemical events linking the toxin–target interaction to the symptom may vary widely. We can visualize this concept as follows. Assume that the length of the arrows in Figure 4 represents the complexity of the events between the primary interaction and the symptom, and that toxins can act at the tail of the arrows. In example A only one toxin can induce the symptom and there are only a few events in the intervening linkage. Example B introduces the concept that more than one toxin can affect the same pathway, but at different points. This is further complicated in example C where different toxins can act on one or both of two independent pathways leading to the same symptom. In example D, the toxin is acting at a common 'shuttle' site, for instance the alternate oxidation–reduction of NAD, which links two different pathways biochemically.

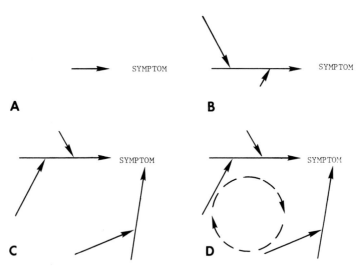

Figure 4. Diagrammatic representations of the different metabolic complexities that may exist between the toxin–target interaction(s) and the resulting symptom. The arrows represent the physico-chemical events in the host leading to the symptom. Toxins can act at the tail of the arrows

Very little is known about these physico-chemical events linking the primary interaction to a symptom (Durbin, 1982). An understanding of this linkage would have value for several reasons besides its intrinsic interest. It would represent a validation of the primary mechanism of action proposed for that toxin. By this is meant that even although a specific toxin–target interaction has already been identified, the crucial question is whether the biochemical consequences of this interaction can fully explain the effect observed on the host. Secondly, it offers the potential opportunity of obviating the symptom, and perhaps controlling the disease by altering the course of some reaction in the linkage. It also has the potential for providing new insights into the normal functioning of plants, and even of pathogenic microorganisms.

Besides visual symptoms, toxins induce ultrastructural abnormalities which ultimately may or may not lead to visual effects (Hanchey, 1981). Generally, there is some correspondence between the subcellular response and the symptom elicited. For example, chlorosis is indicative of chloroplast damage whereas water soaking is closely related to membrane damage. A study of ultrastructural abnormalities can be very informative because they are evident much earlier than visual symptoms—sometimes only a matter of minutes or seconds after treatment. Thus, they can provide valuable clues as to what the primary interaction may be.

The common ultrastructural responses to toxins include (a) the formation of appositions, blister-like areas between the cell wall and plamalemma, which contain callose, (b) the collapse of cell walls and cell plasmolysis, particularly in resistant tissues, (c) injury to the tonoplast and plasmalemma, and (d) damage to specific organelles, such as mitochondria, chloroplasts, nuclei, ribosomes, and the endoplasmic reticulum. In many instances the ultrastructural effects elicited by a toxin resemble the effects of the pathogen that produces it. This is particularly true for the host-specific toxins. However, in other cases, the two do not induce similar effects so that caution in interpretation is required. Further, the primary effects may be obscured by secondary events, for any cellular injury will set in motion a web of non-specific, degradative reactions.

6. MODE OF ACTION

Most toxins appear to have some degree of specificity for their target. They can be divided into two classes depending on whether this target is restricted to higher plants or whether it also occurs in microorganisms (Table 1). A further subdivision can be made within the former class. In the first subclass the targets are broadly, perhaps ubiquitously, distributed in plants, as for example chloroplasts. Members in the second subclass comprise the very interesting group commonly termed host-specific toxins, whose targets are confined to selected cultivars within single plant species (Wheeler and Luke,

Table 1. A partial listing of toxins classified according to the
distribution of their targets

I. Target present in higher plants and microorganisms
   Enniatins (*Fusarium* spp.)[a]
   Fomannosin (*Fomes annosus*)
   Graminin A (*Cephalosporium gramineum*)
   Helminthosporal (*Helminthosporium sativum*)
   Naphthazarines (*Fusarium* spp.)
   Phaseolotoxin (*Pseudomonas syringae* pv. *phaseolicola*)
   Syringomycin (*P. syringae* pv. *syringae*)
   Tabtoxin (*P. syringae* pathovars)

II. Target restricted to higher plants
   A. Target broadly distributed[b]
      Colletotrichins (*Colletotrichum* spp.)
      Coronatine (*Pseudomonas syringae* pv. *atropurpurea*)
      Cotylenins (*Cladosporium* sp.)
      Fusicoccins (*Fusicoccum amygdali*)
      Marasmins (*Pyrenophora teres*)
      Phyllostine (*Phyllosticta* sp.)
      Piricularin (*Pyricularia oryzae*)
      Tagetitoxin (*Pseudomonas syringae* pv. *tagetis*)
   B. Target restricted to specific species or cultivars
      AK-toxin (*Alternaria kikuchiana*)
      AM-toxin (*A. mali*)
      HC-toxin (*Helminthosporium carbonum* race 1)
      HS-toxin (*H. sacchari*)
      PC-toxin (*Periconia circinata*)
      PM-toxin (*Phyllosticta maydis*)
      T-toxin (*Helminthosporium maydis* race T)
      Tentoxin (*Alternaria alternata*)
      Victorin (*Helminthosporium victoriae*)

---

[a] Name of producing pathogen.
[b] Provisional grouping. Some toxins listed here could belong in Class
I, but as yet toxicity tests on microorganisms have been negative.

1963; Pringle and Scheffer, 1964; Scheffer, 1976). There are now 14
examples of this subclass known, all produced by the Fungi Imperfecti within
the genera *Alternaria* (6), *Corynespora* (1), *Helminthosporium* (4), *Periconia*
(1), *Phyllosticta* (1), and *Rhynchosporium* (1). Undoubtedly, other valid
examples will be found; however, to date none from bacteria have been proposed except for amylovorin from *Erwinia amylovora*, the causal agent of fire
blight of rosaceous plants. The proposal has now been shown to be invalid,
however (see Chapter 4).

   Until recently, research on many of the host-specific toxins has been seriously hampered by the unavailability of purified toxins, complete structural

analyses, and, in some cases, chemically stable preparations (Macko *et al.*, 1981). In no case do we known what their specific target is, but reports that one of the earliest effects they have is on cellular membrane systems may be prophetic (Scheffer, 1976; Daly, 1981; Hanchey, 1981). Conceivably, membranes could be their general site of action (see also Chapter 16). The individual toxins would then react with specific membrane components coded for by the genes in that cultivar conditioning susceptibility. The events following this primary interaction could all be initially different but ultimately lead to a loss of membrane function, much like example C in Figure 4. However, it is still not clear whether they have only one site of action, for several apparently diverse functions are quickly affected, among them mitochondrial respiration, electrolyte leakage, and dark CO_2 fixation.

An exact description at the molecular level of toxin–target interactions is known for only a few cases. The most extensive of these studies have been done with three bacterial toxins, produced by *P. syringae* pv. *phaseolicola* (phaseolotoxin), pv. *tabaci* (tabtoxin) and *Rhizobium japonicum* (rhizobitoxine), and three fungal toxins, produced by *Alternaria alternata* (tentoxin), *Fusicoccum amygdali* (fusicoccin), and *Helminthosporium maydis* race T (T-toxin).

Interestingly, the three bacterial toxins act by irreversibly binding to specific enzymes, thereby blocking their catalytic activity. At least with phaseolotoxin and rhizobitoxine, which have been more thoroughly studied, the interaction occurs by a mechanism termed Kcat (Giovanelli *et al.*, 1971; Rando, 1974; Kwok *et al.*, 1979). An important characteristic of this mechanism is its high degree of specificity, which results from a specific chemical activation of the toxin by its target, followed by a specific alkylation of the target by the activated toxin. It is too early to determine if this mechanism is generally involved in other bacterial toxin–target interactions.

Tabtoxin itself has no biological activity but first must be hydrolysed to yield the bioactive component, tabtoxinine-β-lactam (Figure 5A). This 'activation' is carried out by non-specific peptidases, probably of plant origin (Uchytil and Durbin, 1980). The tabtoxinine-β-lactam irreversibly binds to the enzyme glutamine synthetase (GS) (Thomas *et al.*, 1983). This enzyme is crucial for ammonia metabolism in plants and microorganisms, being central to inorganic nitrogen assimilation, the mobilization of stored nitrogen, and the reassimilation of ammonia released in photorespiration. As a consequence of GS inhibition, the ammonia builds up rapidly and causes an uncoupling of the carbon and energy fixation components of photosynthesis. In addition, the ammonia, by an unknown mechanism, selectively destroys the internal membrane system of the chloroplast.

Phaseolotoxin specifically inhibits ornithine carbamoyltransferase (OCTase), an enzyme of the ornithine (urea) cycle which converts ornithine and carbamoyl phosphate to citrulline. Kinetic analysis has shown that the

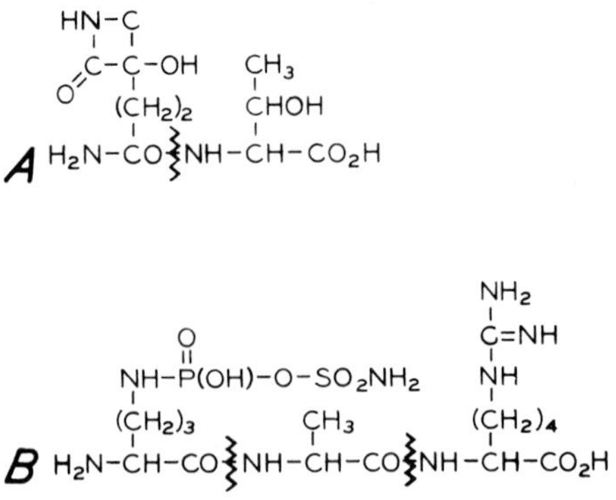

$$
\begin{array}{c}
\text{HN-C} \\
\underset{O}{\overset{}{\diagdown}}\text{C-C-OH} \quad \text{CH}_3 \\
\text{(CH}_2)_2 \quad \text{CHOH} \\
\textbf{A}\ \text{H}_2\text{N-CO}\{\text{NH-CH-CO}_2\text{H}
\end{array}
$$

$$
\begin{array}{c}
\qquad\qquad\qquad\qquad \text{NH}_2 \\
\text{O} \qquad\qquad\qquad\qquad \text{C=NH} \\
\text{NH-P(OH)-O-SO}_2\text{NH}_2 \quad \text{NH} \\
\text{(CH}_2)_3 \qquad \text{CH}_3 \qquad \text{(CH}_2)_4 \\
\textbf{B}\ \text{H}_2\text{N-CH-CO}\{\text{NH-CH-CO}\{\text{NH-CH-CO}_2\text{H}
\end{array}
$$

Figure 5. Effect of plant enzymes on tabtoxin (A) and phaseolotoxin (B). The wavy lines indicate the position of the peptide bonds that are hydrolysed. The biologically active portion of the molecules after hydrolysis is on the left side of the lines

toxin interferes with the binding of carbomoyl phosphate, but not ornithine, to the enzyme (Patil, 1974). As a result of OCTase inhibition, ornithine accumulates in the affected tissue; the carbamoyl phosphate, on the other hand, is diverted into nucleic acid synthesis (Jacques and Sung, 1978). Applications of citrulline, arginine, or orotic acid to bean leaves will counteract toxin-induced chlorosis. From this it has been postulated that the toxin causes chlorosis because of (a) an arginine deficiency which leads to a significant reduction in protein synthesis (Patil, 1974) or (b) a citrulline deficiency which, in turn, depresses pyrimidine synthesis via a reduction in carbamoyl-L-aspartate level (Rudolph, 1976). Both proposals have serious drawbacks, however, and more work is needed to clarify the effect(s) of phaseolotoxin on the carbon flux through the ornithine cycle and its related reactions together with their regulation.

Rhizobitoxine reacts with at least two different pyridoxal phosphate-linked enzymes. One, β-cystathionase, is an important enzyme in sulphur-containing amino acid biosynthesis (Giovanelli et al., 1971). The exact nature of the other enzyme is unknown except that it is involved in ethylene formation (Lieberman, 1979). At present, it is unclear whether the pathological effects of rhizobitoxine are due its inhibition of one or both of these enzymes or, alternatively, other reactions at present unknown.

Tentoxin interacts with coupling factor 1 (CF_1) (Steele et al., 1976). CF_1 is the key enzyme in the chloroplast which couples the energy generated from a transmembrane proton electrochemical gradient to ATP synthesis. The enzyme is structurally complex, being composed of multiple copies of five different kinds of protein subunits. Tentoxin tightly binds to two of them: the $\alpha-\beta$ complex. How this occurs is unknown. We do know, though, that the binding is very strong (affinity constant of $2 \times 10^8\ M^{-1}$) and that only one molecule of tentoxin is needed to bind to each CF_1 molecule. In the case of most plant species which are unaffected by tentoxin, this high-affinity binding site is lacking. From the pattern of enzyme inhibition, it appears that the interaction involves a portion of CF_1 not directly concerned with its catalytic function. Inhibition of CF_1 causes an immediate inhibition of light-driven, but not ATP-driven, protein and RNA synthesis in the chloroplast (Bennett, 1976). Soon thereafter many other energy-requiring reactions are adversely affected, and catabolic reactions then predominate.

An interesting and unusual aspect of the genetic basis of tentoxin reaction is that the structural genes which code for one or both of the α and β subunits are contained in the chloroplast itself rather than the nucleus (Burk and Durbin, 1978). This means that the response of plants to tentoxin is maternally inherited, since in almost all species the chloroplasts of an F_1 plant are derived only from the female parent. Thus, a cross of resistant $\times$ sensitive parental types produces all resistant progeny, whereas the reciprocal cross gives all sensitive progeny.

Fusicoccin causes an increase in water loss from diseased almond and peach to such an extent that wilting, desiccation, and eventual death of infected leaves and shoots ensues. This increase occurs by several means: (a) an exaggerated opening of the stomata and a loss of endogenous control of their movement so that they remain open and (b) a decrease of plasmalemma resistance to water movement resulting in increased cuticular water loss. The toxin also has auxin- and cytokinin-like activities (see also Chapter 19). It is held that all of these actions result from the stimulation by fusicoccin of a membrane-bound, K^+-activated ATPase which links K^+ influx to proton extrusion, causing a negative transmembrane potential (hyperpolarization).

Studies on the mechanism of action of T-toxin vividly illustrate the advantages of structurally defining the toxin prior to mechanistic studies: (a) pure toxin, rather than partially purified preparations with their attendant experimental problems, is available, (b) chemical properties such as stability can be ascertained, (c) toxin concentrations can be precisely defined, and (d) knowledge may be gained on how the toxin might be acting. It should be pointed out, though, that most structural analyses have provided very little insight into what site(s) or mechanism(s) of action a toxin might have.

T-toxin has been found to consist of a complex of linear polyketols of lengths C_{37}–C_{43} with varying numbers of hydroxy and ketone functions. The evidence to date supports the view that the mitochondria are the major site of action of T-toxin (Daly, 1981). However, other sites, most notably the plasmalemma and chloroplast, have been suggested, and it may be that the best criterion for defining this toxin's site(s) of action may be to determine at what toxin concentration the site is affected. Although, superficially, T-toxin appears to act by uncoupling oxidative phosphorylation in mitochondria from T, but not N, maize plants, there are still many uncertainties which need to be elucidated. For instance, oxidation of NADH and succinate, under some conditions, is stimulated and uncoupled in T but not N mitochondria. However, malate oxidation is inhibited and still coupled in both T and N mitochondria. Daly (1981) has suggested that T-toxin, as well as certain other host-specific toxins, may be acting as ionophores at several energy-transducing membrane sites in the cell. Interference with their electrochemical and proton fluxes could severely affect many aspects of cellular metabolism, and result in the range of biochemical lesions observed with these toxins.

Significant progress has also been made in understanding the modes of action of toxins synthesized by the fungi *Alternata mali* (AM-toxins), *Periconia circinata* (PC-toxin), *H. victoriae* (victorin) and *Ceratocystis ulmi* (cerato-ulmin), and the bacteria *P. syringae* pv. *tagetis* (tagetitoxin) and pv. *syringae* (syringomycin). However, since final answers are not at hand, these will not be discussed further. Additional information can be obtained by consulting the references (see especially Scheffer, 1976; Yoder, 1980; Daly, 1981).

Not all toxins are biochemically specific. Organic acids, fatty acid esters, and phenolics, for example, react non-specifically with a variety of plant constituents, resulting in rapid death of the affected cell. These so-called necrogenic toxins are produced most commonly by facultative parasites which essentially parasitize dying and dead tissues. Biotrophs, at the other end of the spectrum, most probably synthesize toxins which interact in a highly specific manner with targets whose disruption would not lead to immediate cell death. For instance, they may be concerned with subtly facilitating pathogen nutrition or cell penetration. Unfortunately, so far very little research has been carried out in this area, chiefly because of our inability to grow most of these pathogens in artificial culture.

7. RESISTANCE

The suggestion has been made from time to time that plant tissues resistant to the host-specific toxins metabolize them more rapidly than do susceptible tissues (Scheffer, 1976; Wheeler, 1981). However, there is still no body of positive evidence for any system to support this suggestion. Rather, the best interpretation of the data is that resistant tissues simply lack the target found in susceptible tissues. The host selectivity of tentoxin can be explained by this hypothesis: it binds tightly to the $\alpha-\beta$ subunit complex of CF_1 from sensitive plants but not to comparable complexes from resistant plants (Steele *et al.*, 1976). The molecular differences in the complexes determining whether binding will occur are unknown.

Another hypothesis put forward to explain resistance is that resistant tissues have a self-repair mechanism which is activated by the toxin. However, we have no conclusive evidence to support this possibility either. Finally, as stated earlier, there is some evidence that toxins suppress phytoalexin production in susceptible but not resistant plants. Again, the experimental evidence for this is still weak.

The response of different plant species to many of the non-host-specific toxins varies. Some species require comparatively much higher dosages of toxin than do others to give equivalent responses. We do not know if this differential response is a reflection of molecular differences at the site of action, or whether the explanation might involve such phenomena as a differential rate of toxin absorption, transport, and/or detoxification. In any case, many of these species-response tests are artificial, in that the toxins affect a much broader range of plant species than do the pathogens that produce them (i.e. factors other than toxins determine host specificity).

Earlier a subclass of pathogens was categorized which produce toxins that act on cellular targets common to both higher plants and microorganisms, including potentially the producers themselves (Table 1). Obviously these pathogens must have some means of protecting themselves from the toxin(s) they produce. Expressed another way, what keeps them from committing suicide? Unfortunately, very little information is available on this subject, and that which is available comes from studies on only two pathovars of the bacterium *Pseudomonas syringae:* pv. *phaseolicola* and pv. *tabaci*. In both instances several self-protection mechanisms appear to be operative. In the case of *Pseudomonas syringae* pv. *tabaci*, firstly the biosynthetic pathway to tabtoxin may act as a protective mechanism since only the tabtoxinine-β-lactam portion has biological activity (Figure 5A). Conceivably during tabtoxin synthesis, the lactam is not present in the free state, but rather is a bound intermediate. Secondly, neither tabtoxinine-β-lactam nor tabtoxin appreciably binds to or is taken up by the bacterium, suggesting that once the toxin is released, inward transport is inhibited (Durbin, unpublished data). Lastly, tabtoxin may be spatially separated in the bacterium from hydrolytic enzymes that can release tabtoxinine-β-lactam.

A similar series of mechanisms operate in *P. syringae* pv. *phaseolicola*. Here, though, the expected target in the bacterium, ornithine carbamoyltransferase, is resistant rather than being sensitive like the enzyme from the host. Interestingly, this occurs only when the bacterium is grown at temperatures permitting toxin production (Staskawicz *et al.*, 1980). At non-permissive temperatures for toxin production, the form of the enzyme produced is inhibited like that in the host. However, at these non-permissive temperatures, bacterial growth is still not inhibited by exogenous phaseolotoxin application. This indicates that one or more additional resistance mechanisms must be involved.

One of these potential mechanisms, but one for which there is no experimental evidence, assumes that phaseolotoxin is synthesized on a protein template. One of the properties of this synthetic system is that intermediates are tightly bound and not freely exchangeable. This would provide a self-protection mechanism since it would effectively keep the toxic intermediates from interacting with other cellular components. Since many of these systems are membrane-bound, it is conceivable that the final product is not released internally, but rather is immediately transported to the cell's exterior. A third possibility, for which evidence has been presented, suggests that phaseolotoxin is not readily absorbed by the cells of pv. *phaseolicola* (Staskawicz and Panopoulos, 1980). Obviously, this mechanism would be even more effective if the preceding possibility were true.

The multiple nature of the protective mechanisms is particularly fascinating, for if this pattern broadly holds, it suggests several things worthy of note. Firstly, it provides strong support for the argument that toxins do indeed have significant functions, even though we cannot always specify what they are. It seems unlikely that the genetic information required for both synthesis and multiple self-protection mechanisms would be evolutionarily conserved unless toxin production had some adaptive value for the pathogen. Secondly, it suggests that obtaining stable genetic resistance to this type of toxin would be difficult since more than one protective mechanism appears to be required. Perhaps individually these mechanisms are either 'leaky' (i.e. do not totally protect), or are subject to a high mutation rate. An interesting finding of potential significance is the connection between protection mechanisms and toxin production: in both pv. *tabaci* and pv. *phaseolicola*, Tox strains are sensitive to the toxins (Gasson, 1980; Staskawicz *et al.*, 1980). It is not known whether both toxin production and self-protection are lost simultaneously, or whether the protective mechanisms are lost later, once the Tox mutants have been isolated. In any event, such a linkage would result in a strong selection pressure against the loss of toxin production. This may account for the rare occurrence of Tox⁻ strains in natural populations.

8. METABOLISM

In general, we have tacitly assumed that toxins, once synthesized by the pathogen, are not altered before they interact with their targets. However, since most experiments have used living plants, plant parts, or complex sub-cellular components rather than simple, purified systems, we cannot be certain, without direct evidence, that the toxins are not being altered before they interact with their target. Actually, instances are known in which the host plants are capable of degrading toxins. Unfortunately, research in this area has been hampered by a lack of suitably 'tagged' toxins or highly specific and sensitive assays. Based on evolutionary arguments and the fact that toxins are secondary metabolites, one might presume. (a) that *in planta* many toxins would be unchanged, (b) that if chemical modifications do occur, they do not appreciably diminish toxicity, or (c) that if modifications do result in detoxification, they are quantitatively unimportant. Of course, the effective toxicity can be diminished physically by transport of the toxin away from the target.

The consequence of structural alterations varies from examples in which post-synthetic metabolism by the host and/or parasite is mandatory for biological activity of the toxin to those in which similar chemical modifications result in a large decrease or loss of bioactivity. The most interesting case involves tabtoxin which, as described, does not inhibit glutamine synthetase until plant peptidases cleave the peptide bond, releasing the biologically active portion, tabtoxinine-β-lactam (Figure 5A). Phaseolotoxin is also hydrolysed by plant peptidases but in this instance the parent compound is as toxic as either of its two products (Figure 5B). A third example, which probably is more generally characteristic of what happens to toxins, is the conversion of fusaric acid by plant enzymes into several, less toxic, derivatives.

The presumption that toxins are relatively unaffected by non-specific, degradative enzymes is supported by an examination of their structural features. Some of these features which could be responsible for biochemical stability include the presence of (a) D-amino acids, (b) cyclized carbon skeletons, (c) sterically hindered structures, (d) unusual and modified amino acids, (e) a high degree of structural modification, and (f) relatively few structural requirements for bioactivity. In addition, most toxins appear to be chemically stable under physiological conditions.

There also may be biological mechanisms for avoiding losses in toxicity, although at present we have no direct evidence for them. A few potential possibilities might include pathogens that exist for a major portion of their disease cycle outside of living cells (e.g. in intercellular spaces or the vascular system) where the potential for toxin degradation is less. Another spatial avoidance mechanism may be illustrated by toxins which interact with the plasmalemma. Here again, these toxins would have less opportunity to come into contact with the cytoplasm, the greatest source of degradative enzymes. Toxins which bind to their targets either irreversibly or tightly (i.e. $K_i < 1$ nM) also would not be as available to degradative enzymes as loosely bound toxins. Although no body of quantitative binding data is available, many toxins which are active at very low concentrations would probably fall into the former category. Lastly, pathogens, by rapidly altering the internal pH, may create conditions unfavourable for toxin-degrading enzymes.

9. FUTURE OUTLOOK

Research on pathogen-produced toxins has greatly accelerated, especially during the last 10–15 years. Although this increased activity has resulted in the generation of a large amount of significant information, we still do not have a full understanding of the role played by any one toxin in disease causation. As I have remarked, to do this will require 'that we go beyond just proving complicity, as difficult as even this task is to accomplish. . . . a complete understanding will demand a determination of the structure of the toxin and which portion or portions constitute the active site(s); an identification of the primary receptor or receptors in the host; a kinetic study of the toxin–target interaction; an elucidation of how this primary interaction is related to the subsequent physiological alterations in the host cell that ultimately culminate in symptom expression; and what the significance is, in a quantitative sense, of these alterations to the disease as a whole. In addition, it will be necessary to know if and how the host transforms the toxin and the manner in which the toxin interacts in a coordinated fashion with the other attack mechanisms of the pathogen' (Durbin, 1981).

It seems likely that in the relatively near future these goals will be realized in selected cases. An increasing number of researchers are now entering the field, many of whom are working as part of mutli-disciplinary teams studying the same host–parasite systems, and this, together with the recent advances in structural analysis and separation techniques, portends a bright future. As I hope is obvious from this presentation, there are many exciting and meaningful problems which need attention. Our ultimate challenge, as we become more knowledgeable about the intricacies of toxin action, will be to develop novel approaches to disease control which will either prevent the production of toxins or alleviate their effects.

REFERENCES

Babczinski, P., Matern, V., and Strobel, G. A. (1978). Serinol phosphate as an intermediate in serinol formation in sugarcane, *Plant Physiol.*, **61**, 46–49.

Baker, K. F., and Cook R. J. (1974). *Biological Control of Plant Pathogens*, Freeman, San Francisco, 433 pp.

Ballio, A. (1981). Structure–activity relationships, in *Toxins in Plant Disease* (Ed. R. D. Durbin), Academic Press, New York, pp. 295–441.

Bennett, J. (1976). Inhibition of chloroplast development by tentoxin, *Phytochemistry*, **15**, 263–265.

Burk, L. G., and Durbin, R. D. (1978). The reaction of *Nicotiana* species to tentoxin: a new technique for identifying the cytoplasmic parent, *J. Hered.*, **69**, 117–120.

Coffman, F. A. (1961). *Oats and Oat Improvement*, American Society of Agronomy, Madison, WI, 650 pp.

Comai, L., and Kosuge, T. (1980). Involvement of plasmid deoxyribonucleic acid in indoleacetic acid synthesis in *Pseudomonas savastanoi, J. Bacteriol.*, **143**, 950–957.

Daly, M. (1981). Mechanisms of action, in *Toxins in Plant Disease* (Ed. R. D. Durbin), Academic Press, New York, pp. 331–394.

Dimond, A. E., and Waggoner, P. E. (1953). On the nature and role of vivotoxins in plant disease, *Phytopathology*, **43**, 229–235.

Drew, S. W., and Demain, A. L. (1977). Effect of primary metabolites on secondary metabolism. *Annu. Rev. Microbiol.*, **31**, 343–356.

Durbin, R. D. (Ed.) (1981). *Toxins in Plant Disease*, Academic Press, New York, 515 pp.

Durbin, R. D. (1982). Sites of action of disease determinants in relation to sympton expression, in *The Physiological and Biochemical Basis of Plant Infection* (Eds. Y. Asada, W. R. Bushnell, S. Ouchi, and C. P. Vance), Japan Scientific Societies Press, Tokyo.

Durbin, R. D., and Steele, J. A. (1979). What are thou, O specificity?, in *Recognition and Specificity in Plant Host–Parasite Interactions* (Eds J. M. Daly and I. Uritani), Japan Scientific Societies Press, Tokyo, pp. 115–131.

Elwell, L. P., and Shipley, P. L. (1980). Plasmid-mediated factors associated with virulence of bacteria to animals. *Annu. Rev. Microbiol.*, **34**, 465–496.

Gasson, M. J. (1980). Indicator technique for antimetabolic toxin production by phytopathogenic species of *Pseudomonas, Appl. Envir. Microbiol.*, **39**, 25–29.

Gäumann, E. (1954). Toxins and plant diseases, *Endeavour*, **13**, 198–204.

Giovanelli, J., Owens, L. D., and Mudd, H. (1971). Mechanism of inhibition of spinach β-cystathionase by rhizobitoxine, *Biochim. Biophys. Acta*, **227**, 671–684.

Gnanamanickam, S. S., and Patil, S. S. (1976). Bacterial growth, toxin production, and levels of ornithine carbamoyltransferase in resistant and susceptible cultivars of bean inoculated with *Pseudomonas phaseolicola, Phytopathology*, **66**, 290–294.

Graniti, A. (1972). The evolution of the toxin concept in plant pathology, in *Phytotoxins in Plant Diseases* (Eds R. K. S. Wood, A. Ballio, and A. Graniti), Academic Press, London, pp. 1–18.

Hanchey, P. (1981). Ultrastructural effects, in *Toxins in Plant Disease* (Ed. R. D. Durbin), Academic Press, New York, pp. 449–475.

Horsfall, J. G., *et al.* (Eds) (1972). *Genetic Vulnerability of Major Crops, National Academy of Sciences*, Washington, DC, 307 pp.

Jacques, S. L., and Sung, Z. R. (1978). Effects of halo blight toxins on pyrimidine and arginine biosynthesis in cell culture of *Daucus carota* L, *Proc. 4th Int. Cong. Plant Pathol. Bacteria*, **2**, 657–661.

Kwok, O. C. H., Ako, H., and Patil, S. S. (1979). Inactivation of bean ornithine carbamoyltransferase by phaseotoxin: effect of phosphate, *Biochem. Biophys. Res. Commun.*, **89**, 1361–1368.

Liebermann, M. (1979). Biosynthesis and action of ethylene, *Annu. Rev. Plant. Physiol.*, **30**, 533–591.

Ludwig, R. A. (1960). Toxins, in *Plant Pathology: An Advanced Treatise* Vol. II. (Eds J. G. Horsfall, and A. E. Dimond), Academic Press, New York, pp. 315–357.

Macko, V., Goodfriend, K., Wachs, T., Renwick, J. A. A., Addin, W. and Arrigoni, D. (1981). Characterization of the host-specific toxins produced by *Helminthosporium sacchari*, the causal organism of eyespot disease of sugarcane, *Experientia*, **37**, 923–924.

Mitchell, R. E. (1981). Structure: bacterial, in *Toxins in Plant Disease* (Ed. R. D. Durbin), Academic Press, New York, pp. 259–293.

Nester, E. W., and Kosuge, T. (1981). Plasmids specifying plant hyperplasias, *Annu. Rev. Microbiol.*, **35**, 351–365.

Ownes, L. D., and Wright, D. A. (1965). Rhizobial-induced chlorosis in soybeans: isolation, production in nodules, and varietal specificity of the toxin, *Plant Physiol.*, **40**, 927–930.

Patil, S. S. (1974). Toxins produced by phytopathogenic bacteria, *Annu. Rev. Phytopathol.*, **12**, 259–279.

Patil, S. S., and Gnanamanickam, S. S. (1976). Suppression of bacterial-induced hypersensitive reaction and phytoalexin accumulation in bean by phaseotoxin, *Nature (London)*, **259**, 486–487.

Perlman, D., and Bodanszky, M. (1979). Biosynthesis of peptide antibiotics, *Annu. Rev. Biochem.*, **40**, 449–464.

Pringle, R. B. (1973). Abolishment of specific toxicity of host-specific toxin of *Helminthosporium carbonum* by electrolytic reduction, *Plant Physiol.*, **53**, 403–404.

Pringle, R. B., and Scheffer, R. P. (1964). Host-specific plant toxins, *Annu. Rev. Phytopathol.*, **2**, 133–156.

Rando, R. R. (1974). Chemistry and enzymology of Kcat inhibitors, *Science*, **185**, 320–324.

Rich, D. H. (1981). Chemical synthesis, in *Toxins in Plant Disease* (Ed. R. D. Durbin), Academic Press, New York, pp. 295–329.

Rudolph, K. (1976). Non-specific toxins, in *Physiological Plant Pathology* (Eds R. Heitfuss, and P. H. Williams), Springer-Verlag, Berlin, N.S. 4, pp. 270–315.

Sands, D. C., *et al.* (1978). Plasmids as virulence markers in *Pseudomonas syringae*, *Proc. 4th Int. Conf. Plant Pathol. Bacteria*, **1**, 39–45.

Scheffer, R. P. (1976). Host-specific toxins, in *Physiological Plant Pathology* (Eds R. Heitfuss, and P. H. Williams), Springer-Verlag, Berlin, N.S. 4, pp. 270–315.

Shaw, P. D. (1981). Production and isolation, in *Toxins in Plant Disease* (Ed. R. D. Durbin), Academic Press, New York, pp. 21–44.

Silverman, W. (1960). A toxin extract from Marquis wheat infected by race 38 of the stem rust fungus, *Phytopathology*, **50**, 130–136.

Staskawicz, B. J., and Panopoulos, N. J. (1980). Phaseolotoxin transport in *Escherichia coli* and *Salmonella typhimurium* via the oligopeptide permease, *J. Bacteriol.*, **142**, 474–479.

Staskawicz, B. J., Panopoulos, N. J., and Hoogenroad, N. J. (1980). Phaseolotoxin-insensitive ornithine carbamoyltransferase of *Pseudomonas syringae* pv. *phaseolicola*: bases for immunity to phaseolotoxin, *J. Bacteriol.*, **142**, 720–723.

Steele, J. A., Uchytil, T. F., Durbin, R. D., Bhatnagar, P., and Rich, D. (1976). Chloroplast coupling factor 1: a species-specific receptor for tentoxin, *Proc. Natl. Acad. Sci. USA*, **73**, 2245–2248.

Strobel, G. A. (1977). Bacterial phytotoxins, *Annu. Rev. Microbiol.*, **31**, 205–224.

Thomas, M., Langston, P. J., and Durbin, R. D. (1983). Inhibition of pea glutamine synthetase by tabtoxin-β-lactam produced by *Pseudomonas syringae* pv. *tabaci*, *Plant Physiol.*, in press.

Uchytil, T. F., and Durbin, R. D. (1980). Hydrolysis of tabtoxins by plant and bacterial enzymes, *Experientia*, **36**, 301–302.

Wheeler, H., and Luke, H. H. (1963). Microbial toxins in plant disease, *Annu. Rev. Microbiol.*, **17**, 223–242.

Wheeler, H. E. (1981). Role in pathogenesis, in *Toxins in Plant Disease* (Ed. R. D. Durbin), Academic Press, New York, pp. 477–494.

Woltz, S. S. (1978). Nonparasitic plant pathogens, *Annu. Rev. Phytopathol.*, **16**, 403–430.

Yoder, O. C. (1972). Host-specific toxins as determinants of successful colonization by fungi, in *Phytotoxins in Plant Diseases* (Eds R. K. S. Wood, A. Ballio, and A. Graniti), Academic Press, London, pp. 457–463.

Yoder, O. C. (1980). Toxins in pathogenesis, *Annu. Rev. Phytopathol.*, **18**, 103–129.

9

The Host–Parasite Interface and Nutrient Transfer in Biotrophic Parasitism

JOHN M. MANNERS and JOHN L. GAY

Department of Biochemistry, University of Queensland, St. Lucia, Brisbane, Queensland, Australia 4067, and Department of Pure and Applied Biology, Imperial College of Science and Technology, Prince Consort Road, London SW7 2BB, UK

1. INTRODUCTION

In biotrophic parasitism, infected cells and tissues remain alive and active, often for extensive periods. Many economically important plant pathogens establish this type of association which always entails a precise and intimate relationship of the two organisms. Studies of nutrient transport thus demand prior consideration of the interfaces between particular pairs of organisms. However, it should be recognized that contacts are effected during a sequence of stages some of which (e.g. appressoria) are more concerned with the establishment of the ultimate nutritional interface than directly with nutrition. In this chapter, attention will be concentrated on interfaces known or likely to be concerned with nutrient transfer and the other interfaces will be considered elsewhere (Chapter 13). A comprehensive review and classification of interfaces has been prepared by Bracker and Littlefield (1973).

The least specialized nutritional interface occurs between host cells and hyphae which occupy intercellular spaces as in Gramineae infected with *Claviceps* spp (ergot) or peach with *Taphrina deformans* (leaf-curl). In many biotrophic associations the interface additionally comprises haustoria which establish an intimate relationship with the cytoplasm of individual host cells. This class includes the rust fungi, downy mildews, some smuts and two genera of powdery mildews. In most powdery mildews the haustoria provide the only nutritional interface, the mycelium and sporing structures lying on the leaf or stem surface and not within the tissue. In diseases caused by members of the Plasmodiophorales and some of the Chytridiales the interface in the established infection is entirely intracellular, each fungal unit being enclosed by the host protoplast.

2. STRUCTURE AND COMPOSITION OF INTERFACES

2.1. Intercellular hyphae

Intercellular hyphae do not possess any features which, when compared with those of their saprophytic relatives, clearly indicate especial involvement in nutrient transfer. The same conclusion is reached from examination of axenic cultures of *Claviceps purpurea* (Voříšek *et al.*, 1974) and rust fungi (Coffey, 1975) and comparison with parasitic mycelia. Pores about 100 nm or less in diameter occur in the septa of the Ascomycetes and rust fungi and these presumably allow nutrient translocation but in many instances the pores are occluded (Littlefield and Heath, 1979). Nutrient reserves such as lipids and, except in Oomycetes, glycogen also, are common features of intercellular hyphae.

The walls of rust hyphae have an amorphous outer layer and it has been suggested that this may be concerned with nutrient exchange or isolation of the hyphae from host toxins (Littlefield and Heath, 1979). The position of intercellular hyphae is often suggestive of nutrient exchange. In section, the hyphae commonly show angular profiles making close contact with host cell walls. In *Taphrina deformans*, Syrop (1975) showed amorphous matrical material filling the interstices around intercellular hyphae. In ergot infections, hyphae penetrate between host cells and also replace them at the base of the ovary where a compact fungal tissue is formed (foot region) (Luttrell, 1980; Shaw and Mantle, 1980). In *Claviceps purpurea* the former author additionally claimed that hyphae develop in the xylem immediately beneath the foot and there lie adjacent to cells of the bundle sheath and phloem. Clearly these hyphae are in a highly advantageous position to obtain nutrients.

2.2. Haustoria

The term haustorium, which literally means 'drawer of water', was introduced for fungal structures by Visiani in 1851. A recent definition (Bushnell, 1972) is much more guarded on the functional role and refers to 'a specialised organ which is formed inside a living host cell as a branch of an extracellular (or intercellular) hypha or thallus, which terminates in that host cell and which probably has a role in the interchange of substances between host and fungus.' The terminal region is usually swollen into a 'body' connected to the hypha (or appressorium) by a 'neck' which is especially narrow where the plant cell wall is penetrated. In the downy mildews (Oomycetes) the haustorial body is spherical (*Albugo* spp. Figures 1 and 2) or lobed (*Peronospora* spp.) and similar shapes are formed in rust infections derived from aeciospores and urediospores. However, rust infections arising from basidiospores have filamentous, unbranched haustoria and these are sometimes referred to as intracellular hyphae. Powdery mildews (Ascomycetes) are unique in having haustoria whose surface area–volume ratios approach or exceed those of filamentous hyphae. This is achieved by large numbers of narrow extensions arising in clusters at the proximal (neck–body junction) and distal ends of the haustorial bodies. In *Erysiphe graminis* the extensions project outwards and are conspicuous but in all other genera they are curled around the body so that, by light microscopy, they appear as a granular investment (Figures 3 and 8). Powdery mildew haustoria are uninucleate and have a perforate septum, typical of Ascomycetes, in the neck. In other fungal groups the haustoria are not so discrete. In pathogenic Oomycetes each is a continuation of the coenocytic intercellular hyphal system and may contain no or several nuclei. In rust fungi, the haustorium and the intercellular haustorial mother 'cell' (which functions as an appressorium) have no septum between them and in aeciospore- and urediospore-derived infections they share the pair of dikaryotic nuclei. The cytoplasm of haustoria include the organelles characteristic of the rest of the fungus except that vacuoles are seldom formed. Mitochondria are frequently observed and are abundant in the fingers of powdery mildew haustoria. Lipid droplets are often present and glycogen occurs in some rust and powdery mildew haustoria (Figures 1 and 5).

Fine structure and other studies have shown that although haustoria enter

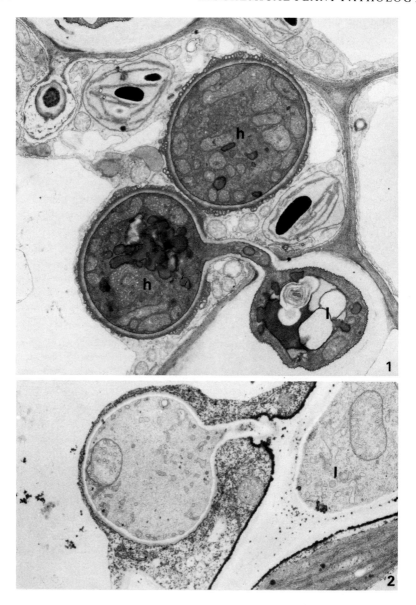

Figures 1 and 2. *Albugo candida* infections. Figure 1 shows two haustoria (h) in a mesophyll cell; the lower haustorium is connected by a slender neck to an intercellular hypha (I). The haustorial cytoplasms include lipid droplets and many mitochondria. × 10 800. Figure 2 is from material processed to show sites of ATPase activity. The plasmalemmas of the two host cells show dense deposits indicating activity where they line the walls but the extrahaustorial region and the fungal plasmalemma [in the haustorium and the intercellular hypha (I)] are inactive. × 13 500. (Reproduced by permission of A. M. Woods, Imperial College, London)

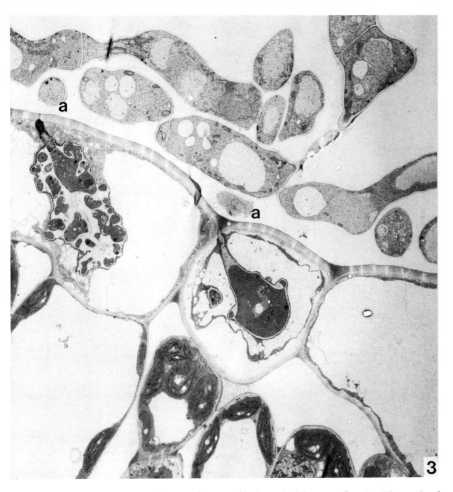

Figure 3. Section of surface tissues of pea leaf infected with *Erysiphe pisi*. The ends of mesophyll cells are shown in the lower part of the plate and two epidermal cells contain haustoria. The haustorium on the left has been cut so that the lobes are conspicuous and their number indicates that the haustorium is older than the other. Sections of superficial hyphae are shown above the epidermis and appressoria (a) are present immediately above the haustorial necks. × 2600. (Reproduced by permission of M. Martin, Imperial College, London)

individual cells they do not breach the host plasmalemma but occupy an invagination developed by its proliferation. Furthermore, material differing in structure from that of the normal plant or fungal cell wall is developed adjacent to this portion of the plasma membrane. Its structure tends to be characteristic of particular groups of fungi and a variety of names, including outer haustorial wall have been applied to the region. Because it invests the haustorium, the term 'extrahaustorial matrix' is the most appropriate and is now

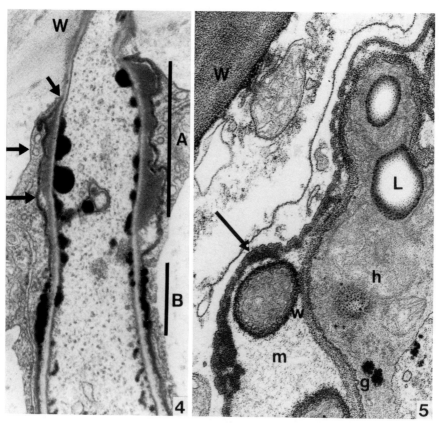

Figures 4 and 5. Sections of haustoria of *Erysiphe pisi*. Figure 4 shows the neck where it emerges from the epidermal cell wall (W) and continuity of the host plasmalemma over the haustorial neck can be traced (arrows). A and B indicate the neckbands where the host and haustorial plasmalemmas are attached to the neck wall. × 40 000 Figure 5 shows part of an haustorial body (h) near the plant cell wall (W) in a section processed to stain polysaccharides. The plant cell wall, the elaborately folded extrahaustorial membrane (arrowed), haustorial walls (w), and glycogen granules (g) are stained. Lipid droplets (L) have been partly degraded by the processing. Note the amorphous matrix (m) between the haustorial walls and the extrahaustorial membrane. × 33 000. (Reproduced by permission of M. Martin, Imperial College, London.)

generally used for this region. Few studies have been made on its nature. In pea powdery mildew, electron microscopy after periodic acid–silver treatment or enzyme digestion has indicated that it is rich in polysaccharide and specific fluorescent staining indicates that this is β-linked (Gil, 1976; Gil and Gay, 1977). It does not contain callose. Polysaccharides were also found in the extrahaustorial matrix around haustoria of *Peronospora pisi* (Hickey and Coffey, 1978) who also combined specific enzyme digestion and staining

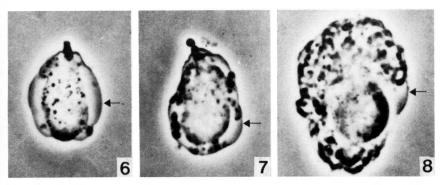

Figure 6–8. Living haustorial complexes isolated from pea infected with *E. pisi*. They are arranged in order of ascending age and are mounted in water so that the extrahaustorial membrane (arrowed) is distended and the haustorial lobes are conspicuous. × 1650. [Reproduced from Gil and Gay (1977) by permission of Academic Press Inc. (London) Ltd.]

with electron microscopy. The presence of cellulose, β1–3-glucan, and protein were indicated.

The term extrahaustorial membrane has been introduced for the region of the host plasmalemma invaginated around haustoria and there is considerable evidence of its distinctive character. The first clear indication for this was obtained by Bracker (1968), who observed that it was thicker and stronger than the wall-lining region of the plasmalemma of barley epidermis infected with *Erysiphe graminis*. Subsequently, Littlefield and Bracker (1972), investigating flax rust infections, found that the extrahaustorial region lacked both particular staining properties and structural features characteristic of normal plasma membranes. The transition occurred at an osmiophilic ring or neckband which had been seen much earlier in rust haustoria (Rice, 1927). Studies of powdery mildew of pea have done much to elucidate the significance of both features. Using methods employed in sub-cellular fractionation, Gil and Gay (1977) isolated and characterized structures which were named 'haustorial complexes'. Each comprised an haustorium with extrahaustorial matrix still enveloped in the extrahaustorial membrane (Figures 6, 7, and 8). The haustorial cytoplasm was retained by closure of the pore in the septum in the neck (Figure 9). Since the haustorial complex comprises the whole parasitic interface and can be obtained in sufficient numbers for population studies *in vitro*, its isolation has provided an essential technique for studies of *in vivo* nutrient transport in powdery mildews.

When living haustorial complexes were observed in water and aqueous solutions they swelled and shrunk, respectively (Figures 10 and 11). In the swollen condition the matrix was expanded but the extrahaustorial membrane remained firmly attached to the haustorial neck at a distinct dark neckband

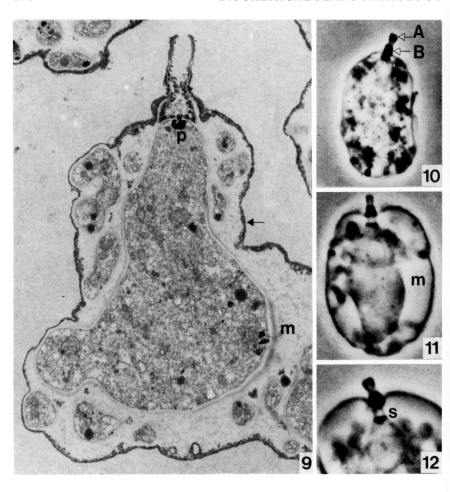

Figures 9–12. Haustorial complexes isolated from pea infected with *Erysiphe pisi*. The section (Figure 9) shows the haustorial cytoplasm retained by material (p) plugging the pore in the septum at the junction with the neck. Lobes of the haustorium project into the extrahaustorial matrix (m), which is bounded by the extrahaustorial membrane (arrowed). × 8000. (Reproduced by permission of P. T. N. Spencer-Phillips, Imperial College, London) Figure 10 shows a living complex mounted in 100 mM sucrose solution where the extrahaustorial membrane follows the contour of the haustorium. × 2000. Figure 11: mounted in water, the extrahaustorial matrix (m) expands and distends the extrahaustorial membrane, which remains attached to the neck at the B neckband. × 2000. Figure 12 shows details of the membrane attachment and the septum (s) in the neck. × 3000. [Reproduced from Gil and Gray (1977) by permission of Academic Press Inc. (London) Ltd.]

(B band) (Figure 12). A second annulus (the A band; Figures 4, 9, and 10) was also distinguished and, for both, electron microscopy demonstrated the close association of the extrahaustorial membrane and fungal plasmalemma with the surface of the neck wall. From these studies it was concluded that the invaginated portion of the host cell plasmalemma was semi-permeable, and that the attachments of plasma membranes to the haustorial neck were instrumental in preventing the escape of solutes from the extrahaustorial matrix. It would also preclude the ingress of host metabolites along the wall of the neck. Thus, the haustorial surface is isolated from the general apoplast of the leaf and it was proposed that metabolites are directed along a pathway through the host cytoplasm. Recent observation using substances which do not pass through plasma membranes has confirmed this conclusion. When epidermes from infected pea were placed in a fluorescent stilbene (P. Spencer-Phillips, personal communication), and isolated haustorial complexes were treated with uranyl acetate and examined by electron microscopy after precipitating the uranyl ion (Gay and Manners, 1983) neither substance passed the neckband region. The uranyl experiment also confirmed that the extrahaustorial membrane was semi-permeable; the matrix was not entered unless the membrane was damaged in preparation. The implications of these results are shown in Figure 13 and it will be noted that the haustorium is, in functional terms, an intracellular structure. The investigation of pea mildew (Gil and Gay, 1977) also showed the extent to which the invaginated portion of the plasmalemma differs from the region lining the cell wall. It is over twice as thick, highly convoluted, and contains polysaccharide in unusually large amounts (Figure 5). When swelling was induced, it distended slowly and did not rupture as plasma membranes normally do in hypotonic media. Even after detergent treatment it remained intact although its distension rate increased. However, the thickness of the membrane was reduced, and swelling rates substantially increased, when cell wall-degrading enzymes were applied. As in the rust infection (Littlefield and Bracker, 1972), the extrahaustorial membrane lacked the intramembrane particles in freeze-fracture preparations (Gil, 1976).

It is thus seen that in cells infected by rust and powdery mildew haustoria the plasmalemma is highly differentiated where it is adjacent to the haustorium while the rest of it remains normal. Transition occurs at the neckbands and thus these structures, in addition to controlling apoplastic diffusion, act as 'domain delimiters' which prevent the interchange of components of the two membrane domains. The position of the modified membrane indicates that differentiation is induced by the haustorium and if the domains were not so delimited lateral diffusion of membrane components would occur as after fusion of antibody-labelled animal cells where an averaged situation arises

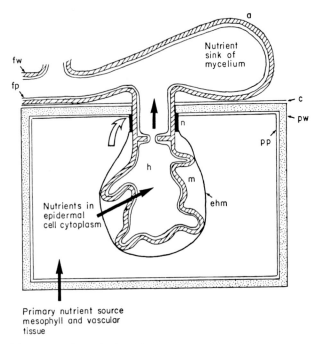

Figure 13. Diagram to show the interfaces, and to illustrate the pathways and barriers to nutrient flux, between *Erysiphe pisi* and *Pisum sativum*. The haustorium occupies an invagination (extrahaustorial membrane, ehm) of the epidermal plasmalemma (pp). When the mycelium is removed from the surface of the leaf, or the haustorium is released from the cell during isolation, the neck (n) is broken at the host cell wall (open arrow) and the extrahaustorial membrane remains attached at the neckbands (shown here as one) so that it and the enclosed matrix (m) are retained with the haustorium as a structural unit (haustorial complex). The haustorial cytoplasm (h) is retained by closure of the pore in the neck septum. The solid arrows indicate the postulated nutrient pathway. a, appressorium; c, cuticle; fp, fungal plasmalemma; fw, fungal cell wall; pw, plant cell wall. [Reproduced from Manners and Gray (1978) by permission of Academic Press Inc. (London) Ltd.]

(Frye and Edidin, 1970). It will be explained below that the polarised differentiation of infected cells is a key feature in haustorial function.

Haustorial complexes have been isolated from several powdery mildew infections including *Erysiphe graminis* (Dekhuijzen, 1966; Gil and Gay, 1977; Manners and Gay, 1977) and the distinctive structural characteristics of the invaginated domain of the host plasmalemma, and the neckbands maintaining the domains and plasmalemma attachment are common features. Also, in cow pea infected with *Uromyces phaseoli* var. *vignae* and corn with *Puccinia sorghi*, restriction of apoplastic diffusion by neckbands has been demonstrated by introducing diffusion indicators into infected tissues (Heath, 1976). The reagents did not pass the neckband to enter the extrahaustorial matrix. Silicon and possibly ferric pyrophosphate have been shown in the neckbands of a rust by Chong and Harder (1980) using X-ray micro-analysis. The former element has also been implicated in restricting diffusion of water in cell walls near infection sites (Sargent and Gay, 1977). However, in basidiospore infections neither neckbands nor differentiated extrahaustorial membranes have been reported. Nor have neckbands been recognized in downy mildew infections but there are several reports of structural transitions in the neck wall and recently clear evidence of differentiation of the extrahaustorial membrane has been obtained (Woods and Gay, in press). It is perhaps significant that the parasitic Oomycetes are considered as a series ranging from necrotrophs to biotrophs and in some biotrophy is brief (Ingram, 1981).

2.2.1. Other features of cells infected by haustoria

Most haustoria have collar-like structures around their necks. These are deposited outside the host plasmalemma. They vary greatly in size and may be completely separate from the neck wall whereas in other instances they are contiguous. The latter situation suggests that the collar developed from a papilla which commonly forms before or during host wall penetration. The collar may subsequently extend at its rim so that the whole haustorium and the extrahaustorial membrane is enveloped. Such encasement, like the papillae, may provide the host with a resistance mechanism (Heath, 1971; Heath and Heath, 1971). It is suggested that the presence of the impermeable polysaccharide callose precludes diffusion of metabolites to the haustorium.

There are many records of host nuclei lying adjacent to rust haustoria. The cytoplasm usually shows evidence of high metabolic activity, e.g. increased ribosome frequency, and often extensive endoplasmic reticulum (e.r.) is developed near haustoria. This is conspicuous adjacent to the extrahaustorial membrane of downy mildew infections and occasional continuity can be traced. Sometimes smooth e.r. (s.e.r.) is organized as complex lattices around young haustoria of *Puccinia graminis* f. sp. *tritici* and occasionally their membranes also connect with the extrahaustorial membrane (Harder *et al.*, 1978). The connections are probably transient. Similar lattice forms of s.e.r. occur in host cytoplasm near haustoria of *Erysiphe graminis* (Gay and Sargent, unpublished work) and *Peronospora pisi* (Hickey and Coffey, 1977). High activity of Golgi dictyosomes is also indicated in haustorium-infected cells and it seems possible that their vesicles fuse and augment the extrahaustorial membrane. The contents of e.r. cisternae and Golgi vesicles may thus contribute to the extrahaustorial matrix but there is no proof of this hypothesis.

2.2.2. Sites of haustoria

In rust and downy mildew infections the haustoria are commonly in the mesophyll. Thus, host cells are engaged in photosynthesis and have direct plasmodesmatal connections with each other. Some rusts, e.g. *Puccinia poarum*, also infect cells of the bundle sheath (in *Poa*) and at the pycnial–aecial stage on coltsfoot, all living cell types in the vascular tissue, including the transfer cells but excluding the sieve tubes, are additionally infected (Al-Khesraji *et al.*, 1980). The powdery mildew *Phyllactinia corylea* exclusively infects bundle sheath cells (Spencer-Phillips and Gay, in preparation) but, as noted above, most genera of the group produce haustoria which are confined to the epidermis which has few or no chloroplasts.

Thus, in many instances, haustoria infect cells which have special roles in the transport systems of their hosts.

2.3. Intracellular fungi

Plasmodia of *Plasmodiophora brassicae* are like almost all intracellular parasites in being enclosed by a structure believed to be derived from the host plasma membrane. The plasmodia have a multilayered envelope 23–24 nm thick. This is interpreted as two closely adpressed membranes of host and parasite origin respectively (Williams and McNabola, 1970). No structural counterparts of walls or matrix associated with haustoria are present. The envelope is a stable structure remaining intact after the isolation of plasmodia which may thus be considered equivalent to haustorial complexes. Further evidence inviting comparison has been obtained from freeze–etch studies by Aist (1974) who showed that particles were absent from the outer membrane of the envelope of young stages but occurred, as expected, in other membranes of the host and parasite.

Olpidium brassicae is exceptional among eukaryotic parasites in having a single plasma membrane between the cytoplasms of host and parasite (Bracker and Littlefield, 1973).

3. TRANSPORT AT INTRACELLULAR OR INTERCELLULAR INTERFACES?

Because haustoria and intercellular hyphae are intimately associated with host cells it has long been assumed that they are instrumental in nutrient absorption. Experimental evidence has now been obtained to confirm this role for powdery mildew haustoria whilst the dependence of rusts, downy mildew, and smut fungi on haustoria or extracellular hyphae for nutrient uptake remains obscure. Two approaches have been successful in assessing the role of intracellular and intercellular fungal structures. These are studies of the onset of nutrient transfer during primary infection and direct sampling of isotopes in haustoria.

3.1. Initiation of nutrient transfer

In an elegant series of experiments on the primary infection processes of cereal powdery mildew fungi, Ellingboe and co-workers (Mount and Elling-boe, 1969; Slesinski and Ellingboe, 1971) have shown that the transfer of ^{32}P- and ^{35}S-labelled compounds from host to pathogen depended on the successful establishment of the primary haustorium. Experiments on the transfer of the dye acridine orange between epidermal strips of barley and spores of the powdery mildew fungus led Kunoh and Ishizaki (1981) to suggest that some exchange may be mediated at a much earlier stage by the non-appressorial germ tube formed 2–3 h after inoculation. Communication prior to the inception of haustoria is also indicated by the double inoculation experiments by Ouchi et al., (1979) but in the absence of evidence for the identification of moving nutrients at these stages of infection it seems likely that transfer is limited to informational molecules.

Experiments with downy mildews and rusts have indicated less dependence on haustoria. Using autoradiography to locate assimilated tritiated isotopes, Andrews (1975) showed that *Bremia lactucae* infecting lettuce cotyledons could utilize glucose prior to penetration and throughout its growth, whilst utilization of leucine was delayed until haustoria were established. Similarly in crown rust-infected oat, label from ^{14}CO$_2$ or [^{3}H]glucose was incorporated into fungal structures prior to haustorium formation whilst tritiated pyrimidine nucleosides were utilized subsequently (Onoe et al., 1973). Both sets of experiments suggest that in these pathogens, the haustorial interfaces have specialized transport processes.

3.2. Detection of isotopes in haustoria

Investigations using high-resolution autoradiography have shown that radioisotopes initially fed to the host are incorporated into both haustoria and mycelia of rust fungi (Onoe et al., 1973; Manocha, 1975; Mendgen, 1977) and downy mildews (Andrews, 1975; Takahashi et al., 1977). However, it cannot be concluded from these studies that the isotope in the haustorium entered directly from the host or whether entry was indirect through intercellular hyphae.

A direct demonstration of nutrient absorption by haustoria of the powdery mildew fungus *Erysiphe pisi* has been achieved by extracting haustorial complexes from infected leaves photosynthesizing in the presence of ^{14}CO$_2$ (Manners and Gay, 1978, 1980a). In these experiments isotope readily entered the complexes even when the mycelium was detached prior to ^{14}CO$_2$ feeding.

Proof that isotope had entered the haustorium, and was not restricted to the extrahaustorial matrix, was obtained by collecting $^{14}CO_2$ respired during a subsequent *in vitro* incubation of labelled fractions. The distribution of label assimilated by the complexes has been further explored by high-resolution autoradiography (Spencer-Phillips and Gay, 1980). After exposure of infected leaves to $^{14}CO_2$ for 2 h, approximately 70% of the insoluble label in the complexes had been incorporated into the haustorial body and lobes and 30% into the extrahaustorial matrix and membrane.

Rigorous control experiments have shown that between 80 and 90% of the radioactivity in haustorial complex fractions isolated in these short-term $^{14}CO_2$-feeding experiments was present in the complexes and contaminants of host origin were virtually unlabelled (Manners and Gay, 1978, 1980a, 1982b; Spencer-Phillips and Gay, 1980). Powdery mildew-infected pea thus represents a model system for biochemical studies of biotrophic transport. The structure of the interface which mediates transport is clearly defined and has been extensively documented, and direct techniques are available for its biochemical analysis and *in vitro* studies of its physiological properties.

4. TRANSPORT AND METABOLISM OF PHOTOSYNTHATES

So far biochemical investigations of the transfer of photosynthates to the pathogen have concentrated on host sugars and amino acids.

4.1. Sugars

Numerous biochemical investigations of the nutrition of various biotrophs have emphasized the role of carbohydrates which appear to be the major translocates from host to fungus (Smith *et al.*, 1969; Scott, 1972; Lewis, 1976; Bushnell and Gay, 1978) and several distinctive features of the transport of these compounds have been resolved. Firstly, movement of carbohydrate from unaffected areas is stimulated and sometimes this is accompanied by an increase in polysaccharide synthesis and accumulation of host sugars in and around infection sites. Imported carbohydrates, together with those produced by photosynthesis in infected cells, are then released, absorbed by the fungus, and subsequently used for growth, or are converted to fungal storage products. Of these processes, it is those of nutrient release from host cells and uptake by the fungus which are most pertinent to the functions of the host–parasite interface. Other aspects of carbohydrate behaviour in diseased plants have been reviewed elsewhere (Smith *et al.*, 1969; Lewis, 1976; Bushnell and Gay, 1978).

When infected leaves are offered $^{14}CO_2$ in the light, a considerable proportion of the photoassimilates become converted to soluble carbohydrates which are characteristic of fungi; these are usually sugar alcohols or trehalose

(Smith *et al.*, 1969; Lewis, 1976; Bushnell and Gay, 1978). The sugars sucrose, glucose, and fructose are the major products of photosynthesis in most higher plants and a large flux of at least one of them to the parasite probably occurs. For most biotrophic associations between higher plants and fungi studied so far, experimental evidence indicates that sucrose is released from host cells at the interface and this is consistent with sucrose being the compound most commonly transported between cells of higher plants (Geiger, 1975).

4.1.1. *Powdery mildews*

Research on carbohydrate transfer in powdery mildew of pea and barley has indicated that the mechanism of sucrose assimilation by the fungus is unlike that of most other biotrophs. Efflux of sucrose from infected epidermal cells at the extrahaustorial membrane has been demonstrated directly by analysis of haustorial complexes isolated from powdery mildewed pea leaves in $^{14}CO_2$ pulse-chase labelling experiments (Manners and Gay, 1982b). The results of these experiments are summarized in Table 1. After 30 min exposure of infected leaves to $^{14}CO_2$, sucrose and glycerol were the predominantly labelled soluble carbohydrates in the isolated complexes. As noted above, sucrose is a characteristic host compound and thus was probably transported directly across the extrahaustorial membrane into the extrahaustorial matrix. In host tissues, the quantity of label in glucose and fructose exceeded the label in sucrose, unlike that in the haustorial complexes. This suggests specificity in the movement of sugars from the photosynthesizing cells to the haustorium.

Table 1. Distribution of ^{14}C in sugars and polyols of the superficial mycelium and haustorial complexes isolated from powdery mildew-infected pea leaves after a 30-min 'pulse' exposure to $^{14}CO_2$ or after a similar pulse of $^{14}CO_2$ followed by a 20-h chase in $^{12}CO_2$; both treatments were in the light. Radioactivity is expressed as dpm in fungal structures from 100 mm^2 of leaf

	Mycelium		Haustorial complexes	
	Pulse	Chase	Pulse	Chase
Sucrose	650	Trace[a]	2.023	0.501
Glucose	370	Trace	0.646	1.054
Fructose	300	Trace	0.416	0.365
Mannitol	12 100	376	1.037	4.836
Arabitol	160	100	0.204	0.4845
Erythritol	30	140	—	—
Glycerol	220	Trace	1.572	0.272
Others	980	115	1.233	3.188

[a]Trace: <5 dpm.

Alternatively, the transit or rate of metabolism of hexoses in haustorial complexes occurred more rapidly than in the host. Thus, although the results do not preclude the transport of hexoses they do demonstrate a flux of sucrose at the interface.

Two lines of evidence indicate that sucrose is absorbed intact and translocated to the mycelium by powdery mildew haustoria. The first is the kinetics of labelling of metabolites in the mycelium during $^{14}CO_2$-feeding of infected leaves. This has been described in detail for barley powdery mildew, where labelled sucrose accumulated initially in the mycelium and the proportion of total radioactivity contained in the compound eventually declined indicating its assimilation into fungal metabolites (Edwards and Allen, 1966). Similarly, labelled sucrose was detected in the mycelium of *Erysiphe pisi* (Table 1; Manners and Gay, 1982b) immediately after the $^{14}CO_2$ pulse and its isotopic content declined with chase indicating metabolism in the mycelium. A second line of evidence for the location of sucrose catabolism has emerged from studies of invertase in powdery mildew-infected pea leaves (Manners, 1979). When the mycelia or haustorial complexes were homogenized in distilled water almost all of the invertase activity was solubilized, thus suggesting an intracellular location for this enzyme. The pH optimum differed markedly from that of the host invertase. Calculation of the invertase activity per unit area of leaf showed that the mycelium contained 25 times more activity than the haustorial complexes and thus was the most probable site of sucrose hydrolysis. No significant increases were detected in host (leaf minus mycelium) invertase.

In the $^{14}CO_2$ feeding experiments of Manners and Gay (1982b), the labelling pattern of a compound tentatively identified as glycerol was similar to that of sucrose. It was suggested that although this compound may be a translocate at the interface it was more likely to be an intermediate of fungal metabolism. Glycerol may be involved in the synthesis of lipids which constituted a major sink for photosynthates in the haustorial complexes. Glycerol was detected in only minor quantities in the host and thus if it were transported into the complexes it would necessitate specialized processes induced at close proximity to the haustorial interface. In several symbiotic associations between algae and fungi or animals it has been shown that a carbohydrate not normally abundant in the alga is often released to the biotrophic partner (Smith *et al.*, 1969). Thus in future experiments the possibility that glycerol is transported across the extrahaustorial membrane of powdery mildew-infected pea also should be entertained.

In the mycelia of *Erysiphe pisi* and *E. graminis* f. sp. *hordei* most of the sucrose and possibly other carbohydrates assimilated by the fungus were rapidly converted to mannitol which acts as a primary sink for photosynthates. In *E. pisi* the quantity of label in mannitol declined during the chase period and isotope accumulated in glycogen, cell walls, and other macromolecules. Thus mannitol is probably translocated to spores which contain glycogen as a major reserve (Martin and Gay, in press) and other sites of growth where it is reutilized. This sequence of transformation is likely to play a major role in maintaining a suitable concentration gradient for continued nutrient flux from host to parasite at the interface (Smith et al., 1969). The importance of the metabolic sink of the mycelium for transport of the host–haustorial interface of E. pisi and Pisum sativum has been demonstrated directly by Manners and Gay (1978). The rate of transport of ^{14}C-labelled photosynthates across the extrahaustorial membrane in the absence of the mycelium was only 0.2% of that when the mycelium was left attached. Manners and Gay (1982b) also recognized an additional sink for ^{14}C-labelled photosynthates in a glucose-rich polysaccharide at the host–parasite interface and this is consistent with the abundance of polysaccharides detected in haustorial complexes by cytochemical methods (Gil, 1976; Gil and Gay, 1977) and above (Section 2.2).

4.1.2. Other pathogenic biotrophic fungi

Methods available for probing the nature and metabolism of specific translocates at the interface of other biotrophic fungi are less direct than those described for powdery mildews. Most extensively investigated is the association of Puccinia poarum with coltsfoot (Tussilago farfara). This is a good model system as intercellular hyphae are compact and well developed and further, as described above, the haustoria in these basidiospore-derived infections lack neckbands. Thus, the fungus may obtain its nutrients entirely from the leaf apoplastic fluids. Because of these considerations, it has been possible to apply to this system the 'inhibition technique' originally devised by Drew and Smith (1967) for identification of mobile carbohydrate in lichens. The technique is essentially one of isotopic dilution and involves labelling host photosynthates by feeding $^{14}CO_2$ and simultaneously or subsequently an unlabelled form of the putative mobile carbohydrate is added to the ambient medium. This results in leakage of radioactive compounds from the infected tissue into the medium. Results have suggested that sucrose is released from infected cells but in this instance it is hydrolysed in the apoplast prior to uptake by the fungus (Smith et al., 1969; Lewis, 1976). Application of the inhibition technique to uredial infections has yielded similar results (So and Thrower, 1976). However, because neckbands are present on the necks of the haustoria in this association the results are unlikely to indicate processes at the host-haustorial interface.

An experimental system analogous to that in Puccinia poarum–Tussilago farfara occurs in rye ovaries infected with the ergot fungus which is entirely intercellular. Dickerson et al. (1976) noted that sucrose was the first compound to become labelled in fresh honeydew exuded from infected ovaries after exposure to $^{14}CO_2$. The composition of freshly exuded honeydew presumably closely resembles that of the intercellular spaces as the two are in a continuous aqueous phase. Again, hydrolysis of sucrose in the intercellular spaces seemed to occur as the proportion of total label in sucrose declined as that in glucose and fructose increased. Significantly, the specific activity of sucrose was greater than glucose and fructose in freshly exuded honeydew, indicating that sucrose was the precursor of the hexoses.

It is well documented for saprophytic fungi that some invertase is either cell wall bound or secreted to the medium (Sutton and Lampen, 1962; Maruyama *et al.*, 1979) and a similar invertase location has been demonstrated for ergot, smut, and rust fungi when growing in axenic culture (Bassett *et al.*, 1972; Callow *et al.*, 1980; Hankin and McIntyre, 1980; Maclean, 1982). These results imply that in these fungi sucrose hydrolysis occurs before uptake. That this is of significance to these fungi growing parasitically is further indicated by large increases in invertase activity associated with the infection of tissues (Lunderstadt, 1966; Long *et al.*, 1975; Callow *et al.*, 1980).

The origin of the additional invertase activity has been investigated in maize smut infections. The increase in neoplastic infections of maize leaf sheath tissue is associated with the appearance of new invertase isozymes (Callow *et al.*, 1980). The major new isozyme was distinguished from those in uninoculated controls by gel filtration and gel electrophoresis but it showed a similar molecular weight, although different electrophoretic mobility to one species of fungal invertase and these authors tentatively suggested that the new isozyme may be of fungal origin. A different conclusion was drawn by Billett *et al.* (1977) using smut-infected leaf bases of more advanced maize seedlings. On the basis of electrophoretic mobility alone these authors proposed that a new invertase of host origin was produced in these tissues. Application of immunological techniques are now required to resolve this problem. Similarly, some attempt should be made to locate this enzyme at the ultrastructural level in smut, rust, and downy mildew infections and again in the absence of a cytochemical stain for invertase activity an immunological approach would be rewarding. Invertase is present in the honeydew exuded from ergot-infected rye ovaries and this presumably is also at the parasitic interface (Dickerson *et al.*, 1976). Honeydew invertase is of the transferase type producing oligosaccharides in addition to hexoses and, since *Claviceps purpurea* has an enzyme of similar catalytic properties (Dickerson, 1972), fungal enzymes are probably involved. In each of the biotrophic interactions described above hydrolysis of sucrose in the apoplast separating host and parasite would be a primary step in maintaining a concentration gradient promoting further sucrose efflux from host cells.

Increased invertase activity in tissues infected with biotrophs is not universal (Hewitt and Ayres, 1976; Mitchell and Cooke, 1976; Roberts and Mitchell, 1979). It is probable in these instances that carbohydrates other than sucrose are mobilized or that pre-existing invertase levels are adequate. The dependence of increased invertase on high infection densities in wheat stem rust and crown rust of oat (Mitchell *et al.*, 1978) suggests that increased enzyme activity is related to the pathogen's nutritional demands.

Rye ovaries infected with the ergot fungus contain another enzyme concerned with carbohydrate transfer. The enzyme, β-1–3-glucanase, is thought to be secreted by the fungus to prevent callose deposition in the phloem and thus maintains the flow of nutrients to the infected tissue (Dickerson *et al.*, 1978). Callose degradation presumably occurs in many instances when biotrophic pathogens penetrate host cell wall oppositions (papillae) formed during infection but the possibility that fungal β-1–3-glucanases have a role in preventing further callose deposition which would then prevent nutrient transfer has not been considered for such interactions. However, it was noted above (Section 2.2) that callose has not been reported in extrahaustorial matrices and its absence may be significant in this context.

4.2. Amino acids

The transfer of amino acids between host and parasite has not been investigated in the diverse range of biotrophic associations examined for carbohydrate transfer. Rust fungi have received most attention and some important aspects have been elucidated. Firstly, studies in axenic culture have indicated a requirement for sulphur-containing amino acids for growth and presumably these are supplied by the host to the fungus growing parasitically (Howes and Scott, 1972; Maclean, 1982).

Transfer of amino acids at the parasitic interface has been inferred from the ^{3}H : ^{14}C ratios present in alanine isolated from stem rust spores and uninfected areas of the same wheat leaf after offering to it [1-^{3}H]glucose and [6-^{14}C]alanine simultaneously (Reisener *et al.*, 1970). The results indicated that the fungus absorbs this amino acid intact and carries out some synthesis of alanine from host sugars. Comparisons of the specific activities of spore and host alanine, lysine, arginine, glutamic acid, or glycine after offering each ^{14}C-labelled amino acid to infected leaves indicated that quantitatively only a small proportion of these amino acids were synthesized by the fungus (Jager and Reisener, 1969). However, the results of these authors also showed that glutamic acid, alanine, and glycine had been metabolized prior to incorporation into spores and thus the form in which their carbon skeletons were transferred to the fungus could not be positively identified.

The 'inhibition technique' described above for study of the transfer of carbohydrates to the fungus was adapted by Burrell and Lewis (1977) to investigate the nature of mobile amino acids in rust-infected coltsfoot. The results indicated that alanine and serine were absorbed from the host more readily than glutamic acid, glutamine, or aspartic acid.

The possibility that amino acids undergo rapid interconversions after uptake as observed for host carbohydrates (Smith *et al.*, 1969; Manners and Gay, 1982b; Manners *et al.*, 1982) has yet to be explored and continued transport of these compounds may rely entirely on active transport mechanisms (see below) at the fungal plasmalemma.

5. MECHANISMS OF TRANSPORT

5.1. Haustoria

It is instructive to consider first transport mechanisms at the host–haustorial interface of powdery mildews where so much is currently known of the structural and physiological properties.

A common feature of biotrophy is that the transport of photoassimilates from autotroph to heterotroph is both rapid and extensive (Smith *et al.*, 1969; Smith, 1974, 1975). Powdery mildews are no exception, and in $^{14}CO_2$-feeding experiments 20–30% of fixed $^{14}CO_2$ was transferred to the fungus with little lag (Edwards and Allen, 1966; Manners and Gay, 1978; Manners, 1979). This represents approximately half of the fixed carbon normally exported from an uninfected leaf (Geiger and Fondy, 1979) and thus transport mechanisms which maintain this flux have an efficiency which rivals that of the plant's own translocation system.

In powdery mildews, photoassimilates transferred to the fungus are directed through infected epidermal cells (Figure 3). In infected pea leaves, frequent plasmodesmata connect adjacent epidermal cells but epidermal–mesophyll connections are rare (Bushnell and Gay, 1978) and thus solute transfer to the epidermis is analogous to the processes of phloem-loading which are thought to be mediated via the apoplast (Geiger, 1975; Giaquinta, 1976). The concentration of solutes in the apoplast is uncertain because no comprehensive study has yet been made of cell permeability in powdery mildew-infected leaves. Experiments monitoring cell leakage has indicated increased permeability to ions and soluble nitrogenous compounds (Ayres, 1977; Bushnell and Gay, 1978) whereas studies of ^{14}C-labelled photosynthates suggested a slight decrease in apoplastic concentrations (Manners, 1979). As transfer via the apoplast is involved in both phloem loading and transfer to infected cells, an overall increase in cell permeability in infected leaves is probably not essential for solute flux to the pathogen. A mechanism involving redirection of solutes seems more likely and this is supported by studies of translocation from infected leaves (Bushnell and Gay, 1978).

It was explained above (Section 2.2, Figure 13) that the extrahaustorial matrix is isolated from the leaf apoplast by the haustorial neckbands and thus the infected cell is functionally polarized with a net solute efflux occurring across the extrahaustorial membrane and a net influx at the uninvaginated region of the plasma membrane. These structural features of infected cells are an integral part of the transport system and recent research has thus concentrated on elucidating physiological properties of the individual regions of the plasma membrane of infected cells and a clear picture of how transport to the pathogen is maintained is now emerging.

Evidence that the *in vivo* transport of sugars across the extrahaustorial membrane is carrier-mediated is indicated from estimates of fluxes of labelled photoassimilates at the interface (Manners, 1979). From the rate of ^{14}C accumulation in the superficial mycelium during $^{14}CO_2$-feeding experiments, the frequency of haustoria and the mean area of the extrahaustorial membrane enclosing each haustorium, Manners (1979) calculated the flux of labelled assimilates across this membrane to be 2.4 ng $cm^{-2}s^{-1}$. This rate must be considered as minimal as it does not allow for any reduction of the specific activity of labelled assimilates by unlabelled compounds already present in the leaf and photosynthetic rates were not optimal. Nonetheless, the value obtained is several orders of magnitude greater than the passive effluxes of sucrose from plant cells down large concentration gradients (Edelman, *et al.*, 1971). It is similar to the ATP-coupled sucrose influx for maize scutellum (Humphreys, 1978), and is within one order of magnitude of the sucrose flux in sieve-tube loading [18 ng $cm^{-2}s^{-1}$ (Sovonick *et al.*, 1974)], and of glucose influx calculated for *Neurospora* [16.2 ng $cm^{-2}s^{-1}$ (Jennings, 1976)]. Manners thus proposed that transport across the interface was limited by an active or facilitated process.

The possibility that the extrahaustorial membrane is freely permeable to small molecules, as is the outer membrane of the mitochondrion and chloroplast envelope (Heldt and Sauer, 1971), can be discounted because uranyl ions do not enter isolated haustorial complexes unless the membrane is broken or treated with detergents or polysaccharide-degrading enzymes (Gay and Manners, 1983). Additionally, the complexes swell in water and shrink in sucrose solutions (Gil and Gay, 1977) and labelled sugars, amino acids, and other small hydrophilic molecules are either excluded or enter isolated complexes very slowly (Manners, in Bushnell and Gay, 1978; Manners and Gay, 1980b, 1982a). Isolated complexes cannot accumulate sucrose against a concentration gradient and although glucose enters slightly faster, passive processes are indicated because uptake is insensitive to 2,4-dinitrophenol and pH, and is proportional to the external glucose concentration. Neither sucrose nor glucose at concentrations up to 400 mM had any effect on the respiration rate of isolated complexes. These results indicate that isolation of haustorial complexes from the host impairs their ability to take up solutes. Unfortunately, a large proportion of the isolated complexes have damaged membranes (Gay and Manners, 1983) and it is likely that severance of the metabolic sink provided by the mycelium and loss of certain membrane constituents by cold osmotic shock and glycerol treatment (Rubinstein *et al.*, 1977) during the isolation have adverse effects. Thus, future *in vitro* characterization of inter-face transport processes will require the isolation of haustorial complexes of greater integrity and physiological activity.

It may be argued alternatively that the impairment of uptake is due to removal of the host cell rather than damage to the complex itself (Bushnell and Gay, 1978). Recent studies of the distribution of ATPase activity in membranes of infected epidermal cells of pea (Spencer-Phillips and Gay, 1981) have added greatly to our understanding of transport mechanisms and membrane function. The region of the plasmalemma lining the epidermal wall showed the high level of ATPase activity characteristic of these cells but the extrahaustorial membrane was completely devoid of any phosphatase activity. The transition occurred precisely where the invagination began and coincided with the edge of the A neckband described above (Figure 14). Thus, the two structural domains recognized in the plasmalemma by Gil and Gay (1977) are enzymatically and, therefore, functionally distinct. The haustorial plasmalemma showed ATPase and other phosphatase activity.

This information made it possible to propose models for the mechanism of assimilate transport at the interface. Entry into the epidermal cells is pre-sumed to be almost exclusively by transmembrane import, and the high ATPase activity (similar to that of phloem transfer cells in the same leaf) shown by the wall-lining section of the plasmalemma suggests that the epid-ermal cells scavenge solutes from the apoplast, thereby increasing the con-centration of solutes within their cytoplasms. The ATPase activity of the haustorial plasmalemma is assumed to deplete the solutes in the extrahaustor-ial matrix simultaneously and thus, a high concentration gradient is main-tained across the extrahaustorial membrane so that efflux from the host cell may then proceed by facilitated diffusion (Fig. 15).

This hypothesis is based on current theory that plant and fungal plasma membrane ATPases activate nutrient transport by maintaining an electro-chemical gradient across their plasma membranes by extrusion of protons or other ions (Baker, 1978). As shown for bacteria, the influx of organic compounds can then be coupled to the equilibration of this gradient by a carrier-mediated ion symport or antiport mechanism (West, 1980). In some instances, application of uncouplers of these electrochemical gradients to bacteria or higher plant cells reverses transport processes and leads to a net solute efflux (Humphreys, 1977). Similarly, it seems likely that the absence of ATPase on the extrahaustorial membrane means that normal transport polar-ity cannot be maintained on this region of the plasmamembrane.

Spencer-Phillips and Gay (1981) also pointed out that the two membranes with ATPase activity have the same polarity with regard to both their cyto-plasms and the direction of transport, and therefore they further suggested that the ionic products of the ATPase pumps may complement one another, interacting at the intervening extrahaustorial membrane to activate the efflux of assimilates through it.

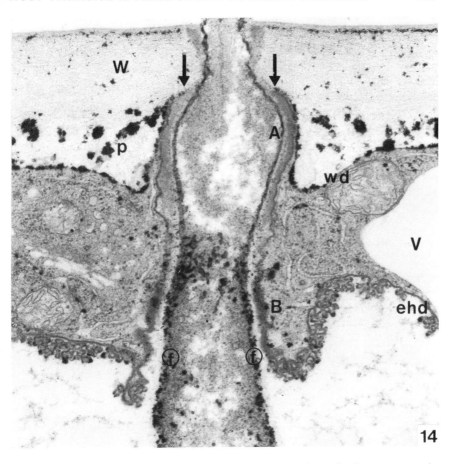

Figure 14. Section through haustorium of *Erysiphe pisi* where the fungus enters the epidermal cell. The material has been processed to show ATPase activity, which is indicated by the dense deposits over the fungal plasma membrane (f) and the plant plasma membrane (wd) where it lines the cell wall. No activity above control level is shown by the invaginated domain of the plant membrane (ehd, extrahaustorial membrane) and the transition is arrowed. Deposits in the region (p, papilla) confluent with the cell wall (W) represent the positions of sequestered membranes. A and B are the neckbands where the invagination of the plasmalemma is sealed to the neck wall. V = host vacuole. × 37 500. (Reproduced from Spencer-Phillips and Gay, 1981, by permission of the trustees of the *New Phytologist*)

Sugar efflux from mesophyll cells of wheat and tobacco is probably mediated by a K^+ symport mechanism (Huber and Moreland, 1981) and the possibility of an externally generated gradient of this ion at the extrahaustorial membrane should be considered. Fungi can generate high membrane potentials (-200 mV for *Neurospora crassa* (Slayman, 1965) which are generally in excess of those measured for plant cells (Higinbotham, 1973;

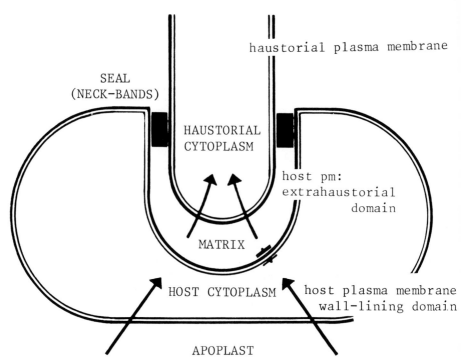

Figure 15. Diagram to show the proposed mechanisms of solute transport through the plasma membranes of pea epidermal cells and infecting haustoria of *Erysiphe pisi*. The arrows indicate ATPase-activated influxes through the wall lining domain of the host and the haustorial plasma membranes. It is proposed that, because the matrix is maintained as a closed compartment by the seal between host and haustorial plasma membranes, efflux through the invaginated domain of the host plasma membrane is effected by either a high concentration difference of the transported species (left) or interaction of complementary ionic products of electrogenic pumps at the ATPase active membranes (right). The seal also maintains the segregation of the two domains of the host plasma membrane. (Reproduced from Spencer-Phillips and Gay, 1981, by permission of the trustees of the *New Phytologist*)

Rubinstein, 1978) and it is thus possible that a continuous electrochemical gradient may exist from the host apoplast through the infected cell to the haustorial cytoplasm. This is clearly a challenging area for future electrophysiological research.

The essential features of these proposals for transport into haustoria arise from the observation of two unique circumstances, viz. the functional differentiation of two domains of one plasmalemma, and the interpolation of a sealed compartment between particular regions of the plasmalemmas of two cells. Active influx occurs into both cells and thus the plant cell is presumed to cooperate with the fungus to maintain the flux. Such cooperation would explain the observations that the integrity of the host cell is essential for hyphal growth by *E. graminis* (Sullivan *et al.*, 1974). Thus, it seems that by means of the extreme structural and functional modification of the host plasma-lemma in the region which surrounds the haustorium, the fungus controls efflux from the host to its own nutritional advantage. This is probably a key factor in the parasitism.

At present there is no evidence to indicate whether ATPase is absent from the invaginated region of the plasma membrane or whether enzyme activity is inhibited. In freeze-fracture studies the absence of 10-nm particles from this region suggests that it may be the former. Whichever is true, it seems certain that, as recognised by Gil and Gay (1977), the A band maintains the distinctive characteristics of the two domains.

Other features are likely to be conducive to higher rates of transport. Because powdery mildew haustoria have long finger-like lobes, the haustorial plasmalemma is extensive and in rose mildew it has been calculated to be twice the area of the extrahaustorial membrane (Perera and Gay, 1976). Autoradiography suggests that the relative rate of assimilate absorption by the lobes, is twice that of the haustorial body (Spencer-Phillips and Gay, 1980). Additionally, the rapid conversion of solutes in the fungus would also maintain concentration gradients. As lipophilic molecules can traverse cell membranes relatively freely it seems unlikely that their transport is related to the ATPase activities discussed above. It is, however, highly probably that lipid precursors traverse the interface of powdery mildew-infected pea, as evidenced by the rapid synthesis of triglycerides in the haustorium (Manners and Gay, 1982b). Further, isolated haustorial complexes accumulate long-chain fatty acids (lauric and linoleic) and other lipophilic compounds against very large concentration gradients (Manners and Gay, 1982a) and it is probable that these are partitioned into the haustorial lipids. Coupled with esterification by glycerol, which is also present in the complexes, this would maintain a steep concentration gradient of fatty acids towards the lipids in the haustorial cytoplasm. Linoleic and other fatty acids occur in superficial cells of plant tissues and are probably cuticle precursors (Kolattukudy, 1977). Lipid droplets are a common feature of biotrophic fungi growing on their hosts (Brennan and Lösel, 1978) and the mechanism outlined above may also be common.

The haustoria formed by fungal pathogens of other taxonomic groups are usually in mesophyll cells and the additional presence of intercellular hyphae makes it difficult to assess their absorptive function. In rust fungi the haustorial interface shares several features with that in powdery mildews (Section 2.2); the invaginated region of the plasma membrane is attached to the surface of the neck and, thus as in *E. pisi*, apoplastic transport along the fungal wall in this region is precluded (Heath, 1976). The invagination lacks the staining and freeze-fracture characteristics of the wall-lining region (Littlefield and Bracker, 1972) and also, as in *E. pisi*, ATPase activity is absent from it in french bean infected with *Uromyces appendiculatus* (Spencer-Phillips and Gay, 1981). Absence of ATPase from the extrahaustorial membrane has also been found in an Oomycete infection (Woods and Gay, in press) (Figure 2). In these mesophyll infections the ATPase activity of the wall-lining region is slight compared with that of the epidermal cells. This probably reflects the presence of chloroplasts and photosynthetic activity in mesophyll cells and, thus, a lower influx than occurs in epidermal cells. The similarity in the functional aspects of these polarised cells infected with different classes of fungi indicates, however, that a general mechanism exists.

Our knowledge of the degree to which the metabolism of the infected cell is altered by an haustorial pathogen is so far limited. There may be considerable changes analogous to those which the malaria parasite induces in mammalian erythrocytes. In this system, the parasite is completely enclosed in a membrane derived by invagination of the host's plasma membrane in which the polarities of ATPase and NADH oxidase are reversed (Langreth, 1977). The uninvaginated region also undergoes extreme modifications which cause gross changes in rates of solute transport and specificity (Dunn, 1969; Hansen *et al.*, 1980; Hoelzl-Wallach and Schmidt-Ullrich, 1980). In the symbiotic biotrophic associations of vesicular–arbuscular mycorrhizas and root nodules an intracellular interface is also formed with the invading microorganism enclosed in either a partial or complete membrane invagination. Presumably solute movement in both directions occurs in these associations and it is notable that here ATPase activity is present on the invagination (Verma *et al.*, 1978; Marx *et al.*, 1982). It is not known whether the activity is related to active solute efflux or influx processes.

5.2. Transport by intercellular hyphae

The concentration of photosynthates in the apoplast of uninfected plants is probably high enough to account for the growth of biotrophic fungal parasites (Hancock and Huisman, 1981). As sugars efflux relatively freely from mesophyll protoplasts (Guy *et al.*, 1980; Huber and Moreland, 1981) and such efflux is essential for phloem loading (Giaquinta, 1976) it seems likely that intercellular hyphae have access to normal host nutrient transport pathways. The early plasmolysis experiments of Thatcher (1942) and more recent leakage experiments (Hoppe and Heitefuss, 1974) have indicated an increase in host cell permeability in rust-infected leaves, particularly at late stages of disease when the parasite's demand is high. Unfortunately, there has been no recent systematic study of cell membrane permeability in smut or downy mildew infections and the precise nature and cause of increased permeability are not yet known. It has recently been reported that phosphate sequestration by the parasite and storage in intracellular polyphosphate pools may be involved (D. H. Lewis, in Smith, 1981). Polyphosphates are abundant in rust and powdery mildew fungi (Bennett, 1972) and if the host is deprived of orthophosphate for ATP synthesis, it will lose its ability to retain solutes (cf. Humphreys, 1977).

Little is known of the activity of intercellular hyphae. Although techniques are available for the isolation of intercellular hyphae of rusts (Dekhuijzen *et al.*, 1967) axenic cultures present a more amenable system for *a priori* assessment of fungal metabolism. These have been employed in investigations of metabolic pathways of sugar assimilation (Manners *et al.*, 1982) and transport mechanisms (Manners, Maclean and Scott, unpublished work) in *Puccinia graminis*. Transport of the non-metabolizable sugar analogue, 3-*O*-methylglucose, is an active process as evidenced by accumulations against a concentration gradient and sensitivity to metabolic inhibitors. Complex kinetics of sugar uptake indicate the presence of more than one carrier system for glucose and other hexoses. At lower concentrations of glucose (<20 mM), the analogues 2-deoxyglucose and β-thioglucose showed a greater affinity for glucose carriers than glucose itself, and these may be useful in future as potent inhibitors of transport in biotrophic interactions.

No account of biotrophic transport processes can be complete without mention of the work of Smith and co-workers on lichens (Smith, 1974; 1975). Here intercellular contact is predominant and carbohydrate efflux probably proceeds by means of facilitated diffusion (Chambers *et al.*, 1976). This interaction may not be entirely comparable to those between higher plants and fungi as in lichens the carbohydrate transferred is not abundant in the free-living alga and its efflux is specific to the biotrophic association (Smith *et al.*, 1969). Biochemical plant pathologists, however, should take note of the experimental approaches adopted by lichenologists whose understanding of transport has come from the following techniques: accurate estimates of fluxes at the host and fungal plasma membranes (Collins and Farrar, 1978); isolation of host cells and characterization of their transport processes (Smith, 1974); and, importantly, the use of inhibitors specific to the transport processes of autotroph and heterotroph (Chambers *et al.*, 1976). The application of such approaches to interactions of fungi with higher plants where contact is principally intercellular should be rewarding.

What, then, are the specific advantages of haustorial over intercellular associations? The establishment of an isolated apoplast and modification of membrane function at a localized region of the host cell permits stimulation of solute efflux over an area closely subtending the pathogen. Solutes cannot be lost by diffusion or uptake by adjacent cells; thus a tight coupling of efflux from the host and influx to the pathogen is achieved. Further, the influx to the pathogen is almost certainly assisted and may even be dependent on unimpaired pumps in the unaffected part of the host plasma membrane. Future research must concentrate on the mechanisms which initiate the structural and functional changes observed in host cells and ultimately the identification of fungal and host molecules involved in the processes.

REFERENCES

Aist, J. R. (1974). A freeze–etch study of membranes of *Plasmodiophora*—infected and non-infected cabbage root hairs, *Can. J. Bot.*, **52**, 1441–1449.

Al-Khesraji, T. O., Lösel, D. M., and Gay, J. L. (1980). The infection of vascular tissue in leaves of *Tussilago farfara* L. by pycnial–aecial stages of *Puccinia poarum* Niel, *Physiol. Plant Pathol.*, **17**, 193–197.

Andrews, J. H. (1975). Distribution of label from ^{3}H-glucose and ^{3}H-leucine in lettuce cotyledons during the early stages of infection with *Bremia lactucae, Can. J. Bot.*, **53**, 1103–1115.

Ayres, P. G. (1977). Effects of powdery mildew *Erysiphe pisi* and water stress upon the water relations of pea, *Physiol. Plant Pathol.*, **10**, 139–145.

Baker, D. A. (1978). Proton co-transport of organic solutes by plant cells, *New Phytol.*, **81**, 485–497.

Bassett, R. A., Chain, E. B., Corbett, K., Dickerson, A. G. F., and Mantle, P. G. (1972). Comparative metabolism of *Claviceps purpurea in vitro* and *in vivo, Biochem. J.*, **127**, 3P–4P.

Bennett, J. (1972). Macromolecular aspects of fungal plant disease, *PhD Thesis*, University of Queensland, Brisbane, Australia.

Billett, E. E., Billett, M. A., and Burnett, J. H. (1977). Stimulation of maize invertase activity following infection by *Ustilago maydis, Phytochemistry,* **16**, 1163–1166.

Bracker, C. E. (1968). Ultrastructure of the haustorial apparatus of *Erysiphe graminis* and its relationship to the epidermal cell of barley, *Phytopathology*, **58**, 12–30.

Bracker, C. E., and Littlefield, L. J. (1973). Structural concepts of host–pathogen interfaces, in *3rd Long Ashton Symposium, 1971, Fungal Pathogenicity and the Plant's Response* (Eds R. J. W. Byrde and C. V. Cutting), Academic Press, London, pp. 159–318.

Brennan, P. J., and Lösel, D. M. (1978). Physiology of fungal lipids, *Adv. Microbiol. Physiol.*, **17**, 47–179.

Burrell, M. M., and Lewis, D. H. (1977). Amino acid movement from leaves of *Tussilago farfara* L. to the rust, *Puccinia poarum* Neil, *New Phytol.,* **79**, 327–333.

Bushnell, W. R. (1972). Physiology of fungal haustoria, *Annu. Rev. Phytopathol.*, **10**, 151–176.

Bushnell, W. R., and Gay, J. L. (1978). Accumulation of solutes in relation to the structure and function of haustoria in powdery mildews, in *The Powdery Mildews* (Ed. D. M. Spencer), Academic Press, London, pp. 183–235.

Callow, J. A., Long, D. E., and Lithgow, E. D. (1980). Multiple molecular forms of invertase in maize smut infections, *Physiol. Plant Pathol.*, **16**, 93–107.

Chambers, S., Morris, M., and Smith, D. C. (1976). Lichen physiology. XV. The effect of digitonin and other treatments on biotrophic transport of glucose from alga to fungus in *Peltigera polydactyla, New Phytol.*, **76**, 485–500.

Chong, J., and Harder, D. E. (1980). Ultrastructure of haustorium development in *Puccinia coronata avenae*. I. Cytochemistry and electron probe X-ray analysis of the haustorial neck ring, *Can. J. Bot.*, **58**, 2496–2505.

Coffey, M. D. (1975). Obligate parasites of higher plants, particularly rust-fungi, *Symp. Soc. Exp. Biol.*, **29**, 297–323.

Collins, C. R., and Farrar, J. F. (1978). Structural resistances to mass transfer in the lichen *Xanthoria parietina, New Phytol.*, **81**, 71–83.

Dekhuijzen, H. M. (1966). The isolation of haustoria from cucumber leaves infected with powdery mildew, *Neth. J. Plant Pathol.*, **72**, 1–11.

Dekhuijzen, H. M., Singh, H., and Staples, R. C. (1967). Some properties of hyphae isolated from bean leaves infected with the bean rust fungus, *Contrib. Boyce Thompson Inst.*, **23**, 367–372.

Dickerson, A. G. (1972). A β-D-fructo-furanosidase from *Claviceps purpurea, Biochem. J.*, **129**, 263–272.

Dickerson, A. G., Mantle, P. G., and Nisbet, L. J. (1976). Carbon assimilation by *Claviceps purpurea* growing as a parasite, *J. Gen. Microbiol.*, **97**, 267–276.

Dickerson, A. G., Mantle, P. G., Nisbet, L. J., and Shaw, B. I. (1978). A role for β-glucanases in the parasitism of cereals by *Claviceps purpurea, Physiol. Plant Pathol.*, **12**, 55–62.

Drew, E. A., and Smith, D. C. (1967). Studies in the physiology of lichens. VIII. Movement of glucose from alga to fungus during photosynthesis in the thallus of *Peltigera polydactyla, New Phytol.*, **66**, 389–400.

Dunn, M. J. (1969). Alteration of red blood cell sodium transport during malaria infection, *J. Clin. Invest.*, **48**, 674–684.

Edelman, J., Schoolar, A. I., and Bonnor, W. B. (1971). Permeability of sugar-cane chloroplasts to sucrose, *J. Exp. Bot.*, **22**, 534–545.

Edwards, H. H., and Allen, P. J. (1966). Distribution of the products of photosynthesis between powdery mildew and barley, *Plant Physiol.*, **41**, 683–688.

Frye, L. D., and Edidin, M. (1970). The rapid intermixing of cell surface antigens after formation of mouse–man heterokaryons, *J. Cell Sci.*, **7**, 319–335.

Gay, J. L., and Manners, J. M. (1981). Transport of host assimilates to the pathogen, in *Effects of Disease on the Physiology of the Growing Plant* (Ed. P. G. Ayres), Cambridge University Press, Cambridge, pp. 85–100.

Gay, J. L., and Manners, J. M. (1983). Permeability of the parasitic interface in powdery mildews, *Physiol. Plant Pathol.*, in press.

Geiger, D. R. (1975). Phloem loading, in *Encyclopedia of Plant Physiology, New Series. Vol. 1. Transport in Plants. I. Pholem Transport* (Eds M. H. Zimmerman and J. A. Milburn). Springer-Verlag, Berlin, New York, pp. 395–431.

Geiger, D. R., and Fondy, B. R. (1979). A method for continuous measurement of export from a leaf, *Plant Physiol.*, **64**, 361–365.

Giaquinta, R. T. (1976). Evidence for phloem loading from the apoplast, *Plant Physiol.*, **57**, 872–875.

Gil, F. (1976). Ultrastructural and physiological properties of haustoria of powdery mildews and their host interfaces, *PhD Thesis*, University of London.

Gil, F., and Gay, J. L. (1977). Ultrastructural and physiological properties of the host interfacial components of haustoria of *Erysiphe pisi in vivo* and *in vitro, Physiol. Plant Pathol.*, **10**, 1–12.

Guy, M., Reinhold, L., and Rahat, M. (1980). Energisation of the sugar transport mechanism in the plasmalemma of isolated mesophyll protoplasts, *Plant Physiol.*, **65**, 550–553.

Hancock, J. G., and Huisman, O. C. (1981). Nutrient movement in host–pathogen systems, *Annu. Rev. Phytopathol.*, **19**, 309–331.

Hankin, L., and McIntyre, J. L. (1980). Hydrolytic and transferase activities of invertases from physiological races of *Phytophthora parasitica* var. *nicotianae, Mycologia*, **72**, 749–758.

Hansen, B. D., Sleeman, H. K., and Pappas, P. W. (1980). Purine base and nucleoside uptake in *Plasmodium berghei* and host erythrocytes, *J. Parasitol.*, **66**, 205–212.

Harder, D. E., Rohringer, R., Samborski, D. J., Kim, W. K. and Chong, J. (1978), Electron microscopy of susceptible and resistant near isogenic (sr6/Sr6) lines of wheat infected by *Puccinia graminis tritici*. I. The host pathogen interface of the compatible (sr6/P6) interaction, *Can. J. Bot.*, **56**, 2955–2966.

Heath, M. C. (1971). Haustorial sheath formation in cowpea leaves immune to rust infection, *Phytopathology*, **61**, 383–388.

Heath, M. C. (1976). Ultrastructural and functional similarity of the haustorial neckbank of rust fungi and the Casparian strip of vascular plants, *Can. J. Bot.*, **54**, 2484–2489.

Heath, M. C., and Heath, I. B. (1971). Ultrastructure of an immune and a susceptible reaction of cowpea leaves to rust infection, *Physiol. Plant Pathol.*, **1**, 277–287.

Heldt, H. W., and Sauer, F. (1971). The inner membrane of the chloroplast envelope as the site of specific metabolite transport, *Biochim. Biophys Acta*, **234**, 83–91.

Hewitt, H. G., and Ayres, P. G. (1976). Effect of infection of *Microsphaera alphitoides* (powder mildew) on carbohydrate levels and translocation in seedlings of *Quercus robur, New Phytol.*, **77**, 379–390.

Hickey, E. L., and Coffey, M. D. (1977). A fine-structural study of the pea downy mildew fungus *Peronospora pisi* in its host *Pisum sativum, Can. J. Bot.*, **55**, 2845–2858.

Hickey, E. L., and Coffey, M. D. (1978). A cytochemical investigation of the host-parasite interface in *Pisum sativum* infected by the downy mildew fungus *Peronospora pisi, Protoplasma*, **97**, 201–220.

Higinbotham, N. (1973). Electropotentials of plant cells, *Annu. Rev. Plant Physiol.*, **24**, 25–46.

Hoelzl-Wallach, D. F., and Schmidt-Ullrich, R. (1980). Cellular membranes and the host–parasite interaction, in *The Host Invader Interplay* (Ed. H. Van den Bossche), Elsevier-North Holland Biomedical Press, Amsterdam, pp. 3–14.

Howes, N. K., and Scott, K. J. (1972). Sulphur nutrition of *Puccinia graminis* f. sp. *tritici* in axenic culture, *Can. J. Bot.*, **50**, 1165–1170.

Hoppe, H. H., and Heitefuss, R. (1974). Permeability and membrane lipid metabolism of *Phaseolus vulgaris* infected with *Uromyces phaseoli*. I. Changes in the efflux of cell constitutents, *Physiol. Plant Pathol.*, **4**, 5–9.

Huber, S. C., and Moreland, D. E. (1980). Efflux of sugars across the plasmalemma of mesophyll protoplasts, *Plant Physiol.*, **65**, 560–562.

Huber, S. C., and Moreland, D. E. (1981). Co-transport of potassium and sugars across the plasmalemma of mesophyll protoplasts, *Plant Physiol.*, **67**, 163–169.

Humphreys, T. (1977). Dinitrophenol-induced efflux of sucrose from maize scutellum cells, *Phytochemistry*, **16**, 1359–1364.

Humphreys, T. E. (1978). A model for sucrose transport in the maize scutellum, *Phytochemistry*, **17**, 679–684.

Ingram, D. S. (1981). Physiology and biochemistry of host–parasite interaction, in *The Downy Mildews* (Ed. D. M. Spencer), Academic Press, London, pp. 143–163.

Jager, K., and Reisener, H. J. (1969). Untersuchungen über Stoffwechselbeziehungen zwischen Parasit und Wirt am Bespiel von *Puccinia graminis* var. tritici auf Weizen, *Planta*, **85**, 57–72.

Jennings, D. H. (1976). Transport and translocation in filamentous fungi, in *The Filamentous Fungi*, Vol. 2 (Eds J. E. Smith and D. R. Berry), Edward Arnold, London, pp. 32–64.

Kolattukudy, P. E. (1977). Biosynthesis and degradation of lipid polymers, in *Lipids and Lipid Polymers in Higher Plants* (Eds M. Tevini and H. K. Lichenthaler), Springer-Verlag, New York, pp. 271–292.

Kunoh, H., and Ishizaki, H. (1981). Cytological studies of early stages of powdery mildew in barley and wheat. VII. Reciprocal translocation of a fluorescent dye between barley coleoptile cells and conidia, *Physiol. Plant Pathol.*, **18**, 207–211.

Langreth, S. G. (1977). Electron microscope cytochemistry of host–parasite membrane interactions in malaria, *Bull, WHO*, **55**, 171–178.

Lewis, D. H. (1976). Interchange of metabolites in biotrophic symbioses between angiosperms and fungi, in *Perspectives in Experimental Biology, Vol. 2, Botany* (Ed. N. Sunderland), Pergamon Press, Oxford, pp. 207–219.

Littlefield, L. J., and Bracker, C. E. (1972). Ultrastructural specialisation of the host–pathogen interface in rust-infected flax, *Protoplasma*, **74**, 271–305.

Littlefield, L. J., and Heath, M. C. (1979). *Ultrastructure of Rust Fungi*, Academic Press, New York.

Long, D. E., Fung, A. K., McGee, E. E. M., Cooke, R. C. and Lewis, D. H. (1975). The activity of invertase and its relevance to the accumulation of storage polysaccharides in leaves infected with biotropic fungi, *New Phytol.*, **74**, 173–182.

Lunderstädt, J. (1966). Effects of rust infection on hexokinase activity and carbohydrate dissimilation in primary leaves of wheat, *Can. J. Bot.*, **44**, 1345–1364.

Luttrell, E. S. (1980). Host–parasite relationships and development in the ergot sclerotium in *Claviceps purpurea, Can. J. Bot.*, **58**, 942–958.

Maclean, D. J. (1982). Axenic culture and metabolism of rust fungi, in *The Rust Fungi* (Eds K. J. Scott and A. K. Chakravorty), Academic Press, London, pp. 38–120.

Manners, J. M. (1979). Physiology of fungal haustoria (Erysiphales), *PhD Thesis*, University of London.

Manners, J. M., and Gay, J. L. (1977). The morphology of haustorial complexes isolated from apple, barley, beet and vine infected with powdery mildews, *Physiol. Plant Pathol.*, **11**, 261–266.

Manners, J. M., and Gay, J. L. (1978). Uptake of ^{14}C photosynthates from *Pisum sativum* by haustoria of *Erysiphe pisi, Physiol. Plant Pathol.*, **12**, 199–209.

Manners, J. M., and Gay, J. L. (1980a). Autoradiography of haustoria of *Erysiphe pisi, J. Gen. Microbiol.*, **116**, 529–533.

Manners, J. M., and Gay, J. L. (1980b). Fluxes and accumulation of systemic fungicide (ethirimol) in haustoria of *Erysiphe pisi* and protoplasts of *Pisum sativum, Ann. Appl. Biol.*, **96**, 283–293.

Manners, J. M., and Gay, J. L. (1982a). Accumulation of systemic fungicides and other compounds by haustorial complexes isolated from *Pisum sativum* infected with *Erysiphe pisi, Pestic. Sci.*, **13**, 195–203.

Manners, J. M., and Gay, J. L. (1982b). Transport, translocation and metabolism of ^{14}C photosynthates at the host–parasite interface of *Pisum sativum* and *Erysiphe pisi, New Phytol.*, **91**, 221–244.

Manners, J. M., Maclean, D. J., and Scott, K. J. (1982). Pathways of glucose assimilation in *Puccinia graminia, J. Gen. Microbiol.*, **128**, 2621–2630.

Manocha, M. S. (1975). Autoradiography and fine structure of host–parasite interface in temperature-sensitive combinations of wheat stem rust, *Phytopathol. Z.*, **82**, 207–215.

Martin, M., and Gay, J. L. (1983). Ultrastructure of conidium development in *Erysiphe pisi, Can. J. Bot.* (in press).

Maruyama, Y., Onodera, K., Seto, H., and Funahashi, S. (1979). Surface localization of enzymes in mycelia and microconidia of *Fusarium oxysporum, J. Gen. Appl. Microbiol.*, **25**, 307–313.

Marx, C., Dexheimer, J., Gianinazzi-Pearson, V., and Gianinazzi, S. (1982). Enzymatic studies on the metabolism of vesicular-arbuscular mycorrhizas. IV. Ultracytoenzymological evidence (ATPase) for active transfer processes in the host–arbuscule interface, *New Phytol.*, **90**, 37–43.

Mendgen, K. (1977). Reduced lysine uptake by bean rust haustoria in a resistant reaction, *Naturwissenschaften*, **64**, 438.

Mitchell, D. T., and Cooke, R. C. (1976). Carbohydrate composition and activity of invertase in infected cabbage root tissues during club root development, *Trans. Br. Mycol. Soc.*, **67**, 344–349.

Mitchell, D. T., Fung, A. K., and Lewis, D. H. (1978). Changes in the ethanol-soluble carbohydrate composition and acid invertase in infected first leaf tissues susceptible to crown rust of oat and wheat stem rust, *New Phytol.*, **80**, 381–392.

Mount, M. S., and Ellingboe, A. H. (1969). ^{32}P and ^{35}S transfer from susceptible wheat to *Erysiphe graminis* f. sp. *tritici* during primary infection, *Phytopathology*, **59**, 235.

Onoe, T., Tani, T., and Haito, N. (1973). The uptake of labelled nucleosides by *Puccinia coronata* grown in susceptible oat leaves, *Rep. Tottori Mycol. Inst.*, **10**, 303–312.

Ouchi, S., Hibino, C., Oku, H., Fujiwara, M., and Nakabayashi, H. (1979). The induction of resistance or susceptibility, in *Recognition and Specificity in Plant Host–Parasite Interactions* (Eds J. M. Daly and I. Uritani), Japan Scientific Societies Press, Tokyo, and University Park Press, Baltimore, pp. 49–65.

Perera, R., and Gay, J. L. (1976). The ultrastructure of haustoria of *Sphaerotheca pannosa* (Wallroth ex Fries) Leveille and changes in infected and associated cells of rose, *Physiol. Plant Pathol.*, **9**, 57–65.

Reisener, H. J., Ziegler, E., and Prinzing, A. (1970). Zum Stoffwechsel des Mycels von *Puccinia graminis* var. *tritici* auf der Weizenpflanze, *Planta*, **92**, 355–357.

Rice, M. A. (1927). The haustoria of certain rusts and the relation between host and pathogene, *Bull. Torrey Bot. Club*, **54**, 63–153.

Roberts, S. M., and Mitchell, D. T. (1979). Carbohydrate composition and invertase activity in poplar leaf tissues infected by *Melampsora aecidioides* (D.C.) Schroeter, *New Phytol.*, **83**, 499–508.

Rubinstein, B. (1978). Use of lipophilic cations to measure the membrane potential of oat leaf protoplasts, *Plant Physiol.*, **62**, 927–929.

Rubinstein, B. P., Mahar, T. A., and Tattar, T. A. (1977). Effects of osmotic shock on some membrane regulated events of oat coleoptile cells, *Plant Physiol.*, **59**, 365–368.

Sargent, C., and Gay, J. L. (1977). Barley epidermal apoplast structure and modification by powdery mildew contact, *Physiol. Plant Pathol.*, **11**, 195–205.

Scott, K. J. (1972). Obligate parasitism by phytopathogenic fungi. *Biol. Rev.*, **47**, 537–572.

Shaw, B. I., and Mantle, P. G. (1980). Host infection by *Claviceps purpurea, Trans. Br. Mycol. Soc.*, **75**, 77–90.

Slayman, C. L. (1965). Electrical properties of *Neurospora crassa*. Effect of external cation on the intracellular potential, *J. Gen. Physiol.*, **49**, 69–92.

Slesinski, R. S., and Ellingboe, A. H. (1971). Transfer of ^{35}S from wheat to the powdery mildew fungus with compatible and incompatible parasite/host genotypes, *Can. J. Bot.*, **49**, 303–310.

Smith, D. C. (1974). Transport from symbiotic algae and symbiotic chloroplasts to host cells, *Symp. Soc. Exp. Biol.*, **28**, 437–508.

Smith, D. C. (1975). Symbiosis and the biology of lichenised fungi, *Symp. Soc. Exp. Biol.*, **29**, 373–405.

Smith, D., Muscatine, L., and Lewis, D. H. (1969). Carbohydrate movement from autotroph to heterotroph in parasitic and mutualistic symbiosis, *Biol. Rev.*, **44**, 17–90.

So, M. L., and Thrower, L. B. (1976). The host–parasite relationship between *Vigna sesquipedalis* and *Uromyces appendiculatus*. III. Uptake of sugars by the parasite and respiration of the host–parasite combination, *Phytopathol. Z.*, **86**, 302–309.

Sovonick, S. A., Greiger, D. R., and Fellows, R. J. (1974). Evidence for active phloem loading in the minor veins of sugar beet, *Plant Physiol.*, **54**, 886–891.

Spencer-Phillips, P. T. N., and Gay, J. L. (1980). Electron microscope autoradiography of ^{14}C photosynthate at the haustorium–host interface in powdery mildew of *Pisum sativum, Protoplasma*, **103**, 131–154.

Spencer-Phillips, P. T. N., and Gay, J. L. (1981). Plasma membrane ATPase domains and transport through infected cells, *New Phytol.*, **89**, 393–400.

Sullivan, T. P. Bushnell, W. R., and Rowell, J. B. (1974). Relations between haustoria of *Erysiphe graminis* and host cytoplasm in cells opened by microsurgery, *Can. J. Bot.*, **52**, 987–998.

Sutton, O. D., and Lampen, J. O. (1962). Localisation of sucrose and maltose fermenting systems in *Saccharomyces cerevisiae, Biochim. Biophys. Acta*, **56**, 303–312.

Syrop, M. (1975). Leaf curl disease of almond caused by *Taphrina* deformans (Berk.) Tul. II. An electron microscope study of the host–parasite relationship, *Protoplasma*, **85**, 57–69.

Takahashi, K., Inaba, T., and Kajiwara, T. (1977). Distribution of ^{14}C assimilated from $^{14}CO_2$ in cucumber leaves infected with downy mildew, *Physiol. Plant Pathol.*, **11**, 255–259.

Thatcher, F. S. (1942). Further studies of osmotic and permeability relations in parasitism, *Can. J. Res.*, **C21**, 283–311.

Verma, D. P. S., Kazazian, V., Zogbi, V., and Bal, A. K. (1978). Isolation and characterisation of the membrane envelope enclosing the bacteriods in soya bean root nodules, *J. Cell Biol.*, **78**, 919–936.

Vořišek, J., Ludvík, J., and Řeháček, Z. (1974). Morphogenesis and ultrastructure of *Claviceps purpurea* during submerged alkaloid formation, *J. Bacteriol.*, **120**, 1401–1408.

West, I. C. (1980). Energy coupling in secondary active transport, *Biochim. Biophys. Acta*, **604**, 91–126.

Williams, P. H., and McNabola, S. S. (1970). Fine structure of the host-parasite interface of *Plasmodiophora brassicae* in cabbage, *Phytopathology*, **60**, 1557–1561.

Woods, A. M., and Gay, J. L. (1983). Evidence for neckband delimiting structural and physiological regions of the host plasma membrane around the haustoria of *Albugo candida, Physiol. Plant Pathol.*, in press.

Further Reading

Bracker, C. E., and Littlefield, L. J. (1973). Structural concepts of host–pathogen interfaces, in *3rd Long Ashton Symposium, 1971, Fungal Pathogenicity and the Plant's Response* (Eds R. J. W. Byrde and C. V. Cutting), Academic Press, London, pp. 159–318.

Bushnell, W. R. (1972). Physiology of fungal haustoria, *Annu. Rev. Phytopathol.*, **10**, 151–176.

Bushnell, W. R., and Gay, J. L. (1978). Accumulation of solutes in relation to the structure and function of haustoria in powdery mildews, in *The Powdery Mildews* (Ed. D. M. Spencer), Academic Press, London, pp. 183–235.

Gay, J. L., and Manners, J. M. (1981). Transport of host assimilates to the pathogen, in *Effects of Disease on the Physiology of the Growing Plant* (Ed. P. G. Ayres), Cambridge University Press, Cambridge, pp. 85–100.

Hancock, J. G., and Huisman, O. C. (1981). Nutrient movement in host–pathogen systems, *Annu. Rev. Phytopathol.*, **19**, 309–331.

Ingram, D. S. (1981). Physiology and biochemistry of host–parasite interaction, in *The Downy Mildews* (Ed. D. M. Spencer), Academic Press, London, pp. 143–163.

Lewis, D. H. (1976). Interchange of metabolites in biotrophic symbioses between angiosperms and fungi, in *Perspectives in Experimental Biology, Vol. 2. Botany* (Ed. N. Sunderland), Pergamom Press, Oxford, pp. 207–219.

Littlefield, L. J., and Heath, M. C. (1979). *Ultrastructure of Rust Fungi*, Academic Press, New York.

Maclean, D. J. (1982). Axenic culture and metabolism of rust fungi, in *The Rust Fungi* (Eds. K. J. Scott and A. K. Chakravorty), Academic Press, London, pp. 38–120.

III. Specificity and Resistance

Biochemical Plant Pathology
Edited by J. A. Callow
© 1983 John Wiley & Sons Ltd

10

Molecular Genetics of Plant Disease

Brian J. Staskawicz

International Plant Research Institute, San Carlos, CA 94070, USA

1. INTRODUCTION

Recent advances in the methodologies of molecular genetics now allow the phytopathologist to approach the study of plant disease from an entirely new perspective. The use of recombinant DNA technologies will be paramount in this effort in that it will allow the identification and cloning of genes that are involved in such processes as host–pathogen specificity, pathogen virulence functions, synthesis and regulation of toxin production, and ultimately the regulation of symptom expression in the host plant.

One of the major limitations in studying the above processes is the development of genetic systems in both the pathogen and the host that are amendable to molecular analysis. Thus, one of the major efforts to date has been to develop DNA cloning systems in both bacterial and fungal phytopathogens as a prelude to a detailed analysis of pathogenicity. The scope of this chapter will be limited to a few studies where progress has been made by employing molecular genetics in the analysis of plant disease.

Finally, an effort has been made to point out areas where molecular genetics may advance our basic understanding of disease processes or where such systems can be used to clarify and prove a specific biochemical hypothesis.

2. A MOLECULAR APPROACH TOWARDS UNDERSTANDING PLANT DISEASE

The events leading to the development of disease symptoms result from the expression of genes in both the host and the pathogen. Since this is an extra-ordinarily complex situation, initial studies have concentrated on bacterial phytopathogens because of the ease of genetic manipulations and small genome size.

A molecular geneticist usually approaches such a problem by first isolating mutants involved in pathogenicity. This defines the numbers of genes involved in such a process. In parallel experiments, a genomic library of pathogen DNA is made by isolating DNA fragments and cloning them into a suitable vector that can replicate in both *Escherichia coli* and the pathogen of interest. The library of DNA is usually made in *E. coli* first because of its ease of manipulation and high transformation rate.

Once the library of DNA has been constructed, recombinant plasmids containing wild-type DNA can be introduced back into the various mutants either by transformation or conjugation to determine which clones complement pathogenicity. In this manner, genes involved in pathogenicity can be isolated and further analysed for gene structure and function. If a suitable transformation system is available in the pathogen, the wild-type DNA can be directly reinserted into the mutant and scored for pathogenicity.

An alternative strategy to isolate genes involved in interactions with the host is to use transposon mutagenesis (Kleckner *et al.*, 1977; Kleckner, 1981).

Figure 1. Cloning a gene by transposon mutagenesis. (A) A transposon is introduced into a bacterium by conjugation via a suicide plasmid. (B) The transposon inserts into the genome at random sites. The physical insertion of this element into a gene causes a mutation. Since Tn5 encodes for kanamycin resistance (Km^R), this element can be followed in subsequent manipulations. Colonies that are Km^R are then screened for a change in phenotype, e.g. growth on minimal media, inability to cause disease, altered host range, or any trait one can test. Once a Km^R colony has been found to be altered in the phenotype of interest, total DNA is isolated from that single purified colony and cut to completion with enzyme EcoR1. Since Tn5 does not contain EcoR1 restriction sites, the first recognition sites that flank the transposon to the left and right will be recognized. (C) The plasmid pBR325 contains a single recognition site for the enzyme EcoR1 and cloning foreign DNA into this site can be detected by insertion into the chloramphenicol gene. In the next series of steps, the pBR325 is cut with EcoR1 and mixed together with the total EcoR1-cut DNA isolated in step B. These DNA molecules are then ligated together and transformed into *E. coli*. By selecting for

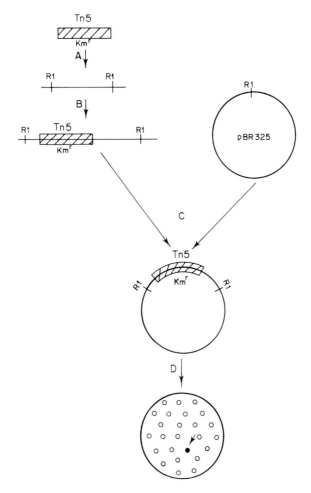

Km^R, Amp^R, and Cm^S a colony can be obtained that contains the plasmid with a single EcoR1 insert. This insert will consist of the entire Tn5 element and the flanking sequences of the gene that was inactivated which was shown to be altered in the phenotype tested. Thus, at this point the mutated gene has been cloned. (D) In order to isolate the non-mutated version of this gene the following series of steps may be employed. The clone isolated in step C is grown up and plasmid DNA is purified. In parallel experiments a library of wild-type DNA has been constructed in an appropriate vector. On a single Petri dish an entire library may be maintained in 500–1000 individual cosmid clones. The plasmic DNA is then labelled with ^{32}P by a process known as nick translation. The nick-translated ^{32}P probe is then used to screen the wild-type library for a colony containing a cosmid clone that has DNA sequence homology. This isolate is then purified and tested for its biological activity by introducing the cosmid clone back into the Tn5-mutagenized strain. If the wild-type gene is present on the clone the mutation should be complemented and the phenotype restored to the original wild-type condition

Transposons are defined segments of DNA capable of moving from one DNA replicon to another, independent of homologous recombinational systems (Calos and Miller, 1980). Upon the insertion of a transposon into DNA a mutation is effected by insertional inactivation of the gene of interest. Most importantly, this method of mutagenesis is useful because of the improbability of the transposon being inserted simultaneously at two unique sites. In addition, transposons usually carry a selectable antibiotic resistance marker that allows one to follow such elements during genetic manipulations. Once a mutant has been identified by these procedures, the transposon and flanking sequences of the inactivated gene can be cloned directly into *E. coli* using the antibiotic resistance marker for a direct selection. Using the flanking sequences as a probe, the wild-type gene may be identified and cloned from a wild-type library (Figure1).

3. DEVELOPMENT OF DNA CLONING SYSTEMS

The ability to isolate DNA from an organism, cut with a restriction enzyme, and clone into a suitable vector has become a routine technique in molecular genetics. The main premise behind the development of cloning systems is that relatively small segments of DNA can be maintained in a vector which can facilitate the structural and functional analysis of the gene in question.

DNA cloning vectors usually consist of plasmids, bacteriophages, or a hybrid of a phage and a plasmid called a cosmid. Cosmids are defined as plasmids which contain the *cos* site from the bacteriophage λ and can be packaged *in vitro*; their use is discussed by Hohn (1980).

The development of DNA cloning vectors for bacterial plant pathogens has been greatly advanced by the work of Panopoulos (1981). The plasmid RSF1010 has shown considerable promise as a wide host range vector that is capable of directly transforming several pathovars of *Pseudomonas syringae* (Panopoulos *et al.*, 1979). Improved cloning derivatives have been constructed and their utility proved (Panopoulos, 1981). Derivatives of the wide host range plasmid pRK2 have also been employed in both *Agrobacterium tumefaciens* and various *Rhizobium spp*. Ditta *et al.* (1980), have used the plasmid pRK290 to construct a gene bank of *Rhizobium meliloti* DNA. More recently, a cosmid derivative of pRK290, referred to as pLAFRI, has been constructed by cloning the *cos* site from pHC 79, into the single Bg II site of pRK290. This vector has been successfully used in the construction of a library of *Rhizobium meliloti* DNA (Friedman *et al.*, 1982). Wide host range cosmid cloning vectors are generally more efficient for library construction because of the large inserts necessary to be packaged in the bacteriophage lambda (Collins and Hohn, 1978). Furthermore, most colonies that grow in *E. coli* are bonafide clones because an insert of 16–31 kilobases is usually required for *in vitro* packaging in the cosmid pLAFRI. An exception is the packaging of two pLAFRs resulting in a dimer. By altering the ratio of insert DNA to vector DNA an optimum may be obtained to increase the percentage of true inserts. Thus, a complete library of DNA of any Gram-negative bacterium can be contained in several hundred to a thousand individual colonies of *E. coli*. In this particular instance, the cosmid vector pLAFRI is capable of replication in both *E. coli* and *Rhizobium meliloti* and can be shown to be mobilized from the primary library stored in *E. coli* back to *Rhizobium meliloti* by the helper plasmid pRK2013 (Ditta *et al.*, 1980). Using these techniques, Long *et al.* (1982) have identified a gene involved in the early stages of nodule formation by mobilizing a wild-type cosmid library from *E. coli* into an appropriate nodulation mutant of *Rhizobium meliloti, en masse.* Employing the plant as a selection they were able to isolate colonies of *Rhizobium meliloti* that had regained normal nodulation functions by the acquisition of a cosmid clone containing the wild-type allele. The development of similar cloning systems in phytopathogenic bacteria should allow the isolation of genes involved in processes such as cultivar specificity, and other genes involved in pathogenicity and virulence.

4. MOLECULAR ANALYSIS OF HOST–PATHOGEN SPECIFICITY

Although the area of host–pathogen specificity has been the subject of many previous investigations, the basic molecular mechanisms involved have yet to be discovered. Past strategies have often employed a comparative biochemical approach in which a pathogen is typically inoculated on near isogenic lines of a host plant and a difference in a biochemical process between the susceptible and resistant host plant is studied. Even though this has led many researchers to hypothesize the role of compounds such as phytoalexins and lectins as determinants of host–pathogen specificy, unequivocal proof remains to be shown. Thus, we find a situation in which most researchers are looking for the gene products that determine specificity and have ignored the physical isolation of the genes that encode for these products.

The early work of Flor (1947) clearly demonstrated the relationship between specific genes in the pathogen and specific genes in the host plant. The gene-for-gene hypothesis suggests that incompatibility in a host–pathogen interaction results from the interaction of a dominant avirulence gene product in the pathogen and a dominant resistance gene product in the host (Ellingboe, 1976). Therefore, it has been suggested that incompatibility arises from the interplay of positive gene functions in the pathogen and positive gene functions in the host and compatibility is due to a lack of recognition for incompatibility. If this is true, we also must assume that there are basic gene functions of the parasite which are necessary to induce disease on a susceptible cultivar, and that avirulence genes mask the expression of these basic compatibility genes (Ellingboe, 1976). If we assume the avirulence genes encode for specific gene products and that the expression of these genes precludes the expression of basic compatibility genes, then it should be possible to obtain induced mutants in these genes with a concomitant change to compatibility.

Mutants have been obtained by nitrosoguanidine mutagenesis to increase virulence in several host–pathogen combinations. Gabriel *et al.* (1982) have demonstrated that mutants with increased virulence can be obtained in the *Erysiphe graminis*–wheat host–pathogen interaction. Temperature-sensitive mutants of avirulence genes have also been obtained, suggesting that the primary gene product of an avirulence gene is a protein (Gabriel *et al.*, 1979). Furthermore, mutants with increased virulence have been obtained in host–parasite interactions between bacteria and host plants. Dahlbeck and Stall (1979) clearly showed by fluctuation analysis that *Xanthomonas vesicatoria*, the causal agent of bacterial blight on tomato and peppers, was spontaneously changing to virulent races in host lines that were different in a single allele. Bacterial mutants of *Pseudomonas syringae* pv. *phaseolicola* to increased virulence have been obtained by EMS mutagenesis by Staskawicz and Panopoulos (unpublished work). In this particular instance, a Race 1 strain of *P. syringae* pv. *phaseolicola* was mutagenized and inoculated into bean plants of the cultivar Red Mexican. Several colonies were isolated and were shown to be stably changed to the compatible phenotype.

Genetic molecular studies have shown that genes controlling host range are pTi plasmid-encoded properties in *Agrobacterium tumefaciens* (Loper and Kado, 1979; Thomashow *et al.*, 1980). These studies have been useful in that they have clearly delineated the genes involved in host–pathogen specificity as assayed by the phenotype of the disease, but evidence for specific gene products is still lacking.

In our laboratory we are studying the pathogen *Pseudomonas syringae* pv. *glycinea*. This pathogen causes a leaf-spot disease on the cultivated soybean, *Glycine max*. We have chosen this host–pathogen combination because of the natural variability which exists in the pathogen when inoculated on several differential cultivars of soybean. To date, ten races have been identified (Cross *et al.*, 1966; Fett and Sequeira, 1981). As an initial attempt to isolate the specific genes involved in host–pathogen specificity, we have constructed gene libraries of several races of the pathogen on the cosmid cloning vector pLAFRI. As mentioned previously, a cosmid cloning vector is extremely efficient in library constructions because of the large insert necessary for *in vitro* packaging. We have also demonstrated that the library of DNA stored in *E. coli* can be mobilized to *P. syringae* pv. *glycinea* by conjugation with the plasmid pRK2013 in triparental matings (Ditta *et al.*, 1980). Table 1 depicts the known reactions of races 1, 5, and 6 on the soybean cultivars Harasoy 63, Acme, and Flambeau.

We have concentrated initially on cloning DNA from Races 1 and 6

Table 1. Reactions of races 1, 5, and 6 of *Pseudomonas syringae* pv. *glycinea* on three cultivars of soybean

Race	Soybean cultivar		
	Harasoy 63	Acme	Flambeau
1	−	+	+
5	+	−	−
6	−	−	+

in the vector pLAFRI. If Race 1 or 6 DNA is introduced into strain of Race 5, we should be able to recognize a change from virulence to avirulence when the appropriate DNA fragment is present. The DNA fragment containing the avirulence gene can then be physically isolated. The same transconjugants can be tested for a change from avirulence to virulence when inoculated on the cv. Flambeau. Once the clone has been identified we can then determine its gene product and begin a detailed study on the genes involved in race specificity. In essence, the tools of recombinant DNA now allow us to analyse host–pathogen interactions on the level of the gene.

5. CLONING OF VIRULENCE GENES

The molecular identification of virulence genes or basic compatibility genes will add greatly to our basic knowledge of plant disease and the evolution of pathogenicity. Significant advances along these lines have been made in *Agrobacterium tumefaciens*; as they are presented in Chapter 5 they will not be considered further here. On the other hand, the bacterial pathogen *Pseudomonas savastanoi*, the causal agent of oleander and olive knot, has been the subject of recent investigations by Comai and Kosuge (1980, 1982). Earlier work by Smidt and Kosuge (1978) has demonstrated that mutants of *Pseudomonas savstanoi* resistant to α-methyltryptophan were altered in their ability to produce indoleacetic acid (IAA). Two main classes of mutants were obtained, those in which strains failed to synthesize IAA and those in which twice the quantity of IAA was synthesized. More importantly, it was demonstrated that the mutants which failed to synthesize IAA were no longer able to incite gall formation, although they exhibited similar growth patterns in infected tissue. The other class of mutants accumulated twice the amount of IAA and produced larger galls. Thus, we find a situation in which a strong correlation exists between the ability to synthesize IAA and the ability to induce galls, but other interpretations could be inferred. This type of correlation is strong but only suggestive evidence for the primary role of IAA in gall formation. Previous work by Comai and Kosuge (1980) had determined that the gene for tryptophan 2-monooxygenase was located on a resident 34 mdalton plasmid (pIAAI) and that strains cured of this plasmid were IAA–and avirulent. The reintroduction of this plasmid back to acridine orange-cured strains restored IAA production and virulence. This type of experimental rigour is essential in assigning definitive functions to plasmid genes involved in virulence. Subsequently Comai and Kosuge (1982) have reported the cloning of the tryptophan 2-monooxygenase gene into the cloning vectors RSF1010 and pBR328. Introduction of a 2.75 kilobase Eco R1 fragment from pIAA1 cloned in RSF1010 (pLUC1) into a pIAA1-less strains of *P. savastanoi* conferred indoleacetamide production and the isolate was restored to virulence. The work by Comai and Kosuge (1980, 1982) is a good example of how a molecular genetics approach has led to the assigning of specific genes and gene products involved in pathogenicity in a particular host–pathogen combination.

   Pseudomonas solanacearum causes a destructive wilt disease in many sol-

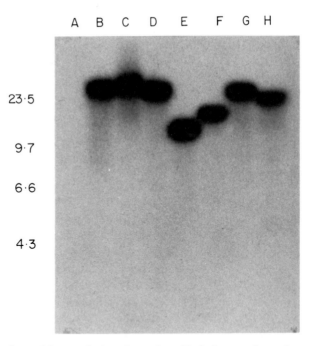

Figure 2. Southern blot analysis of random Tn5 inserts into the genome of *Pseudomonas solanacearum*. Total DNA was isolated from wild-type *P. solanacearum* (lane A) and seven random Km^R transconjugants (lanes B–H). The DNA was cut with the enzyme EcoR1 and blotted on to nitrocellulose and probed with a nick translated ColE1 : : Tn5 (pRZ102) plasmid. The control in lane A shows that neither ColE1 sequences nor Tn5 sequences share homology with the *P. solanacearum* genome. Each strain contains a single Tn5 insert at a different location in the genome. as revealed by the various mobilities

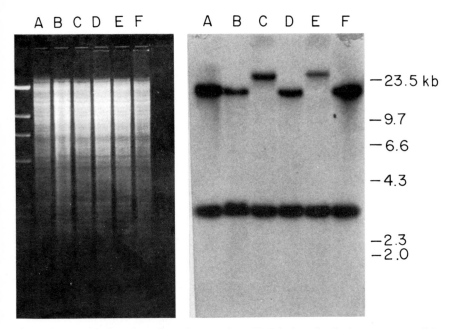

Figure 3. EcoR1 digestion of total DNA from Tn5-induced avirulent mutants of *P. solanacearum*. The left figure represents an agarose gel containing DNA from six Tn5-induced avirulent mutants cut with the enzyme EcoR1. The gel was stained with ethidium bromide and photographed under UV light. The right figure is an auto-radiogram of the DNA from the left figure blotted on to nitrocellulose and probed with a nick-translated ColE1 : : Tn5 probe. See text for further explanation

anaceous plants and has been the subject of many physiological and biochemical experiments. Early work by Kelman (1954) had clearly demonstrated a strong correlation between a fluidal colony morphology and the ability to cause wilt. Mutants to the non-fluidal phenotype were shown to be avirulent, thus suggesting a role for the extracellular polysaccharide (EPS) as a virulence determinant. Since spontaneous revertants back to the fluidal phenotype have not been reported (Sequeria, personal communication), evidence for the role of EPS in the virulence of *P. solanacearum* is still only a correlation. The development of genetic systems in *P. solanacearum* have been greatly advanced by Boistard and Boucher (1978). A transformation system is available and transposon mutagenesis has been reported (Boucher *et al.*, 1981; Staskawicz *et al.*, unpublished work).

In our laboratory we have been using transposon mutagenesis in an attempt to isolate mutants to the non-fluidal phenotype. Our strategy has been to employ the suicide plasmid pJB4JI (Beringer *et al.*, 1978) to introduce the transposon Tn5 into a Race 3 strain of *P. solanacearum*. As mentioned previously, transposon mutagenesis causes mutations by insertional inactivation

A   B

33

3.2

Figure 4. Genomic analysis of a fluidal revertant of *P. solanacearum*. Total DNA was isolated from a Tn5-induced non-fluidal, avirulent mutant (lane A) and from a fluidal, virulent revertant (lane B). The DNA was cut with the enzyme EcoR1, transferred to nitrocellulose and probed with a ColE1 : : Tn5 (pRZ102) probe. Note that the revertant no longer contains the small 3.2 kb hydridizable sequence

(Figure 1). Once a mutant has been isolated it should be possible to isolate the gene into which the transposon has been inserted by cloning the transposon and flanking sequences.

We have been successful in inserting the transposon Tn5 into the genome of *P. solanacearum*. The majority of the inserts were single inserts into unique sites in the genome (Figure 2). Auxotrophic mutations were isolated at a frequency of 0.5%. In addition, we have isolated avirulent non-fluidal mutants by this method. Preliminary analysis of some of the non-fluidal mutants has revealed an association with a 3.2 kb hybridizable sequence to a ColE1::Tn5 probe (Figure 4). These results were unexpected since the minimum size of any fragment of *P. solanacearun* DNA cut with the enzyme EcoR1 should be in excess of the size of Tn5 (5.7 kb). These mutants also contained an additional larger sequence greater than 5.7 kb in addition to the 3.2 kb fragment (Figure 3). Furthermore, we have isolated revertants back to the fluidal virulent phenotype that have lost this small hybridizable sequence (Figure 4). We are currently attempting the cloning of this fragment to ascertain its possible role in virulence. If this sequence is involved in virulence, it should be possible to isolate the wild-type gene and introduce it back into avirulent non-fluidal mutants to test if virulence is restored. Once the gene has been cloned, the product of this gene can be determined by *in vitro* transcription and translation systems.

6. ROLE OF INDIGENOUS INSERTION SEQUENCES IN THE MODULATION OF VIRULENCE

Insertion sequences (IS) are discrete units of DNA that are capable of transposing from one location in a genome to another site in the genome independent of homologous recombination systems in the host (Calos and Miller, 1980). The most thoroughly studied are those in *E. coli* and were first detected by their physical insertion into functional genes in *E. coli* resulting in an insertional inactivation of a specific gene product. Subsequently, naturally occurring IS elements have been detected in the bacterial phytopathogens *Agrobacterium tumefaciens* (Garfinkel and Nester, 1980), *Pseudomonas syringae* pv. *savastanoi* (Comai, personal communication), and *Pseudomonas syringae* pv. *glycinea* (Staskawicz, Dahlbeck, Miller and Damm, unpublished work), and the bacterial endosymbiont *Rhizobium meliloti* (Meade *et al.*, 1982).

In *Agrobacterium tumefaciens*, Garfinkel and Nester (1980) discovered insertion sequences that caused alterations in the ability of *Agrobacterium tumefaciens* to cause galls. One insertion was mapped in the T-DNA and was shown to be 1.0 mdalton in size. In another plant pathogen, *P. syringae* pv. *savastanoi*, Comai and Kosuge (personal communication) have detected mutants in indoleacetic acid production (IAA−). Physical characterization of this mutant by Southern blotting revealed a 1.2 kb insertion into the gene encoding for indoleacetamide *iaaM*. Definitive evidence that these elements are true IS elements still remains to be proved, but the evidence for IS activity is strongly suggestive, since these elements can cause polar mutations. The evidence for IS elements is less definitive for *P. syringae* pv. *glycinea* and is based on physical hybridization data of the *P. syringae* pv. *savastanoi* IS elements (PsIS1) with total genomic DNA of various races of *P. syringae* pv. *glycinea* (Figure 5). The copy number of this presumptive IS element is between 7

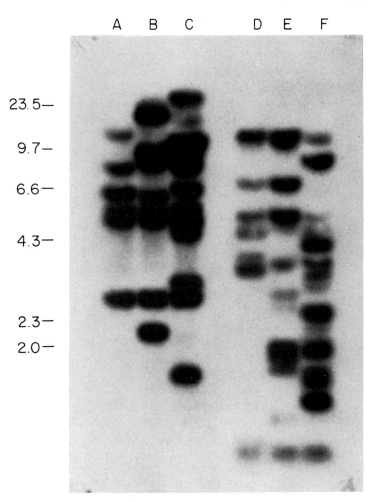

Figure 5. Presence of hybridizable sequences in the genome of *Pseudomonas syringae* pv. *glycinea* to an indigenous insertion sequence in *Pseudomonas savastanoi* (PsIS1). Total DNA was isolated from three races of *P. syringae* pv. *glycinea*, cut with the restriction enzymes EcoR1 (lanes A–C) and Sal I (lanes D–F), blotted on to nitrocellulose and probed with PsIS1. Lanes A and D, race 1; lanes B and E, race 2; lanes C and F, race 5

and 12 and is present in all races of *P. syringae* pv. *glycinea* tested. It is interesting that each race contains common hybridizable bands in addition to unique bands. Purified plasmid DNA has also been shown to contain hybridizable sequences with the PsIS1. It will be interesting to determine if a correlation in race specificity can be found with the insertion of an IS element into unique sites in the genome. If this is the case,the role of IS may be a genetic mechanism enabling a bacterium to mutate to a new race. This could best be explained by hypothesizing that endogenous IS change race specificity of a pathogen by inactivating genes for avirulence, allowing the pathogen to become virulent. This type of strategy by the pathogen would allow the pathogen to modulate its virulence by insertionally inactivating a single locus. The remarkable outcome of this strategy is that the gene can be restored at a latter point in time by the precise excision of this element, thus restoring the original gene function.

Finally, IS have been discovered in *Rhizobium spp.* Meade *et al.* (1982) have described the insertion of RmIS1 into a nif gene in the megaplasmid of *Rhizobium meliloti*. This was confirmed by Southern blotting, using a cloned probe pRmR2 for the nif region in *R. melioti*. It can also be hypothesized that other pathogens, both bacterial and fungal, will contain IS and that the uncovering of their role in gene regulation may lead to new insights into the evolution of host–pathogen interactions.

7. CONCLUSION AND FUTURE PERSPECTIVES

To date, only a handful of plant pathogens have been analysed at the gene level. It is hoped that a detailed analysis of gene functions in plant pathogens will allow us to conclude the role of specific hypothesized gene products in pathogenicity. The isolation of genes involved in host–pathogen specificity will also determine the gene products that are involved in the highly evolved recognitional systems inherent to many plant pathogens. Once the gene products have been determined in the pathogen, one might be able to hypothesize or even isolate the gene products of the host that encode for resistance genes. In many host–parasite combinations, genetic crosses have shown that host resistance genes are usually single dominant genes. Once these genes have been cloned and identified it may be possible to introduce these genes into host plants by transformation systems when they become available. In this manner, genetically engineered plants containing unique combinations of resistance genes may be achieved that could not be accomplished by normal sexual crosses. Finally, we shall be better able to design novel strategies for plant disease control with the aid of the tools of molecular genetics.

8. ACKNOWLEDGEMENTS

The expert technical assistance of Doug Dahlbeck, Judy Miller, and Debby Damm is greatly appreciated. The stimulating conversations and ideas of Nick Panopoulos and Al Ellingboe are acknowledged. Thanks are due to Luca Comai for the clone of PsIS1 and results prior to publication.

REFERENCES

Beringer, J. E., Beynon, J. L., Buchanan-Wollaston, A. V., and Johnston, A. W.B. (1978). Transfer of the drug-resistance transposon Tn5 to *Rhizobium, Nature (London)*, **276**, 633–634.

Boistard, P., and Boucher, C. (1978). P-1 group plasmids as tools for genetic study of phytopathogenic bacteria, *Proc. 4th Int. Conf. Plant Pathol., Angers, France*, pp. 17–30.

Boucher, C., Message, B., Debieu, D., and Zischek, C. (1981). Use of P-1 incompatibility group plasmids to introduce transposons into *Pseudomonas solanacearum, Phytopathology*, **71**, 639–642.

Calos, M., and Miller, J. (1980). Transposable elements, *Cell*, **20**, 579–596.

Collins, J., and Hohn, B. (1978). Cosmids: a type of plasmid gene-cloning vector that is packageable *in vitro* in bacteriophage heads, *Proc. Natl. Acad. Sci. USA*, **75**, 4242–4246.

Comai, L., and Kosuge, T. (1980). Involvement of plasmid deoxyribonucleic acid in indoleacetic acid synthesis in *Pseudomonas savastanoi, J. Bacteriol*, **143**, 950–957.

Comai, L., and Kosuge, T. (1982). Cloning and characterization of *iaaM*, a virulence determinant of *Pseudomonas savastanoi, J. Bacteriol*, in press.

Cross, J. E., Kennedy, B. W., Lambert, J. W., and Cooper, R. L. (1966). Pathogenic races of the bacterial blight pathogen of soybeans, *Pseudomonas glycinea, Plant Dis. Rep.*, **50**, 557–560.

Dalhbeck, D., and Stall, R. (1979). Mutations for change of race in cultures of *Xanthomonas vesicatoria, Phytopathology*, **69**, 634–636.

Ditta, G., Stanfield, S., Corbin, D., and Helinski, D. R. (1980). Broad host range cloning system for Gram-negative bacteria: construction of a gene bank of *Rhizobium meliloti, Proc. Natl. Acad. Sci. USA*, **77**, 7393–7351.

Ellingboe, A. H. (1976). Genetics of host-parasite interactions, in *Encyclopedia of Plant Physiology*, New Series, Vol. 4, *Physiological Plant Pathology* (Eds R. Heitefuss and P. H. Williams), Springer-Verlag, Heidelberg, pp. 761–778.

Fett, W. F., and Sequeira, L. (1981). Further characterization of the physiological races of *Pseudomonas glycinea, Can. J. Bot.*, **59**, 283–287.

Flor, H. H. (1947). Inheritance of reaction to rust in flax, *J. Agric. Res.*, **74**, 241–262.

Friedman, A. M., Long, S., Brown, S., Buikema, W., and Ausubel, F. (1982). Construction of a broad host range cosmid cloning vector and its use in the genetic analysis of *Rhizobium* mutants, *Gene*, in press.

Gabriel, D., Ellingboe, A., and Rossman, E. (1979). Mutations affecting virulence in *Phyllosticta maydis. Can. J. Bot.*, **57**, 2639–2643.

Gabriel, D., Lisker, N., and Ellingboe, A. (1982). The induction and analysis of two classes of mutations affecting pathogenicity in an obligate parasite, *Phytopathology*, in press.

Garfinkel, D., and Nester, E. (1980). *Agrobacterium tumefaciens* mutants affected in crown gall tumorigenesis and octopine catabolism, *J. Bact.* 144, 732–743.

Hohn, B. (1981). *In vitro* packaging of cosmid DNA, *Methods Enzymol.*, **68**, 299–309.

Kelman, A. (1954). The relationship of pathogenicity in *Pseudomonas solanacearum* to colony appearance on a tetrazolium medium, *Phytopathology*, **44**, 693–695.

Kleckner, N. (1981). Transposable elements in prokaryotes, *Annu. Rev. Genet.*, **15**, 341–404.

Kleckner, N., Roth, J., and Botstein, D. (1977). Genetic engineering *in vivo* using transposable drug-resistance elements. New methods in bacterial genetics, *J. Mol. Biol.*, **116**, 125–159.

Loper, J. E., and Kado, C. I. (1979). Host range conferred by the virulence-specifying plasmid of *Agrobacterium tumefaciens, J. Bacteriol,* **139,** 591–596.

Long, S. R., Buikema, W., and Ausubel, F. (1982). Cloning of *Rhizobium meliloti* nodulation genes by direct complementation of nod mutants, *Nature (London),* **298,** 485–488.

Meade. H., Long, S., Ruvkun, G., Brown, S., and Ausubel, F. (1982). Physical and genetic characterization of symbiotic and auxotrophic mutants of *Rhizobium meliloti* induced by transposon Tn5 mutagenesis, *J. Bacteriol.,* **149,** 114–122.

Panopoulos, N. J. (1981). Emerging tools for *in vitro* and *in vivo* genetic manipulations in phytopathogenic Pseudomonas and other non-enteric Gram-negative bacteria, in *Genetic Engineering in the Plant Sciences* (Ed. N. J. Panopoulos), Praeger Publishers, New York, pp. 163–186.

Panopoulos, N. J., Staskawicz, B. J., and Sandlin, D. E. (1979). Search for plasmid-associated properties and for a cloning vector in *Pseudomonas phaseolicola,* in *Plasmids of Medical, Environmental and Commercial Importance* (Eds K. N. Timmis and A. B. Puhler), Elsevier/North Holland Biomedical Press, Amsterdam, pp. 365–371.

Smidt, M., and Kosuge, T. (1978). The role of indole-3-acetic acid accumulation by alphamethy tryptophan-resistant mutants of *Pseudomonas savastonoi* in gall formation on oleanders, *Physiol. Plant Pathol.,* **13,** 203–214.

Thomashow, M., Panogopoulos, C., Gordon, M., and Nester, E. (1980). Host range of *Agrobacterium tumefaciens* is determined by the Ti-plasmid, *Nature (London),* **283,** 794–796.

Further Reading

Molecular Biology of The Gene (Ed. J. D. Watson), W. A. Benjamin, Menlo Park, CA (1976).

Molecular Genetics: An Introductory Narrative (Eds S. Stent and R. Calander), W. H. Freeman, San Francisco (1978).

Cold Spring Harbor Symposia On Quantitative Biology, Vol. XLV, Movable Genetic Elements, Parts 1 and 2, Cold Spring Harbor Laboratory (1981).

Genetic Engineering: Principles and Methods (Eds J. K. Setlow and A. Hollaender), Plenum Press, New York (1980).

Advanced Bacterial Genetics: A Manual For Genetic Engineering (Eds R. W. Davis, D. Botstein, and J. E. Roth), Cold Spring Harbor Laboratory, New York (1980).

Biochemical Plant Pathology
Edited by J. A. Callow
© 1983 John Wiley & Sons Ltd

11

Cell Walls and Other Structural Barriers in Defence

J. P. RIDE

Department of Microbiology, University of Birmingham, P.O. Box 363, Birmingham B15 2TT, UK

1. INTRODUCTION

The polysaccharide-based walls which surround plant cells act as natural barriers to infection by microorganisms. Thus the relative infrequence of bacterial entry into plant cells may be related to the comparative inability of these organisms to penetrate the protective walls. The frequent requirement for vectors in the transmission of viruses must also be dependent in part on the barrier posed by the outer walls. The strength of the wall may be naturally enhanced by the presence of materials such as cutin, suberin, lignin, calcium, and silicon. The presence of lignin in xylem vessels, for instance, may help explain why penetration of this tissue by pathogens is relatively uncommon.

The barrier presented by the walls is rarely complete, in that natural openings (e.g. stomata, lenticels) and intercellular spaces provide easy routes of progress for those parasites capable of finding them and surviving in the extracellular environment. Nevertheless, most parasites must at some stage breach the walls, either to allow entry and colonization, or to release nutrients essential for continued growth in the intercellular spaces. Many fungi, both parasites and saprophytes, produce cell wall-degrading enzymes in artificial culture (see Chapter 7) and hence it is perhaps surprizing that the cell walls of living plants are not breached more often. There are obviously many reasons for the failure of potential parasites to enter host cells, but one possibility which has recently been attracting more attention is that the host cell walls are modified to a more resistant form on approach of a potential parasite. It is becoming increasingly apparent that the walls of plants do not just present a static barrier to infection but are capable of altering dynamically, sometimes extremely quickly, on contact with potential pathogens. Alterations may involve changes to the existing wall, the deposition of new wall-like material on the outside or inside of the cells, or the production of a completely new barrier by cell division.

Numerous reports in the literature, many based on microscopical observations, suggest that wall alterations occur on attempted infection, notably in response to fungi. However, knowledge of the molecular composition of the modified walls, how the changes are induced, and their importance in determining resistance and specificity is very slight. Much of the identification of molecular components has depended on the use of a few traditional histochemical tests and the application of modern microscopical and analytical techniques is really only just beginning. Many of the crucial questions concerning a role in resistance remain unanswered: for instance, does an alteration occur at the right time, in the right place, and in sufficient concentration to explain the observed cessation in fungal growth? These are not easy questions to answer. Both altered and virgin cell walls may not just present a structural barrier to physical penetration by the invading microorganism, but may also be resistant to enzyme attack, either due to shielding of susceptible components in the wall or by chemical alteration of the substrates to forms unsuitable for the parasite's enzymes. Walls may contain components which directly inhibit fungal growth; they may contain inhibitors of enzymes, or toxins, of the parasite, and they might reduce 'molecular interchange' between host and parasite, i.e. inhibit the movement of nutrients and water from host to parasite and the movement of enzymes and toxins from parasite to host.

This somewhat confusing combination of possibilities is made more complex by the knowledge that several different defence mechanisms, some of which are probably currently unknown, are likely to be acting in a concerted fashion against the invading organism. Sorting out the degree of importance of any one mechanism is thus very difficult and further complicated by the possibility that a given plant may respond in several different ways depending on the nature of the pathogen, the nature and physiological state of the affected tissue, and the environmental conditions.

Despite these difficulties, it is becoming clear that the role of the plant cell wall in defence is not just that of a static physical barrier to penetration. It is a dynamic, multifunctional, and possibly very important component in the integrated battery of defence mechanisms controlled by the plant. In the same way that phytoalexin accumulation involves the biosynthesis of many novel compounds, the cell walls of challenged plants may contain many new constituents of which we are currently unaware. The area is ripe for investigation.

2. PRE-EXISTING STRUCTURAL DEFENCES

The primary cell wall is a natural physical barrier to those microorganisms which are incapable of producing the appropriate degradative enzymes. In addition to acting as a direct impedance to the invading organism it may also sieve out or inhibit potentially harmful enzymes of the parasite (see Chapter 7). The basic polysaccharide and glycoprotein structure of the primary cell wall (Chapter 7) may also be naturally strengthened and made considerably more resistant to microbial enzymes by overlayering or impregnation with materials such as cutin, suberin, waxes, lignin, silicon, and calcium. Detailed aspects of the biochemistry of cutin and suberin, their microbial breakdown and potential role in resistance are covered in Chapter 6.

Lignin is a term covering a group of aromatic phenolic polymers which are related in being formed by peroxidase-initiated dehydrogenative polymerization of three hydroxycinnamyl alcohols: *trans-p*-coumaryl, *trans*-conferyl, and *trans*-sinapyl alcohols (Figure 1). The intermonomer linkages between these C_6–C_3 (phenylpropanoid) units can vary since the final free-radical coupling reactions are not enzymatically controlled. The proportion of the monomer units can also vary and thus it is not possible to pronounce an exact structure for the polymer. Lignin occurs naturally throughout the middle lamella, primary and secondary walls of xylem tissues, often being at highest concentrations in the middle lamella at cell corners (Wardrop, 1971). It may also be found in other cells such as sclerenchyma fibres, bundle sheath cells, and occasionally, epidermal cells. Lignin is deposited in the matrix of the wall between the cellulose microfibrils, and its complex aromatic network provides a rigidity and resistance to compressive forces as well as considerable resistance to microbial degradation.

The properties of lignin indicate a potential role in resistance to pathogens and an association between the presence of preformed lignin and the limitation of fungal growth has been noted for some diseases. Thus Hursh (1924) noted that *Puccinia graminis* f.sp. *tritici* was restricted in wheat stems to the chlorenchyma, being unable to colonize the adjacent lignified sclerenchyma fibres. Wheat varieties in which the chlorenchyma was divided into small bundles by the fibres exhibited smaller individual lesions on infection than those in which the chlorenchyma was more continuous. Similarly, the smaller lesions caused by *Colletotrichum lindemuthianum* on older hypocotyls of bean is associated with an apparent inability of the mycelium to penetrate lignified fibres and xylem elements (Griffey and Leach, 1965). Observations such as these suggest a role for naturally lignified structures in disease resistance but the possibility that components other than lignin are more important cannot be excluded.

Silicon is a natural component of the epidermis and woody tissues of many plants, although the distribution pattern depends on the plant (e.g. Scurfield *et al.*, 1974; Postek, 1981). The element exists in soil solution as monosilicic acid (H_4SiO_4) and is transported as such in plants. It is generally deposited as amorphous hydrated silica ($SiO_2.nH_2O$), its form varying from thin layers around and within the cell walls to solid masses filling the cell lumina. It may function to strengthen plant organs, reduce transpiration, and perhaps increase resistance to pathogens. The silica content of rice is apparently related to its resistance to blast (*Pyricularia oryzae*) with soil supplementations of silicate enhancing both silica content and resistance (Volk *et al.*, 1958; Akai and Fukutomi, 1980).

The levels of divalent cations such as Ca^{2+} and Mg^{2+} have also been associated with resistance in some cases. Calcium is important in cell wall structure in that it forms cross-links, particularly between adjacent carboxylic groups of the pectic component. Bateman and Lumsden (1965) noted that the increase in resistance with age of bean hypocotyls to *Rhizoctonia solani* could be correlated with an increase in calcium content and a decrease in methoxyl content of the pectin. The corresponding increase in calcium cross-links was proposed as an explanation for the observed increase in the resistance of tissue to maceration by fungal polygalacturonases, and hence the increase in resistance to disease. The ability of *Sclerotium rolfsii* to attack older bean hypocotyls was apparently related to its capacity to produce oxalic acid which, by complexing calcium, allowed enzymic hydrolysis of the pectic fraction (Bateman and Beer, 1965). However, alterations in wall structure other than an increase in calcium might be involved in the resistance to fungal enzymes and the potential importance of other defence mechanisms, such as phytoalexins, further complicates the issue (Bateman *et al.*, 1969).

In addition to the resistance provided by the primary and strengthened cell walls described above, there have been numerous attempts to correlate other morphological features, e.g. form and arrangement of stomata, leaf hairs, lenticels, and xylem vessels, with resistance. For a review of this area where numerous correlations but little conclusive evidence exist, see Royle (1976).

3. POST-INFECTIONAL STRUCTURAL DEFENCES

Structural defences induced by potential pathogens may be artificially classified according to the physical appearance or chemical nature of the alteration.

3.1. Alterations to existing cell walls

Lignification apparently occurs in many plants in response to infection or attempted infection by fungi, bacteria and occasionally viruses (Vance *et al.*, 1980). In some instances it is also a response to wounding. Identification of the polymer has relied heavily on histochemical tests, e.g. the phloroglucinol–HC1 test for cinnamaldehyde side-chains, which can be misleading. Degradative chemical tests (acidolysis, nitrobenzene oxidation, permanganate–peroxide oxidation) are not only more reliable indicators of the presence of lignin but the quantities and proportions of the products can also give an idea of the quantity and structure of the original polymer.

Many different lignins can be formed from the three hydroxycinnamyl alcohol precursors (Figure 1): lignins derived principally from coniferyl alcohol units are termed guaiacyl lignins, whilst those derived from both coniferyl and sinapyl units are termed guaiacyl-syringyl lignins. The proportion derived from *p*-coumaryl alcohol (giving *p*-hydroxyphenyl residues) is usually small. In cases where the composition of infection-induced lignin has been at least partially examined it is clear that the induced polymers may be quite different from the polymers in healthy plants. Thus a guaiacyl lignin is formed in Japanese radish roots in response to *Peronospora parasitica* while the polymer in healthy roots is of a guaiacyl-syringyl nature (Asada and Matsumoto, 1972). Similarly in muskmelons protected against *Colletotrichum lagenarium* by prior inoculation with the same pathogen there is an increase in the synthesis of guaiacyl, and later *p*-hydroxyphenyl, but not syringyl lignin units (Touzé and Rossignol, 1977). Conversely, the lignin synthesized in the leaves of wheat seedlings in response to non-pathogenic fungi has a very much higher syringyl and *p*-hydroxyphenyl content than the predominantly guaiacyl lignin from healthy leaves. The polymer also differs in its staining reactions and ultraviolet spectrum (Ride, 1975).

Although in some cases lignification may be a response to wounding or general damage, in many cases it is clearly not so. Thus, wounding wheat leaves, or treating them with a range of deleterious compounds known to induce phytoalexins in other plants, does not generally induce lignification. Lignification can, however, be elicited by a range of filamentous fungi, this specificity of induction in itself suggesting that the response has an important function (Pearce and Ride, 1980). The chemical nature of the fungal components that induce lignification is unknown, although the polysaccharide chitin, which is a common constituent of fungal cell walls, is known to induce lignification in wheat (Pearce and Ride, 1982). Thus wall fragments, possibly released by host polysaccharases, may be the relevant triggers. However, the possibility that in some situations specific lignification elicitors do not exist cannot be ignored. The lignification-inducing factor isolated from infected radish roots is interesting in this context in that it is thought to be a host product (Asada *et al*., 1979), possibly being produced in response to general 'stress' caused by the parasite.

The biosynthesis of induced lignin and its control has not been studied in depth for any host-parasite system. The probable pathway for the biosynthesis of lignin from phenylalanine in wheat is shown in Figure 1 and this route would be essentially the same in most higher plants (Gross, 1980). The initial conversion of phenylalanine to *trans*-cinnamic acid is an irreversible step catalysed by the enzyme phenylalanine ammonia-lyase (PAL) with a small amount of carbon in the Gramineae also entering from tyrosine via tyrosine ammonia-lyase (TAL) (see also Chapter 17). The conversion of phenylalanine to cinnamic acid may lead to the biosynthesis of many phenolic compounds apart from lignin (Chapter 17) and hence problems arise in interpreting the importance of increased PAL (and TAL) activity in infected tissues. The possibility that infection-induced lignification might involve the utilization of existing pools of phenolic precursors further complicates the issue. Nevertheless, the fact that increases in PAL activity have been spatially and temporally associated with lignification in infected tissues (Maule and Ride, 1976; Vance and Sherwood, 1976a), together with the observed incorporation of radiolabelled phenylalanine into induced lignin (Touzé & Rossignol, 1977; Maule and Ride, 1982), suggests that PAL activity is important and that a major portion of the carbon for lignification is derived directly from phenylalanine on infection.

In addition to increases in ammonia-lyase, lignifying tissues have also shown increases in cinnamate-4-hydroxylase, hydroxycinnamate:CoA ligase, *O*-methyltransferase, and peroxidase activities (Maule and Ride, 1976; Maule, 1977; Vance and Sherwood, 1976a; Vance *et al*., 1976). The activities of other enzymes in the pathway await determination but it is clear that induced lignification is an active process involving the rapid increase in activities of numerous enzymes (possibly as multienzyme complexes), a process which probably involves RNA and protein synthesis since it is inhibited by cycloheximide, actinomycin D, and a range of other antimetabolites (Vance and Sherwood, 1976b; Asada *et al*., 1979; Bird and Ride, 1981; see also Chapter 17). At present we know nothing about the way fungi or their products induce these changes, although it is possible that in some cases the response is mediated by natural plant hormones such as ethylene (Vance *et al*., 1980).

The factors regulating the often different structure of induced lignin are also largely unexplored. The increased syringyl content of induced lignin in wheat cannot be explained by changes in the specificity of either the *O*-methyltransferase or hydroxycinnamate:CoA ligase activity (Maule and Ride, 1976; Maule, 1977). The CoA ligase from healthy and infected wheat exhibits a preference for *p*-coumaric acid and little or no affinity for sinapic acid. In this it resembles the hydroxycinnamate:CoA ligase from many other plants (Gross, 1977), although an isoenzyme with a high affinity for sinapate has been associated with the natural biosynthesis of syringyl lignin in older tissues of *Petunia* (Ranjeva *et al.*, 1976). Changes in the specificity of the cinnamoyl-CoA reductase or the cinnamyl alcohol dehydrogenase may be important in regulating the structure of induced lignins, but this has yet to be examined. Ohguchi and Asada (1975) have suggested that changes in the levels of specific peroxidase isoenzymes may account for changes in the guaiacyl to syringyl ratio in infected radish roots. However, it seems perhaps unlikely that control would be exerted at such a late stage of biosynthesis when, presumably, the precursors had already been synthesized in the protoplasm and exported to the wall. Logically, perhaps, the point most likely to be important in the control of the guaiacyl to syringyl ratio is that at which feruloyl-CoA is converted to either coniferyl aldehyde or 5-hydroxy-feruloyl-CoA, and studies on the relative activities of the enzymes catalysing these steps might be profitable.

The cellular location of induced lignification can vary according to the host–parasite combination, although the factors which control the site of deposition are unknown. In Japanese radish lignification follows a relatively normal pattern with lignin deposition being restricted to the wall and greatest in the middle lamella (Asada and Matsumoto, 1971). Presumably the precursors are synthesized and 'packaged' in the cytoplasm prior to export to the wall where peroxidase and H_2O_2 initiate polymerization. In infected wheat, however, lignification may also occur throughout the protoplasm of cells, particularly those which are responding hypersensitively to fungal attack (Beardmore *et al.*, 1982; Maule and Ride, 1982). In this case the increasing disorganization of the cytoplasm involved in hypersensitivity presumably causes the release of precursors normally destined for the wall into the cytoplasm to be polymerized by cytoplasmic peroxidase.

Conclusive evidence that induced lignification is important in resistance, and hence in determining specificity, is difficult to find. Some of the most convincing evidence relates to the lignification of newly formed papillae and is discussed in the next section. However, some correlations exist between the induced lignification of existing structures and resistance, and these indicate a possible role of defence. Thus in wounded wheat leaves lignification occurs rapidly in response to non-pathogenic fungi, the polymer being deposited around the wound in advance of the hyphae (Ride, 1975). The response occurs more slowly in response to pathogenic *Septoria* species (Ride, 1975), *Fusarium graminearum*, and a *Penicillium* species (Ride, 1983). Prior inoculation of wounds with non-pathogenic fungi or chitin induces lignification and protects the leaves to a large degree against subsequent inoculation with pathogens (Ride, 1975, 1983). Lignification is also apparently involved in the resistance of wheat to stem rust (*Puccinia graminis* f.sp. *tritici*). Cultivars carrying the *Sr*-5 and *Sr*-6 alleles for resistance exhibit rapid lignification of

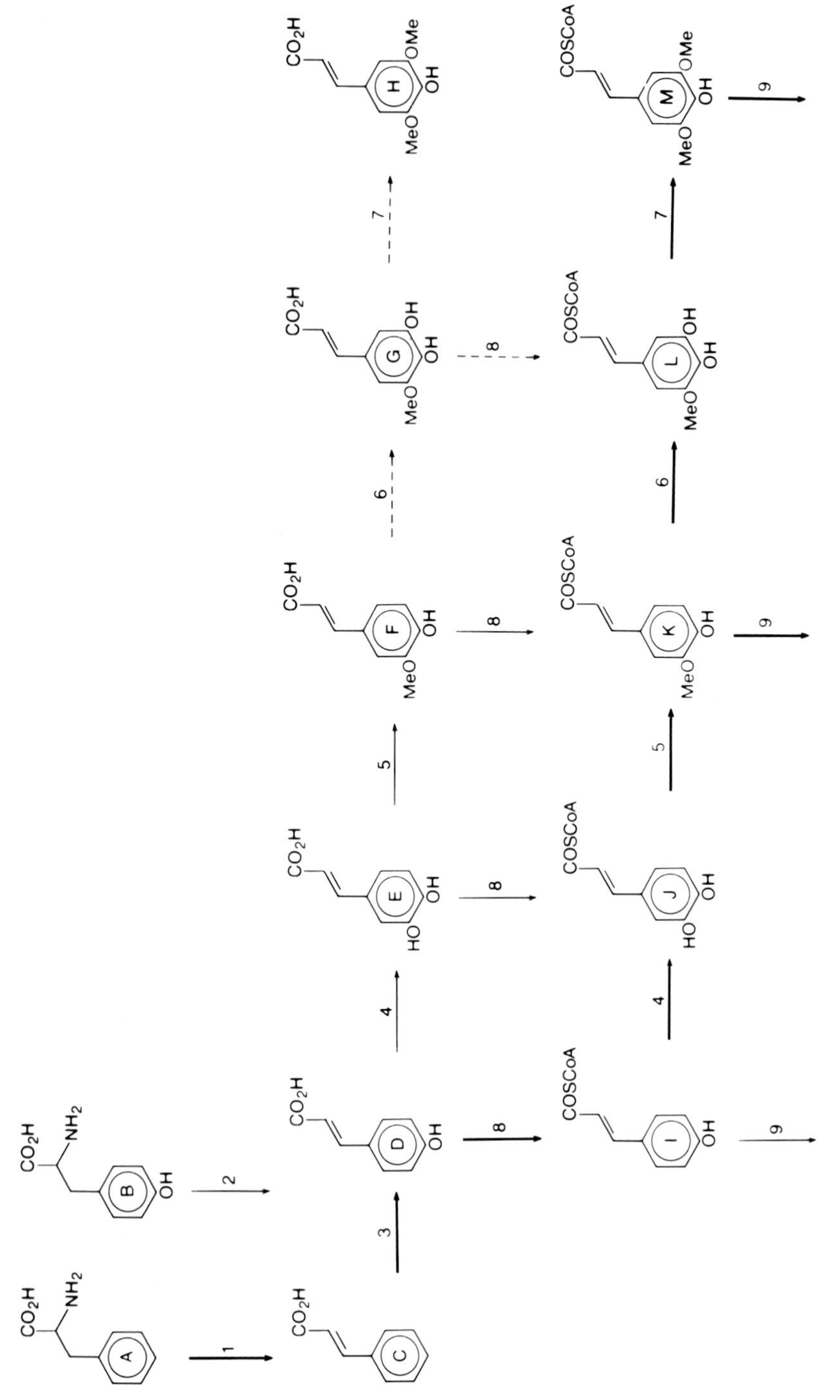
CO2H
NH2
OH
MeO
OMe
HO
COSCoA
A
B
C
D
E
F
G
H
I
J
K
L
M
1
2
3
4
5
6
7
8
9

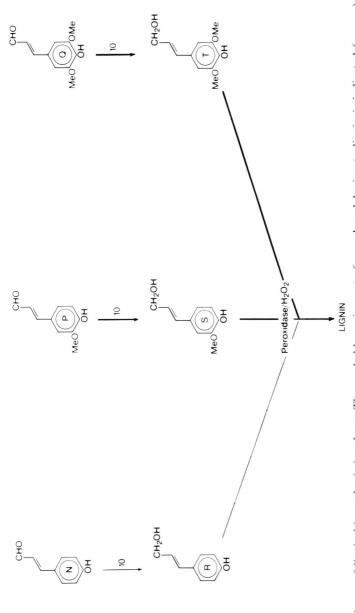

Figure 1. Pathway of lignin biosynthesis in wheat. The probable main route from phenylalanine to lignin is indicated (——), together with pathways of lesser significance (– – –). The intermediates are (A) phenylalanine, (B) tyrosine, (C) *trans*-cinnamic acid, (D) *p*-coumaric acid, (E) caffeic acid, (F) ferulic acid, (G) 5-hydroxyferulic acid, (H) sinapic acid, (I–M) the CoA esters of the acids (D–H), (N) *p*-coumarylaldehyde, (P) coniferylaldehyde, (Q) sinapylaldehyde, (R) *p*-coumaryl alcohol, (S) coniferyl alcohol, and (T) sinapyl alcohol. The enzymes are (1) phenylalanine ammonia-lyase, (2) tyrosine ammonia-lyase, (3) cinnamate 4-hydroxylase, (4) *p*-coumarate 3-hydroxylase, (5) caffeate *O*-methyltransferase, (6) ferulate 5-hydroxylase (hypothetical), (7) 5-hydroxyferulate *O*-methyltransferase, (8) hydroxycinnamate: CoA ligase, (9) cinnamoyl-CoA reductase, (10) cinnamyl alcohol dehydrogenase

hypersensitively responding cells (Beardmore *et al.*, 1983). Conversely, cultivars of wheat with differing resistance to *Septoria nodorum* do not show differences in the rate of induced lignification (Bird and Ride, 1981).

Difficulties in determining the importance of lignification in resistance stem in part from insufficient knowledge of the ways in which the process might hinder microbial growth and the quantities of lignin required to prevent growth. Lignification could hinder fungal progress in a number of ways. Firstly, the polymer may make the walls more resistant to mechanical penetration. There is, as yet, no direct evidence that this is the case with induced lignins, or that such resistance would be an important factor in defence. Secondly, lignification may make the host walls more resistant to degradation by fungal enzymes, either by shielding the polysaccharide substrates or by chemically altering them to make them less suitable subtrates. Certainly the induced lignin in wheat leaves makes the walls very resistant to enzyme attack and fungal degradation (Ride, 1975; 1980) and such resistance may prevent fungal penetration in addition to preventing host-cell death and nutrient leakage via the action of pectic enzymes. Thirdly, the relatively impermeable nature of lignin might result in restricted molecular interchange between host and parasite, i.e. limit the diffusion of enzymes and toxins from the fungus to the host, and conversely nutrients and water from the host to the fungus. This possibility has not been tested experimentally. Fourthly, the low molecular weight phenolic precursors and free radicals produced during polymerization may be toxic to the fungus or inactivate important fungal enzymes. In this context it is interesting that increased levels of two antifungal lignin precursors, conferyl alcohol and conferyl aldehyde, have been associated with the resistance of flax to *Melampsora lini* (Keen and Littlefield, 1979). Lastly, it is conceivable that the tips of hyphae that are in close contact with lignifying host material might themselves become lignified and lose the plasticity necessary for growth. There is little direct evidence that this occurs although it is known that chitin, cellulose, and hydroxyproline-rich proteins, all of which occur in fungal walls, can serve as matrices for the *in vitro* polymerization of lignin precursors.

Some insight into the importance of lignification might be provided in the future by the use of specific inhibitors. For example, the relatively specific inhibitors of PAL, α-aminooxyacetic acid and α-aminooxy-β-phenylpropionic acid, have already been shown to inhibit lignification in virus-infected tobacco, with a corresponding increase in virus multiplication (Massala *et al.*, 1980; see also Chapter 17). The results are not conclusive since inhibition of PAL activity will affect the biosynthesis of many phenolics apart from lignin, and also the possibility of secondary effects arising from the lower rate of phenolic synthesis cannot be ruled out. Hopefully, more specific inhibitors of lignin biosynthesis will be available in the future.

In addition to the formation of 'traditional' lignins, the walls of some plants are modified on infection by phenolic deposits of different nature. Thus a lignin-like material is formed in potato tubers with major gene resistance on inoculation with incompatible races of *Phytophthora infestans* (Friend, 1981). Much of this material appears to be *p*-coumaric and ferulic acids esterified probably to carbohydrates in the cell wall. Friend (1976) has suggested that modification of the wall in this way may chemically alter the polysaccharides sufficiently so that they are no longer suitable substrates for fungal enzymes. Oxidative cross-linking between wall-bound phenolics may further aid resistance (Friend, 1981). However, the alterations appear to occur much later than the hypersensitive response and the ensuing cessation of hyphal growth, and so are possibly later, secondary features of resistance.

The oxidation and polymerization of phenolic compounds, typified by the production of brown pigments, is a common feature of the response of many plants to infection, often occurring in the walls and protoplasm of cells undergoing a hypersensitive response to an incompatible pathogen. The structure and origin of the oxidized material is often difficult to determine. In some cases the quinones produced on phenol oxidation may be polymerized to melanins, which are analogous in many ways to lignin and may function in a similar way in defence. Quinones and melanins may also be relatively non-specific enzyme inhibitors, and thus inhibit the important wall-degrading enzymes of potential pathogens (e.g. Byrde, 1957). Quantitative estimates of melanins are scarce, although Langcake and Wickens (1975) demonstrated that a faster rate of deposition of a melanoid pigment occurred in rice leaves in response to *Pyricularia oryzae* when the plants had been previously treated with dichlorocyclopropane derivatives, compounds which enhance the natural resistance of rice to this fungus. However, phytoalexins also appear to be active in this enhanced resistance (Cartwright *et al.*, 1980), illustrating the complementary nature of defence mechanisms which is probably a feature of many host–parasite interactions (Bell, 1980); e.g. phytoalexin accumulation may slow down fungal growth sufficiently to allow time for structural modifications to be effective or *vice versa*.

Suberization is also a relatively frequent response to infection or wounding and it has been associated with resistance in some cases (e.g. Walker and Wade, 1978). The deposition of suberin is often associated with renewed cell division and is thus considered in Section 3.3.

The accumulation of calcium and sometimes other ions can also occur in some infections, e.g. *Rhizoctonia* infections of bean hypocotyls (Bateman, 1964), and this may make pectate in the wall more resistant to enzymic degradation. The use of electron probe (energy-dispersive X-ray) micro-analysis for the location and quantitation of elements such as calcium and silicon in infected tissues will hopefully lead to a better understanding of the role of such inorganic accumulations in resistant reactions. For instance, this technique has been used to show that an electron-opaque material that is deposited on and within mesophyll cell walls in bean leaves after infection by the cowpea rust contains silicon (Heath, 1979). Interestingly, injection of bean rust exudates, or extracts of bean rust-infected leaves, into healthy bean tissues suppressed the formation of these silicon-containing deposits on subsequent inoculation with cowpea rust and increased the frequency of haustoria (Heath, 1981). This suggests that the ability to suppress the formation of these deposits is an important element in the success of bean rust as a bean parasite.

3.2. Deposition of new wall-like material

The deposition of wall-like material on the internal or external surface of the cell wall is a common response of plants to infection. Deposits on the interior of the wall, termed papillae, have been noted in many plant–fungus interactions, particularly in relation to the resistance of epidermal cells to penetration. The attempted penetration of many plants, but particularly members of the Gramineae, by fungi, results in the formation of a papilla between the plasmalemma and the host cell wall, at a point directly opposite the fungal appressorium and penetration peg. The adjacent epidermal wall is sometimes also altered to produce a disc-shaped area frequently termed a halo. The formation of a papilla can be very rapid and is often preceded by an aggregation of the cytoplasm at the point of attempted penetration (Figure 2).

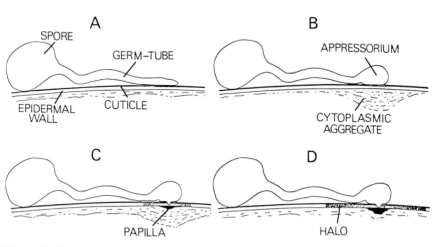

Figure 2. General processes observed, in some plants, at sites of attempted penetration of epidermal cells by non-infecting plant-parasitic fungi. During the development of the infection structures of the fungus, i.e. the adhesive appressorium and fine penetration peg, the cytoplasm of the host aggregates and deposits a lump of new wall-like material, the papilla. The adjacent epidermal wall may also be affected to produce a 'halo'

Papillae vary enormously in their appearance, from homogeneous to multi-textured or multi-layered structures, from small hemispherical deposits to elongated encasements around the penetrating hyphae. They may be a response to the physical damage caused by the fungal penetration peg and/or to some form of chemical encitement (Aist, 1976).

Identification of the chemical components of papillae has relied almost entirely on histochemical tests. The commonest constituent appears to be callose, a $\beta 1 \rightarrow 3$ glucan usually identified by its fluorescence after staining with aniline blue. Lignin is a common strengthening component, with suberin, silicon, cellulose, protein, and autofluorescent materials also having been reported. The composition may depend not only on the plant species but also on the nature of the pathogen, the type of cell infected and the age of the papilla, etc.

Many correlations exist between the production of papillae and resistance to fungal penetration (Aist, 1976). Thus, for example, inoculation of reed canary grass leaves with non-pathogenic fungi incites the formation of lignified papillae and haloes underneath the appressoria at sites of attempted penetration (Sherwood and Vance, 1976). Even with the pathogen *Helminthosporium catenarium* about 99% of attempted penetrations fail, with approximately 78% of appressoria being opposed by a well developed, lignified papilla. Penetration occurs only when papillae are absent or poorly developed. Treatment of leaves with cycloheximide, to inhibit host protein synthesis, blocks papilla formation and enhances penetration (Vance and Sherwood, 1976).

Lignified papillae and haloes also occur in wheat leaves and coleoptiles in response to attempted penetration by non-pathogenic fungi (Young, 1926; Ride and Pearce, 1979). In leaves, the papillae and haloes are formed shortly after appressorial development, and lignification can be detected by its auto-fluorescence (Figure 3), histochemical reactions, and by autoradiography following feeding with ^{3}[H] phenylalanine or ^{14}[C] cinnamic acid (Figure 4). Lignified papillae and haloes are resistant to wall-degrading enzymes, a resistance which is destroyed by chemical delignification (Ride and Pearce, 1979). They are also resistant to *in vitro* degradation by a range of fungal species, wheat pathogens generally being no more capable of degrading the structures than non-pathogens (Ride, 1980). Papillae and haloes taken from leaves as early as 12 h after inoculation show substantial resistance to degradation, demonstrating the possible importance of relatively small quantities of lignin, quantities which are undetectable by most traditional histochemical tests.

Correlations alone do not prove conclusively the importance of a factor in resistance and the possibility that other resistance mechanisms, operating prior to or during papilla development, are equally or more important cannot be ruled out. Papilla formation might then be viewed as perhaps a secondary response to a weakened parasite (Aist, 1976). Experiments with general

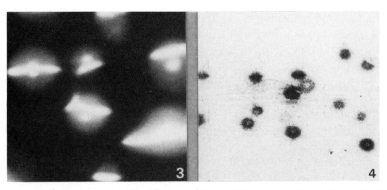

Figure 3. Autofluorescence of lignified papillae and haloes on the surface of wheat leaves 24 h after inoculation with a non-pathogenic isolate of *Botrytis cinerea*. The fungus has been removed and in this surface view papillae appear as bright central dots in the roughly circular haloes, with the lateral (anticlinal) walls also being lignified where they pass through the haloes. ×480

Figure 4. Surface autoradiogram of a wheat leaf harvested 12.5 h after inoculation with *B. cinerea* following a 2-h exposure to [^{14}C]cinnamate. Note the marked incorporation of the labelled lignin precursor into the halo areas. The leaf was pre-extracted with 70% ethanol, water, and hot dilute alkali to remove soluble and wall-esterified phenolics. ×144

inhibitors such as cycloheximide cannot solve the problem since many defence mechanisms may be simultaneously inhibited. Thus in most host–parasite systems where they occur the role of papillae is still in doubt. Powdery mildew (*Erysiphe graminis*) infections of barley and wheat provide an interesting example. Here the production of papillae has been investigated with respect to both race-specific incompatibility and the general resistance of compatible cultivars, often with conflicting results. Some workers have concluded that the structures are associated with race-specific incompatibility (Lin and Edwards, 1974; Johnson *et al.*, 1979), while others conclude that they are not involved, except perhaps as a general defence mechanism (Bushnell and Bergquist, 1975; Hyde and Colhoun, 1975), or play only a minor role (Stanbridge *et al.*, 1971).

Aist and co-workers have experimentally manipulated papilla formation in barley coleoptiles inoculated with a compatible race of mildew in an attempt to determine the role of papillae in resisting fungal ingress. Thus, for example, inoculated, live coleoptiles were centrifuged sufficiently to produce a cytoplasm-rich and cytoplasm-poor zone within each cell (Waterman *et al.*, 1978). Papilla deposition occurred at interaction sites in cytoplasm-rich zones only. Penetration efficiency was, however, no higher in cytoplasm-poor regions than in non-centrifuged coleoptiles, implying that the penetration failures that occur normally in this host–pathogen combination are not due to papillae. However, penetration efficiency was much lower in the cytoplasm-rich areas, apparently due to papilla production. Other artificially induced, oversize papillae were also capable of preventing mildew penetration (Aist *et al.*, 1979). Acoustic microscopy of such papillae suggests that they are more elastic, viscous, or dense than the normal papillae and hence may provide mechanical resistance to fungal penetration (Israel *et al.*, 1980).

Much of the confusion surrounding the role of papillae in resistance to cereal mildews could be resolved by knowledge of the chemical composition of the structures. Papillae, even in the same leaf, may have crucially different chemical compositions. Thus papillae associated with penetration failures in barley contain more basic-staining material (Lin and Edwards, 1974) and are more autofluorescent (Koga *et al.*, 1980) than those which are associated with successful penetration. It must also be remembered that each gene for incompatibility may present more than one 'hurdle' for the parasite and that only some individuals from the parasite population will surmount each hurdle (Ellingboe, 1972). These hurdles may be numerous (Johnson *et al.*, 1979) and the importance of any one hurdle (e.g. papilla formation) may depend on many factors, e.g. age of the plant, tissue type, genotype, and environmental conditions, thus explaining the apparently conflicting nature of some results.

3.3. Barrier formation by renewed cell division

The formation of a zone of 'barrier' cells by renewed cell division is a common response of plants, particularly woody plants and storage roots, to wounding or infection. Such cambial activity separates the infected zone from the healthy cells and may eventually result in the sloughing off, or at least the desiccation, of the infected cells. Thus the phenomenon of lesion rejection in infections of potato tubers by *Phoma exigua* is associated with periderm formation underneath the infected region (Walker and Wade, 1976). Cells in such a barrier zone commonly contain suberin, lignin, or other unidentified strengthening components, although the importance of these constituents is frequently unknown. However, in wounded oaks the deposition of suberin appears crucial to the resistance of the barrier zone to invasion (Pearce and Rutherford, 1981). Living trees commonly respond to wounding by the production of a barrier of cells [termed 'wall 4' in the 'compartmentalization of decay in trees' (CODIT) hypothesis (Shigo and Marx, 1977)], laid down by the vascular cambium in the vicinity of the wound. In oaks this parenchyma barrier is suberized when the underlying sapwood is colonized by the decay fungus *Stereum gausapatum*, but suberization does not occur where colonization is absent. *S. gausapatum* does not pass through the suberized barrier. The importance of suberin deposition in this limitation of the fungus is suggested by the much greater resistance of suberized cells to *in vitro* degradation by *S. gausapatum* (Figures 5 and 6), a resistance which is destroyed by the prior extraction of suberin (Pearce and Rutherford, 1981).

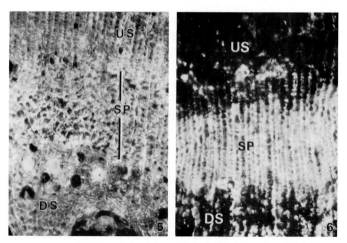

Figure 5. Transverse section showing the extent of the suberized parenchyma (SP) formed in wounded oak trees between sapwood decayed by *Stereum gausapatum* (DS) and the undecayed sapwood laid down after wounding (US). Dark field illumination. ×100. [Reproduced with permission from Pearce and Rutherford (1981). Copyright: Academic Press Inc. (London) Ltd.]

Figure 6. A section similar to that in Figure 5, but after 7 weeks of incubation *in vitro* with *Stereum gausapatum*. Both the decayed (DS) and undecayed (US) sapwood have been extensively degraded by the fungus, only the suberized tissues (SP) retaining their integrity. Dark field illumination. ×100. [Reproduced by permission from Pearce and Rutherford (1981). Copyright: Academic Press Inc. (London) Ltd.]

3.4. Vascular occlusion

Host responses which result in the blockage of xylem elements are crucial in defence against fungi which penetrate the vascular system, whether through wounds or by direct penetration of the roots. Fungi which successfully colonize xylem vessels, such as the vascular wilt fungi *Fusarium oxysporum* and *Verticillium albo-atrum*, spread through plants primarily as spores which are carried upwards in the transpiration stream. Mechanisms which trap spores are thus likely to be important in physically limiting the spread of the fungus. They may also allow the accumulation of antifungal chemicals and, by sealing off the infected vessels, prevent the upward movement of fungal enzymes and toxins.

A vessel is not a continuous tube but a series of cells joined end to end. Spores are thus naturally trapped at vessel end walls or perforation plates and successful fungi must grow through these barriers and produce secondary spores. Vascular occlusion commonly occurs above these natural trapping sites and usually involves the formation of tyloses and gel or gum plugs (Beckman and Talboys, 1981).

Vascular gels, in banana at least, appear to be composed primarily of pectin and hemicellulose materials (Beckman and Zaroogian, 1967) and arise by swelling of the primary walls and middle lamellae of pit membrances, perforation plates, and end walls (VanderMolen et al., 1977). The production of gels appears to be a host response rather than a simple consequence of the action of fungal enzymes (Beckman and Talboys, 1981).

Tyloses are formed by the bulging of adjacent parenchyma cells through pit membranes into the vessels. The plasticizing of the wall necessary for this process may involve an infusion of H^+ and removal of Ca^{2+} (Beckman, 1971) and a similar process may be involved in gel formation.

The rapid formation of gels and tyloses has frequently been associated with the resistance of plants to vascular pathogens, with susceptibility being correlated with a poor or delayed response (Beckman and Talboys, 1981). A delayed response allows the escape of secondary spores from the initial trapping site and thus systemic spread of the pathogen, followed sometimes by widespread tylosis which in itself may damage the plant.

Both gels and tyloses are frequently strengthened by the infusion and polymerization of phenolic compounds (e.g. Beckman et al., 1974), resulting in the brown discolouration of the xylem vessels that is so common in vascular diseases. The phenolics and their oxidation products may inhibit fungal enzymes and form a permanent seal around the infected zone. Vascular occlusions may also become infused with phytoalexins. In cotton the synthesis of terpenoid aldehyde phytoalexins occurs simultaneously with tylose formation near sites inoculated with V. albo-atrum. Above the site tylose formation slightly precedes phytoalexin formation (Mace, 1978). Infusion with antibiotic materials may in fact be crucial to the stability of the occlusive mechanisms (Bell and Mace, 1981). Knowledge of the interdependence of 'biochemical' and 'structural' defence mechanisms would be of value for most host–parasite systems.

CONCLUSIONS

Many different types of cell wall alterations occur in infected plants. The precise nature and importance of most alterations is not known; some may involve novel biochemical pathways and hence have passed so far unnoticed. The detailed chemical nature and associated biochemistry of most alterations deserve more attention.

The site and timing of wall modifications in relation to the development of potential pathogens has not been accurately determined for most systems. The quantity of a compound in gross tissue extracts is relatively meaningless in this context since the timing of the response at specific locations is more important. The sensitivity of the detection technique is also crucial: conclusions that an alteration is absent, or produced too late to be important in resistance, are invalid unless the sensitivity of the technique, particularly in relation to the quantity of material required to inhibit microbial progress, is known. A critical problem here lies in assessing not only the quantity at a specific location but also its likely effect on microbial growth. The concept of 'minimum lethal dose' is obviously not appropriate. Nevertheless, much useful information might be obtained by suitable *in vitro* tests (cf. Ride, 1980; Pearce and Rutherford, 1981) or by challenging artificially induced alterations (cf. Aist *et al.*, 1979). Knowledge of the effect of alterations on the permeability of walls to water, nutrients, toxins, and enzymes will also be helpful.

There is, then, still much to be discovered about the role of plant cell walls in defence. Since many of the alterations apparently involve the deposition of materials resistant to microbial degradation, it seems unlikely that they play no part at all in resistance. However, their precise functions and importance and their relation to other defence mechanisms await experimentation.

REFERENCES

Aist, J. R. (1976). Papillae and related wound plugs of plant cells, *Annu. Rev. Phytopathol.*, **14**, 145–163.

Aist, J. R., Kunoh, H., and Israel, H. W. (1979). Challenge appressoria of *Erysiphe graminis* fail to breach preformed papillae of a compatible barley cultivar, *Phytopathology*, **69**, 1245–1250.

Akai, S., and Fukutomi, M. (1980). Preformed internal physical defenses, in *Plant Disease. An Advanced Treatise. Vol. 5. How Plants Defend Themselves* (Eds J. G. Horsfall and E. B. Cowling), Academic Press, London, pp. 139–159.

Asada, Y., and Matsumoto, I. (1971). Microspectrophotometric observations on the cell walls of Japanese radish (*Raphanus sativus*) root infected by *Peronospora parasitica, Physiol. Plant Pathol.*, **1**, 377–383.

Asada, Y., and Matsumoto, I. (1972). The nature of lignin obtained from downy mildew-infected Japanese radish root, *Phytopathol. Z.*, **73**, 208–214.

Asada, Y., Ohguchi, T., and Matsumoto, I. (1979). Induction of lignification in response to fungal infection, in *Recognition and Specificity in Plant Host–Parasite Interactions* (Eds J. M. Daly and I. Uritani), Japan Scientific Societies Press, Tokyo, pp. 99–115.

Bateman, D. F. (1964). An induced mechanism of tissue resistance to polygalacturonase in *Rhizoctonia*-infected hypocotyls of bean, *Phytopathology*, **54**, 438–445.

Bateman, D. F., and Beer, S. V. (1965). Simultaneous production and synergistic action of oxalic acid and polygalacturonase during pathogenesis by *Sclerotium rolfsii, Phytopathology*, **55**, 204–211.

Bateman, D. F., and Lumsden, R. D. (1965). Relation of calcium content and nature of pectic substances in bean hypocotyls of different ages to suscepibility to an isolate of *Rhizoctonia solani, Phytopathology*, **55**, 734–738.

Bateman, D. F., Van Etten, H. D., English, P. D., Nevins, D. J., and Albersheim, P. (1969). Susceptibility to enzymatic degradation of cell walls from bean plants resistant and susceptible to *Rhizoctonia solani* Kuhn, *Plant Physiol.*, **44**, 641–648.

Beardmore, J., Ride, J. P., and Granger, J. W. (1983). Cellular lignification as a factor in the hypersensitive resistance of wheat to stem rust, *Physiol. Plant Pathol.*, in press.

Beckman, C. H. (1971). The plasticizing of plant cell walls and tylose formation—a model, *Physiol. Plant Pathol.*, **1**, 1–10.

Beckman, C. H., and Talboys, P. W. (1981). Anatomy of resistance, in *Fungal Wilt Diseases of Plants* (Eds M. E. Mace, A. A. Bell, C. H. Beckman), Academic Press, London, pp. 487–521.

Beckman, C. H., and Zaroogian, G. E. (1967). Origin and composition of vascular gel in infected banana roots, *Phytopathology*, **57**, 11–13.

Beckman, C. H., Mueller, W. C., and Mace, M. E. (1974). The stabilization of artificial and natural cell wall membranes by phenolic infusion and its relation to wilt disease resistance, *Phytopathology*, **64**, 1214–1220.

Bell, A. A. (1980). The time sequence of defense, in *Plant Disease. An Advanced Treatise. Vol. 5. How Plants Defend Themselves* (Eds J. G. Horsfall and E. B. Cowling), Academic Press, London, pp. 53–73.

Bell, A. A., and Mace, M. E. (1981). Biochemistry and physiology of resistance, in *Fungal Wilt Diseases of Plants* (Eds M. E. Mace, A. A. Bell, and C. H. Beckman), Academic Press, London, pp. 431–486.

Bird, P. M., and Ride, J. P. (1981). The resistance of wheat to *Septoria nodorum*: fungal development in relation to host lignification, *Physiol. Plant Pathol.*, **19**, 289–299.

Bushnell, W. R., and Bergquist, S. E. (1975). Aggregation of host cytoplasm and the formation of papillae and haustoria in powdery mildew of barley, *Phytopathology*, **65**, 310–318.

Byrde, R. J. W. (1957). The varietal resistance of fruits to brown rot. II. The nature of resistance in some varieties of cider apple, *J. Hort. Sci.*, **32**, 227–238.

Cartwright, D. W., Langcake, P., and Ride, J. P. (1980). Phytoalexin production in rice and its enchancement by a dichlorocyclopropane fungicide, *Physiol. Plant Pathol.*, **17**, 259–267.

Ellingboe, A. H. (1972). Genetics and physiology of primary infection by *Erysiphe graminis, Phytopathology*, **62**, 401–406.

Friend, J. (1976). Lignification in infected tissue, in *Biochemical Aspects of Plant–Parasite Relationships* (Eds J. Friend and D. R. Threlfall), Academic Press, London, pp. 291–303.

Friend, J. (1981). Plant phenolics, lignification and plant disease, *Prog. Phytochem.*, **7**, 197–261.

Griffey, R. T., and Leach, J. G. (1965). The influence of age on the development of bean anthracnose lesions, *Phytopathology*, **55**, 915–918.

Gross, G. G. (1977). Biosynthesis of lignin and related monomers, in *Recent Advances in Phytochemistry, Vol. 11, The Structure, Biosynthesis and Degradation of Wood* (Eds F. A. Loewus and V. C. Runeckles), Plenum Press, New York, pp. 141–184.

Gross, G. G. (1980). The biochemistry of lignification, *Adv. Bot. Res.*, **8**, 25–63.

Heath, M. C. (1979). Parial characterization of the electron-opaque deposits formed in the non-host plant, French bean, after cowpea rust infection, *Physiol. Plant Pathol.*, **15**, 141–148.

Heath, M. C. (1981). The suppression of the development of silicon-containing deposits in French bean leaves by exudates of the bean rust fungus and extracts from bean rust-infected tissue, *Physiol. Plant Pathol.*, **18**, 149–155.

Hyde, P. M., and Colhoun, J. (1975). Mechanisms of resistance of wheat to *Erysiphe graminis* f. sp. *tritici, Phytopathol. Z.*, **82**, 185–206.

Hursh, C. R. (1924). Morphological and physiological studies on the resistance of wheat to *Puccinia graminis tritici* Erikss. and Henn., *J. Agric. Res.*, **27**, 381–411.

Israel, H. W., Wilson, R. G., Aist, J. R., and Kunoh, H. (1980). Cell wall appositions and plant disease resistance: acoustic microscopy of papillae that block fungal ingress, *Proc. Nat. Acad. Sci. USA*, **77**, 2046–2049.

Johnson, L. E. B., Bushnell, W. R., and Zeyen, R. J. (1979). Binary pathways for analysis of primary infection and host response in populations of powdery mildew fungi, *Can. J. Bot.*, **57**, 497–511.

Keen, N. T., and Littlefield, L. J. (1979). The possible association of phytoalexins with resistance gene expression in flax to *Melampsora lini*, *Physiol. Plant Pathol.*, **14**, 265–280.

Koga, H., Mayama, S., and Shishiyama, J. (1980). Correlation between the deposition of fluorescent compounds in papillae, and resistance in barley against *Erysiphe graminis hordei*, *Can J. Bot.*, **58**, 536–541.

Langcake, P., and Wickens, S. G. A. (1975). Studies on the action of the dichlorocyclopropanes on the host–parasite relationship in the rice blast disease, *Physiol. Plant Pathol.*, **7**, 113–126.

Lin, M.-R., and Edwards, H. H. (1974). Primary penetration process in powdery mildewed barley related to host cell age, cell type, and occurrence of basic staining material, *New Phytol.*, **73**, 131–137.

Mace, M. E. (1978). Contributions of tyloses and terpenoid aldehyde phytolexins to *Verticillium* wilt resistance in cotton, *Physiol. Plant Pathol.*, **12**, 1–11.

Massala, R., Legrand, M., and Fritig, B. (1980). Effect of α-aminooxyacetate, a competitive inhibitor of phenylalanine ammonia-lyase, on the hypersensitive resistance of tobacco to tobacco mosaic virus, *Physiol. Plant Pathol.*, **16**, 213–226.

Maule, A. J. (1977). Biochemical and ultrastructural alterations in wheat leaves inoculated with *Botrytis cinerea*, *PhD Thesis*, University of Birmingham.

Maule, A. J., and Ride, J. P. (1976). Ammonia-lyase and O-methyltransferase activities related to lignification in wheat leaves infected with *Botrytis*, *Phytochemistry*, **15**, 1661–1664.

Maule, A. J., and Ride, J. P. (1982). Ultrastructure and autoradiography of lignifying cells in wheat leaves wound-inoculated with *Botrytis cinerea*, *Physiol. Plant Pathol.*, **20**, 235–241.

Ohguchi, T., and Asada, Y. (1975). Dehydrogenation polymerization products of p-hydroxycinnamyl alcohols by isoperoxidase obtained from downy mildew-infected roots of Japanese radish (*Raphanus sativus*), *Physiol. Plant Pathol.*, **5**, 183–192.

Pearce, R. B., and Ride, J. P. (1980). Specificity of induction of the lignification response in wounded wheat leaves, *Physiol. Plant Pathol.*, **16**, 197–204.

Pearce, R. B., and Ride, J. P. (1982). Chitin and related compounds as elicitors of the lignification response in wounded wheat leaves, *Physiol. Plant Pathol.*, **20**, 119–123.

Pearce, R. B., and Rutherford, J. (1981). A wound-associated suberized barrier to the spread of decay in the sapwood of oak (*Quercus robur L.*), *Physiol. Plant Pathol.*, **19**, 359–369

Postek, M. T. (1981). The occurrence of silica in the leaves of *Magnolia grandiflora* L., *Bot. Gaz.*, **142**, 124–134.

Ranjeva, R., Boudet, A. M., and Faggion, R. (1976). Phenolic metabolism in Petunia tissue. IV. Properties of p-coumarate coenzyme A ligase isoenzymes, *Biochimie*, **58**, 1255–1262.

Ride, J. P. (1975). Lignification in wounded wheat leaves in response to fungi and its possible role in resistance, *Physiol. Plant Pathol.*, **5**, 125–134.

Ride, J. P. (1980). The effect of induced lignification on the resistance of wheat cell walls to fungal degradation, *Physiol. Plant Pathol.*, **16**, 187–196.

Ride, J. P. (1983). The effects of treatments inducing or suppressing lignification on the resistance of wheat to fungi, *Physiol Plant Pathol.*, submitted for publication.

Ride, J. P., and Pearce, R. B. (1979). Lignification and papilla formation at sites of attempted penetration of wheat leaves by non-pathogenic fungi, *Physiol. Plant Pathol.*, **15**, 79–92.

Royle, D. J. (1976). Structural features of resistance to plant diseases, in *Biochemical Aspects of Plant–Parasite Relationships* (Eds J. Friend and D. R. Threlfall), Academic Press, London, pp. 161–193.

Scurfield, G., Anderson, C. A., and Segnit, E. R. (1974). Silica in woody stems, *Aust. J. Bot.*, **22**, 211–229.

Sherwood, R. T., and Vance, C. P. (1976). Histochemistry of papillae formed in reed canarygrass leaves in response to noninfecting pathogenic fungi, *Phytopathology*, **66**, 503–510.

Shigo, A. L., and Marx, H. G. (1977). Compartmentalization of decay in trees, United States Department of Agriculture, Forest Service, Agricultural Information Bulletin No. 405.

Stanbridge, B., Gay, J. L., and Wood, R. K. S. (1971). Gross and fine structural changes in *Erysiphe graminis* and barley before and during infection, in *Ecology of Leaf Surface Micro-organisms* (Eds T. F. Preece and C. H. Dickinson), Academic Press, London, pp. 367–379.

Touzé, A., and Rossignol, M. (1977). Lignification and the onset of premunition in muskmelon plants, in *Cell Wall Biochemistry Related to Specificity in Host–Plant Pathogen Interactions* (Eds B. Solheim and J. Raa), Universitetsforlaget, Oslo, pp. 289–293.

Vance, C. P., and Sherwood, R. T. (1976a). Regulation of lignin formation in reed canarygrass in relation to disease resistance, *Plant Physiol.*, **57**, 915–919.

Vance, C. P., and Sherwood, R. T. (1976b). Cycloheximide treatments implicate papilla formation in resistance of reed canarygrass to fungi, *Phytopathology*, **66**, 498–502.

Vance, C. P., Anderson, J. O., and Sherwood, R. T. (1976). Soluble and cell wall peroxidases in reed canarygrass in relation to disease resistance and localized lignin formation, *Plant Physiol.*, **57**, 920–922.

Vance, C. P., Kirk, T. K., and Sherwood, R. T. (1980). Lignification as a mechanism of disease resistance, *Annu. Rev. Phytopathol.*, **18**, 259–288.

VanderMolen, G. E., Beckman, C. H., and Rodehorst, E. (1977). Vascular gelation: a general response phenomenon following infection, *Physiol. Plant Pathol.*, **11**, 95–100.

Volk, R. J., Kahn, R. P., and Weintraub, R. L. (1958). Silicon content of the rice plant as a factor influencing its resistance to infection by the blast fungus, *Piricularia oryzae*, *Phytopathology*, **48**, 179–184.

Walker, R. R., and Wade, G. C. (1976). Epidemiology of potato gangrene in Tasmania, *Aust. J. Bot.*, **24**, 337–347.

Walker, R. R., and Wade, G. C. (1978). Resistance of potato tubers (*Solanum tuberosum*) to *Phoma exigua* var. *exigua* and *Phoma exigua* var. *foveata, Aust. J. Bot.*, **26**, 239–251.

Wardrop, A. B. (1971). Lignins in the plant kingdom. Occurrence and formation in plants, in *Lignins. Occurrence, Formation, Structure and Reactions* (Eds K. V. Sarkanen and C. H. Ludwig), Wiley, New York, pp. 19–41.

Waterman, M. A., Aist, J. R., and Israel, H. W. (1978). Centrifugation studies help to clarify the role of papilla formation in compatible barley powdery mildew interactions, *Phytopathology*, **68**, 797–802.

Young, P. A. (1920). Penetration phenomena and facultative parasitism in *Alternaria, Diplodia*, and other fungi, *Bot. Gaz.*, **81**, 259–279.

Further Reading

Aist, J. R. (1976). Papillae and related wound plugs of plant cells, *Annu. Rev. Phytopathol.*, **14**, 145–163.

Beckman, C. H., and Talboys, P. W. (1981). Anatomy of resistance, in *Fungal Wilt Diseases of Plants* (Eds M. E. Mace, A. A. Bell, and C. H. Beckman), Academic Press London, pp. 487–521.

Gross, G. G. (1980). The biochemistry of lignification, *Adv. Bot. Res.*, **8**, 25–63.

Royle, D. J. (1976). Structural features of resistance to plant diseases, in *Biochemical Aspects of Plant-Parasite Relationships* (Eds J. Friend and D. R. Threlfall), Academic Press, London, pp. 161–193.

Vance, C. P., Kirk, T. K., and Sherwood, R. T. (1980). Lignification as a mechanism of disease resistance, *Annu. Rev. Phytopathol.*, **18**, 259–288.

Biochemical Plant Pathology
Edited by J. A. Callow
© 1983 John Wiley & Sons Ltd

12

Antimicrobial Compounds

JOHN W. MANSFIELD

Biological Sciences Department, Wye College (University of London), Ashford, Kent, UK

1. INTRODUCTION AND HISTORICAL PERSPECTIVE

Most microorganisms are unable to colonize living plant tissue. Bacteria and fungi that cause plant disease are usually able to colonize only one species and are restricted to a particular tissue within the susceptible host plant. The specific nature of parasitism to crop plants is further demonstrated by the occurrence of disease-resistant varieties and races of pathogens which are differentiated by their success or failure to colonize cultivars possessing certain genes for resistance. The different levels of specificity in plant–microbe interactions are reflected in the following terms which are used to classify the processes of disease resistance: (1) *non-host resistance*, referring to interactions between plants and saprophytic organisms or pathogens of other hosts; (2) *race-non-specific resistance*, which describes interactions between pathogens and cultivars which show some resistance to all races or cases where such resistance develops with age or in certain tissues; (3) *race-specific resistance*, which acts only against certain races of a pathogen.

The failure of microorganisms to colonize plants has often been attributed to the presence of inhibitory compounds within challenged tissues (Ingham, 1973; Deverall, 1977). Antimicrobial compounds isolated from plants generally fall into two categories: (1) *constitutive* compounds, which are present in healthy plants, and (2) *induced* compounds synthesized from remote precursors following infection. The term *constitutive* includes compounds which are released from inactive precursors following tissue damage, for example the release of toxic hydrogen cyanide from cyanogenic glycosides (Millar and Higgins, 1970).

The first compounds found to have a role in disease resistance were two constitutive inhibitors, the water-soluble phenolics protocatechuic acid and catechol (Figure 1), which occur in the outer pigmented scales of onion bulbs resistant to the smudge disease caused by *Colletotrichum circinans*. The phenolic compounds diffuse from the dead cells in these scales into overlying inoculum droplets and inhibit spore germination at the bulb surface (Link *et al.*, 1929; Link and Walker, 1933). Some time before the discoveries of Walker and colleagues, Bernard (1909, 1911) had provided the first convincing experimental evidence for the induced production of antifungal compounds by plant tissues. He found that when surface-sterilized pieces of orchid tuber were placed on gelatin plates and inoculated with the mycorrhizal fungus *Rhizoctonia repens*, the fungus grew over the medium except in a zone around the pieces of tuber. In contrast, the fungus grew over extracts of crushed tuber or tuber pieces that had been treated for 35 min at 55 °C. Bernard proposed that live cells in the intact pieces of tuber responded to fungal secretions and synthesized compounds which diffused into the medium and inhibited fungal growth.

The concept of induced resistance was further developed by Müller and Börger (1940) in a classic paper describing studies on the resistance of potato

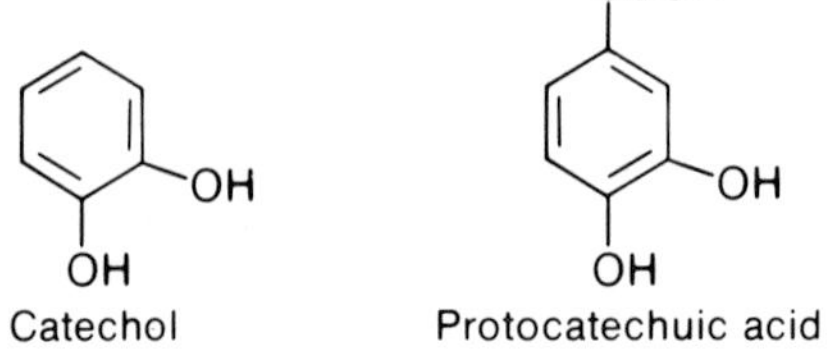

Catechol   Protocatechuic acid

Figure 1. Constitutive antimicrobial compounds from pigmented onion bulb scales

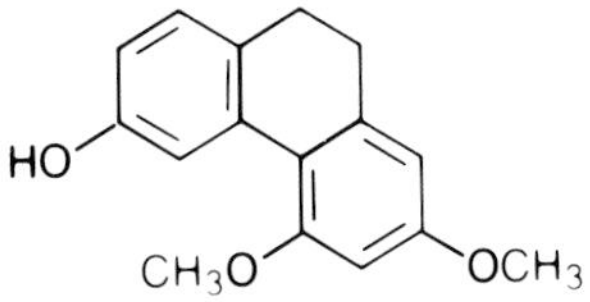

Figure 2. Orchinol

tuber tissue to *Phytophthora infestans*. Their most important experiments were performed with a cultivar which was resistant to one race of the late-blight fungus but susceptible to another. The compatible race rapidly colonized slices of tuber whereas growth of the incompatible race was restricted within one or two dead cells at the cut tuber surface. Compatible races of *P. infestans* were unable to colonize tissue previously inoculated with zoospores of an incompatible race, the protective effect was highly localized around inoculation sites and was active against other pathogens, such as *Fusarium caeruleum*. Müller and Börger concluded that protection was caused by the production of a fungitoxic substance by potato cells in response to the initial inoculation. They termed this substance a phytoalexin (from the Greek, *phyton* = plant and *alexin* = protecting substance). Interestingly, one of the first compounds to be identified as a phytoalexin was the phenanthrene orchinol (Figure 2), which was isolated from orchid tubers challenged by *Rhizoctonia repens* (Gäumann and Kern, 1959; Hardegger *et al*., 1963). It was not until 1968 that the phytoalexin from potato tubers was identified as the sesquiterpene rishitin (Figure 3; Katsui *et al*., 1968).

During the past 20 years the production of phytoalexins has become the most extensively studied process of disease resistance in plants. Phytoalexins may now be defined as 'low molecular weight antimicrobial compounds that are both synthesized by and accumulate in plants which have been exposed to microorganisms.' An important feature of this definition is that it restricts phytoalexins to compounds which are synthesized from remote precursors, probably through *de novo* synthesis of enzymes (Mansfield and Bailey, 1982).

Fungal growth within resistant plants may be restricted at one of a number of different sites: (1) within the partially degraded walls of epidermal cells (for example *Botrytis* spp. in non-host plants); (2) intracellularly, either within the epidermis (*Colletotrichum* spp in non-host plants or resistant cultivars)

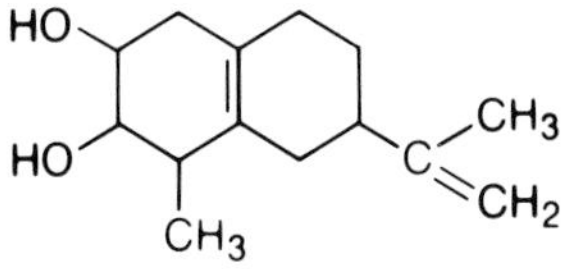

Figure 3. Rishitin

or in mesophyll cells (restricted development of haustoria of rust fungi); (3) in intercellular spaces (*Cladosporium fulvum* in leaves of resistant tomato cultivars); and (4) within xylem vessels (*Verticullium* and *Fusarium* spp. in wilt-resistant plants). Bacterial multiplication is usually restricted within intercellular spaces. In order to prove whether or not inhibition of microbial growth at these sites is caused by antimicrobial compounds, it would be necessary to measure the concentrations of inhibitors to which hyphae and bacteria are exposed at the time they stop growing and also to examine the activity of what may be a mixture of antibiotics at the site of exposure. An essential requirement

Table 1. Examples of constitutive antimicrobial compounds reported to have a role in resistan

Plant family	Species	Compound	Active form[a]
Aceraceae	*Acer platanoides* (Norway maple)	Gallic acid	—
Amaryllidaceae	*Allium cepa* (onion)	Catechol and protocatechuic acid	—
Araliaceae	*Hedera helix* (English ivy)	Hederasaponin C	Hederin
Berberidaceae	*Mahonia* spp. (barberry)	Berberine	—
Cruciferae	*Brassica* spp.	Mustard oil glucoside	Mustard oil
Gramineae	*Avena sativa* (oat)	Avenacin	—
	Hordeum vulgare (barley)	Hordatines	—
	Triticum aestivum (wheat)	Dihydroxymethoxybenzoxazinone (DIMBOA) glucoside	DIMBOA
	Zea mays (maize)	DIMBOA glucoside	DIMBOA
Lauraceae	*Persea americana* (avocado)	Borbonol	—
Leguminoseae	*Lotus corniculatus* (birds-foot trefoil)	Linamarin	Hydrogen cyanid
Liliaceae	*Tulipa gesneriana* (tulip)	Tuliposides A and B	Tulipalins A and
Rosaceae	*Malus sylvestris* (apple)	Phloridzin and phloretin	*o*-Quinones
	Pyrus communis (pear)	Arbutin	Hydroquinone
Solanaceae	*Lycopersicon esculentum* (tomato)	Tomatine	—
	Solanum tuberosum (potato)	α-Solanine and α-chaconine	—
		Caffeic and chlorogenic acids	—

[a]—indicates that the compound is antimicrobial as it occurs in the plant and, for example, does not requir hydrolytic conversion to an active form.

for obtaining such a direct proof is that the biochemical changes occurring within infected tissues must be closely associated with the biology of infection development. These criteria should always be borne in mind when assessing the role of antimicrobial compounds in disease resistance.

2. CONSTITUTIVE LOW MOLECULAR WEIGHT COMPOUNDS

Some examples of constitutive inhibitors thought to have a role in resistance are given in Table 1. The compounds are chemically diverse and are found in

hemical group of active form	Microorganism studied	Reference
henolic	Various fungi	Dix (1979)
henolic	*Colletotrichum circinans*	Walker and Stahmann (1955)
iterpenoid saponin	Various fungi	Schlösser (1973)
lkaloidal saponin	*Phymatotrichum omnivorum*	Greathouse and Watkins (1938)
othiocyanate iterpenoid saponin	*Fusarium oxysporum* *Gaeumannomyces graminis*	Davis (1964) Turner (1961)
oumaroylagmatine derivatives	*Helminthosporium sativum*	Stoessl and Unwin (1970)
yclic hydroxamate	*Septoria nodorum*	Baker and Smith (1977)
yclic hydroxamate	Various bacteria	Hartman *et al.* (1975)
actone	*Phytophthora cinnamomi*	Zaki *et al.* (1980)
	Stemphylium loti	Millar and Higgins (1970)
actone	*Fusarium oxysporum*	Bergman and Beijersbergen (1968)
	Venturia inaequalis	Raa (1968)
	Erwinia amylovora	Powell and Hildebrand (1970)
lkaloidal saponin	*Botrytis cinerea*	Verhoeff and Liem (1975)
lkaloidal saponins	*Helminthosporium carbonum*	Allen and Kuć (1968)
henolic	,, ,,	Kuć *et al.* (1956)

a wide range of plant families. Their presence in plant tissues has been reported to confer non-host or race non-specific resistance; there are no convincing reports that differences in levels of constitutive antimicrobial compounds explain patterns of race-specific resistance.

Most constitutive inhibitors are considered to be effective following penetration rather than by inhibiting microbial development at the plant surface. Exceptions to this general rule are catechol and protocatechuic acid from onion bulb scales (as already discussed) and other phenolic compounds, including gallic acid, which are thought to regulate fungal growth on the surface of leaves of *Acer platanoides* (Dix, 1979). Unidentified compounds associated with epicuticular waxes have also been shown to influence spore germination on beetroot and broad bean leaves (Blakeman and Atkinson, 1976; Rossall and Mansfield, 1980; see also Chapter 6). Although chlorogenic acid and the steroid glycoalkaloid solanine are found in healthy potato tissues, they have also been shown to accumulate around wounds and after fungal infection (Kuć, 1982). There are few other examples of compounds which can be classified as both constitutive and induced inhibitors. Only avenacin, cyanogenic glucosides, and tuliposides will be considered in more detail here, but the reader is urged to examine thoroughly the evidence for the involvement in disease resistance of other compounds listed in Table 1.

2.1. Avenacin

The fungus *Gaeummannomyces graminis* (previously *Ophiobolus graminis*) causes take-all disease of wheat, barley, and several grasses but does not cause a lasting infection of oats. Oat roots are, however, susceptible to *G. graminis* var. *avenae*. As part of an investigation of the mechanisms underlying the success or failure of *G. graminis* var. *avenae* and *G. graminis* to colonize oat roots, Turner (1953, 1960) examined the growth of fungi in root extracts. Whereas *G. graminis* var. *avenae* grew in the extracts, the type species was inhibited, particularly in extracts from the crown roots and root tips which are very resistant to colonization. The extract from one root tip in 2 cm^3 of water was sufficient to cause a 50% reduction in growth of *G. graminis*. The major toxic principle was subsequently isolated and characterized as the saponin avenacin (Figure 4; Tscheche *et al.*, 1973).

Turner (1961) found that *G. graminis* var. *avenae* was tolerant of avenacin because the fungus produced an induced extracellular glycosidase which removed the terminal sugar from the carbohydrate chain (Figure 4) and thereby detoxified the molecule. Five other fungi tested, including *G. graminis*, failed to produce 'avenacinase.' Detoxification of avenacin by the removal of a sugar residue contrasts with other glycosides where the aglycone is often more toxic than the parent compound (see Table 1).

Figure 4. Avenacin. R = β-D-glucose, 1,4-β-D-glucose, 1,2-α-L-arabinose-1,3

Turner's hypothesis that constitutive inhibitors in oat roots have a role in disease resistance was recently supported by results obtained by Holden (1980). He found that growth on oat seedlings of varieties of *G. graminis*, *Phialophora radicicola*, and *Leptosphaeria narmasi* was closely paralleled by their growth *in vitro* in oat tissue extracts.

2.2. Cyanogenic glycosides

Cyanogenic glycosides have been found in over 800 plant species representing about 70 families (Eyjólfsson, 1970). The glycosides are generally thought to be present in cell vacuoles but convincing evidence for their subcellular localization is not always available. The presence of dhurrin, the cyanogenic glucoside from sorghum (*Sorghum bicolor*), in vacuoles has been confirmed by microautoradiography (Saunders *et al.*, 1977). Following the loss of compartmentation associated with tissue injury, the glycosides become exposed to hydrolytic enzymes and are degraded releasing the highly toxic gas hydrogen cyanide (HCN). For example, liberation of cyanide from linamarin, the glucoside in birds-foot trefoil (*Lotus corniculatus*), is a two-step process involving the cytoplasmic enzymes β-glucosidase and oxynitrilase (see Figure 5).

Several studies have shown that HCN is released from cyanogenic plants during their attack by necrotrophic fungi (Millar and Higgins, 1970; Myers and Fry, 1978). Most cyanogenic plants have a remarkable ability to generate HCN, a typical example being sorghum, which produces 400 μmol of HCN per gram dry weight of leaf tissue (Myers and Fry, 1978). It has been proposed that the levels of HCN released at infection sites are sufficiently high to kill or at least inhibit the growth of invading hyphae, but there are no reports of attempts to determine the concentrations of HCN around invading microorganisms.

Results obtained by Lüdtke and Hahn (1953) and Trione (1960) suggest that race-specific resistance in cyanogenic plants cannot be explained by differences in glycoside contents. Support for a role for cyanogenic glycosides in

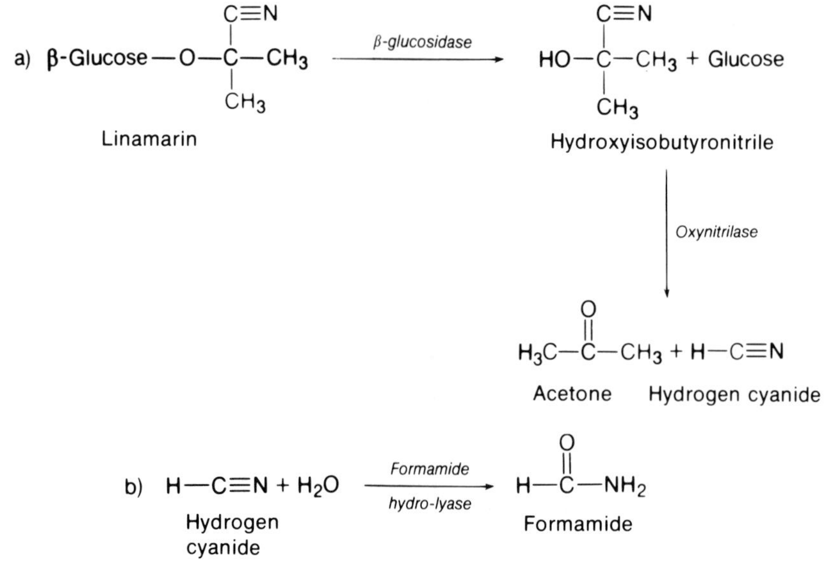

a) β-Glucose—O—$\overset{\overset{\displaystyle C\equiv N}{|}}{\underset{\underset{\displaystyle CH_3}{|}}{C}}$—CH₃ $\xrightarrow{\textit{β-glucosidase}}$ HO—$\overset{\overset{\displaystyle C\equiv N}{|}}{\underset{\underset{\displaystyle CH_3}{|}}{C}}$—CH₃ + Glucose

        Linamarin   Hydroxyisobutyronitrile

                                                      $\downarrow$ *Oxynitrilase*

$$H_3C-\overset{\overset{\displaystyle O}{\|}}{C}-CH_3 + H-C\equiv N$$

                  Acetone   Hydrogen cyanide

b) H—C≡N + H₂O $\xrightarrow[\textit{hydro-lyase}]{\textit{Formamide}}$ $H-\overset{\overset{\displaystyle O}{\|}}{C}-NH_2$

    Hydrogen   Formamide
    cyanide

Figure 5. (a) Hydrolysis of the cyanogenic glucoside, linamarin to yield glucose, acetone and hydrogen cyanide. (b) Detoxification of hydrogen cyanide by conversion to formamide by the enzyme formamide hydro-lyase

non-host resistance comes from the discovery that fungal pathogens of cyanogenic plants are tolerant of HCN (Millar and Higgins, 1970; Fry and Millar, 1971). Tolerance appears to be due to the induced production of the enzyme formamide hydro-lyase, which converts HCN into harmless formamide (Figure 5). Fry and Evans (1977) examined 31 species of fungi and found that high or moderate levels of formamide hydro-lyase were produced by all of the eleven pathogens of cyanogenic plants tested and four of the fourteen pathogens of non-cyanogenic plants, but not by any of the six saprophytes studied. In the last group only *Neurospora crassa* possessed any formamide hydro-lyase activity. Two of the lyase-producing fungi found to be non-pathogens of cyanogenic plants were previously reported to be pathogenic to sorghum (Nelson and Kline, 1962). Bearing this in mind the association between the level of induced formamide hydro-lyase activity and virulence reported by Fry and Evans (1977) is striking. The presence of the tolerance mechanism in pathogenic fungi provides circumstantial evidence that cyanogenic glucosides may contribute to non-host resistance. It is clear that the glycosides are not the only mechanism of resistance in cyanogenic plants, otherwise a fungus with high formamide hydro-lyase activity might be expected to be pathogenic to all cyanogenic plants. This is not the case: the growth of sorghum pathogens is restricted in *L. corniculatus* and *vice versa*. Leaves of *L. corniculatus* are known to produce the phytoalexins sativan and vestitol (Ingham, 1982).

In contrast to the close relationship found between HCN tolerance and fungal pathogenicity, Rust *et al.* (1980) reported that bacterial pathogens of cyanogenic plants were not distinctly more tolerant of HCN than other bacteria. The role of cyanogenic glycosides in resistance to bacteria requires further examination.

2.3. Tuliposides

Extracts of all parts of the healthy tulip plant possess some antibacterial and antifungal activity (Schönbeck, 1967). The active principles in extracts have been identified as the butyrolactones tulipalin A and B (Tscheche *et al.*, 1968). The active lactones are formed during extraction from unstable tuliposides A and B, which are the esters of glucose with α-methylene-γ-hydroxybutyric acid and the β-hydroxy derivative of the acid, respectively.

The tuliposides are thought to be stored in the vacuoles of tulip cells and the concentrations and proportions of each tuliposide vary between different tissues, tuliposides A and B occurring in particularly large quantities in the white skin of tulip bulbs and in pistils, respectively. Tuliposides are unstable at pH > 5.0 and readily convert to the active lactones via intermediate acids (Figure 6). Liberation of the lactones can occur under the direct influence of pH, slowly at pH 5.5, rapidly at pH 7.5, or very quickly by the action of enzymes present in extracts of tulips at pH 5.0 (Schönbeck, 1967; Tscheche *et al.*, 1968; Beijersbergen and Lemmers, 1972). Some antifungal activity has been attributed to the tuliposides themselves (Schönbeck, 1967; Tscheche *et al.*, 1968) but tests on the activity of the esters are complicated by their conversion to the very active lactones during bioassays (Beijersbergen and Lemmers, 1972). If tuliposides have any role to the resistance of the tulip plant to fungal colonization they must be released from vacuoles and converted to lactones in order to inhibit invading hyphae.

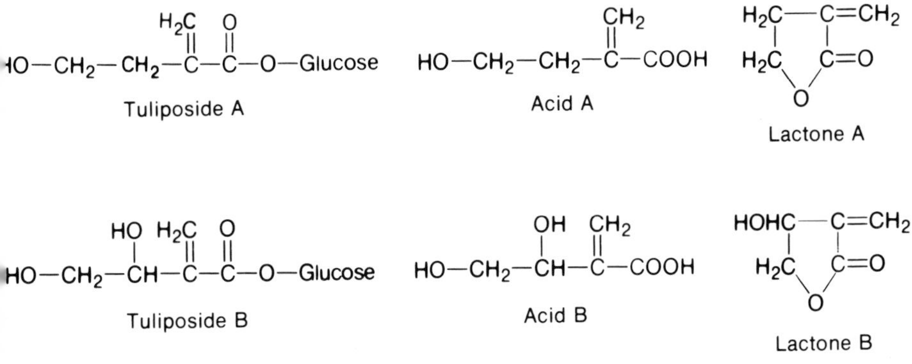

Figure 6. Structures of tuliposides and their derivatives

Unsaturated lactones occur as glycosides in many plants notably members of the Liliaceae, Ranunculaceae, and Rosaceae (Slob *et al.*, 1975; Schönbeck and Schlösser, 1976).

The resistance of tulip bulbs to *Fusarium oxysporum* f. sp. *tulipae* and of pistils to *Botrytis cinerea* has been attributed to their tuliposide content. The antifungal activity of extracts from these tissues can be striking. For example, Beijersbergen and Lemmers (1972) found that extracts from pistils and outer fleshy bulb scales were fungicidal to *F. oxysporum* when diluted in potato dextrose agar to one tenth of their fresh weight concentration.

The growing tulip bulb is attacked by *F. oxysporum* only during the last few weeks before harvest even if the mother bulb planted in the autumn was infected or contaminated with conidia. During the short period of susceptibility before lifting, the outermost white bulb scale, which previously contained high levels of tuliposides, turns into a papery brown husk which lacks antimicrobial substances. The underlying white scales which are temporarily susceptible to colonization by *F. oxysporum* also have a low tuliposide content. However, within a few days after harvesting tuliposide concentrations increase rapidly so that yields of more than 200 μg of tulipalin A per gram of fresh tissue can be recovered from the outer layers of the white scales which become resistant to infection. Tulipalin A is fungicidal to mycelium of *F. oxysporum* at concentrations between 200 and 300 μg cm^{-3}. The coincidence of increasing susceptibility of bulb scales with the drop in tuliposide concentrations leads to the conclusion that the constitutive inhibitors normally protect the white scales and therefore the growing bulb against infection by *F. oxysporum* f.sp. *tulipae* (Bergman and Beijersbergen, 1968).

Tulip pistils are susceptible to colonization by *Botrytis tulipae* the cause of tulip fire disease but resistant to *B. cinerea* the ubiquitous grey-mould fungus. Schönbeck and Schroeder (1972) found that *B. tulipae* was not only less sensitive to tulipalin than other *Botrytis* spp. but also converted tuliposides and their derivatives to stimulatory products.

The role of tuliposides in the resistance of other parts of the tulip plant has not been examined. The need to postulate some mechanism of resistance other than tuliposides in leaves is suggested by the results of Rutter *et al.* (1977), who found low levels of the compounds in the epidermis (first challenged by most fungi) compared with the mesophyll. Resistance of leaves to *Botrytis* is associated with lignification at inoculation sites (Mansfield and Hutson, 1980). Although good correlations have been established between tuliposide concentration and the resistance of bulbs and pistils to fungal colonization, the localized distribution of the glucosides in the tissues and their requirement for conversion to the active lactones raise doubts about the exposure of invading hyphae to fungitoxic concentrations of the inhibitors. Measurements of changes in concentrations of tuliposides and lactones occurring within infected tissues coupled with microscopical studies on the timing and sites of restriction of fungal growth would allow more definite conclusions to be drawn on the role of tuliposides in disease resistance.

There have been few reports of the use of experimental approaches for examining the role of constitutive inhibitors in resistance. Susceptibility to colonization can be induced in many plants by antimetabolites or sub-lethal heat treatments but manipulation of the host–parasite interaction seems to have been attempted only with plant tissues which lack constitutive inhibitors. For example, susceptibility of soybeans to colonization by incompatible races of *Phytophthora megasperma* is induced by treatment of hypocotyls with the transcription inhibitor actinomycin D and also the protein synthesis inhibitor blasticidin S (Yoshikawa, 1978; see also Chapters 2 and 13). Similarly, heating French bean pods at 44 °C for 2 h immediately prior to inoculation decreases their resistance to the normally avirulent fungi *Botrytis cinerea* and *Sclerotinia fructigena* (Jerome and Müller, 1958). Demonstrations of induced susceptibility in soybean and French bean have been used to support a role for phytoalexins in disease resistance in these species. It would be interesting to carry out similar experiments with plants containing constitutive inhibitors. The use of antimetabolites with tulip tissues should allow an assessment to be made of the relative importance of tuliposides and induced lignification in disease resistance.

An interesting genetical approach has been adopted by Défago and Kern (1983) and Défago *et al.* (1983) to examine the role of the saponin tomatine in the resistance of tomato fruits to fungal colonization. They obtained an isolate of *Fusarium solani* which rotted roots, stems and red fruits but was unable to colonize green tomato fruits which contain very high concentrations of tomatine. Growth of the wild-type isolate of *F. solani* was inhibited *in vitro* by 100 μg cm^{-3} of tomatine. Treatment of spores with mutagens allowed the selection of mutants able to grow on media containing 800 μg cm^{-3} of tomatine. The tomatine-tolerant mutants were found to have low sterol content (sterols in the cell membrane being the site of action of the saponin), and were able to rot green fruits. Analysis of the progeny from crosses between tolerant and wild-type strains showed that virulence to green fruits, low sterol content, and insensitivity to tomatine were always inherited together—convincing evidence that tomatine is an important resistance factor.

3. INDUCED LOW MOLECULAR WEIGHT COMPOUNDS—PHYTOALEXINS

3.1. Distribution, chemistry and biosynthesis

Phytoalexins have been isolated and characterized from fifteen families of angiosperms, the Leguminoseae and Solanaceae being the most intensively studied. There are few well documented examples of phytoalexins from gymnosperms and no evidence is available on their presence in lower plants (Shain, 1967; Ingham, 1973; Coxon, 1982). The chemical diversity of phytoalexins is remarkable; they range from hydroxyflavans in *Narcissus* bulbs to polyacetylenic compounds in safflower stems (Coxon, 1982). Ingham and Harborne (1976) have proposed that phytoalexin production may be used as a novel approach to the study of systematic relationships among higher plants. Two main conclusions may be drawn from studies on the chemical diversity of phytoalexins: (1) members of plant families usually produce chemically similar types of phytoalexin, for example isoflavonoids from the Leguminoseae and terpenoids from the Solanaceae; and (2) although accumulation of one compound may predominate in a particular species, most plants produce several closely related phytoalexins. *Vicia faba* provides a notable exception to these general rules; like most other legumes it produces an isoflavonoid phytoalexin, in this case medicarpin, but the principal induced antimicrobial compounds are furanoacetylenic wyerone derivatives. Eight furanoacetylenes have been recognized as phytoalexins in *Vicia faba* (Ingham, 1982). Some examples of plants with multicomponent phytoalexin responses are given in Table 2.

The interactions between primary and secondary metabolism leading to phytoalexin biosynthesis, and the source of important substrates such as phenylalanine, acetyl-CoA, malonyl-CoA, and mevalonic acid, are outlined in Figure 7.

Several mechanisms have been proposed which might lead to the diversion of normal metabolism to the accumulation of phytoalexins. For example, Stoessl (1982) suggested the imposition or removal of metabolic blocks relatively late on in the normal metabolic route or greatly stimulated synthesis of early, general biosynthetic precursors. Development of an understanding of the metabolic control of phytoalexin biosynthesis requires investigation of the properties and activities of the enzymes catalysing steps in the pathways involved. The most detailed information available on these matters concerns the isoflavonoid phytoalexin, phaseollin, but increases in enzyme activities have also been recorded in tissue accumulating furanoterpenoid, sesquiterpene, and stilbene phytoalexins (Bailey, 1982a). Enzymes studied include those controlling the synthesis of early precursors, for example phenylalanine ammonia-lyase (PAL), which has a key role in the biosynthesis of phenylpropanoid phytoalexins (Figure 7), and enzymes directly responsible for the formation of phytoalexins, for example furanosesquiterpene reductase, which catalyses the formation of ipomeamarone from dehydroipomeamarone (see Table 2).

The mechanisms leading to increased PAL activity have recently been examined in cell suspension cultures of *Phaseolus vulgaris*. Lamb *et al*. (1980; see also Chapter 17) used the elegant technique of comparative density label-

Plant family	Species	Phytoalexins	Chemical group	Reference
Amaryllidaceae	*Narcissus pseudonarcissus* (daffodil)	7,4'-Dihydroxyflavan 7,4'-Dihydroxy-8-methylflavan	Flavanoid Flavanoid	Coxon *et al.* (1980)
Compositae	*Carthamus tinctorius* (safflower)	7-Hydroxyflavan Dehydrosafynol Safynol	Flavanoid Polyacetylene Polyacetylene	Allen and Thomas (1971)
Convolvulaceae	*Ipomoea batatas* (sweet potato)	Dehydroipomeamarone Ipomeamarone Ipomeamoronol	Furanosesquiterpene Furanosesquiterpene Furanosesquiterpene	Oguni and Uritani (1974)
Gramineae	*Oryza sativa* (rice)	Momilactone A Momilactone B	Diterpene Diterpene	Cartwright *et al.* (1980)
Leguminoseae	*Glycine max* (soybean)	Glyceocarpin Glyceollins I–IV	Isoflavonoid Isoflavonoid	Ingham *et al.* (1981)
	Phaseolus vulgaris (French bean)	Kievitone Phaseollidin Phaseollin Phaseollinisoflavan	Isoflavonoid Isoflavonoid Isoflavonoid Isoflavonoid	Bailey and Burden (1973)
	Vicia faba (broad bean)	Dihydrowyerol Dihydrowyerone Dihydrowyerone acid Wyerol Wyerone Wyerone acid Wyerone epoxide Medicarpin	Furanoacetylene Furanoacetylene Furanoacetylene Furanoacetylene Furanoacetylene Furanoacetylene Furanoacetylene Isoflavonoid	Mansfield *et al.* (1980)
Solanaceae	*Lycopersicon esculentum* (tomato)	Falcarindiol Falcarinol Rishitin	Polyacetylene Polyacetylene Sesquiterpene	Hargreaves *et al.* (1976) De Wit and Kodde (1981a)
	Solanum tuberosum (potato)	Lubimin Phytuberin Rishitin	Sesquiterpene Sesquiterpene Sesquiterpene	Coxon *et al.* (1974)
Vitaceae	*Vitis vinifera* (grape)	Pterostilbene α-Viniferin ε-Viniferin	Stilbene Stilbene oligomer Stilbene oligomer	Langcake *et al.* (1979)

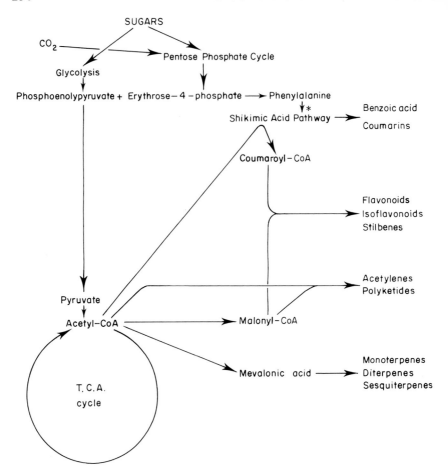

Figure 7. Relationships between primary metabolism and synthesis of phytoalexins (modified from Bailey, 1982). Note that the enzyme phenylalanine ammonia lyase catalyses the conversion of phenylalanine to cinnamic acid (reaction marked with an asterisk)

ling to demonstrate that *de novo* synthesis of PAL occurred within 2 h after treatment of cultured cells with elicitors of phaseollin accumulation. The important general implication of this work is that phytoalexin production may involve the coordinated activation of biosynthetic pathways by *de novo* enzyme synthesis. If experiments on other enzymes and phytoalexins lend support to this conclusion it is possible that phytoalexin biosynthesis may become a useful model for studies on the control of gene expression in plant cells.

Phytoalexins are not necessarily the end-products of plant metabolism. Under certain conditions phytoalexin concentrations may decrease due to conversions catalysed by the plant. As discussed by VanEtten *et al.* (1982), most of the evidence for this comes from experiments in which phytoalexins were added to plant tissues and a problem with this approach is that observed transformations may not be typical of those occurring *in situ*. However, Ishiguri *et al.* (1978) have isolated from potato tuber tissues enzymes catalysing the metabolism or rishitin. It is probable that in many plants the observed accumulation of phytoalexins is the net result of their biosynthesis and metabolism by plant tissues. This aspect is discussed further in Chapters 13 and 17.

3.2. Role of phytoalexins in resistance

Phytoalexin accumulation is thought to have a role in the resistance of plants to colonization by bacteria and nematodes as well as fungi. The involvement of phytoalexins in disease resistance is not restricted to a particular interaction between genotypes, for example race-specific or non-host resistance, but is more closely associated with morphologically similar types of response. Most data supporting a role for phytoalexin accumulation as the cause of the inhibition of microbial growth in resistant plants come from interactions in which resistance is expressed following penetration and is associated with the necrosis of plant cells. Such a local lesion response, often described as a hypersensitive reaction, can occur during the expression of non-host, race-specific and race-non-specific resistance.

The more detailed coordinated biochemical and microscopical studies designed to examine the role of phytoalexins in resistance have been carried out with fungi. Two examples are studies on interactions between *Phaseolus vulgaris* and *Colletotrichum lindemuthianum*, the cause of anthracnose disease, and between *Vicia faba* and *Botrytis* spp., including *B. fabae*, the cause of chocolate spot disease.

3.2.1. Phaseolus vulgaris *and* Colletotrichum lindemuthianum

The anthracnose fungus exists as several physiological races which can be differentiated by their reaction with the hypocotyls of various cultivars of French bean. Differences between cultivars occur following penetration into epidermal cells. In susceptible cultivars, developing intracellular hyphae grow between the cell wall and plasmalemma of epidermal and cortical cells without causing any observable host response. Biotrophic growth can continue for several days leading to extensive colonization of symptomless tissues. After this period infected cells die, the tissue collapses, and a brown lesion develops. Extensive necrotrophic fungal growth may then occur particularly when tissues are incubated at lower temperatures (<17 °C). Race-specific resistance to *C. lindemuthianum* is expressed by the hypersensitive reaction of cells to infection by incompatible races. The initially penetrated cell and perhaps one or two adjacent cells die and turn brown soon after infection, appearing as scattered flecks beneath inoculum droplets. Growth of the incompatible race is usually restricted to the cells initially penetrated. In susceptible cultivars resistance can be induced by transferring hypocotyls from 16 to 25 °C, 3 days after inoculation. Transfer to the higher temperature leads to premature death of invaded cells, the development of localized lesions, and inhibition of the short intracellular hyphae established during the brief period of biotrophic growth at 16 °C (Landes and Hoffman, 1979; Bailey, 1982b).

Phaseollin is the main phytoalexin produced by French bean hypocotyls challenged by *C. lindemuthianum*. Accumulation of phaseollin is closely associated with cell death and browning in both resistant and susceptible hypocotyls. Bailey and Deverall (1971) were unable to detect the phytoalexin during the biotrophic period of colonization by a compatible race but recorded accumulation of phaseollin after the death of infected tissues and formation of lesions. At susceptible sites where spreading lesions coalesced low concentrations of less than 5 μg g^{-1} fresh weight were recovered from colonized tissues. Phaseollin accumulated rapidly following the death of cells during the hypersensitive reaction to incompatible races. Yields of more than 150 μg g^{-1} fresh weight were recovered from excised inoculation sites bearing isolated flecked lesions. The elegant experiments of Hargreaves and Bailey (1978) have demonstrated that phaseollin and other isoflavonoid phytoalexins are synthesized in tissues around necrotic cells and are absorbed and accumulate in the dead tissue containing intracellular hyphae. The concentrations of phaseollin to which incompatible hyphae are exposed within cells undergoing hypersensitive reactions are therefore probably much higher than those recorded in excised pieces of tissue, which mostly contain unaffected cells. Assuming that phytoalexins accumulated within dead cells, Bailey and Deverall (1971) calculated that local concentrations approached 3000 μg cm^{-3} almost 300 times the minimum concentration completely restricting growth of germ-tubes *in vitro* by all races of *C. lindemuthianum*.

Although the rapid accumulation of phaseollin during hypersensitive reactions provided circumstantial evidence that the phytoalexin causes the inhibition of incompatible races, the timing of phytoalexin accumulation and cessation of hyphal growth could not be examined. Bailey and Deverall (1971) found that the intracellular hyphae produced were too small to determine precisely when they were restricted. More recently, however, Bailey *et al.* (1980) have used the resistance induced by transfer of infected hypocotyls from 16 to 25 °C to examine in detail the timing of fungal growth inhibition, host cell death, and phytoalexin accumulation. Fungal growth during this induced resistant response is much greater than during normal hypersensitivity, enabling measurements of the lengths of intracellular hyphae to be made throughout the resistance process. The results obtained showed that death of infected cells occurred several hours before phytoalexins formed at inoculation sites and that inhibition of hyphal growth occurred shortly after phaseollin and phaseollinisoflavan began to accumulate. These findings are consistent with the view that the accumulated phytoalexins cause the restriction of fungal growth during the resistance of beans to *C. lindemuthianum*. Conversely, the unrestricted colonization of hypocotyls occurs when the fungus avoids exposure to the phytoalexins by establishing a prolonged biotrophic phase of growth. This leads to the formation of coalescing lesions, within and around which cells remain alive for insufficient time to synthesize the amounts of inhibitors required to reach fungitoxic concentrations.

3.2.2. Vicia faba *and* Botrytis *spp.*

Recent microscopical studies have shown that *B. fabae* is a totally necrotrophic pathogen of broad bean leaves. Following inoculation on to the abaxial epidermis of field-grown leaves the chocolate spot fungus penetrates the cuticle and grows rapidly through epidermal walls which swell in advance of fungal colonization. Degradation of cell walls seems to lead to the death of affected cells around penetration points. Areas of dead cells soon coalesce and the fungus ramifies through dead epidermal and mesophyll tissues producing black lesions which spread rapidly away from the site of inoculation. Non-pathogens such as *B. cinerea*, *B. elliptica*, and *B. tulipae* also penetrate into the epidermis but they kill far fewer cells than *B. fabae* during the early stages of infection and their hyphae stop growing within epidermal walls by 12 h after inoculation. Lesions which develop at sites inoculated with non-pathogens are confined to the tissue beneath inoculum droplets (Mansfield and Hutson, 1980).

In early studies on the changes in phytoalexin concentrations occurring during infection development, wyerone derivatives were extracted from leaf discs cut from beneath inoculum droplets. The results obtained demonstrated the rapid accumulation of phytoalexins (in particular wyerone acid) during limited lesion formation by *B. cinerea* but they provided little information on the accumulation of inhibitors around invading hyphae (Hargreaves *et al.*, 1977). More recently, wyerone acid and wyerone concentrations were measured in epidermal strips collected from infected leaves. The phytoalexins were found to accumulate in the epidermis before fungal growth was restricted and quickly reached concentrations which were fungicidal *in vitro* towards *B. cinerea* (Mansfield, 1982). When spreading lesions were examined, an initial increase in phytoalexin concentrations within the mixture of live and dead cells at inoculation sites was followed by a decrease as tissues became completely necrotic and colonized by *B. fabae*. Very low concentrations of inhibitors (if any) were present in totally necrotic tissue or in inoculum droplets containing *B. fabae* (Hargreaves *et al.*, 1977). These results suggest that phytoalexin accumulation restricts the growth of *B. cinerea* and that *B. fabae* is able to metabolize and detoxify the inhibitors to which it is exposed and thereby to prevent their accumulation to fungitoxic concentrations around invading hyphae.

Botrytis fabae has been shown to be much less sensitive than other *Botrytis* spp. to wyerone derivatives and is able to detoxify the phytoalexins by reducing them to less fungitoxic products. An additional important factor in the virulence of *B. fabae* to *V. faba* is the ability of the pathogen to kill bean cells during the early stages of infection, before phytoalexins accumulate. The 'bulldozer' effect of *B. fabae*, killing numerous cells quickly, reduces the synthetic activity of infected tissue and thus suppresses phytoalexin production, particularly if inoculum droplets contain high numbers of spores (Hutson and Mansfield, 1980). The interaction between *Botrytis* and leaves of *V. faba* may, therefore, be envisaged as a balance between phytoalexin production by the plant and phytoalexin degradation by the fungus (Figure 8). Modification of any of the factors contributing to this balance could affect the outcome of the

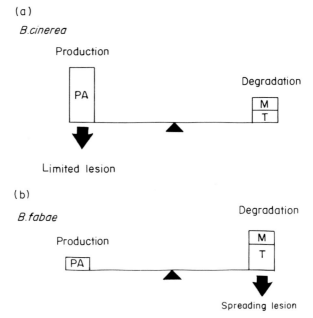

Figure 8. Interaction between *Botrytis* and leaf and pod tissues of *Vicia faba* expressed as a balance between phytoalexin production (PA) by the plant and phytoalexin degradation by the fungus. The degradation weighting incorporates both metabolic capacity to detoxify the inhibitors (M) and a tolerance factor (T) which accounts for the differential sensitivity of *B. cinerea* and *B. fabae* to wyerone derivatives. (From Mansfield, 1980)

interaction. Thus, decreasing spore numbers leads to reduced development of *B. fabae*, probably by reducing the numbers of plant cells killed at inoculation sites and thereby increasing phytoalexin production and accumulation around invading hyphae (Mansfield and Hutson, 1980; Mansfield, 1982).

3.3. Phytoalexins, recognition, and elicitors

Anthracnose of French bean is representative of many examples of race-specific resistance in which incompatible races (of bacteria or fungi) seem to be recognized by invaded plant cells which then undergo a hypersensitive reaction leading to phytoalexin accumulation. By contrast, cells fail to recognize the presence of compatible races which grow biotrophically at least during the early stages of infection. Important additional examples are late blight of potatoes caused by *Phytophthora infestans* (Tomiyama *et al.*, 1979) and halo blight of French bean caused by the bacterium *Pseudomonas phaseolicola* (Lyon and Wood, 1975). The concept of recognition is included in Figure 9, which summarizes the means by which virulent pathogens may avoid exposure to inhibitory concentrations of phytoalexins.

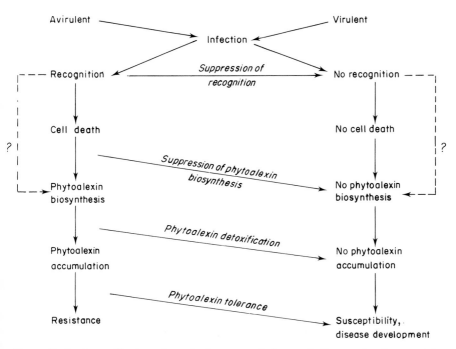

Figure 9. Factors affecting phytoalexin accumulation and disease development in plants following inoculation with avirulent and virulent pathogens. (From Mansfield, 1982. Reproduced by permission of Blackie & Son)

Bailey (1982) proposed that phytoalexin biosynthesis is activated by endogenous elicitors released from cells undergoing the hypersensitive reaction or dying as a result of exposure to various phytotoxic materials, for example fungal glycoproteins (De Wit and Kodde, 1981b) or Triton surfactants (Hargreaves, 1981). It is envisaged that the endogenous elicitors diffuse from the dying or dead cells into surrounding live cells which are induced to synthesize phytoalexins. If this hypothesis is correct, then the biochemical basis underlying differentiation between compatible and incompatible races should lie in the presence within incompatible races of molecules triggering hypersensitive cell death. In Figure 9, however, an alternative route for triggering phytoalexin biosynthesis which avoids cell death has been included, as it has been suggested that recognition may involve direct and specific elicitation of phytoalexin biosynthesis (Bruegger and Keen, 1979; Keen and Legrand, 1980). The topics of host–parasite recognition and phytoalexin elicitors are considered in more detail in Chapter 13.

3.4. Phytoalexin detoxification

Following the onset of phytoalexin accumulation, continued microbial growth depends upon the ability of the plant to maintain high concentrations of inhibitors and the toxicity of the phytoalexins to the invading organism—the balance illustrated in Figure 8 comes into play. *Botrytis fabae* provides a good example of a totally necrotrophic pathogen that is able to detoxify its host's phytoalexins and minimize their accumulation by rapidly killing cells within infected tissue.

Detoxification of phytoalexins by pathogens has frequently been reported (VanEtten *et al.*, 1982). Various conversions have been described but in only one case, detoxification of kievitone by *Fusarium solani* f.sp *phaseoli*, has the enzyme concerned been isolated. Kievitone is converted to kievitone hydrate by an extracellular enzyme secreted into culture fluids and infected tissues (see Figure 10).

VanEtten *et al.* (1980) have used a genetical approach to analyse the relationship between phytoalexin (pisatin) tolerance and the virulence of *Nectria haematococca* on pea. Studies on an isolate of *N. haematococca* pathogenic to pea showed that it was highly tolerant of pisatin and was able to detoxify the phytoalexin by demethylation (VanEtten and Stein, 1978). *Nectria haematococca* can be found in a number of habitats other than diseased pea tissue, indicating that a great deal of natural variability exists within the species. VanEtten *et al.* (1980) examined a total of 59 isolates with diverse origins and found that virulence was closely correlated with tolerance to and ability to demethylate pisatin. Crosses between isolates with widely differing characteristics further confirmed the close relationships between virulence and pisatin tolerance which was always associated with an ability to metabol-

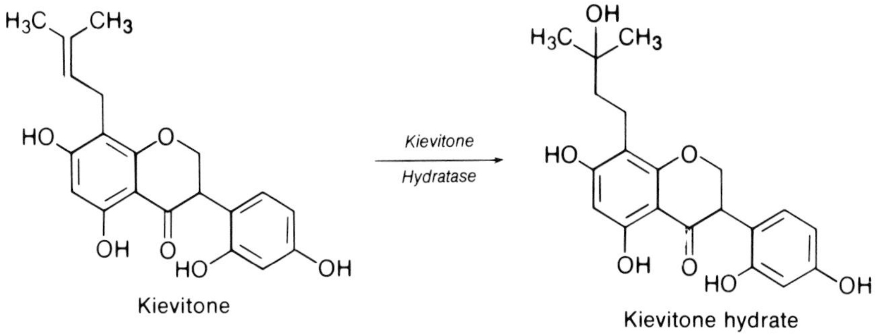

Figure 10. Detoxification of kievitone by *Fusarium solani* f.sp. *phaseoli* (Kuhn and Smith, 1979)

ize the phytoalexin. These results provide convincing evidence for the importance of pisatin detoxification in pathogenicity and also for the role of pisatin in disease resistance.

4. HIGH MOLECULAR WEIGHT COMPOUNDS

The role of high molecular weight phenolic polymers and lignin as antimicrobial compounds in plants has been described in Chapter 11. Other high molecular weight compounds thought to have a role in disease resistance are glycoproteins and proteins which have been shown to bind to bacteria, fungal cells, and cell wall-degrading enzymes produced by invading microorganisms. The binding factors are often described as lectins but many of the compounds reported have not been studied in sufficient detail to fulfil the definition of lectin proposed by Callow (1975), which requires that lectins bind to structures such as cell surfaces through specific carbohydrate-containing receptor sites. Lectins and other agglutinins generally may be considered as constitutive inhibitors.

Albersheim and Anderson (1971) have isolated from plant cell walls proteins which inhibit polygalacturonases produced by some necrotrophic fungi. Although the proteins are not fungitoxic they may contribute to resistance by preventing invading fungi from establishing their normal nutritional relationship within infected tissues. Such an indirect mode of inhibition has also been suggested for glycoproteins found in sweet potato roots which agglutinate germinated spores of *Ceratocystis fimbriata* (Kojima and Uritani, 1974, 1978). An interesting feature of this agglutinin is that, in the presence of certain unidentified low molecular weight factors from sweet potato tissue, it binds specifically to avirulent strains of *C. fimbriata*; virulent strains are not agglutinated.

Fungitoxic activity has been reported for a lectin from wheat embryos. The wheat germ agglutinin (WGA) binds specifically to carbohydrates containing N-acetylglucosamine. Chitin, an important constituent of some fungal cell walls, is a polymer of this amino sugar. Mirelman *et al*. (1975) found that WGA bound to the growing hyphal tips of *Trichoderma viride* where chitin was exposed, and inhibited further fungal growth. The authors suggested that the lectin might protect the germinating grain from fungal attack but further evidence is needed to confirm that WGA is active under natural conditions.

The search for factors which agglutinate bacteria was encouraged by the observation that bacteria appeared to be bound to cell walls of resistant plants (Sequeira *et al*., 1977; Hildebrand *et al*., 1980). There is some evidence that tobacco lectin may be localized in cell walls and bind to avirulent but not to virulent cells of *Pseudomonas solanacearum* (Sequeira, 1978). An agglutinin isolated from various legumes by Jasalavich and Anderson (1981) also showed a degree of specificity, it bound to saprophytic pseudomonads but not to plant pathogenic bacteria or other genera. It is not clear if agglutination *per se* prevents bacterial multiplication *in vivo*.

The role of agglutinating factors in disease resistance remains uncertain; it is possible that their agglutinating activity is fortuitous and that their main function is not to act as antimicrobial compounds but to control the process of recognition leading to the triggering of other mechanisms of resistance as outlined in Figure 9.

5. CONCLUSION

As emphasized earlier in this chapter, proof that an antimicrobial compound causes the restriction of microbial growth requires not only measurement of the concentration of the inhibitor to which the invading microorganism is exposed but also determination of the toxicity of the compound under the conditions found within plant tissues. Attempts to determine localized concentrations of inhibitors have included rather crude dissection experiments involving, for example, analysis of thin sections of tissue (Yoshikawa *et al*., 1978). More critical data need to be obtained, possibly by use of microdissection coupled with very sensitive methods of quantitative analysis, or by the development of immunohistochemical techniques. Precise localization of antimicrobial compounds presents considerable technical difficulties but is a less intractable problem than attempting to reproduce the conditions which occur in infected plants during experimental assessments of antimicrobial activity. Because of the difficulties involved in the direct approach to proving the role of compounds in resistance future developments are likely to come from increased application of experimental methods and in particular use of genetic analyses as pioneered by VanEtten *et al*. (1980) and Défago *et al*. (1983).

There is convincing circumstantial evidence that the failure of certain bacteria and fungal species to colonize inoculated tissues is caused by the presence or accumulation of antimicrobial compounds. For example, in the interactions between French bean hypoctyls and *C. lindemuthianum* and between tulip bulbs and *F. oxysporum*, resistance can be attributed to the accumulation of one phytoalexin, phaseollin, and the presence of one constitutive inhibitor, tulipalin A, respectively. In other plants, neither constitutive nor induced inhibitors may alone account for resistance but the combined effects of the two categories of antimicrobial compound may be sufficient to prevent disease development. Evidence for such a coordinated mechanism of defence is provided by the work of Kojima and Uritani (1978), who proposed that agglutinating factors and furanoterpenoid phytoalexins act synergistically to restrict colonization of sweet potato roots by avirulent strains of *C. fimbriata*.

In view of the remarkable range of antimicrobial compounds isolated from plants and the variety of microorganisms they are thought to counteract, it is difficult to make worthwhile generalizations about the role in disease resistance of the different categories of inhibitor. Evidence suggests, however, that unlike phytoalexin accumulation and other induced processes of resistance, constitutive inhibitors do not have a role in race-specific resistance. Constitutive inhibitors seem to be involved only in the processes of non-host and race non-specific resistance.

Paleobotanical evidence suggests that plants have co-existed with fungal parasites for over 400 million years (Swain, 1978). Probably because of the influence of plant pathologists, studies on antimicrobial compounds have largely been confined to the angiosperms and in particular to economically important crop plants. Studies on the occurrence of constitutive and induced inhibitors in gymnosperms and non-vascular plants should not only lead to the discovery of novel chemicals with antimicrobial activity, but also provide valuable clues to the involvement of the compounds in the evolution of plants and their parasites.

REFERENCES

Albersheim, P., and Anderson, A. J. (1971). Host pathogen interactions. III. Proteins from plant cell walls inhibit polygalacturonases secreted by plant pathogens, *Proc. Natl. Acad. Sci. USA*, **68**, 1815–1819.

Allen, E. H., and Kuć, J. (1968). α-Solanine and α-chaconine as fungitoxic compounds in extracts of Irish potato tubers, *Phytopathology*, **58**, 776–781.

Allen, E. H., and Thomas, C. A. (1971). A second antifungal polyacetylene compound from *Phytophthora*-infected safflower, *Phytopathology*, **61**, 1107–1109.

Bailey. J. A. (1982a). Mechanisms of phytoalexin accumulation, in *Phytoalexins* (Eds J. A. Bailey and J. W. Mansfield), Blackie, Glasgow, pp. 289–318.

Bailey, J. A. (1982b). Physiological and biochemical events associated with the expression of resistance to disease, in *Active Defence Mechanisms in Plants* (Ed. R. K. S. Wood), Plenum Press, New York, London, pp. 39–65.

Bailey, J. A., and Burden, R. S. (1973). Biochemical changes and phytoalexin accumulation in *Phaseolus vulgaris* following cellular browing caused by tobacco necrosis virus, *Physiol. Plant. Pathol.*, **3**, 171–177.

Bailey, J. A., and Deverall, B. J. (1971). Formation and activity of phaseollin in the interaction between bean hypocotyls (*Phaseolus vulgaris*) and physiological races of *Colletotrichum lindemuthianum, Physiol. Plant Pathol.*, **1**, 435–449.

Bailey, J. A., Rowell, P. M., and Arnold, G. M. (1980). The temporal relationship between infected cell death, phytoalexin accumulation and the inhibition of hyphal development during resistance of *Phaseolus vulgaris* to *Colletotrichum lindemuthianum, Physiol. Plant Pathol.*, **17**, 329–339.

Baker, E. A., and Smith, I. M. (1977). Antifungal compounds in winter wheat resistant and susceptible to *Septoria nodorum, Ann. Appl. Biol.*, **87**, 67–73.

Beijersbergen, J. C. M., and Lemmers, C. B. G. (1972). Enzymic and non-enzymic liberation of tulipalin A (α-methylene butyrolactone) in extracts of tulip, *Physiol. Plant Pathol.*, **2**, 265–270.

Bergman, B. H. H., and Beigersbergen, J. C. M. (1968). A fungitoxic substance extracted from tulips and its possible role as a protectant against disease, *Neth. J. Plant Pathol.*, **74**, (suppl. 1), 157–162.

Bernard, N. (1909). L'évolution dans la symbiose, les orchidées, et leurs champignons commensaux. *Trans. Brit. Mycol. Soc.*, **75**, 97–105.

Bernard, N. (1911). Sur la fonction fungicide des bulbes d'Ophrydees, *Ann. Sci. Nat. (Bot.)*, **14**, 221–234.

Blakeman, P. J., and Atkinson, P. (1976). Evidence for a spore germination inhibitor co-extracted with wax from leaves, in *Microbiology of Aerial Plant Surfaces* (Eds C. H. Dickinson and T. F. Preece), Academic Press, London, New York, pp. 441–449.

Bruegger, B., and Keen, N. T. (1979). Specific elicitors of glyceollin accumulation in the *Pseudomonas glycinea*—soybean host–parasite system, *Physiol. Plant Pathol.*, **15**, 43–51.

Callow, J. A. (1976). Plant lectins, *Curr. Adv. Plant Sci.*, **18**, 181–193.

Cartwright, D. W., Langcake, P., and Ride, J. P. (1980). Phytoalexin production in rice and its enhancement by a dichlorocyclopropane fungicide, *Physiol. Plant Pathol.*, **17**, 259–267

Coxon, D. T. (1982). Phytoalexins from other plant families, in *Phytoalexins* (Eds J. A. Bailey and J. W. Mansfield), Blackie, Glasgow, pp. 106–129.

Coxon, D. T., Curtis, R. F., Price, K. R., and Howard, B. (1974). Phytuberin: a novel antifungal terpenoid from potato, *Tetrahedron Lett.*, **27**, 2363–2366.

Coxon, D. T., O'Neill, T. M., Mansfield, J. W., and Porter, A. E. A. (1980). Identification of three hydroxyflavan phytoalexins from daffodil bulbs, *Phytochemistry*, **19**, 889–891.

Davis, D. (1964). Host fungitoxicity in selective pathogenicity of *Fusarium oxysporum, Phytopathology*, **54**, 290–293.

Défago, G., and Kern, H. (1983). Induction of *Fusarium solani* mutants insensitive to tomatine, their pathogenicity and aggressiveness to tomato fruits, *Physiol. Plant Pathol.*, in press.

Défago, G., Kern, H., and Sedlar, L. (1983). Genetic analysis of tomatine insensitivity, sterols content and pathogenicity for green tomato fruits in mutants of *Fusarium solani, Physiol. Plant Pathol.*, in press.

Deverall, B. J. (1977). *Defence Mechanisms of Plants*, Cambridge University Press, Cambridge.

De Wit, P. J. G. M., and Kodde, E. (1981a). Induction of polyacetylenic phytoalexins in *Lycopersicon esculentum* after inoculation with *Cladosporium fulvum* (syn. *Fulvia fulva), Physiol. Plant Pathol.*, **18**, 143–148.

De Wit, P. J. G. M., and Kodde, E. (1981b). Further characterisation and cultivar specificity of glycoprotein elicitors from culture filtrates and cell walls of *Cladosporium fulvum* (syn. *Fulvia fulva*), *Physiol. Plant Pathol.*, **18**, 297–314.

Dix, N. J. (1979). Inhibition of fungi by gallic acid in relation to growth on leaves and litter, *Trans. Br. Mycol. Soc.*, **73**, 329–336.

Eyjólfsson, R. (1970). Recent advances in the chemistry of cyanogenic glycosides, *Fortschr. Chem. Org. Naturst.*, **28**, 74–108.

Fry, W. E., and Evans, P. E. (1977). Association of formamide hydrolyase with fungal pathogenicity to cyanogenic plants, *Phytopathology*, **67**, 1001–1006.

Fry, W. E., and Millar, R. L. (1971). Development of cyanide tolerance in *Stemphlium loti, Phytopathology*, **61**, 501–506.

Gäumann, E., and Kern, H. (1959). Uber chemische Abwehrreacktionen bei Orchiden, *Phytopathol. Z.*, **36**, 1–26.

Greathouse, G. A., and Watkins, G. M. (1938). Berberine as a factor in the resistance of *Mahonia trifoliata* and *Mahonia swaseyi* to *Phymatotrichum* root rot, *Am. J. Bot.*, **25**, 743–748.

Hardegger, E., Biland, H. R., and Corrodi, H. (1963). Synthese von 2, 4-Dimethoxy-6-hydroxyphenanthren und Konstitution des Orchinols, *Hevl. Chim. Acta*, **46**, 1354–1360.

Hargreaves, J. A. (1981). Accumulation of phytoalexins in cotyledons of French bean (*Phaseolus vulgaris* L.) following treatment with Triton (t-octylphenol polyethoxyethanol) surfactants, *New Phytol.*, **87**, 733–741.

Hargreaves, J. A., and Bailey, J. A. (1978). Phytoalexin production by hypocotyls of *Phaseolus vulgaris* in response to constitutive metabolites released by damaged cells, *Physiol. Plant Pathol.*, **13**, 89–100.

Hargreaves, J. A., Mansfield, J. W., and Coxon, D. T. (1976). Identification of medicarpin as a phytoalexin in the broad bean plant (*Vicia faba* L.), *Nature (London)*, **262**, 318–319.

Hargreaves, J. A., Mansfield, J. W., and Rossall, S. (1977). Changes in phytoalexin concentration in tissues of the broad bean plant (*Vicia faba* L.) following inoculation with species of *Botrytis, Physiol. Plant Pathol.*, **11**, 227–242.

Hartman, J. R., Kelman, A., and Upper, C. D. (1975). Differential inhibitory activity of a corn extract to *Erwinia* spp. causing soft rot, *Phytopathology*, **65**, 1082–1088.

Hildebrand, D. C., Alosi, M. C., and Schroth, M. N. (1980). Physical entrapment of *Pseudomonads* in bean leaves by films formed at air–water interfaces, *Phytopathology*, **70**, 98–109.

Holden, J. (1980). Relationship between pre-formed inhibitors in oats and infection by *Gaeumannomyces graminis* and *Phialophora radicicola, Trans. Br. Mycol. Soc.*, **75**, 97–105.

Hutson, R. A., and Mansfield, J. W. (1980). A genetical approach to the analysis of mechanisms of pathogenicity in *Botrytis/Vicia faba* interactions, *Physiol. Plant Pathol.*, **17**, 309–317.

Ingham, J. L. (1973). Disease resistance in higher plants. The concept of pre-infectional and post-infectional resistance, *Phytopathol. Z.*, **78**, 314–335.

Ingham, J. L. (1982). Phytoalexins from the Leguminoseae, in *Phytoalexins* (Eds J. A. Bailey and J. W. Mansfield) Blackie, Glasgow, pp. 21–80.

Ingham, J. L., and Harborne, J. B. (1976). Phytoalexin production as a new dynamic approach to the study of systematic relationships among higher plants, *Nature (London)*, **260**, 241–243.

Ingham, J. L., Keen, N. T., Mulheirn, L. J., and Lyne, R. L. (1981). Inducibly formed isoflavonoids from leaves of soybean, *Phytochemistry*, **20**, 795–798.

Ishiguri, Y., Tomiyama, K., Doke, N., Murai, A., Katsui, N., Yagihashi, F., and

Masamune, T. (1978). Induction of rishitin-metabolizing activity in potato tuber tissue disks by wounding and identification of rishitin metabolites, *Phytopathology*, **68**, 720–725.

Jasalavich, C. A., and Anderson, A. J. (1981). Isolation from legume tissues of an agglutinin of saprophytic pseudomonads, *Can. J. Bot.*, **59**, 264–271.

Jerome, S. M. R., and Müller, K. O. (1958). Studies on phytoalexins. II. Influence of temperature on resistance of *Phaseolus vulgaris* towards *Sclerotinia fructicola* with reference to phytoalexin output, *Aust. J. Biol. Sci.*, **11**, 301–314.

Katsui, N., Murai, A., Takasugi, M., Imaizumi, K., Musamune, T., and Tomiyama, K. (1968). The structure of rishitin, a new antifungal compound from disease potato tubers, *Chem. Commun.*, **1968**, 43–44.

Keen, N. T., and Legrand, M. (1980). Surface glycoproteins: evidence that they may function as the race specific phytoalexin elicitors of *Phytophthora megasperma* f. sp. *glycinea*, *Physiol. Plant Pathol.*, **17**, 175–192.

Kojima, M., and Uritani, I. (1974). The possible involvement of a spore agglutinating factor(s) in various plants in establishing host specificity by various strains of black rot fungus, *Ceratocystis fimbriata, Plant Cell Physiol.*, **15**, 733–737.

Kojima, M., and Uritani, I. (1978). Isolation and characterisation of factors in sweet potato root which agglutinate germinated spores of *Ceratocystic fimbriata*, black rot fungus, *Plant Physiol.*, **62**, 751–753.

Kuć, J. (1982). Phytoalexins from the Solanaceae, in *Phytoalexins* (Eds J. A. Bailey and J. W. Mansfield), Blackie, Glasgow, pp. 81–100.

Kuć, J., Henze, R. E., Ullstrup, A. J., and Quackenbush, F. W. (1956). Chlorogenic and caffeic acids as fungistatic agents produced by potatoes in response to inoculations with *Helminthosporium carbonum, J. Am. Chem. Soc.*, **78**, 3123–3125.

Kuhn, P. J., and Smith, D. A. (1979). Isolation from *Fusarium solani* f. sp. *phaseoli* of an enzymic system responsible for kievitone and phaseollidin detoxification, *Physiol. Plant Pathol.*, **14**, 179–190.

Lamb, C. J., Lawton, M. A., Taylor, S. J., and Dixon, R. A. (1980). Elicitor modulation of phenyl ammonia-lyase in *Phaseolus vulgaris, Ann. Phytopathol.*, **12**, 423–433.

Landes, M., and Hoffmann, G. M. (1979). Ultrahistological investigations of the interaction in compatible and incompatible systems in *Phaseolus vulgaris* and *Colletotrichum lindemuthianum, Phytopathol. Z.*, **96**, 330–350.

Langcake, P., Cornford, C. A., and Pryce, R. J. (1979). Indentification of pterostilbene as a phytoalexin from *Vitis vinifera* leaves, *Phytochemistry*, **18**, 1025–1027.

Link, K. P., and Walker, J. C. (1933). The isolation of catechol from pigmented onion scales and its significance in relation to disease resistance in onions, *J. Biol. Chem.*, **100**, 379–383.

Link, K. P., Dickson, A. D., and Walker, J. C. (1929). Further observations on the occurrence of protocatechuic acid in pigmented onion scales and its relation to disease resistance in onions, *J. Biol. Chem.*, **84**, 719–725.

Lüdtke, M., and Hahn, H. (1953). Uber den Linamaringehalt gesunder und von *Colletotrichum lini* befallener junger Leinpflanzen, *Biochem. Z.*, **325**, 433–442.

Lyon, F. M., and Wood, R. K. S. (1975). Production of phaseollin, coumestrol and related compounds in bean leaves inoculated with *Pseudomonas* spp., *Physiol. Plant Pathol.*, **6**, 117–124.

Mansfield, J. W. (1980). Mechanisms of resistance to *Botrytis*, in *The Biology of Botrytis* (Eds J. R. Coley-Smith, K. Verhoeff, and W. R. Jarvis), Academic Press, London, pp. 181–218.

Mansfield, J. W. (1982). The role of phytoalexins in disease resistance, in *Phytoalexins* (Eds J. A. Bailey and J. W. Mansfield), Blackie, Glasgow, pp. 253–282.

Mansfield, J. W., and Bailey, J. A. (1982). Phytoalexins: current problems and future prospects, in *Phytoalexins* (Eds J. A. Bailey and J. W. Mansfield), Blackie, Glasgow, 319–322.

Mansfield, J. W., and Hutson, R. A. (1980). Microscopical studies on fungal development and host responses in broad bean and tulip leaves inoculated with five species of *Botrytis, Physiol. Plant Pathol.*, **17**, 131–145.

Mansfield, J. W., Porter, A. E. A., and Smallman, R. V. (1980). Dihydrowyerone derivatives as components of the furanoacetylenic phytoalexin response of tissues of *Vicia faba, Phytochemistry*, **19**, 1057–1061.

Millar, R. L., and Higgins, V. J. (1970). Association of cyanide with infection of birdsfoot trefoil by *Stemphylium loti, Phytopathology*, **60**, 104–110.

Mirelman, D., Galun, E., Sharon, N., and Lotan, R. (1975). Inhibition of fungal growth by wheat germ agglutinin, *Nature (London)*, **256**, 414–416.

Müller, K. O., and Börger, H. (1940). Experimentelle Untersuchungen uber die *Phytophthora*—Resistenz der Kartoffel—zugleich ein Beitrag zum Problem der erworbenen Resistenz im Pflanzenreich, *Arb. Biol. Anst. Reichsanst (Berl.)*, **23**, 189–231.

Myers, D. F., and Fry, W. E. (1978). Hydrogen cyanide potential during pathogenesis of sorghum by *Gloecercospora sorghi* or *Helminthosporium sorghicola, Phytopathology*, **68**, 1037–1041.

Nelson, R. R., and Kline, D. M. (1962). Intraspecific variation in pathogenicity in the genus *Helminthosporium* to gramineous species, *Phytopathology*, **52**, 1045–1049.

Oguni, I., and Uritani, I. (1974). Dehydroipomeamarone as an intermediate in the biosynthesis of ipomeamarone, a phytoalexin from the sweet potato root infected with *Ceratocystis fimbriata, Plant Physiol.*, **53**, 649–652.

Powell, C. C., and Hildebrand, D. C. (1970). Fire-blight resistance in *Pyrus:* involvement of arbutin oxidation, *Phytopathology*, **60**, 337–340.

Raa, J. (1968). Polyphenols and natural resistance of apple leaves against *Venturia inaequalis, Neth. J. Plant Pathol.*, **74**, (suppl 1), 37–45.

Rossall, S., and Mansfield, J. W. (1980). Investigation of the causes of poor germination of *Botrytis* spp. on broad bean leaves (*Vicia faba* L.), *Physiol. Plant Pathol.*, **16**, 369–382.

Rust, L. A., Fry, W. E., and Beer, S. V. (1980). Hydrogen cyanide sensitivity in bacterial pathogens of cyanogenic and non-cyanogenic plants, *Phytopathology*, **70**, 1005–1008.

Rutter, J. C., Johnston, W. R., and Willmer, C. M. (1977). Free sugars and organic acids in the leaves of various and plant species and their compartmentation between the tissues. *J. Exp. Bot.*, **28**, 1019–1028.

Saunders, J. A., Conn, E. E., Chin, H. L., and Stocking, C. R. (1977). Subcellular localization of the cyanogenic glucoside of sorghum by autoradiography, *Plant Physiol.*, **59**, 647–652.

Schlösser, E. (1973). Role of saponins in antifungal resistance. II. The hederasaponins in leaves of English ivy (*Hedera helix* L.), *Z. Pflanzenkr. Pflanzenschutz*, **80**, 704–710.

Schönbeck, F. (1967). Untersuchungen uber Bluteninfektionen. V. Untersuchungen an Tulpen, *Phytopathol. Z.*, **59**, 205–224.

Schönbeck, F., and Schlösser, E. (1976). Preformed substances as potential protectants, in *Encyclopedia of Plant Physiology, Vol. 4, Physiological Plant*

Pathology (Eds R. Heitefuss and P. H. Williams), Springer-Verlag, Berlin, pp. 653–678.

Schönbeck, F., and Schroeder, C. (1972). Role of antimicrobial substances (tuliposides) in tulips attacked by *Botrytis spp., Physiol. Plant Pathol.*, **2**, 91–99.

Sequeira, L. (1978). Lectins and their role in host–pathogen specificity *Annu. Rev. Phytopathol.*, **16**, 453–481.

Sequeria, L., Gaard, G., and de Zoeten, G. A. (1977). Attachment of bacteria to host cell walls: its relation to mechanisms of induced resistance, *Physiol. Plant Pathol.*, **10**, 43–50.

Shain, L. (1967). Resistance of sapwood in stems of loblolly pine to infection by *Fomes annosus, Phytopathology*, **57**, 1034–1045.

Slob, A., Jekel, B., de Jong, B., and Schlatmann, E. (1975). On the occurrence of tuliposides in the Liliiflorae, *Phytochemistry*, **14**, 1997–2005.

Stoessl, A. (1982). Biosynthesis of phytoalexins, in *Phytoalexins* (Eds J. A. Bailey and J. A. Mansfield), Blackie, Glasgow, pp. 133–180.

Stoessl, A., and Unwin, C. H. (1970). The antifungal factors in barley, V. Antifungal activity of the hordatines, *Can. J. Bot.*, **48**, 465–470.

Swain, T. (1978). Plant–animal co-evolution; a synoptic view of the paleozoic and mezozoic, in *Biochemical Aspects of Plant and Animal Coevolution* (Ed. J. B. Harborne), Academic Press, London, pp. 3–19.

Tomiyama, K., Doke, N., Nozue, M., and Ishiguri, Y. (1979). The hypersensitive response of resistant plants, in *Recognition and Specificity in Plant Host–Parasite Interactions* (Eds J. M. Daly and I. Uritani), Japan Scientific Societies Press, Tokyo, pp. 317–333.

Trione, E. J. (1960). The HCN content of flax in relation to flax wilt resistance, *Phytopathology*, **50**, 482–486.

Tscheche, R., Kammerer, F.-J., Wulff, G., and Schönbeck, F. (1968). Uber die antibiotisch wirksamen substanzen der Tulpe (*Tulipa geseriana*), *Tetrahedron Lett.*, **6**, 701–706.

Tscheche, R., Chandra Jha, H., and Wulff, G. (1973). Uber Triterpene—Zur Struktur des Avenacins, *Tetrahedron*, **29**, 629–633.

Turner, E. M. C. (1953). The nature of the resistance of oats to the take-all fungus, *J. Exp. Bot.*, **4**, 264–271.

Turner, E. M. C. (1960). The nature of the resistance of oats to the take-all fungus. III. Distribution of the inhibitor in oat seedlings, *J. Exp. Bot.*, **11**, 403–412.

Turner, E. M. C. (1961). An enzymic basis for pathogenic specificity in *Ophiobolus graminis, J. Exp. Bot.*, **12**, 169–175.

VanEtten, H. D., and Stein, J. I. (1978). Differential response of *Fusarium solani* isolates to pisatin and phaseollin, *Phytopathology*, **68**, 1276–1283.

VanEtten, H. D., Matthews, P. S., Tegtmeier, K. J., Dietert, M. F., and Stein, J. I. (1980). The association of pisatin tolerance and demethylation with virulence on pea in *Nectria heamatococca, Physiol. Plant Pathol.*, **16**, 257–268.

VanEtten, H. D., Matthews, D. E., and Smith, D. A. (1982). Metabolism of phytoalexins, in *Phytoalexins* (Eds J. A. Bailey and J. W. Mansfield), Blackie, Glasgow, pp. 181–217.

Verhoeff, K., and Liem, J. I. (1975). Toxicity of tomatine to *Botrytis cinerea* in relation to latency, *Phytopath. Z.*, **82**, 333–338.

Walker, J. C., and Stahmann, M. A. (1955). Chemical nature of disease resistance in plants, *Annu. Rev. Plant Physiol.*, **6**, 351–366.

Yoshikawa, M. (1978). *De novo* messenger RNA and protein synthesis are required for phytoalexin-mediated disease resistance in soybean hypocotyls, *Plant Physiol.*, **61**, 314–317.

Yoshikawa, M., Yamauchi, K., and Masago, H. (1978). Glyceollin: its role in restricting fungal growth in resistant soybean hypocotyls infected with *Phytophthora megasperma* var. *sojae, Physiol. Plant Pathol.*, **12**, 73–82.

Zaki, A. I., Zentmeyer, G. A., Pettus, J., Sims, J. J., Keen, N. T., and Sing, V. O. (1980). Borbonol from *Persea* spp.—chemical properties and antifungal activity against *Phytophthora cinnamomi, Physiol. Plant Pathol.*, **16**, 205–212.

Further Reading

Commonwealth Mycological Institute (1973). *A Guide to the Use of Terms in Plant Pathology*, Phytopathological Papers No. 17, Commonwealth Agricultural Bureaux, Farnham, Surrey.

Deverall, B. J. (1981). *Fungal Parasitism*, Studies in Biology No. 17, Edward Arnold, London.

Mayama, S., Matsuura, Y., Iida, H. and Tani, T. (1982). The role of avenalumin in the resistance of oat to crown rust, *Puccinia coronata* f. sp *avenae Physiol. Plant Pathol.* **20**, 189–199.

Wheeler, H. (1975). *Plant Pathogenesis*, Springer-Verlag, Berlin, Heidelber, New York.

Wood, R. K. S. (1974). *Disease in Higher Plants*, Oxford Biology Reader No. 57, Oxford University Press, London.

Biochemical Plant Pathology
Edited by J. A. Callow
© 1983 John Wiley & Sons Ltd

13

Macromolecules, Recognition, and the Triggering of Resistance

MASAAKI YOSHIKAWA

Laboratory of Plant Pathology, Faculty of Agriculture, Kyoto Prefectural University, Kyoto 606, Japan

1. INTRODUCTION

Plants in nature are constantly challenged by an immense array of pathogenic microorganisms, but resist almost all of them. In many cases this involves inducible defence mechanisms such as the accumulation of phytoalexins or structural barriers (Chapters 11–13). Failure of plants to invoke such defence reactions may result in disease susceptibility. The induction of these plant defence reactions is presumed to be mediated by an initial recognition process between plants and pathogens (Albersheim and Anderson-Prouty, 1975; Keen and Bruegger, 1977) which involves detection of certain unique structural features of incompatible pathogens by recognitional molecules in plants, thereby setting off a cascade of biochemical events leading ultimately to resistance expression. This idea has been supported by the finding that certain components isolated from pathogenic microorganisms induce biochemical events characteristic of resistance responses in plants. This chapter is concerned with the molecules which may participate in initial plant–pathogen recognition and with the sequence of biochemical events which follows. The evidence to be assessed is primarily concerned with the phytoalexin response in fungal pathogen–plant systems.

2. MOLECULES OF PATHOGEN ORIGIN THAT INDUCE PLANT DEFENCE REACTIONS

Defence in plants is based on diverse and sometimes subtle mechanisms. One of the common features observed in many plant–pathogen interactions, however, is that plant defence is frequently not static but induced *de novo* after microbial infection through metabolically active processes. This may be demonstrated by pre-treatment of plants with metabolic inhibitors or heat treatments that suppress the expression of plant defence and concomitantly result in growth of a normally incompatible pathogen (Heath, 1980). The *de novo* nature of plant defence reactions *a priori* suggests that plants recognize certain unique elements elaborated by pathogens and that this triggers a series of reactions ultimately leading to the expression of disease resistance. Indeed, substances have been isolated from several pathogenic microorganisms that elicit inducible plant defence reactions.

2.1. Phytoalexin elicitors

Phytoalexin accumulation appears to be a widely occurring mechanism of natural disease resistance in plants (e.g. Keen, 1981a, and Chapter 12). Phytoalexin production may also be induced, in the absence of living micro-organisms, by certain microbial components that have been termed elicitors (Keen, 1975; West, 1981). Most of the isolated and characterized elicitors are macromolecules, either carbohydrates or glycoproteins, and frequently they induce tissue browning akin to hypersensitive cell necrosis in addition to phytoalexin accumulation. It is not known, however, whether elicitors are capable of inducing the two symptoms independently, or induce only one of these which then triggers the other. Elicitors may be divided into two groups based on the specificity of phytoalexin elicitation. Specific elicitors are those which have differential elicitor activity in various plant cultivars depending upon the disease resistance genotype, and non-specific elicitors do not have such differential activity. Some structural features of the known elicitors and their possible role in invocation of disease defence reactions will be discussed, with emphasis on investigations carried out in the last few years; for com-prehensive summaries of earlier work the reader is referred to reviews by Albersheim and Anderson-Prouty (1975), Callow (1977), and Keen and Bruegger (1977).

2.1.1. Carbohydrate elicitors

Many fungi contain branched β-glucans as major structural elements of cell walls (Bartnicki-Garcia, 1968) and these glucans appear to be potent phytoalexin elicitors. Ayers *et al.* (1976b,c) extracted elicitors of the soybean phytoalexin glyceollin from mycelial walls of *Phytophthora megasperma* f. sp. *glycinea* (syn. var. *sojae*) by autoclaving, and separated them into four fractions based on affinities for conA-Sepharose and DEAE-cellulose. The four fractions varied in glucose, mannose, and protein contents, but several lines of evidence suggested that elicitor activity of all the fractions resided in the glucan component. Methylation analysis of the elicitor-active glucans indicated that they were highly branched β-1,3-glucans similar to the structural glucans present in the cell walls of the fungus. These extracted glucans were heterogeneous in size, with an average molecular weight of 100 000 daltons. Albersheim's group is currently determining the detailed molecular structure of the active site of the glucan elicitors by preparing the smallest elicitor-active oligosaccharide after partial acid hydrolysis and enzymatic cleavage. The smallest elicitor-active oligosaccharide so far obtained contains nine glucosyl residues composed of two 3-linked, two 6-linked, two 3,6-linked, and three terminal residues (Albersheim and Valent, 1978).

Elicitor-active β-glucans appear to be present in the cell walls of a wide range of fungi. Glucans from *Colletotrichum lindemuthianum* and *Fusarium oxysporum* elicited phaseollin production and hypersensitive tissue browning in green bean (Anderson, 1980a). Glucans from *P. infestans* elicited terpenoid phytoalexin production and tissue browning in potato (Lisker and Kuć, 1977). The structures of the glucans produced by three races of *P. megasperma* f. sp. *glycinea* did not differ (Ayers *et al.*, 1976c). There were, however, some differences in structure of the glucans isolated from different species of fungi. For instance, Anderson (1980a) noted that the linear chain linkage structure of the glucans of *F. oxysporum* was predominantly β-1,3 and β-1,6 whereas those of *C. lindemuthianum* were β-1,3 and β-1,4. It is not known, however, whether the elicitor-active sites of glucan molecules have identical or similar structures among glucans of different fungi. Glucan elicitors are non-specific, since they do not exhibit significant differences when isolated from various pathogen races and assayed on incompatible and compatible host cultivars. Glucans are generally active in plants that are susceptible to those pathogens from which the glucans are isolated, but studies with *P. megasperma* f. sp. *glycinea* showed that its glucans were also active on plants that were not hosts for this fungus, i.e. potato and green bean (*Phaseolus vulgaris*) (Cline *et al.*, 1978). Likewise, green bean responds to glucan elicitors from not only its pathogen *C. lindemuthianum* but also other species of *Colletotrichum* and *F. oxysporum* (Anderson, 1980a) in addition to *P. megasperma* f. sp. *glycinea*. The facts that glucan elicitors are widespread among diverse fungi and many plants respond to them suggest that fungal cell wall glucans may act as elicitors of general or non-host resistance in plants attacked by many pathogenic but incompatible fungi (Albersheim and Valent, 1978).

Glucan elicitors were also detected in culture fluids of *C. lindemuthianum* (Albersheim and Anderson-Prouty, 1975) and *P. megasperma* f. sp. *glycinea* (Ayers *et al.*, 1976a). Since the chemical structures of these elicitors were indistinguishable from those of the cell wall glucans, the extracellular glucans probably originated from the walls, possibly owing to the activity of autolytic enzymes during long-term culturing. An elicitor of glyceollin, which has a structure similar to that of the *Phytophthora* wall glucans, was also isolated from brewers' yeast, a non-pathogen of plants (Hahn and Albersheim, 1978).

Hadwiger and Beckman (1980) observed that chitosan, a polymer of β-1,4-linked glucosamine residues and formerly known as a major constituent in cell walls of Zygomycete fungi, was a minor constituent of spores of *F. solani* and was an active elicitor of pisatin accumulation in pea. Further immunochemical studies (Hadwiger *et al.*, 1981) suggested that chitosan or its fragments were released from *Fusarium* spores on pea pod tissue and detected in the surrounding plant cells. The release of chitosan was probably mediated by hydrolytic enzymes present in the pea tissue. These results provided evidence that chitosan penetrated into and was recognized by plant cells. In addition to acting as a phytoalexin elicitor, chitosan was reported to inhibit directly germination and growth of the fungus.

Yoshikawa *et al.* (1979c, 1981) demonstrated that incubation of cell walls of *P. megasperma* f. sp. *glycinea* with either wounded soybean cotyledons or enzyme preparations from them resulted in the rapid release of high molecular weight but soluble carbohydrate elicitors of glyceollin. Since the elicitors were liberated from the cell walls after incubation for as little as 2 min and were also released from living fungal hyphae (Yoshikawa *et al.*, unpublished work), the process may be important in the initial host–parasite interaction. Later research (Keen and Yoshikawa, unpublished work) showed that the released elicitors accounted for about 4% of the cell wall dry weight and contained approximately equal amounts of glucose and mannose. The enzymatically released carbohydrates were heterogeneous, but a relatively homogeneous fraction with a molecular weight of 40 000 daltons was purified by gel filtration and gave the highest elicitor activity. Two isozymes of *endo-β-1,3*-glucanase with isoelectric points of pH 8.7 and 10.5 were purified to apparent homogeneity from soybean cotyledons and shown to account for the release of the glucomannan elicitor from fungal walls. It is therefore presumed that the released glucomannan elicitors are covalently linked to the *β-1,3*-structural glucans of native cell walls and are liberated when these are enzymatically attacked. The released elicitors appeared to exhibit a degree of race and cultivar specificity for elicitation of glyceollin, but this has not been extensively tested.

2.1.2. Glycoprotein elicitors

Glycoproteins occur universally on the surface of eucaryotic cells and participate in recognition processes in many biological systems. Evidence now indicates that glycoproteins of fungal origin act as phytoalexin elicitors. Stekoll and West (1978) isolated a high molecular weight elicitor of casbene biosynthesis in castor bean (*Ricinus communis*) from culture fluids of *Rhizopus stolonifer*. The most purified elicitor fraction contained both protein and carbohydrate with a molecular weight of approximately 32 000 daltons. The carbohydrate portion comprised about 20% of the elicitor and was composed of 92% mannose and 8% glucosamine. Both protein and carbohydrate moieties were required for elicitor activity, since pronase and periodate treatments reduced elicitor activity. Lee and West (1981a,b) showed that the extensively purified glycoprotein elicitor had *endo*-polygalacturonase activity and that the elicitor activity was dependent upon the catalytic activity of the enzyme. This was consequently the first reported example of a fungal elicitor with catalytic activity. Since polygalacturanase and other pectic enzymes are produced by many plant pathogens, it is possible that these enzymes may act as elicitors in other plant–pathogen systems. It is interesting to question how the *R. stolonifer* polygalacturonase triggers phytoalexin accumulation. One possibility may be that the enzyme releases certain polysaccharides from plant cell walls which act as endogenous elicitors (Albersheim *et al.*, 1980). Although the interaction between castor bean and *R. stolonifer* does not constitute a well defined host–pathogen system, it is nonetheless unique since the enzymes catalysing casbene synthesis have been identified.

Keen and Legrand (1980) extracted glycoproteins from the mycelial walls of nine races of *P. megasperma* f. sp. *glycinea* with mild alkali. The glycoprotein preparations elicited the production of glyceollin in two near-isogenic soybean cultivars with the same relative specificity as the living fungal races from which they were obtained. The carbohydrate but not the protein portion of the glycoproteins appeared to be the elicitor-active moiety. The purified glycoprotein preparations contained approximately equal amounts of glucose and mannose, and were reactive with concanavalin A. Using FITC-labelled concanavalin A, they demonstrated that the extracted glycoproteins were present on the outer surface of the fungal cell walls. The work offers experimental support for the previously formulated specific elicitor hypothesis (Keen, 1975) wherein the surface-borne specific elicitors of the pathogen are determinants of disease specificity.

Anderson (1980b) isolated high molecular weight products from culture fluids of three races of *C. lindemuthianum* which were shown to be composed of glycoproteins when analysed by polyacrylamide gel electrophoresis. Preparations from the three races differed in their carbohydrate and protein compositions and in their ability to elicit phaseollin production and tissue browning in red kidney bean. Preparations from an avirulent race were found to be 100-fold more active in elicitor activity than from a virulent race and 10-fold more active than from a race of intermediate virulence. Although further tests employing more fungal races and plant cultivars are necessary, the results also appear to be consistent with the specific elicitor hypothesis.

Frank and Paxton (1971) extracted a fraction from mycelia of *P. megasperma* f. sp. *glycinea* that contained both carbohydrate and protein with a molecular weight of 10 000–30 000 daltons. The fraction elicited a fluorescing compound in soybean, called PA_k, produced coordinately with glyceollin. Interestingly, elicitor production by the fungus was reported to be stimulated more by plant extracts of incompatible than compatible genotypes.

Virulent and avirulent forms of *F. solani* produce extracellular elicitors that differ in pisatin-inducing potential in pea (Daniels and Hadwiger, 1976). The elicitors were fairly heat-stable but sensitive to pronase treatment, indicating that they might be glycoproteins. The patterns of pisatin-inducing components from virulent and avirulent forms were similar when culture fluids were fractionated on ion-exchange columns; however, the avirulent form produced quantitatively more elicitor components.

Culture fluids, mycelial extracts, and cell walls of *Cladosporium fulvum* contained high molecular weight elicitors of rishitin accumulation in tomato, pisatin in pea, and glyceollin in soybean (De Wit and Roseboom, 1980). Elicitor-active components appeared to be glycoproteins, based on sensitivity to periodination, α-mannosidase, and pronase treatments. The elicitors isolated from two races of the fungus, however, were not race or cultivar specific.

Bruegger and Keen (1979) solubilized elicitor-active components from cellular envelope fractions of five races of a bacterial pathogen of soybean, *Pseudomonas glycinea*, with sodium dodecylsulphate. These contained both carbohydrates and proteins, and elicitor activity was partially destroyed by periodination but not by pronase, thereby suggesting that the activity probably resided in carbohydrates. The solubilized elicitors showed race and cultivar specificity; with one exception, extracts from all incompatible races elicited higher levels of glyceollin in cotyledons of two soybean cultivars than did extracts from compatible races. In recent work, highly purified lipopolysaccharides from the bacteria have been shown to possess elicitor activity when used in the highly soluble triethylamine salt form (Keen, unpublished work).

2.1.3. Elicitors with other chemical properties

The first purified microbial elicitor was a highly water soluble polypeptide with a molecular weight of approximately 8000 daltons isolated from mycelia of *Monilinia fructicola* (Cruickshank and Perrin, 1968). This peptide elicited phaseollin production in green bean, but elicited neither pisatin in pea nor phytoalexins in broad bean (*Vicia faba*).

A lipid fraction from *P. infestans* elicited tissue browning and terpenoid phytoalexin production in potato (Ersek, 1977). Bostock *et al.* (1981) later showed that two fatty acids, arachidonic and eicosapentaenoic acids, isolated from the same fungus were phytoalexin elicitors in potato. These fatty acids appeared to be tightly bound to cell walls of the fungus, and the authors concluded that the elicitor activity of the cell wall fraction was due to the fatty acids and not the glucans. Kurantz and Zacharius (1981), however, reported that neither the lipid nor the glucan fractions from mycelial walls of *P. infestans* were active elicitors when they were separately applied to potato tissue, but combination of the two materials resulted in substantial activity. The reason for the apparent discrepancy in the role of lipids as elicitors is not known.

Many agents not produced by plant pathogens also elicit phytoalexin accumulation in plants. These agents include heavy metal salts, detergents, antibiotics, synthetic peptides (Hadwiger *et al.*, 1971; Hadwiger and Schwochau, 1971; Yoshikawa, 1978), and ultraviolet irradiation (Bridge and Klarman, 1973). Since their significance in pathogenesis is unclear, they will not be discussed further here.

Wounding of plant cells may be a general effector, if not an elicitor, of phytoalexin production. Plant tissues are usually wounded in some manner in elicitor bioassays and elicitor solutions are applied on to the wounded surfaces. Although wounding alone does not cause considerable phytoalexin accumulation, it induces metabolic changes in plants that may be important in phytoalexin accumulation. For instance, wounding of soybean hypocotyls and cotyledons induced considerable synthesis of glyceollin, as measured by [^{14}C]phenylalanine incorporation, which was not detected in the unwounded intact tissues (Yoshikawa, 1978; Yoshikawa et al., 1979b). Partridge and Keen (1977) also noted that the activity of phenyalanine ammonia lyase and two other enzymes involved in glyceollin biosynthesis increased after wounding of soybean hypocotyls. Although glyceollin did not accumulate to a detectable level in the wounded tissues, possibly owing to the presence of constitutive glyceollin degrading activity, the results indicated that wounding can activate the glyceollin biosynthetic pathway. Similar activation of phytoalexin biosynthesis after wounding was reported for rishitin synthesis in potato tuber (Sakai et al., 1979). These observations also suggest that phytoalexins may serve not only as defence substances but also as normal plant metabolites associated with the wound response. Indeed, Yoshikawa et al. (1979a) showed that glyceollin and other several phytoalexins stimulated adventitious root formation in a mung bean bioassay.

Although elicitors stimulate phytoalexin synthesis over the level caused by wounding alone (Yoshikawa, 1978), it has not been tested whether or not wounding is required for phytoalexin accumulation induced by elicitors. Although unwounded suspension-cultured cells of soybean and bean can respond to elicitors and accumulate phytoalexins (Ebel et al., 1976; Hargreaves and Selby, 1978), the metabolic states of the cultured cells may differ from those of normal plant tissues. One of the possible explanations for the observed effects of wounding is that wounding may release certain plant metabolites, normally compartmentalized in intact tissues, that have potential to activate phytoalexin synthesis.

2.2. Substances that induce other plant defence mechanisms

There are several reported instances where certain substances of pathogen origin invoke plant defence reactions which are apparently unrelated to phytoalexin production. Wade and Albersheim (1979) obtained glycoprotein preparations from culture fluids of three races of P. megasperma f. sp. glycinea. The partially purified glycoproteins from incompatible races, but not those from compatible races, protected soybean seedlings against inoculation with living compatible fungal races. The sugar compositions of the carbohydrate portion of the glycoproteins differed among three races, as did the extracellular glycoprotein invertases from the same fungus (Ziegler and

Albersheim, 1977). The glycoprotein preparations were, however, neither potent nor specific elicitors of glyceollin accumulation in soybean hypocotyl and cotyledon bioassays. The authors therefore speculated that the glycoproteins invoke an unknown defence mechanism not involving glyceollin which inhibits growth of the normally compatible fungal races.

Pre-treatment of rice plants with cell-free germination fluids of *Helminthosporium oryzae* induced resistance to subsequent inoculation with the living fungus (Sinha and Das, 1972). In this case the authors suggested that the germination fluids might induce the production of phytoalexin-like resistance factors. Indeed, pre-treatment with certain phytoalexin elicitors has been reported to protect plants from subsequent attack by living fungi (Albersheim and Valent, 1978; Hadwiger and Beckman, 1980). Infiltration with bacterial lipopolysaccharide was shown to induce an unidentified resistance mechanism in tobacco against living bacteria (Graham *et al.*, 1977). A glucan fraction from *P. infestans* also inhibited virus multiplication when the fraction was included in the virus inoculum (Hodgson *et al.*, 1969).

Ishiba *et al.* (1981) found that pre-inoculation of cut ends of cucumber stems with a weakly virulent strain of *F. oxysporum* f. sp. *cucumerinum* systemically protected cucumber leaves from subsequent attack by a foliar pathogen, *Colletotrichum lagenarium*. They also showed that certain fractions from the *Fusarium* strain, notably the insoluble cell wall fraction, were able to mimic the living fungus in inducing systemic protection. Although the mechanism of systemic protection in cucumber has not been established, recent studies have suggested that systemically protected leaves induce rapid lignification upon infection with the challenge fungus (Kuć, unpublished work).

There are a few cases where soluble extracts from incompatible-responding plant tissues induce resistance. Berard *et al.* (1972) showed that bean hypocotyls were protected against compatible races of *C. lindemuthianum* by a factor which diffused into water droplets from incompatible-responding plant tissue. The protection appeared to highly specific for the plant resistance genotype, since the factor protected only the cultivar from which it was obtained. It was not known, however, whether the factor was of pathogen or plant origin, although spore germination fluids or mycelial extracts of the fungus did not protect plants. Prusky *et al.* (personal communication) found that a fraction from crown rust-infected oat leaves containing nucleic acids induced resistance associated with hypersensitive cell necrosis in oat leaves. The protection factor in this case appeared to be of fungal origin, since its production was specific to the fungal virulence genotype.

Plant lignification is also induced in the absence of living pathogens. Matsumoto *et al.* (1978) showed that homogenates from downy mildew-infected Japanese radish roots induced extensive lignification in slices of the root tissue. The lignification-inducing factor was of plant origin and formed in the diseased tissue in response to infection. A later study suggested that the factor was a glycoprotein (Matsumoto *et al.*, 1980). Pearce and Ride (1978) indicated that some filamentous fungi produce factors capable of inducing lignification in wheat (see Chapter 11). Active factors from *Botrytis cinerea* appeared to be mucilage materials and chitin, a cell wall constituent of Ascomycete and Basidiomycete fungi, was also active. Furthermore, pretreatment of wounds with chitin resulted in some protection from attack by *Fusarium gramineum*, thereby indicating that lignification serves as a plant defence mechanism.

2.3. Assessment of the role of elicitors in phytoalexin accumulation in infected tissues

As noted above, many pathogenic fungi produce elicitors of plant defence reactions, especially phytoalexin accumulation. These elicitors mimic living fungi by inducing the symptoms characteristic of disease resistance reactions such as tissue browning or hypersensitive cell necrosis concomitant with phytoalexin accumulation. In addition, some elicitors have been shown to induce accumulation with the same specificity as the living pathogen races. These facts suggest that the observed elicitors may be involved with the invocation of phytoalexin accumulation in pathogen-infected plants. Further evidence supporting this possibility is that pathogen-produced, i.e. biotic, elicitors may lead to phytoalexin accumulation and biochemical effects similar to those occurring in infected tissues, but distinctly different from those caused by abiotic chemical elicitors (Yoshikawa, 1978; Yoshikawa *et al.*, 1979b; see also Section 3.4).

There are, however, several unresolved questions regarding the *in vivo* involvement of elicitors. A major detracting argument arises from the observation that most of the fungal cell wall elicitors can be extracted only by severe treatments such as autoclaving or acid or alkaline treatments, which are unlikely to exist in biological environments. This raises the question of how normally insoluble elicitor molecules on or in fungal walls may be recognized by plant cells during natural infection processes. Furthermore, certain fungi such as *P. megasperma* f. sp. *glycinea* produce many types of extracellular and wall-associated elicitors. Are all of these elicitors active in infected tissues? In addition, *Phytophthora* species contain large amounts of intracellular β-glucans, called mycolaminarans, which also act as non-specific elicitors in certain plant tissues. These intracellular glucans are, however, unlikely to have a physiological role in elicitation of phytoalexin accumulation during the initial host–pathogen interaction. Likewise, cell wall elicitors may not be efficiently detected by plant cells, unless they are exposed at the outer surface of the walls. Thus, unless elicitors can be proved to be surface-borne, extracellularly produced, or detected in some other way by plant cells during pathogenesis, their physiological role remains questionable.

Keen and Legrand (1980) localized race and cultivar-specific glycoprotein elicitors on cell walls of *P. megasperma* f. sp. *glycinea*. Using FITC-labelled concanavalin A that binds to the glycoproteins, they demonstrated that the glycoprotein elicitors were present on the outer surface of the fungal cell walls. Thus, the glycoprotein elicitors are likely to be detected by plant cells during the infection processes. The observation also raises the possibility that the race-non-specific glucan elicitors associated with the cell walls of the fungus (Ayers *et al.*, 1976b,c) may not be detected by plant cells, since they may be masked by the surface glycoproteins.

Another possible way of facilitating the detection of fungal wall elicitors by plant cells has recently been suggested. Enzymes present in soybean and pea tissues rapidly liberated soluble phytoalexin elicitors from cell walls of *P. megasperma* f. sp. *glycinea* (Yoshikawa *et al.*, 1979c, 1981) and *F. solani* (Hadwiger and Beckman, 1980), respectively. The previously discussed soybean *endo-β*-1,3-glucanase activity appeared to be present mostly in the cytosol fraction and therefore the invading fungus would probably be exposed to the enzymes. Furthermore, elicitor release also occurred from the surface of the actively growing fungus after incubation with the soybean enzymes (Yoshikawa *et al.*, unpublished work). These conditions are similar to the natural infection process, and it is accordingly possible that enzyme-mediated liberation of fungal wall elicitors occurs in fungus-infected plants. If so, this would facilitate penetration of the elicitors to host target sites.

Culture fluids of various pathgenic fungi contain elicitor activity. This indicates that they may produce extracellular elicitors. It has not been demonstrated, however, that they are produced during pathogenesis and in sufficient quantity to account for the observed phytoalexin accumulation. For example, Yoshikawa *et al.* (unpublished work) incubated growing mycelia of *P. megasperma* f. sp. *glycinea* with the soybean *endo*-glucanase or with the heat-inactivated enzyme. High elicitor activity was recovered in the cell-free fluids incubated with the active enzyme, but only low levels of elicitor activity were detected with the inactivated enzyme. This suggested that the activity of any extracellular elicitors elaborated by the fungus during the limited time periods of incubation was far lower than that of the enzyme-released wall elicitors.

Critical evaluation of the involvement of elicitors *in vivo* appears to be essential, especially where one fungus produces several extra- and intracellular and wall-associated elicitors, some of which are reported to be race and cultivar specific and some non-specific. One of the possible reasons that various polysaccharides and glycoproteins in a single fungus have elicitor activity may be that they contain common carbohydrate structures accounting for elicitor activity. In addition to such common elicitor-active structures, other portions of elicitor molecules may serve to confer various properties such as race specificity. Assessing the *in vivo* involvement of an individual elicitor is therefore important before interpreting its physiological role in the host–pathogen interaction.

3. SEQUENCE OF BIOCHEMICAL EVENTS LEADING TO PHYTOALEXIN ACCUMULATION AND OTHER DISEASE DEFENCE REACTIONS

3.1. Recognition of phytoalexin elicitor molecules by plant receptors

Elicitors induce a series of metabolic events in plants which eventually result in phytoalexin accumulation. These include specific gene transcription and translation, resulting in *de novo* induction of enzymes of phytoalexin biosynthesis (Yoshikawa, 1978; Zähringer *et al.*, 1979). The initial biochemical event is thought to be the recognition of elicitor molecules by receptors on plant cells. At least two ways in which this might occur have been suggested. The first model proposes that elicitors interact directly with plant DNA, resulting in specific gene transcription required for phytoalexin production (Hadwiger and Schwochau, 1969; Day, 1974). The model is largely based on observations that DNA-intercalating compounds such as actinomycin D and other chemicals with high affinity for DNA can induce pisatin accumulation in pea (Hadwiger and Schwochau, 1971; Hadwiger *et al.*, 1971). Consistent with this idea is the observation that UV irradiation, which induces thymine dimer formation in DNA, also elicits phytoalexin accumulation in certain plants (Bridge and Klarman, 1973). However, the only biotic elicitor which has been shown to interact directly with DNA and to enter the plant nucleus is chitosan, an elicitor of pisatin production present in *Fusarium* spp. (Hadwiger and Beckman, 1980; Hadwiger *et al.*, 1981). Evidence that other biotic elicitors directly interact with plant DNA is lacking. Instead, there are several experimental results which support the alternative model that biotic elicitors have receptors on plant cell membranes.

Glucan elicitors from cell walls of *P. infestans* rapidly altered the electric membrane potential of potato cells (Kota and Stelzig, 1977) agglutinated potato protoplasts (Peters *et al.*, 1978) and disorganized and killed potato protoplasts (Doke and Tomiyama, 1980a). Doke *et al.* (1975) observed that a particulate fraction from zoospores of *P. infestans*, presumably an elicitor fraction, interacted with and precipitated potato membranes. These observations indicate the probable existence of receptors for the glucan elicitors on the cell surface membranes of potato cells. Furthermore, Doke and Tomiyama (1978) reported that the high molecular weight dextran-bound *p*-chloromercuribenzoic acid inhibited the rapid hypersensitive death of potato tuber cells following infection with *P. infestans*. Since the high molecular weight SH-reagent was presumed not to enter the potato cells, the results were interpreted to indicate that the reagent interfered with the recognition process between the plant and incompatible pathogen which probably occurred at the plant cell surface. The SH-reagent, however, did not inhibit the direct binding of the fungal hyphae to host plasma membranes (Nozue *et al.*, 1981), as will be described below.

A tight binding between the surface of *P. infestans* and plasma membranes of potato cells was demonstrated by cytological methods (Nozue *et al.*, 1979). When the fungus-infected potato cells were plasmolysed, parts of plasmolysed protoplasts adhered to the intracellular hyphae. The adhesion appeared to be essential for invocation of hypersensitive cell death and might be mediated by potato lectin, since pretreatment with *N,N′*-diacetyl-D-chitobiose, the potato lectin hapten, inhibited both the adhesion and hypersensitive cell death (Nozue *et al.*, 1980). Garas and Kúc (1981) further showed that the potato lectin indeed bound and precipitated fungal elicitors of terpenoid phytoalexin accumulation and hypersensitive host cell death. These observations suggest that the potato lectin serves as a receptor for elicitor molecules produced by *P. infestans*. The results also suggest that chitobiose-like residues are present in the elicitor molecules and are the site recognized by receptors on potato cells. This is, however, somehow contradictory to the observation that β-methylglucoside specifically inhibited the hypersensitive cell death of potato cells induced by a β-1,3-linked glucan elicitor preparation from *P. infestans* (Marcan *et al.*, 1979).

The above observations with the *P. infestans*–potato system all indicate, albeit indirectly, the existence of receptors for the glucan elicitors on plant membranes. However, heterogeneous elicitor preparations were used in the studies and the possibility accordingly exists that components in the preparations other than elicitor molecules *per se* were responsible for the observed phenomena. Furthermore, the experiments did not employ direct and quantitative binding assay between elicitors and plant membranes

Yoshikawa, Keen, and Wang (unpublished work) recently obtained the first direct evidence suggesting that a biotic phytoalexin elicitor is recognized by a specific receptor on soybean membranes, by employing a ^{14}C-labelled homogeneous fungal elicitor and a direct membrane binding assay. Mycolaminaran, a water-soluble and branched β-1,3-glucan produced by *Phytophthora* spp., was selected as a model elicitor since it is readily isolated as highly homogeneous molecules, an essential requirement for critical binding studies. The exploratory experiments using soybean cotyledon membrane fractions showed that the highest [^{14}C]mycolaminaran specific binding activity was associated with a membrane fraction that contained predominantly plasma membranes. A Scatchard plot of the binding indicated the presence of a single class of binding sites for the glucan elicitor with a K_D value of approximately 10 μmol l^{-1}. The binding was highly specific, based on competition studies with unlabelled ligand, on the fact that no other carbohydrates competed with binding and on the concomitant alterations in both the affinity constant for binding and observed glyceollin elicitor activity of various chemically altered and naturally occurring mycolaminaran derivatives. The results therefore indicate that the observed binding sites are specific receptors for mycolaminaran and may be physiologically involved in the initiation of phytoalexin production in soybean. In addition, the work offers precedence for the possible existence of analogous plant receptors that recognize the race and cultivar specific elicitors, such as the previously discussed glycoprotein and glucomannan elicitors produced by *P. megasperma* f. sp. *glycinea*. A search for such receptors would be of great importance since the specific elicitor–receptor model (Keen, 1975) described previously predicts that plant resistance genes code for such receptors.

3.2. Possible formation of secondary messenger after elicitor–receptor interaction

Provided that the binding of elicitor molecules to receptors on plant plasma membranes is essential for invoking phytoalexin accumulation, this scheme requires the formation of a secondary messenger that transmits signals from the plasma membrane receptors to the nucleus of the plant cell in order to induce specific gene transcription. This model is analogous to animal systems in which cyclic nucleotides, histamine and arachidonic acid, possibly following membrane phospholipid methylation, are produced in response to hormone binding to membrane receptors (Hirata and Axelrod, 1980).

Oguni *et al.* (1976) showed that cyclic-3′,5′-AMP, but not 5′-AMP, induced accumulation of furanoterpenoid phytoalexins and tissue browning in sweet potato tissue. Keen and Kennedy (1974) reported that cyclic AMP alone or with a compatible race of *P. glycinea* greatly stimulated hypersensitive cell necrosis and phytoalexin accumulation in soybean. However, there is still much argument on the existence of cyclic AMP in plants, and further studies are necessary to assess whether cyclic AMP is involved in phytoalexin induction.

Paradies *et al.* (1980) observed that ethylene production increased immediately after elicitor treatment of soybean tissues. However, they concluded that ethylene formation is not directly linked with subsequent glyceollin production, since both a stimulator and an inhibitor of ethylene production did not influence the phytoalexin levels (see also Chapter 19).

Endogenous plant constituents have been described which elicit phytoalexin accumulation. Hargreaves and Bailey (1978) and Hargreaves and Selby (1978) reported that aqueous extracts obtained from bean hypocotyls stimulated phaseollin formation when applied to hypocotyl tissue or cell suspension cultures. The active component in the extracts, termed a constitutive elicitor, was heat stable and dialysable. Based on these observations, it was suggested that the initial interaction between plant cells and an invading pathogen leads to the release of a preformed plant metabolite which then stimulates phytoalexin formation. Albersheim *et al.* (1980) isolated a similar endogenous elicitor from cell walls of soybean hypocotyls and of suspension-cultured cells of tobacco, sycamore, and wheat that elicited glyceollin accumulation in soybean tissue. The elicitor was released from the isolated cell walls by partial acid hydrolysis and purified by ion-exchange and gel filtration chromatography. The elicitor-active fragments were heterogeneous in size, consisting of 10–15 glycosyl residues. The endogenous elicitor appeared to be a pectic polysaccharide. This may be relevant to the observations that polygalacturonases have elicitor activity in certain plants (Lee and West, 1981a; Albersheim et al., 1980). It is possible that exogenously applied enzymes release the endogenous elicitors from plant cell walls, although this has not been demonstrated experimentally (Albersheim et al., 1980). Albersheim et al. (1980) further observed that a freeze–thaw procedure released an enzyme from soybean hypocotyls that elicits phytoalexin accumulation in soybean. The possibility therefore exists that this enzyme, upon activation due to cell damage caused by fungal infection or elicitor treatment, may release the endogenous elicitor which then acts as a secondary messenger and triggers the metabolic alterations leading to phytoalexin production.

Akin to the endogenous elicitors of phytoalexin production is the lignification-inducing factor in Japanese radish roots (Matsumoto et al., 1978) discussed previously (Section 2.2). Further study is necessary to elucidate whether these endogenous substances serve as secondary messengers in plant defence responses.

3.3. *De Novo* DNA transcription and translation

Substantial evidence now suggests that DNA transcription and translation are required for phytoalexin production and the expression of disease resistance. Yoshikawa et al. (1977) observed that soybean hypocotyls inoculated with an incompatible race of *P. megasperma* f. sp. *glycinea* synthesized poly(A)-containing messenger RNA about six times more rapidly than uninoculated control plants as early as 4 h after inoculation. Little activation occurred in hypocotyls inoculated with a compatible race. Yoshikawa (1980) further showed that messenger RNA synthesized during the early incompatible interaction contained new species of messenger RNA which were not detected in uninoculated plants; translation of messenger RNA in an *in vitro* wheat germ system led to the appearance of several new proteins. These observations demonstrated that the incompatible host–pathogen interaction resulted in a rapid quantitative and qualitative alteration of gene transcription. Increases in messenger RNA synthesis at early stage of incompatible interaction have also been noted in crown rust-infected oat leaves (Tani and Yamamoto, 1978).

The importance of *de novo* gene transcription and subsequent translation was further substantiated with the finding that the transcription inhibitor actinomycin D and the translation inhibitor blasticidin S diminished both glyceollin accumulation and resistance expression in soybean, when the inhibitors were applied to the plants at or shortly after inoculation

(Yoshikawa *et al.*, 1978b). In the inhibitor-treated hypocotyls the fungus grew as well as it did in near-isogenic susceptible cultivars. Similar stimulation of fungal growth by application of transcription and translation inhibitors to incompatible host plants was also demonstrated in the crown rust–oat system (Tani and Yamamoto, 1978). In soybean hypocotyls, the inhibitor treatment was also effective in suppressing glyceollin production and resistance expression when plants were inoculated with various *Phytophthora* species that are normally nonpathogenic to soybean. The results therefore indicate that not only cultivar-specific but also general resistance associated with phytoalexin production in soybean is mediated by *de novo* messenger RNA and protein synthesis.

Elicitor-induced phytoalexin production also appears to be mediated by *de novo* gene transcription and translation. Hadwiger and co-workers (Hadwiger *et al.*, 1971; Daniels and Hadwiger, 1976) showed that cordycepin and cycloheximide, which are transcription and translation inhibitors, respectively, inhibited pisatin accumulation induced by culture fluids of *F. solani*. Transcription inhibitors such as 6-methylpurine, α-amanitin, and high concentrations of actinomycin D inhibited pisatin accumulation induced by abiotic agents such as low concentrations of actinomycin D and UV irradiation. Glyceollin production in elicitor-treated soybean cotyledons was also inhibited by cordycepin, 6-methylpurine, blasticidin S, and cycloheximide, but not by actinomycin D (Yoshikawa, unpublished work).

Induced lignification has also been shown to be mediated by *de novo* gene activation. Vance and Sherwood (1976) demonstrated that cycloheximide inhibited the formation of lignified papilla in reed canarygrass in response to infection by several non-pathogenic fungi such as *Helminthosporium avenae* and the plant became susceptible to penetration by these fungi. Matsumoto *et al.* (1978) reported that lignification in Japanese radish induced by a homogenate of the downy mildew-infected tissue was efficiently inhibited by transcription and translation inhibitors.

The role of *de novo* gene induction in hypersensitive cell necrosis appears controversial, depending upon different host–pathogen systems. Tani *et al.* (1975, 1976) reported that blasticidin S diminished the resistance of oat leaves to infection with an incompatible race of the crown rust fungus, but it did not inhibit hypersensitive cell necrosis. Keen *et al.* (1981), however, reported that the same inhibitor negated both resistance expression and hypersensitive cell necrosis in soybean leaves following infection with an incompatible race of *Pseudomonas glycinea*. Doke and Tomiyama (1975) showed that blasticidin S did not inhibit hypersensitive cell necrosis induced in an incompatible potato–*P. infestans* interaction, when the inhibitor was applied to aged tuber discs shortly before fungal inoculation. Nozue *et al.* (1977), however, later reported that when the inhibitor was applied immediately after cutting, hypersensitive cell necrosis did not occur after the fungal inoculation. The results were interpreted to indicate that *de novo* protein synthesis was not required for hypersensitive cell necrosis *per se*, but was necessary for the tuber cells to acquire the potential to respond hypersensitively during the ageing process after cutting.

A consequence of *de novo* gene transcription and translation is the appearance of enzyme activities that may participate in the terminal biosynthetic pathway for the formation of phytoalexins or other defence substances. Oba *et al.* (1976) showed that the enzymes such as 3-hydroxy-3-methylglutaryl CoA reductase and pyrophosphomevalonate decarboxylase, which are presumably involved in biosynthesis of furanoterpene phytoalexins in sweet potato, were activated following fungal inoculation or elicitor treatment, and that cycloheximide treatment inhibited both the enzyme activation and phytoalexin production. Gustine *et al.* (1978) found that fungus-inoculated jackbean cells accumulated medicarpin concomitant with increased activity of an *O*-methyltransferase capable of methylating hydroxylated isoflavonoids. West's group (Dueber *et al.*, 1978) reported that exposure of castor bean seedlings to spores or a partially purified elicitor of *R. stolonifer* resulted in a great increase of casbene synthetase, which catalyses the synthesis of the phytoalexin casbene directly from *trans*-geranylgeranyl pyrophosphate. This is the only reported instance of the complete biosynthesis of a phytoalexin in an inducible cell-free enzyme system.

Zähringer *et al.* (1979) reported that particulate fractions from elicitor-treated soybean cotyledons contained enzymes which catalysed the specific prenylation of 3,6,9a-trihydroxypterocarpan at the 2- or 4- position. Since the trihydroxypterocarpan and its prenylated products appear to be the immediate biosynthetic precursors of glyceollin, the results indicate that the prenyl transferase is involved in glyceollin biosynthesis. No prenyl transferase activity was detected with extracts from cotyledons not treated with the elicitor, thus indicating that the prenyl transferase must be induced following elicitor application. *De novo* synthesis of PAL in French bean cells treated with a phytoalexin elicitor from cell walls of *C. lindemuthianum* has been detected using the sensitive and elegant technique of density-labelling (Dixon and Lamb, 1979; see also Chapter 17).

Enzymes involved in lignin biosynthesis also appear to be activated following infection. Certain, but not all isozymes of peroxidase from Japanese radish, formed lignin-like materials *in vitro* when incubated with *p*-hydroxycinnamyl alcohols, and some of these isozyme activities increased after infection with the downy mildew fungus (Ohguchi and Asada, 1975). Vance *et al.* (1976) also reported that fungus-inoculated reed canarygrass leaves showed increased lignin content and peroxidase activity, and that the increased enzyme activity was attributable to the induction of cathodic isoperoxidases.

Cycloheximide treatment inhibited both the increase of peroxidase activity and lignification, thus indicating that these peroxidases may function in the biosynthesis of lignin at the site of fungal penetration.

3.4. Regulation of phytoalexin accumulation

As described above, the production of phytoalexins following infection or elicitor treatment appears to be due to induced *de novo* biosynthesis, possibly mediated by *de novo* gene transcription and translation. However, there is now evidence, albeit somewhat controversial, that the ultimate levels of phytoalexin accumulation in plant tissues are not regulated solely by the rates of the induced biosynthetic activity since phytoalexins may be degraded in plants. Therefore, levels of phytoalexin accumulation may be regulated by the relative rates of biosynthesis and degradation. Using a pulse-labelling technique with a ^{14}C-labelled precursor, Yoshikawa *et al.* (1979b) observed that glyceollin biosynthesis was apparently triggered by wounding alone. The phytoalexin, however, did not accumulate, owing to the presence of glyceollin degrading activity as demonstrated by a pulse-chase experiment in which [^{14}C]glyceollin synthesized by the wounded hypocotyls rapidly disappeared after a chase with the unlabelled precursor. In addition, sliced soybean hypocotyls possessed substantial glyceollin degrading activity *in vitro*. These observations suggested that there is a dynamic balance between synthesis and degradation in the uninoculated but wounded hypocotyls, such that glyceollin does not accumulate. In addition, phytoalexin metabolizing activity has been observed for capsidiol in pepper (Stoessl *et al.*, 1977), rishitin in potato (Horikawa *et al.*, 1976), and phaseollin in bean (Glazener and VanEtten, 1978).

As in the case of uninoculated, wounded soybean hypocotyls, phytoalexin degrading activity appears to play an important role in determining the ultimate levels of phytoalexin accumulation in infected tissues. In fungus-challenged soybean hypocotyls, glyceollin accumulation is much greater in incompatible than compatible interactions. However, when the rate of glyceollin biosynthesis was estimated by pulse-labelling experiments with [^{14}C]phenylalanine, no significant differences were observed with hypocotyls infected with either an incompatible or compatible fungal race although the rates of the synthesis were greater in the inoculated than uninoculated hypocotyls (Yoshikawa *et al.*, 1979b). In contrast, pulse-chase experiments suggested that glyceollin degrading activity was inhibited more strongly in the hypocotyls infected with the incompatible than the compatible race. The fungus did not appear to contribute directly to the differential glyceollin degrading activity in the infected hypocotyls since neither race degraded glyceollin *in vitro*. Moesta and Grisebach (1981), however, reported that glyceollin synthesis, as measured by $^{14}CO_2$ incorporation, was higher in the incompatible than compatible interaction 14 h after fungal inoculation. Based mainly on this observation, they suggested that the levels of glyceollin accumulation in the incompatible and compatible interactions were determined by its rate of synthesis. In their experiments, however, they failed to detect any difference in the synthetic activity between the two interactions at 8–10 h after inoculation. This is an early and critical time period of the interaction when high, localized levels of glyceollin accumulate only in the incompatible interaction and appear to be responsible for inhibition of fungal growth in the resistant hosts (Yoshikawa et al., 1978a).

Phytoalexin accumulation is induced by a variety of substances including biotic elicitors of pathogen origin and abiotic elicitors such as heavy metal salts and detergents. The finding that phytoalexin accumulation could be dynamically regulated by relative rates of both synthesis and degradation gives a clue to the mechanisms by which such diverse elicitor molecules induce phytoalexin accumulation in plants. Yoshikawa (1978) showed that different biotic elicitors such as cell walls and extracellular metabolites of P. megasperma f. sp. glycinea stimulated glyceollin synthesis in soybean cotyledons to about eight times that in the control, wounded cotyledons, with no significant effect on glyceollin degrading activity. In contrast, various abiotic elicitors caused little or no stimulation of biosynthetic activity over the controls; instead, they all strongly inhibited the degrading activity. The differential effects of biotic and abiotic elicitors on glyceollin degrading activity were confirmed by use of different techniques such as pulse-chase experiments with [14C]phenylalanine and direct assessment of degrading activity with [14C]glyceollin (Yoshikawa et al., 1980). Furthermore, the ability of about 60 tested abiotic chemicals to stimulate glyceollin accumulation was highly correlated with their specific inhibitory effects on degrading activity. These results strongly indicate that the primary regulatory effects of biotic and abiotic elicitors on phytoalexin metabolism are distinctly different.

Moesta and Grisebach (1980) used single biotic and abiotic elicitors on soybean cotyledons and obtained different results from those above. They concluded that the two types of elicitors both function by mainly stimulating glyceollin synthesis. The cause of the apparent discrepancy between the German and Japanese groups (discussed also in Chapter 17) is not known; however the two groups used different pulse-chase reagents [14C]phenylalanine and $^{14}CO_2$ and their unlabelled adducts. Despite the fact that carbon dioxide is a poor chase reagent, similar results would have been expected with the different labelled precursors if the two types of elicitors have the same mode of action. Other problems relating to the interpretation of pulse-chase data are discussed in Chapter 17. Further research such as evaluating the effects of elicitors on in vitro phytoalexin degrading systems or on the induction of enzymes involved in phytoalexin biosynthesis appears to be necessary before their mode of action can be finally established.

4. POSSIBLE MECHANISMS UNDERLYING PLANT DISEASE SPECIFICITY

As described previously, plant disease resistance may be triggered by the initial recognition process between certain pathogen molecules and plant receptors. Such recognition activates a series of biochemical events leading to the expression of resistance as typified by phytoalexin accumulation or structural barrier formation. Susceptible reactions, conversely, are thought to originate from the failure of plants to invoke the active plant defence reactions. Therefore, disease specificity is determined by whether or not defence reactions are invoked. The following is a brief summary of currently proposed mechanisms determining disease specificity, mainly based on the elicitor–phytoalexin mechanism. The reader is also referred to the more comprehensive reviews by Keen (1981b) on this topic.

4.1. Quantitatively different production of elicitors

4.1.1. Differential synthesis of elicitors by pathogens

Frank and Paxton (1971) observed that *P. megasperma* f. sp. *glycinea* produced elicitors when the fungus was incubated with extracts from incompatible but not compatible soybean cultivars. Based on this observation, they developed the double induction hypothesis, which states that specific constitutive substances in incompatible host genotypes interact with the invading fungus such that the pathogen is induced to synthesize a non-specific elicitor. This hypothesis, however, has not been supported by further experimental evidence. Further, the induction period required for synthesis of the elicitors would presumably require several hours, but recognition in the fungus–soybean system occurs very rapidly after initial contact of the fungus and host (Yoshikawa *et al.*, 1978b).

4.1.2. Differential release of elicitors (inhibitors) from pathogen cell walls

Hadwiger and Beckman (1980) found that chitosan, a component of *Fusarium* cell walls, induced pistin accumulation in pea and protected the plant from the fungal infection. Chitosan, in addition, inhibited fungal growth and was deposited on the fungal cell walls during an incompatible host reaction. Similar hexosamine polymers were released from the cell walls during the infection process and by enzyme preparations from pea. Based on these regulatory roles of the hexosamine polymer, Hadwiger and Loschke (1981) proposed a hypothesis that certain unique elements from the cell wall surface of the incompatible fungus interact with plant heterochromatin and result in the synthesis of plant hydrolytic enzymes which in turn release hexosamine polymers from fungal cell walls. The released polymers may then inhibit fungal growth directly and may also activate plant defence reactions. The hypothesis is unique in the sense that the fungal constituents themselves act as inhibitory factors of fungal growth, and is accordingly the converse of the phytoalexin theory. However, several immediate questions arise concerning the hypothesis and further experimental proof is clearly required to establish occurrence of the mechanism. Yoshikawa *et al.* (1979c, 1981) also observed that soybean tissues contained *endo*-glucanases capable of releasing elicitors of glyceollin from fungal cell walls. In this case, however, the initial rates of elicitor release were not affected by enzymes from different plant genotypes, although the possibility remains that the enzyme-mediated release of elicitor may be accelerated during the incompatible interaction owing to further induced synthesis of the enzymes. Again, this is unlikely to occur since recognition takes place very rapidly after infection.

4.2. Production of qualitatively different elicitors or suppressors

4.2.1. Race and cultivar-specific elicitors

The pioneering work by Keen (1975) demonstrated that culture fluids of an incompatible but not compatible race of *P. megasperma* f. sp. *glycinea* contained a race-specific phytoalexin elicitor. The partially purified factor elicited greater production of glyceollin in genetically incompatible than in compatible soybean cultivars. Based on these observations, Keen developed the specific elicitor-specific receptor hypothesis for gene-for-gene host–pathogen systems. In the model, constitutive elements of the incompatible but not compatible pathogen race are recognized by specific receptor molecules in the incompatible host genotype and this initial recognition leads to subsequent invocation of an inducible defence mechanism, possibly involving phytoalexin accumulation. Further support for the hypothesis was obtained with the findings that cell wall surface glycoproteins (Keen and Legrand, 1980) and enzyme-released glucomannans (Keen and Yoshikawa, unpublished work) from *P. megasperma* f. sp. *glycinea* and cellular envelopes from *P. glycinea* (Bruegger and Keen, 1979) acted as race and cultivar-specific elicitors of glyceollin in soybean tissues. Race-specific glycoprotein elicitors were also demonstrated in culture fluids of *C. lindemuthianum* (Anderson, 1980b). The hypothesis is consistent with various biochemical and genetic evidence observed in gene-for-gene systems. The hypothesis also explains the high degree of specificity underlying gene-for-gene systems, since subtle structural changes in elicitor or receptor molecules could alter the affinity of the elicitor–receptor interaction. Additional testing should establish the validity of the model. If the model is correct, pathogen avirulence genes probably encode the primary structures of specific glycosyl transferases which synthesize the carbohydrate moieties of elicitor molecules and host-resistant genes encode the primary structures of receptor molecules (Albersheim and Anderson-Prouty, 1975; Keen, 1981b, 1982). The specific elicitor–receptor model cannot, however, be applied to the general or non-host resistance of plants to various pathogenic microbes or for disease specificity at the species–species level; the model instead suggests that the recognition mechanisms in these cases are different.

4.2.2. Non-specific elicitors and specific suppressors

The suppressor model has been advanced to account for determination of disease reaction at two levels, species–species and race–cultivar specificity. Involvement of suppressors in the former specificity was demonstrated by Oku, Ouchi, and co-workers (Shiraishi *et al.*, 1978). Two fractions, believed to contain low molecular weight peptides, were isolated from germination fluids of *Mycosphaerella pinodes*, a pathogen of pea, and shown to suppress pisatin accumulation in pea leaves induced by high molecular weight elicitors from the fungus. The peptide suppressors were shown to inhibit the biosynthetic pathway to pisatin, resulting in the accumulation of precursors such as cinnamic and coumaric acids in the suppressor-treated tissue (Shiraishi *et al.*, 1980). The suppressors also inhibited the phytoalexin accumulation induced by infection with a non-pathogen, *Stemphylium sarcinaeforme*, and rendered pea leaves susceptible to the normally non-pathogenic fungus. Futhermore, several other leguminous plant species, to which *M. pinodes* is pathogenic, became susceptible to the non-pathogen, *Alternaria alternata*, when plants were treated with the suppressors (Oku *et al.*, 1980). These results therefore indicated that the suppressor of phytoalexin accumulation is a pathogenicity factor of *M. pinodes* and plays an important role in determination of the host range of the fungus.

Suppressors also appear to be involved in overcoming the non-host resistance of certain plants to obligate parasite. Heath (1981) observed that exudates of the bean rust fungus or extracts of bean rust-infected French bean leaves, when injected into uninoculated French bean tissues, increased the frequency of haustoria subsequently produced by *Uromyces phaseoli* var. *vignae*, the cowpea rust fungus. The extracts inhibited the formation of silicon-containing deposits at infection sites, a process associated with prevention of haustorial development.

Suppressors have been strongly implicated as determinants of specificity in the interaction between races of *P. infestans* and potato cultivars. Doke, Kuć, and co-workers (Garas *et al.*, 1979; Doke *et al.*, 1980) isolated water-soluble, low molecular weight glucan molecules from homogenates and germination fluids of several fungal races which functioned as race and cultivar-specific suppressors of the terpenoid phytoalexin accumulation and hypersensitve cell necrosis otherwise caused by high molecular weight non-specific elicitors. It was therefore proposed that the specific suppressors confer race and cultivar specificity in this host–pathogen system. Doke and Tomiyama (1980b) further showed that hypersensitive responses of potato protoplasts to the non-specific elicitors were specifically inhibited by the glucan suppressors from compatible but not incompatible fungal races.

Despite the excellent correlative evidence, there are some uncertainties for the operation of the specific suppressor mechanism in gene-for-gene systems, particularly its conflict with the available genetic evidence. In most of the studied gene-for-gene systems, genes conditioning the expression of resistance (plant resistance and pathogen avirulence genes) are inherited as dominant characters. Indeed, all the known resistance genes in potato were shown to be dominant. If the suppressor mechanism operates, genes for susceptibility should be dominant. The recent proposal by Bushnell and Rowell (1981), however, appears to reconcile the genetic conflict. Their model is based on the proposition that specific suppressors compete with the binding of non-specific elicitors and their plant receptors. Dominant resistance genes may code for the primary structure of the receptors, and dominant pathogen avirulence genes determine the structures of suppressors such that the suppressors cannot bind to the receptors, thereby allowing elicitor–receptor binding. The model is consistent with the observation by Doke *et al.* (1979) that suppressors probably interfere with the recognition between elicitors and plant membranes.

Suppression of phytoalexin production or other defence reactions may be quite general among plant pathogens. For instance, host-specific toxins or pathotoxins may work in precisely this way. Patil and Gnanamanickam (1976) indicated that phaseotoxin produced by *Pseudomonas phaseolicola* suppressed both hypersensitive cell necrosis and phaseollin production in bean leaves otherwise induced by incompatible isolates of the bacterium. A host-specific toxin produced by *Alternaria kikuchiana* was also suggested to suppress resistance responses in toxin-susceptible pear leaves induced by avirulent isolates of the fungus or resistance-inducing factors from the fungus (Hayami *et al.*, 1980).

4.2.3. *Non-specific elicitors and specific protection factors*

This model was developed by Albersheim's group (Ayers *et al.*, 1976c) for the *P. megasperma* f. sp. *glycinea*–soybean system. Based on their observations (Ayers *et al.*, 1976b) that similar levels of glyceollin accumulated during compatible and incompatible interactions and that their elicitor preparations did not show race and cultivar specificity, they indicated that glyceollin accumulation is not the primary factor for inhibition of fungal growth during resistance expression. Instead, they predicted the presence of an unknown primary defence mechanism which is invoked by race and cultivar-specific protection factors elaborated by pathogen races. The predicted presence of such specific protection factors was experimentally demonstrated by Wade and Albersheim (1979). They partially purified glycoproteins from culture fluids of several fungal races and showed that the glycoprotein preparations from incompatible but not compatible races specifically protected soybean plants against inoculation with compatible fungal races. The preparations did not specifically elicit production of glyceollin. However, no direct experimental evidence for the existence of the hypothesized primary defence mechanism has yet appeared. Furthermore, the basis of their model is contradictory to

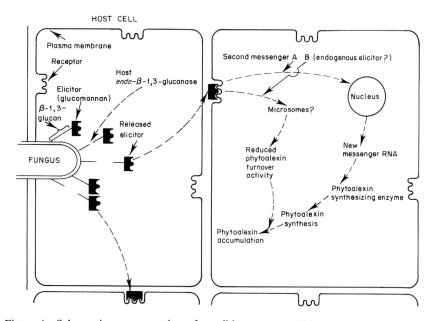

Figure 1. Schematic representation of possible sequence of biochemical events leading to plant defence reactions. The scheme exemplifies phytoalexin accumulation mechanism in the *Phytophthora megasperma* f. sp. *glycinea*–soybean interaction, but it may be applicable to other plant–pathogen systems. Contact of incompatible fungal races with host cells results in rapid release of phytoalexin elicitors from fungal cell wall surface, due to attack by *endo-β-1*, 3-glucanase constitutively present in host cells. The released elicitors then interact with the complementary receptors on plant plasma membrane. This interaction generates second messengers which transmit the signal to the nucleus where *de novo* transcription is invoked. The resulting new messenger RNA leads to the synthesis of enzymes involved in phytoalexin biosynthesis and the phytoalexin is formed. Levels of phytoalexin accumulation are accelerated by simultaneous inhibition of phytoalexin degrading system, which may also result from the elicitor–receptor interaction. Compatible fungal races may either possess elicitors that, upon release, cannot interact with the host receptors, or produce 'suppressors' that interfere with the elicitor–receptor interaction or inhibit one of the subsequent host metabolic processes leading to the phytoalexin accumulation

the finding of several other laboratories that glyceollin accumulation is responsible for inhibition of fungal growth during incompatible interactions (Keen, 1981a; Yoshikawa *et al.,* 1978a).

Several models have thus been proposed as the mechanisms responsible for determining disease specificity. It seems, however, too early at present to conclude that one or the other model is correct. There is a possibility that other specificity mechanisms not yet described may operate. It may also be possible that the molecular basis for specificity is diverse among different plant–pathogen interactions.

As summarized in this chapter (see also Figure 1), extensive research in the last few years has disclosed a sequence of biochemical events that appear to participate in invocation of disease defence reactions. The processes probably involve pathogen-associated molecule (elicitor)–plant receptor interaction, formation of secondary messenger, and gene activation, resulting in *de novo* biosynthesis of enzymes responsible for production of phytoalexins or other defence substances. At present the scheme is still hypothetical in many respects, however, and requires further evaluation of each process before molecular details in the expression of disease resistance are fully understood. A search for physiologically important elicitors as well as their purification and characterization appear to be a crucial aspect. This would be prerequisite for critical experimental analysis of the initial interaction of elicitors with plants and the biochemicals events that follow. Clarification of the regulatory processes involved in the expression of disease resistance, in addition, would eventually help the understanding of the mechanisms of disease specificity.

Disease defence reactions, as exemplified by phytoalexin production or hypersensitivity, constitute one of the most rapid and drastic metabolic responses in plants. Analysis of such drastic responses may disclose certain, as yet unknown, unique regulatory mechanisms in plants that are otherwise difficult to study in uninfected plants, and will contribute to the understanding of many other biologically important problems such as the mechanisms of cell–cell recognition, signal transmission, and gene expression.

5. ACKNOWLEDGEMENTS

The author is very grateful to N. T. Keen for critical comments on the manuscript. Comments made by H. Masago and H. Hamamura are also appreciated. The author's research was supported by grants from the Ministry of Education, Science, and Culture of Japan (Nos. 076178 and 256040).

REFERENCES

Albersheim, P., and Anderson-Prouty, A. J. (1975). Carbohydrates, proteins, cell surfaces, and the biochemistry of pathogenesis, *Annu. Rev. Plant Physiol.*, **26**, 31–52.

Albersheim, P., and Valent, B. S. (1978). Host–pathogen interactions in plants. Plants, when exposed to oligosaccharides of fungal origin, defend themselves by accumulating antibiotics, *J. Cell Biol.*, **78**, 627–643.

Albersheim, P., Darvill, A. G., McNeil, M., Valent, B. S., Hanh, M. G., Sharp, J. K., Desjardins, A. E., Spellman, M. W., Ross, L. M., Robertsen, B. K., Aman, P., and Franzen, L. E. (1980). Structure and function of complex carbohydrates active in regulating plant–microbe interactions, *Proc. Xth Int. Symp. Carbohydrate Chem., Sydney, Australia, July, 1980.*

Anderson, A. J. (1980a). Studies on the structure and elicitor activity of fungal glucans, *Can. J. Bot.*, **58**, 2343–2348.

Anderson, A. J. (1980b). Differences in the biochemical compositions and elicitor activity of extracellular components produced by three races of a fungal plant pathogen, *Colletotrichum lindemuthianum, Can. J. Microbiol.*, **26**, 1473–1479.

Ayres, A. R., Ebel, J., Finelli, F., Berger, N., and Albersheim, P. (1976a). Host–pathogen interactions. IX. Quantitative assays of elicitor activity and characterization of the elicitors present in the extracellular medium of cultures of *Phytophthora megasperma* var. *sojae, Plant Physiol.*, **57**, 751–759.

Ayres, A. R., Ebel, J., Valent, B., and Albersheim, P. (1976b). Host–pathogen interactions. X. Fractionation and biological activity of an elicitor isolated from the mycelial walls of *Phytophthora megasperma* var. *sojae. Plant Physiol.*, **57**, 760–765.

Ayres. A. R., Valent, B., Ebel, J., and Albersheim, P. (1976c). Host–pathogen interactions. XI. Composition and structure of wall-released elicitor fractions, *Plant Physiol.*, **57**, 766–774.

Bartnicki-Garcia, S. (1968). Cell wall chemistry, morphogenesis, and taxonomy of fungi, *Annu. Rev. Microbiol.*, **22**, 87–108.

Bernard, D. F., Kuć, J., and Williams, E. B. (1972). A cultivar-specific protection factor from incompatible interactions of green bean with *Colletotrichum lindemuthianum, Physiol. Plant Pathol.*, **2**, 123–127.

Bostock, R. M., Kuć, J., and Laine, R. A. (1981). Eicosapentaenoic and arachidonic acids from *Phytophthora infestans* elicit fungitoxic sesquiterpenes in the potato, *Science*, **212**, 67–69.

Bridge, M. A., and Klarman, W. L. (1973). Soybean phytoalexin, hydroxyphaseollin, induced by ultraviolet irradiation, *Phytopathology*, **63**, 606–609.

Bruegger, B. B., and Keen, N. T. (1979). Specific elicitors of glyceollin accumulation in the *Pseudomonas glycinea*–soybean host–parasite system, *Physiol. Plant Pathol.*, **15**, 43–51.

Bushnell, W. R., and Rowell, J. B. (1981). Suppressors of defense reactions: a model for roles in specificity, *Phytopathology*, **71**, 1012–1014.

Callow, J. A. (1977). Recognition, resistance and the role of plant lectins in host–parasite interactions, *Adv. Bot. Res.*, **4**, 1–49.

Cline, K., Wade, M., and Albersheim, P. (1978). Host–pathogen interactions. XV. Fungal glucans which elicit phytoalexin accumulation in soybean also elicit the accumulation of phytoalexins in other plants, *Plant Physiol.*, **62**, 918–921.

Cruickshank, I. A. M., and Perrin, D. R. (1968). The isolation and partial characterization of monilicolin A, a polypeptide with phaseollin-inducing activity from *Monilinia fructicola, Life Sci.*, **7**, 449–458.

Daniels, D. L., and Hadwiger, L. A. (1976). Pisatin-inducing components in filtrates of a virulent and avirulent *Fusarium solani* cultures, *Physiol. Plant Pathol.*, **8**, 9–19.

Day, P. R. (1974). *Genetics of Host-Parasite Interaction*, W. H. Freeman, San Francisco.

De Wit, P. J. G. M., and Roseboom, P. H. M. (1980). Isolation, partial characterization and specificity of glycoprotein elicitors from culture filtrates, mycelium, and cell walls, of *Cladosporium fulvum* (syn. *Fulvia fulva*), *Physiol. Plant Pathol., 16*, 391–408.

Dixon, R. A., and Lamb, C. J. (1979). Stimulation of *de novo* synthesis of L-phenylalanine ammonia-lyase in relation to phytoalexin accumulation in *Colletotrichum lindemuthianum* elicitor-treated cell suspension cultures of French bean (*Phaseolus vulgaris*), *Biochem. Biophys. Acta, 586*, 453–463.

Doke, N., and Tomiyama, K. (1975). Effect of blasticidin S on hypersensitive death of potato leaf petiole cells caused by infection with an incompatible race of *Phytophthora infestans., Physiol. Plant Pathol., 6*, 169–175.

Doke, N., and Tomiyama, K. (1978). Effect of sulfhydryl-binding compounds on hypersensitive death of potato tuber cells following infection with an incompatible race of *Phytophthora infestans, Physiol. Plant Pathol., 12*, 133–139.

Doke, N., and Tomiyama, K. (1980a). Effect of hyphal wall components from *Phytophthora infestans* on protoplasts of potato tuber tissues, *Physiol. Plant Pathol., 16*, 169–176.

Doke, N., and Tomiyama, K. (1980b). Suppression of the hypersensitive response of potato tuber protoplasts to hyphal wall components by water soluble glucans isolated from *Phytophthora infestans, Physiol. Plant Pathol., 16*, 177–186.

Doke, N., and Tomiyama, K., Nishimura, N., and Lee, H. S. (1975). *In vitro* interactions between components of *Phytophthora infestans* zoospores and components of potato tissue, *Ann. Phytopathol. Soc., 41*, 425–433.

Doke, N., Garas, N. A., and Kuć, J. (1979). Partial characterization and aspects of the modes of action of a hypersensitive-inhibiting factor (HIF) isolated from *Phytophthora infestans, Physiol. Plant Pathol., 15*, 127–140.

Doke, N., Garas, N. A., and Kuć, J. (1980). Effect on host hypersensitivity of suppressors released during the germination of *Phytophthora infestans* cystospores, *Phytopathology, 70*, 35–39.

Dueber, M. T., Adolf, W., and West, C. A. (1978). Biosynthesis of the diterpene phytoalexin casbene. Partial purification and characterization of casbene synthetase from *Ricinus communis, Plant Physiol., 62*, 598–603.

Ebel, J., Ayers, A. R., and Albersheim, P. (1976). Host–pathogen interactions. XII. Response of suspension-cultured soybean cells to the elicitor isolated from *Phytophthora megasperma* var. *sojae*, a fungal pathogen of soybean, *Plant Physiol., 57*, 775–779.

Ersek, T. (1977). A lipid component from *Phytophthora infestans* inducing resistance and phytoalexin accumulation in potato tubers, *Curr. Topics Plant Pathol. (Symp.), 1977*, 73–76.

Frank, J. A., and Paxton, J. D. (1971). An inducer of soybean phytoalexin and its role in the resistance of soybean to *Phytophthora* rot, *Phytopathology, 61*, 954–958.

Garas, N. A., and Kuć, J. (1981). Potato lectin lyses zoospores of *Phytophthora infestans* and precipitates elicitors of terpenoid accumulation produced by the fungus, *Physiol. Plant Pathol., 18*, 227–237.

Garas, N. A., Doke, N., and Kuć, J. (1979). Suppression of the hypersensitive reaction in potato tubers by mycelial components from *Phytophthora infestans, Physiol. Plant Pathol., 15*, 117–126.

Glazener, J. A., and VanEtten, H. D. (1978). Phytotoxicity of phaseollin to, and alteration of phaseollin by, cell suspension cultures of *Phaseolus vulgaris, Phytopathology, 68*, 111–117.

Graham, T. L., Sequeira, L., and Huang, T. S. (1977). Bacterial lipopolysaccharide as inducers of disease resistance in tobacco, *Appl. Environ. Microbiol., 34*, 424–432.

Gustine, D. L., Sherwood, R. T., and Vance, C. P. (1978). Regulation of phytoalexin synthesis in jackbean callus cultures. Stimulation of phenylalanine ammonia-lyase and O-methyltransferase, *Plant Physiol.*, **61**, 226–230.

Hadwiger, L. A., and Beckman, J. M. (1980). Chitosan as a component of pea–*Fusarium solani* interactions, *Plant Physiol.*, **66**, 205–211.

Hadwiger, L. A., and Loschke, D. C. (1981). Molecular communication in host–parasite interactions: hexosamine polymers (chitosan) as regulator compounds in race-specific and other interactions, *Phytopathology*, **71**, 756–762.

Hadwiger, L. A., and Schwochau, M. E. (1969). Host resistance—an induction hypothesis, *Phytopathology*, **59**, 223–227.

Hadwiger, L. A., and Schwochau, M. E. (1971). Specificity of deoxyribonucleic acid intercalating compounds in the control of phenylalanine ammonia lyase and pisatin levels, *Plant Physiol.*, **47**, 346–351.

Hadwiger, L. A., Jafri, A., von Broembsen, S., and Eddy, R., Jr. (1971). Mode of pisatin induction. Increased template activity and dye-binding capacity of chromatin isolated from polypeptide-treated pea pods, *Plant Physiol.*, **53**, 52–63.

Hadwiger, L. A., Beckman, J. M., and Adams, M. J. (1981). Localization of fungal components in the pea–*Fusarium* interaction detected immunochemically with anti-chitosan and anti-fungal cell wall antisera, *Plant Physiol.*, **67**, 170–175.

Hahn, M. G., and Albersheim, P. (1978). Host–pathogen interactions. XIV. Isolation and partial characterization of an elicitor from yeast extract, *Plant Physiol.*, **62**, 107–111.

Hargreaves, J. A., and Bailey, J. A. (1978). Phytoalexin production by hypocotyls of *Phaseolus vulgaris* in response to constitutive metabolites released by damaged bean cells, *Physiol. Plant Pathol.*, **13**, 89–100.

Hargreaves, J. A., and Selby, C. (1978). Phytoalexin formation in cell suspensions of *Phaseolus vulgaris* in response to an extract of bean hypocotyls, *Phytochemistry*, **17**, 1099–1102.

Hayami, N., Otani, H., Nishimura, S., and Kohmoto, K. (1980). Role of AK-toxin as a suppressor of resistance response in pear tissue, *Ann. Phytopathol. Soc. Jpn.*, **46**, 96.

Heath, M. C. (1980). Reactions of nonsuscepts to fungal pathogens, *Annu. Rev. Phytopathol.*, **18**, 211–236.

Heath, M. C. (1981). The suppression of the development of silicon-containing deposits in French bean leaves by exudate of the bean rust fungus and extracts from bean rust-infected tissue, *Physiol. Plant Pathol.*, **18**, 149–155.

Hirata, F., and Axelrod, J. (1980). Phospholipid methylation and biological signal transmission, *Science*, **209**, 1082–1090.

Hodgson, W. A., Munro, J., Singh, R. P., and Wood, F. A. (1969). Isolation from *Phytophthora infestans* of polysaccharide that inhibits potato virus X, *Phytopathology*, **59**, 1334–1335.

Horikawa, T., Tomiyama, K., and Doke, N. (1976). Accumulation and transformation of rishitin and lubimin in potato tuber tissue infected by an incompatible race of *Phytophthora infestans, Phytopathology*, **66**, 1186–1191.

Ishiba, C., Tani, T., and Murata, M. (1981). Protection of cucumber against anthracnose by a hypovirulent strain of *Fusarium oxysporum* f. sp. *cucumerinum, Ann. Phytopathol. Soc. Jpn.*, **47**, 352–359.

Keen, N. T. (1975). Specific elicitors of plant phytoalexin production: determinants of race specificity in pathogens?, *Science*, **187**, 74–75.

Keen, N. T. (1981a). Evaluation of the role of phytoalexins, in *Plant Disease Control* (Ed. R. C. Staples), Wiley, New York, pp. 155–177.

Keen, N. T. (1981b). Mechanisms conferring specific recognition in gene-for-gene plant–parasite systems, in *Specificity in Plant Disease* (Ed. R. K. S. Wood), Plenum Press, New York, in press.

Keen, N. T. (1982). Specific recognition in gene-for-gene host–parasite systems, in *Advances in Plant Pathology*, Vol. 1 (Eds P. H. Williams and D. Ingram), Academic Press, New York, pp. 35–81.

Keen, N. T., and Bruegger, B. (1977). Phytoalexins and chemicals that elicit their production in plants, *ACS Symp. Ser.*, **62**, 1–26.

Keen, N. T., and Kennedy, B. W. (1974). Hydroxyphaseollin and related isoflavonoids in the hypersensitive resistance reaction of soybeans to *Pseudomonas glycinea, Physiol. Plant Pathol.*, **4**, 173–185.

Keen, N. T., and Legrand, M. (1980). Surface glycoproteins; evidence that they may function as the race specific phytoalexin elicitors of *Phytophthora megasperma* f. sp. *glycinea, Physiol. Plant Pathol.*, **17**, 175–192.

Keen, N. T., Ersek, T., Long, M., Bruegger, B., and Holliday, M. (1981). Inhibition of the hypersensitive reaction of soybean leaves to incompatible *Pseudomonas* spp. by blasticidin S, streptomycin or elevated temperature, *Physiol. Plant Pathol.*, **18**, 325–337.

Kota, D. A., and Stelzig, D. A. (1977). Electrophysiology as a means of studying the role of elicitors in plant disease resistance, *Proc. Am. Phytopathol. Soc.*, **4**, 216–217.

Kurantz, M. J., and Zacharius, R. M. (1981). Hypersensitive response in potato tuber: elicitation by combination of non-eliciting components from *Phytophthora infestans, Physiol. Plant Pathol.*, **18**, 67–77.

Lee, S. C., and West, C. A. (1981a). Polygalacturonase from *Rhizopus stolonifer,* an elicitor of casbene synthetase activity in castor bean (*Ricinus communis* L.) seedlings, *Plant Physiol.*, **67**, 633–639.

Lee, S. C., and West, C. A. (1981b). Properties of *Rhizopus stolonifer* polygalacturonase, an elicitor of casbene synthetase activity in castor bean (*Ricinus communis* L.) seedlings, *Plant Physiol.*, **67**, 640–645.

Lisker, N., and Kuć, J. (1977). Elicitors of terpenoid accumulation in potato tuber slices, *Phytopathology*, **67**, 1356–1359.

Marcan, H., Jarvis, M. C., and Friend, J. (1979). Effect of methyglycosides and oligosaccharides on cell death and browning of potato tuber discs induced by mycelial components of *Phytophthora infestans, Physiol. Plant Pathol.*, **14**, 1–9.

Matsumoto, I., Ohguchi, T., Inoue, M., and Asda, Y. (1978). Lignin induction in roots of Japanese radish by homogenate of downy mildew-infected root tissue, *Ann. Phytopathol. Soc. Jpn.*, **44**, 22–27.

Matsumoto, I., Kumada, T., and Asada, Y. (1980). Studies on lignin formation in infected plants. XXXII. Isolation of a lignin-inducing factor from downy-mildew infected radish roots, *Ann. Phytopathol. Soc. Jpn.*, **46**, 384.

Moesta, P., and Grisebach, H. (1980). Effects of biotic and abiotic elicitors on phytoalexin metabolism in soybean, *Nature (London)*, **286**, 710–711.

Moesta, P., and Grisebach, H. (1981). Investigation of the mechanism of glyceollin accumulation in soybean infected *Phytophthora megasperma* f. sp. *glycinea, Arch. Biochem. Biophys.*, **212**, 462–467.

Nozue, M., Tomiyama, K., and Doke, N. (1977). Effect of blasticidin S on development of potential of potato tuber cells to react hypersensitively to infection by *Phytophthora infestans, Physiol. Plant Pathol.*, **10**, 181–189.

Nozue, M., Tomiyama, K., and Doke, N. (1979). Evidence for adherence of host plasmalemma to infecting hyphae of both compatible and incompatible races of *Phytophthora infestans, Physiol. Plant Pathol.*, **15**, 111–115.

Nozue, M., Tomiyama, K., and Doke, N. (1980). Effect of *N, N* -diacetyl-*D*-

chitobiose, the potato-lectin hapten and other sugars on hypersensitive reaction of potato tuber cells infected by incompatible and compatible races of *Phytophthora infestans, Physiol. Plant Pathol.*, **17**, 221–227.

Nozue, M., Tomiyama, K., and Doke, N. (1981). Effect of some enzyme inhibitors on adherence of host plasmalemma to infecting hyphae of *Phytophthora infestans, Ann. Phytopathol. Soc. Jpn.*, **47**, 189–193.

Oba, K., Tatematsu, H., Yamashita, K., and Uritani, I. (1976). Induction of furano-terpene production and formation of the enzyme system from mevalonate to isopentenyl pyrophosphate in sweet potato root tissue injured by *Ceratocystis fimbriata* and toxic chemicals, *Plant Physiol.*, **58**, 51–56.

Oguni, I., Suzuki, K., and Uritani, I. (1976). Terpenoid induction in sweet potato roots by cyclic-3', 5'-adenosine monophosphate, *Agric. Biol. Chem.*, **40**, 1251–1252.

Ohguchi, T., and Asada, Y. (1975). Dehydrogenation polymerization products of *p*-hydroxycinnamyl alcohols by isoperoxidases obtained from downy mildew-infected root of Japanses radish (*Raphanus sativus*), *Physiol. Plant Pathol.*, **5**, 183–192.

Oku, H., Ishiura, M., Shiraishi, T., and Ouchi, S. (1980). Specificity of the pathogenecity factor produced by *Mycosphaerella pinodes, Ann. Phytopathol. Soc. Jpn.*, **46**, 383.

Paradies, I., Konze, J. R., Elstner, E. F., and Paxton, J. (1980). Ethylene: indicator but not inducer of phytoalexin synthesis in soybean, *Plant Physiol.*, **66**, 1106–1109.

Partridge, J. E., and Keen, N. T. (1977). Soybean phytoalexins: rates of synthesis are not regulated by activation of initial enzymes in flavonoid biosynthesis, *Phytopathology*, **67**, 50–55.

Patil, S. S., and Gnanamanickam, S. S. (1976). Suppression of bacterially-induced hypersensitive reaction and phytoalexin accumulation in bean by phaseotoxin, *Nature (London)*, **259**, 486–487.

Pearce, R. B., and Ride, J. P. (1978). Elicitors of the lignification response of wheat, *Ann. Appl. Biol.*, **89**, 306–307.

Peters, B. M., Cribbs, D. H., and Stelzig, D. A. (1978). Agglutination of plant protoplasts by fungal cell wall glucans, *Science*, **201**, 364–365.

Sakai, S., Tomiyama, K., and Doke, N. (1979). Synthesis of a sesquiterpenoid phytoalexin rishitin in non-infected tissue from various parts of potato plants immediately after slicing, *Ann. Phytopathol. Soc. Jpn.*, **45**, 705–711.

Shiraishi, T., Oku, H., Yamashita, M., and Ouchi, S. (1978). Elicitor and suppressor of pisatin induction in spore germination fluid of pea pathogen, *Mycoshaerella pinodes, Ann. Phytopathol. Soc. Jpn.*, **44**, 659–665.

Shiraishi, T., Hiramatsu, M., Oku, H., and Ouchi, S. (1980). The mechanism of suppression of pisatin accumulation by a factor produced by *Mycosphaerella pinodes, Ann. Phytopathol. Soc. Jpn.*, **46**, 383.

Sinha, A. K., and Das, N. C. (1972). Induced resistance in rice plant to *Helminthosporium oryzae, Physiol. Plant Pathol.*, **2**, 401–410.

Stekoll, M. S., and West, C. A. (1978). Purification and properties of an elicitor of castor bean phytoalexin from culture filtrates of the fungus *Rhizopus stolonifer, Plant Physiol.*, **61**, 38–45.

Stoessl, A., Robinson, J. R., Rock, G. L., and Ward, E. W. B. (1977). Metabolism of capsidiol by sweet pepper tissue: some possible implication for phytoalexin studies, *Phytopathology*, **67**, 111–117.

Tani, T., and Yamamoto, H. (1978). Nucleic acid and protein synthesis in association with the resistance of oat leaves to crown rust, *Physiol. Plant Pathol.*, **12**, 113–121.

Tani, T., Yamamoto, H., Onoe, T., and Naito, N. (1975). Initiation of resistance and host cell collapse in the hypersensitive reaction of oat leaves against *Puccinia coronata avenae, Physiol. Plant Pathol.*, **7**, 231–242.

Tani, T., Yamamoto, H., Kadota, G., and Naito, N. (1976). Development of rust fungi in oat leaves treated with blasticidin S, a protein synthesis inhibitor, *Tech. Bull. Fac. Agric. Kagawa Univ.*, **27**, 95–103.

Vance, C. P., and Sherwood, R. T. (1976). Regulation of lignin formation in reed canarygrass in relation to disease resistance, *Plant Physiol.*, **57**, 915–919.

Vance, C. P., Anderson, J. O., and Sherwood, R. T. (1976). Soluble and cell wall peroxidases in reed canarygrass in relation to disease resistance and localized lignin formation, *Plant Physiol.*, **57**, 920–922.

Vance, C. P., Kirk, T. T., and Sherwood, R. T. (1980). Lignification as a mechanism of disease resistance, *Annu. Rev. Phytopathol.*, **18**, 259–288.

Wade, M., and Albersheim, P. (1979). Race-specific molecules that protect soybeans from *Phytophthora megasperma* var. *sojae, Proc. Natl. Acad. Sci. USA*, **76**, 4433–4437.

West, C. A. (1981). Fungal elicitors of the phytoalexin response in higher plants, *Naturwissenschaften*, **68**, 447–457.

Yoshikawa, M. (1978). Diverse modes of action of biotic and abiotic phytoalexin elicitors, *Nature (London)*, **275**, 546–547.

Yoshikawa, M. (1980). Resistance mechanisms underlying the soybean–*Phytophthora megasperma* f. sp. *glycinea interaction, Plant Prot.*, **34**, 248–256.

Yoshikawa, M., Masago, H., and Keen, N. T. (1977). Activated synthesis of poly(A)-containing messenger RNA in soybean hypocotyls inoculated with *Phytophthora megasperma* var. *sojae, Physiol. Plant Pathol.*, **10**, 125–138.

Yoshikawa, M., Yamauchi, K., and Masago, H. (1978a). Glyceollin: its role in restricting fungal growth in resistant soybean hypocotyls infected with *Phytophthora megasperma* var. *sojae, Physiol. Plant Pathol.*, **12**, 73–82.

Yoshikawa, M., Yamauchi, K., and Masago, H. (1978b). *De novo* messenger RNA and protein synthesis are required for phytoalexin-mediated disease resistance in soybean hypocotyls, *Plant Physiol.*, **61**, 314–317.

Yoshikawa, M., Genma, H., and Masago, H. (1979a). Physiological activity of phytoalexins: rooting cofactor activity of glyceollin, *Ann. Phytopathol. Soc. Jpn.*, **45**, 535–536 (In Japanese).

Yoshikawa, M., Yamauchi, K., and Masago, H. (1979b). Biosynthesis and biodegradation of glyceollin by soybean hypocotyls infected with *Phytophthora megasperma* var. *sojae, Physiol. Plant Pathol.*, **14**, 157–169.

Yoshikawa, M., Yoshizawa, T., Matama, M., and Masago, H. (1979c). Contact of insoluble cell walls of *Phytophthora megasperma* var. *sojae* with wounded soybean cotyledons results in rapid liberation of a phytoalexin elicitor, *Ann. Phytopathol. Soc. Jpn.*, **45**, 535 (In Japanese).

Yoshikawa, M., Imamura, S., and Masago, H. (1980). Differential effects of biotic and abiotic elicitors on glyceolling degrading system in soybean, *Ann. Phytopathol. Soc. Jpn.*, **46**, 89.

Yoshikawa, M., Matama, M., and Masago, H. (1981). Release of a soluble phytoalexin elicitor from mycelial walls of *Phytophthora megasperma* var. *sojae* by soybean tissues, *Plant Physiol.*, **67**, 1032–1035.

Zähringer, U., Ebel, J., Mulheirn, L. J., Lyne, R. L., and Grisebach, H. (1979). Induction of phytoalexin synthesis in soybean, Dimethylallyl pyrophosphate: trihydroxypterocarpan dimethylallyl transferase from elicitor-induced cotyledons, *FEBS Lett.*, **101**, 90–92.

Ziegler, E., and Albersheim, P. (1977). Host–pathogen interactions. XIII. Extracellular invertases secreted by three races of a plant pathogen are glycoproteins which possess different carbohydrate structures, *Plant Physiol.*, **59**, 1104–1110.

Further Reading

Albersheim, P., and Valent, B. S. (1978). Host–pathogen interactions in plants. Plants, when exposed to oligosaccharides of fungal origin, defend themselves by accumulating antibiotics, *J. Cell Biol.*, **78**, 627–643.

Callow, J. A. (1977). Recognition, resistance and the role of plant lectins in host–parasite interactions, *Adv. Bot. Res.*, **4**, 1–49.

Keen, N. T. (1981). Evaluation of the role of phytoalexins, in *Plant Disease Control* (Ed. R. C. Staples), Wiley New York, pp. 155–177.

Keen, N. T. (1982). Specific recognition in gene-for-gene host–parasite systems, in *Advances in Plant Pathology*, Vol. 1 (Eds P. H. Williams and D. Ingram), Academic Press, New York, pp. 35–81.

West, C. A. (1981). Fungal elicitors of the phytoalexin response in higher plants, *Naturwissenschaften*, **68**, 447–457.

Yoshikawa, M., and Masago, H. (1982). Biochemical mechanism of glyceollin accumulation in soybean, in *Plant Infection: The Physiological and Biochemical Basis* (Eds. Y. Asada *et al.*,) Japan Sci. Soc. Press, Tokyo/Springer-Verlag, Berlin, pp. 264–280.

Biochemical Plant Pathology
Edited by J. A. Callow
© 1983 John Wiley & Sons Ltd

14

Recognition of Bacteria by Plants

UMBERTO MAZZUCCHI

Istituto di Patologia Vegetale, Università di Bologna, 40126 Bologna, Italy

1. INTRODUCTION: PLANT–BACTERIA ASSOCIATIONS

All plants live in association with bacteria, and parasitism occurs when these bacteria are commensalistic, mutualistic, and pathosistic (Bateman, 1978). The closest, most specific and specialized bacteria–plant interactions occur in mutualism and pathosism involving endophytic bacteria. In endophytic mutualism of legume root nodules, the bacteria pass the main part of their life as bacteroids in the intracellular state. The bacteroids live enclosed within special vacuoles of the host plant cell and not in direct contact with the cytoplasm. However, during penetration and initial colonization, even the mutualistic, endophytic bacteria have an intercellular phase, either within the infection threads or within the intercellular spaces (Beringer *et al.*, 1979).

In all the other types of association the bacteria usually live in the intercellular spaces and/or in the xylem vessels and tracheids; a few bacteria are also found in the laticiferous canals. Colonization within the plant cells takes place only when these are degenerating or dead. Hence the phytopathogenic bacteria usually have an apoplastic habitat and never come in direct contact with living host plasmalemma.

Many plant–bacteria associations are remarkably specific. It is known, for example, that the nodulation of a legume crop, in an area where legumes have never been grown, usually requires the inoculation of seeds with appropriate strains of *Rhizobium*. For each endophytic, parasitic bacterium the number of non-host plants greatly exceeds that of hosts. Immunity is, therefore, generally the rule, and susceptibility the exception. Only rarely does a bacterium possess all the necessary determinants to establish an endophytic parasitism (under natural conditions) in a given plant. In mutualistic parasitism of legumes by rhizobia it has been postulated that numerous plant genes, at least ten, are involved (Holl and LaRue, 1976; Legocki and Verma, 1980). Plant resistance to pathosism is usually horizontal and polygenic (Van der Planck, 1975, 1978; Robinson, 1976); vertical resistance is rare. There are a few systems in which differential interactions between host plant cultivars and pathogen strains are known (e.g. cotton-*X. malvacearum*, soybean-*P. glycinea*, and broadbean-*P.phaseolicola*) and even in these cases the genetic significance of the interactions is not always clear. As a consequence, Flor's gene-for-gene hypothesis cannot be generally applied to plant–bacteria associations (Van der Planck, 1975). The host specificity of the parasitic bacteria is probably the result of macroevolution (Robinson, 1976). The establishment of an endophytic parasitism requires an unknown series of reciprocal cellular recognitions between the plant and the parasite, in a certain sequence. Each cellular recognition involves one or more processes of molecular recognition between the ligands of one partner and the receptors of the other. The specificity of the association derives, finally, from an appropriate combination in space and time, of the many events of molecular recognition.

In this chapter, the most important molecular and cellular recognitions described in the literature are discussed, with reference to those occurring during the initial period of interaction between the plant and the parasitic bacteria. The subject has also been treated in a number of reviews (Callow, 1977; Sequeira, 1978, 1980, 1981; Lippincott and Lippincott, 1980; Dazzo, 1980; Bauer, 1981; Sölheim and Paxton, 1981; Whatley and Sequeira, 1981).

2. MOLECULAR AND CELLULAR RECOGNITION

Recognition describes an interaction in which molecules or cells discriminate between the materials in their environment (Herbert and Wilkinson, 1977). In molecular recognition, certain sites of a molecule (*ligands*) have the property of being bound to the complementary sites of another molecule, the *receptor* (Curtis, 1981). The binding between the ligand and the receptor may be either specific or relatively unspecific, and the complementary nature is the consequence of physico-chemical interactions of various types. Molecular recognition is a diffusion-controlled process, governed by the laws of thermodynamics, and takes place rapidly (within seconds) (Marchalonis, 1980). The interactions between enzymes and their substrates, between antigens and their antibodies, and between lectins and certain sugars are all examples of molecular recognition.

Cellular recognition presupposes molecular recognition, but is distinguished from it. Cellular recognition occurs once molecular recognition has taken place at the cell surface and the cellular functions are activated, but molecular recognition does not necessarily lead to cellular recognition (Marchalonis, 1980). The ligand, found in the environment, and the receptor on the cell surface responsible for cellular recognition, can be labelled *cognon* and *cognor*, respectively (Burke *et al.*, 1980). The cognor expresses the active recognition part of the system, the cognon the passive part providing the sites to be recognized and functioning as a signal molecule. Cellular recognition implies a series of cellular and biochemical events and is intermediate or slow (from a few minutes to several hours) (Marchalonis, 1980). Cognons can be free or bound and protruding from the surface of the particles, stimulating cellular recognition. Free cognons can give rise to cellular recognition at great distances, but those exposed on the cell surface give rise to cell–cell recognition over much shorter distances when the cognons are released by a cell and transferred to another cell in the immediate vicinity bearing the cognors, or when the two cells come into contact (Clarke and Knox, 1978).

In cellular recognition the molecular recognitions which occur on the surface between cognons and cognors activate the cellular functions. Phagocytosis, for example, is a localized response of a segment of the phagocytic cell to signals generated by the contact between its cytoplasmic membrane and the foreign particle which are transmitted to the cytoplasm. In polymorphonuclear leukocytes the phagocytic process is initiated only 30 s after contact (Boyles and Bainton, 1979). The adhesion of animal cells to a substrate of collagen or artificial material plays an essential role in their replication and differentiation (Kleinman *et al.*, 1981). The adhesion to collagen, for example, stimulates the growth and differentiation of the cells much more than to glass or plastic. In plant–bacteria interactions, the induced protection phenomenon is the result of a cellular recognition following the treatment of the plant cells with bacterial substances (Mazzucchi *et al.*, 1979).

It is clear then, that cellular recognition presupposes the existence of a signal receptor–transmitter system in the cell (Drachman, 1979). The reception of the signal occurs on cognon–cognor binding. The transmission of the signal takes place when it is translated within the cell under the form of some biochemical action. There are extremely few examples in biology in which any details are known about this cellular receptor–transmitter system (Engström and Hazelbauer, 1980; Marchalonis, 1980; Wang and Koshland, 1980; Curtis, 1981; Owen and Crumpton, 1981).

Cell–cell recognition systems may be either negative or positive in operation (Yeoman *et al.*, 1978). In the first case, one type of cell tends to accept and to integrate itself functionally with the other type, unless there has been a signal to indicate incompatibility. In the second case, one cell will tend to regard the other as non-self until a positive signal occurs, enabling it to recognize the other as compatible. In higher plants negative systems are responsible for gametophytic self-incompatibility (Clarke and Gleason, 1981) and for vegetative incompatibility (Yeoman *et al.*, 1978; Moore and Walker, 1981a,b). In both examples the adhesion between two different incompatible cells is successful and the initial course of the interactions is similar to that of compatible interactions. It is only subsequently that an incompatible reaction occurs which blocks any further integration between the two partners.

The recognition of bacteria by plants is a cell–cell recognition in which the bacterium supplies the cognon and the plant the cognor and a cellular effector system (Figure 1). As the bacteria never come into contact with the plasmalemma of the living plant cells, it is difficult to understand how the bacterial cognon is able to bind itself to the plant cognor and how the signal is then translated within the plant cell. In the most widely studied models involving mesophyll cells and root hairs the plant walls form a barrier of several micrometers of water-imbibed polysaccharides. The ways in which cellular recognition can take place in such a situation were recently discussed and various models have been proposed (Heslop-Harrison, 1978; Bauer, 1981; Reisert, 1981; Figure 1). However, with respect to the particular examples of cell–cell recognition of bacteria by plants the actual state of our knowledge is fragmentary and precludes a unequivocal and well-documented interpretation. The available experimental evidence indicates that the bacterial cognons are soluble, low molecular weight substances. It is likely that they do not have a diameter of more than 3–5 nm and a molecular weight of more than about 17 kDa in order to penetrate pores in the plant cell wall (Carpita *et al.*, 1979). The cognons appear to be present in the bacterial envelopes (Mazzucchi and Pupillo, 1976; Graham *et al.*, 1977; Bruegger and Keen, 1979; Mazzucchi *et al.*, 1979; Ozawa and Yamaguchi, 1980; Bauer, 1981; Sölheim and Paxton, 1981) and may be generated by the enzymatic degradation of polysaccharide polymers of the envelopes, such as the O-chain of lipopolysaccharides (LPS) and/or the capsular polysaccharides (CPS). Glycosidase

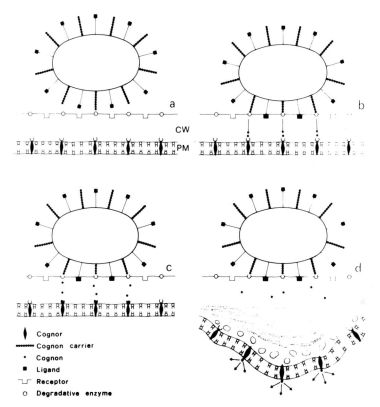

Figure 1. Cell–cell recognition model following the attachment of a bacterium to the plant cell wall (CW). (a) The bacterium in a sorption state, very close to the plant cell wall, bears potential ligands for attachment and cognons for cellular recognition on its surface. (b) The bacterium attaches itself to the wall by means of bonds between its ligands and plant cell receptors. Following enzymatic degradation the cognons are freed from their polymer carriers (probably LPS and/or CPS) and due to their low molecular weight move through the cell wall pores. (c) The cognons bind themselves to cognors of the plasmalemma (PM) inducing a cellular response. (d) The cellular recognition is under way as a cell wall apposition (see Section 4.2.4) localized on the opposite side of the wall from where the bacteria were attached. The cell–cell recognition may also occur when the bacteria, although not attached to the wall, release in the immediate vicinity the polymers carrying the cognons. Alternative models are possible, especially as regards the trans-wall communication (see Reisert, 1981)

enzymes, present on the plant wall, may be responsible for this degradation (Cline and Albersheim, 1981). The existence of small oligosaccharide signal-molecules, responsible for the recognition in plant cells of biotic or abiotic agents, has already been reported in other systems (see Chapter 13). The extracellular polysaccharides (EPS) do not carry cognons or are inefficient carriers: when plant tissues are pre-treated with purified EPS, protection against confluent hypersensitive necrosis is slight (Mazzucchi, El-Banoby, and Rudolph, unpublished work) or not at all (Sequeira, 1981). On the contrary, complete protection can be induced by pre-treatment with LPS even at a 50 μg cm^{-3} dose (Graham et al., 1977).

3. RECOGNITION OF RHIZOBIA BY LEGUME ROOTS

3.1. The attachment

The attachment of rhizobia to the roots of legumes is the first stage in a sequence of interactions which lead to the establishment of the parasitism. Attachment may take place at wounds caused by the emergence of lateral roots, on the root hairs, or on the epidermal cell walls immediately before the formation of the hair (Bhuvaneswari et al., 1980). This attachment is usually polar and occurs within seconds or minutes after inoculation (Chen and Phillips, 1976; Dazzo et al., 1976). The nature of the ligands and the receptors involved has been discussed in recent reviews (Broughton, 1978; Dazzo, 1980; Bauer, 1981; Sölheim and Paxton, 1981). Hamblin and Kent (1973) suggested that the lectins of legumes were responsible for the attachment and Bohlool and Schmidt (1974) showed that soybean lectin bound specifically to strains of rhizobia that can nodulate soybean.

Two models have emerged from the experimental studies. Wall-bound lectin may act as a bridge between its receptor on the plant cell wall and the ligand on the wall of the bacterium, both receptor and ligand being serologically cross-reacting antigens (Dazzo, 1980). Alternatively, the lectin may be released by the plant into the rhizosphere to act as a stabilizing factor when the bacterium attaches itself to the receptor of the plant wall (Raa et al., 1977). In both models, if the lectin acts as a receptor only in homologous interactions, it must be responsible for the specificity of the rhizobium–legume association and the attachment must be the crucial step in the infective process. Specific binding by a lectin has been shown for clover–R. trifolii (Dazzo, 1980) and soybean–R. japonicum (Bohlool and Schmidt, 1974; Stacey et al., 1980). However, for various reasons, it is not easy to accept that there is a specific attachment in all rhizobium–legume associations or that the various lectins are the receptors involved. The lectin of a legume may in some cases bind to non-nodulating rhizobia (Chen and Phillips, 1976; Law and Strijdom, 1977). The attachment of the homologous rhizobia to the wall does not necessarily lead to infection and nodulation, and non-nodulating rhizobia can also attach to the roots, although less frequently than the homologous ones (Broughton, 1978; Dazzo, 1980). Moreover, rhizobia also adhere to inert materials, such as glass and plastic, and to the roots of non-host plants (Tsien and Schmidt, 1977; Shimshick and

Herbert, 1978; Dazzo, 1980). Finally, with the exception of white clover (Dazzo, 1980) it has yet to be shown that lectins are present at the actual infection sites (Sölheim and Paxton, 1981). Nevertheless, the available evidence does strongly indicate that a lectin-mediated attachment is a common occurrence in the legume–rhizobia associations, but it has not been demonstrated in either of the two models mentioned above that the lectin acts as a direct cognor or as a bridging substance for the real cognor of the legume root epidermal cells. There is no proof whatsoever, that the attachment of the rhizobia of their materials to the lectins induces a cellular response.

Dazzo (1980) suggested that non-specific and specific mechanisms are involved in the attachment of the rhizobia to the roots. It is possible that several different receptors are involved. The attachment of the rhizobia in soybean callus cultures was inhibited following treatment of the callus with pectinase (Reporter *et al.*, 1975), suggesting that receptors on the plant wall might also be polygalacturonides or other molecules linked to pectin substances.

The ligands responsible for attachment, exposed on the surface of the rhizobia, appear to be carried by polysaccharides (Bauer, 1981). They may be EPS, CPS, O-chains of LPS, or even glucans. In clover–*R. trifolii* it was shown that the active polysaccharide was capsular and that the determinant involved in the binding with the lectin was 2-deoxyglucose. However, the polysaccharide containing the ligand appeared to be produced only at the beginning of the exponential and stationary phases, but not between these two periods or later in the stationary phase. It has not been shown, conclusively, that the ligand was of capsular origin, because the cellular localization of the active polysaccharide has never been determined. In the soybean–*R. japonicum* interaction there is, on the other hand, convincing evidence that the receptor for the lectin is on the CPS and that it is available *in vivo*. However, the various strains of *R. japonicum* produce CPS with very different compositions. In the 110/130 type CPS, the galactose side-chains are probably the main determinants of the lectin-binding activity, and this type of determinant may be present in all strains of *R. japonicum*.

3.2. Root hair curling

After attachment, the rhizobia may penetrate the legume roots with or without infection threads. During these initial phases of infection, specific cellular recognitions take place involving morphological, ultrastructural, and biochemical responses of the host plant. After the attachment of homologous, but not heterologous, rhizobia to the root hair wall, there is a significant curling of the hair around the bacterium. Rhizobia applied to non-symbiont plants do not induce curling (Haack, 1964), and the capacity of homologous rhizobia to induce curling is positively correlated with infectivity (Yao and

Vincent, 1976). The marked hair curling around the attached, infective rhizobia is, therefore, a cellular recognition characterized by a differential and localized response of the plant cells involving wall biogenesis. Young clover root hairs treated with rhizobia revealed thickening of the walls, which was sometimes associated with arrays of vesicles in the neighbouring cytoplasm (Kumarasinghe and Nutman, 1977). With heterologous strains the thickenings were transient and mainly apical in position, and appeared to be unrelated to attachment sites of the bacteria. However, with the homologous, infective strains they were localized in correspondence with the branching of the root hairs and particularly their curls. Persistent deposits of callose were observable at the point where the infection thread originated and not along its subsequent route.

As the plant cell response is therefore localized in strict relation to the attachment sites of the rhizobia, it is conceivable that the cognons involved in recognition are either substances protruding from the surface of the rhizobia and/or molecules released from them with a limited range of diffusion. In fact, culture filtrates of rhizobia induced the curling and branching of root hairs in both symbiont and non-symbiont plants (Sölheim and Raa, 1973; Yao and Vincent, 1976), but in the latter the response was less marked. The deforming factor in culture filtrates appears to be a heat-stable polysaccharide. Hubbell (1970) treated root hairs of clover with EPS taken from 1 week old culture plates and purified by precipitation in alcohol. Only the higher dose (100 μg cm^{-3}) caused marked curling and deformations compárable to those induced by living rhizobia. Dazzo and Hubbell (1975) treated clover root hairs with CPS (100 μg cm^{-3}) from eight R. trifolii strains and observed that curling of the hairs was only induced by CPS from the infective homologous strains. Yao and Vincent (1976), using partially purified EPS of two R. trifolii strains, were not able to confirm Hubbell's (1970) results. Their material, containing 20% protein and small amounts of LPS and nucleotides, was only able to induce branching and moderate curling, and appeared to be heat-stable and non-dialysable. This discrepancy may be due to the difficulty of comparing the responses of root hairs in bioassays in different laboratories. It is probable that a marked curling, identical with that of natural infections, cannot be obtained in bioassays, for the simple reason that one cannot be sure of a unilateral treatment of the root hairs with purified polysaccharides.

The degree of purity of the polysaccharides used in the above experiments does not permit one to decide conclusively whether the biologically active polysaccharide is capsular or extracellular. The mechanism of synthesis of the two types of polysaccharide may be identical and the distinction between the two is frequently arbitrary (Sutherland, 1977; Jann and Jann, 1977). They can also be excreted by the bacterial cells through common adhesion sites (Bayer, 1981). They may form mixed aggregates, apparently homogeneous for some properties, and the separation of one from the other and their structure analysis presents some difficulties (Bauer, 1981). Also the same bacterium may produce different types of CPS (Jann and Jann, 1977). The biological activity of highly purified EPS of rhizobia has yet to be tested. Dazzo and Hubbell's results and the relative heat stability of the deforming factor indicate that it is a CPS since bacterial CPS are held to be more heat stable than the EPS of the slime (Edwards and Ewing, 1972).

The CPS purified from the rhizobia are high molecular weight, non-dialysable substances and, although they are involved in the attachment of rhizobia to the root hair walls, it is difficult to perceive how they can act as a cognon since it is inconceivable that they pass in intact form through the root hair wall. Ljunggren (1969) suggested that the deforming factor, present in the inoculated root medium, had a molecular weight of about 5000. Sölheim and Raa (1973) demonstrated distinct, biologically active fractions both in culture filtrates and in the inoculated root medium. However, a comparative analysis of the two fractions was not made. Yao and Vincent (1976) admitted that the deforming factor must have both dialysable and non-dialysable forms and recently Fjellheim and Sölheim (cited in Bauer, 1981) reported that extracts of pea and clover roots were capable of selectively degrading radiolabelled CPS preparations (Bauer, 1981). The legume roots seemed to possess enzymes capable of degrading the CPS of their homologous rhizobia (Sölheim and Paxton, 1981). The cognons activating the cellular recognition must therefore be oligosaccharides, generated by enzymatic degradation, and capable of reaching the plant cell cognors through the walls. The CPS is only the cognon carrier. The cognors involved in this cellular recognition leading to curling are unknown; there are no available experimental data to support a role for lectins.

The curling of the root hair around the rhizobia implies a local disturbance in plant cell wall biogenesis. Bauer (1981) has proposed a model based on the localized inhibition of cellulose synthesis in the deepest (β) layer of the emerging hair, induced by an attached *Rhizobium* cell. Recent results (Ozawa and Yamaguchi, 1980) appear to support Bauer's model. Cell envelopes of infective rhizobia added to culture suspensions of soybean cells increased cellulase activity in the medium. Envelopes of a non-infective homologous strain and heterologous strains were ineffective. The response of the soybean cells could be measured after only 30 min and was inhibited by cyclohex-imide. Unfortunately, the treatments were made with only partially purified materials and the exact nature of the cognon is unknown.

4. RECOGNITION OF PHYTOPATHOGENIC BACTERIA

4.1. Bacterial attachment to plant cell walls

Attachment is a crucial step in the bacterial strategy for the colonization of the host plant, and in the induction of resistance responses to avirulent bacteria and must involve molecular recognition (Whatley and Sequeira, 1981).

4.1.1. Attachment of avirulent bacteria

Saprophytic and incompatible phytopathogenic bacteria attach themselves firmly to the plant cell walls (Sequeira, 1980). In the most widely studied systems, however, 2 h after the infiltration of incompatible bacteria, the cells are very close to the plant walls, but not yet in contact. For *P. pisi* and *P. aptata* the distances measured vary between 33 and 68 nm (Goodman *et al.*, 1976, 1977; Politis and Goodman, 1978; Medeghini Bonatti *et al.*, 1979). These distances are those to be expected as secondary minimum values according to the DLVO theory (Marshall *et al.*, 1971; Marshall, 1976). Molecular recognitions responsible for adhesion (molecular contact), occur only at the primary minimum, when the distances between the cells is about 0.1–0.4 nm, that necessary for the various types of bonds (Weiss, 1970; Daniels, 1980). Hence for *P. pisi* and *P. aptata* the situation at 2 h may be comparable to a reversible sorption stage (Marshall and Bitton, 1980). The scarce experimental data concerning the removal of the bacteria from the intercellular spaces by washing with water and buffered saline appears to be contradictory. In the washing technique, however, an arbitrary force is used to dislodge the unattached bacteria and there has been no attempt to quantify the tenacity of attachment (Knox, 1981).

The gap between the bacteria and the plant wall after 2 h may be a mere artifact and so impossible to interpret. Some photomicrographs show bacteria which appear to have been detached from their niches on the walls (Sequeira *et al.*, 1977; Cason *et al.*, 1978). Alternatively, the presence of a crown-shaped water meniscus around one side of the bacterium and also transpirational forces would tend to pull the bacteria towards the wall, but these forces would be opposed by electrostatic repulsion between the electronegative bacteria and electronegative plant cell walls (Marshall, 1976; Läuchli, 1976). At distances close to the secondary minimum the bacteria can only attach themselves to the cell walls by means of filamentous bridging materials, capable of overcoming the repulsion barrier between the primary and secondary minimum to effect molecular contact (Marshall, 1976; Marshall and Bitton, 1980). In general, after 2 h there is no evidence of bridging material.

Sequeira (1981) suggested that for incompatible bacteria, devoid of capsules and EPS, the LPS are the most probable attachment candidates. The bridging polymers must be the O-polysaccharide chains in Gram-negative bacteria and, similarly, the teichoic acid chains in Gram-positive bacteria (Duckworth, 1977). Both of these protrude from the surface of the bacterial walls and constitute a hydrated microcapsule around the cells. The length of the LPS O-chains, usually 20–40 nm (Leive and Davis, 1980) would fit with this interpretation of Sequeira, but these unchanged polysaccharide chains are not stained in the usual electron microscope preparations.

About 3–4 h after inoculation the incompatible bacteria, including *P. sol-*

anacearum, are attached to the plant cell walls (Goodman *et al.*, 1976; Sequeira *et al.*, 1977; Politis and Goodman, 1978; Medeghini Bonatti *et al.*, 1979; Hildebrand *et al.*, 1980). Some photomicrographs, however, still show some bacteria which, although segregated in small niches, are not directly attached.

4.1.2. Attachment of virulent bacteria

The behaviour of compatible, virulent bacteria in the same systems is different. Their cells can be observed at a distance from the walls equal to or higher than the secondary minimum for several hours after infiltration. The distances calculated were as follows: for *E. amylovora* 33–166 nm at 18 h (Goodman and Burkowicz, 1970); for *P. solanacearum* 57–82 nm at 4 h and 24–97 nm at 12 h (Sequeira *et al.*, 1977); and for *P. tabaci* 47–376 nm at 6 h (Goodman *et al.*, 1976) and 30–160 nm until 8 h (Mazzucchi *et al.*, 1982). *P. tabaci*, 5–8 h after infiltration, occasionally appeared to be anchored to the tobacco cell walls with a few fibrils (Goodman *et al.*, 1976; Atkinson *et al.*, 1981; Mazzucchi *et al.*, 1982). Moreover, in bean and tobacco leaves, after 12 and 48 h, respectively, it was observed that the growth of compatible bacteria occurred within a network of fibrillar material, probably EPS, stretched out within the intercellular spaces (Hildebrand *et al.*, 1980; Daub and Hagedorn, 1980; Mazzucchi *et al.*, 1982).

There are systems in which attachment of the compatible bacteria to the plant walls has never been observed (Sequeira *et al.*, 1977; Cason *et al.*, 1978). For the others, a rational attachment model may incorporate the following elements. Initially the bacteria remain adsorbed to the walls close to the secondary minimum. Within a few hours, the bacteria begin to produce bridging substances to attach themselves to the plant cell walls at the primary minimum. Initial attachment occurs at a few points on the bacterial cell and is very fragile. Then new polymers are produced to reinforce the bridging attachment and to create a thin hydrophilic sheath around the rest of the bacterial cell, exposed to the intercellular atmosphere. The bacteria begin to multiply within this sheath. New extracellular material accumulates progressively around the growing microcolonies until it occupies, either partly or completely, the intercellular space. During the anchoring to the wall, the bacteria release the substances carrying the cognons responsible for cellular recognition, inducing a localized response of the plant cell (see Section 4.2.4). These cognons may be part of the bridging materials or associated with them. This model does not exclude a role played by substances of plant origin. To test this model ultrastructurally, staining may be used to reveal polysaccharides (Roth, 1977).

4.1.3. Attachment and role of plant lectins

Molecular recognition between a plant lectin and strains of *P. solanacearum* has been described by Sequeira and Graham (1977). All avirulent strains were strongly agglutinated by potato lectin and all the virulent strains were not agglutinated or only slightly. The avirulent strains of *P. solanacearum* formed rough colonies on nutrient media and did not produce EPS; the virulent strains, however, had fluidal colonies and produced large amounts of EPS. Comparing a virulent strain with its avirulent mutant, it was noted that it was the absence of EPS which permitted the agglutination of the avirulent strain by the lectin; the addition of EPS from the virulent strain inhibited agglutination. When EPS was removed by careful washing, even the virulent strain was agglutinated by the lectin. Therefore, the lectin receptors must have been present in both the virulent and avirulent strains, and it was the presence of EPS which inhibited lectin binding. The bacterial receptor for the lectin appeared to contain *N*-acetylglucosamine, since the addition of chitin oligomers inhibited both the agglutination of the avirulent strain and the precipitation of its LPS. The receptors for the lectin seemed, however, to be available both on EPS and on the LPS of *P. solanacearum* (Sequeira, 1981) and it is not easy to interpret the types of binding which took place *in vitro* or their role *in vivo*.

The ionic strength of the medium in which the interaction occurs appears to play an important role. The lectins might have an active role, *in vivo*, in the attachment of incompatible bacteria to the cell walls of tobacco and potato whereas unattached bacteria are free to multiply in the intercellular spaces. When sections of tissue were treated with specific FITC-labelled antibodies against potato lectin, a localized fluorescence was observable on the plant cell walls. However, the evidence that lectin was exposed on the plant cell wall is not conclusive, because potato lectin contains arabinogalactans, substances commonly present in the walls of plant cells (Sequeira, 1981), and the antiserum may have been reacting to these common immunogenic determinants. The absence of extracellular material between the bacteria and the plant cell walls in the first hours of compatible interaction (Sequeira *et al.*, 1977) does not appear to support the hypothesis that EPS inhibits the attachment of virulent cells. The scarce fibrillar material protruding from the plant walls in the areas, in close proximity to the bacteria, was interpreted as being the result of the disruption of the same plant cell walls.

Avirulent strains of *P. solanacearum*, inducing hypersensitive response (HR) in tobacco leaves, produce LPS which rapidly migrate in SDS polyacrylamide gels and have low xylose/glucose and rhamnose/glucose ratios (Whatley *et al.*, 1980). In contrast, non-HR-inducing avirulent strains produce LPS similar to that of virulent strains with high sugar ratios and low electrophoretic mobility. It is possible to distinguish between non-HR- and

HR-inducing strains using a phage (CH 154) isolated from a lysogenic strain of *P. solanacearum* (Sequeira, 1981), because it is inactivated by the LPS of the former but not the latter. Moreover, the LPS of the avirulent, HR-inducing strains bind much more strongly to the potato lectin than those not inducing HR (Whatley *et al.*, 1980). These results indicate that the structure of the O-side-chain of these two types of LPS must be different. The HR-inducing strains are rough with shorter O-side-chains (incomplete LPS). The non-HR-inducing avirulent strains are smooth and produce longer O-side-chains (complete LPS) like the virulent strains. Although the tobacco lectin possesses a sugar binding activity similar to that of the potato lectin, in the tobacco leaves it is possible to induce protection against HR-inducing strains treating with both types of LPS (Whatley *et al.*, 1980). This indicates that the tobacco lectin may be the receptor responsible for the attachment of HR-provoking strains of *P. solanacearum*, but it does not carry a cellular recognition cognor which can differentiate between them.

4.1.4. *Attachment of Agrobacterium tumefaciens*

Unlike other phytopathogenic bacteria, a rapid, firm attachment to the plant cell walls is, for tumourogenic agrobacteria, essential for pathogenesis and occurs following molecular recognition between bacterial cell wall ligands and plant cell wall receptors (Lippincott and Lippincott, 1980). Tumour initiation was inhibited if the susceptible tissues were treated with certain avirulent strains or with heat-killed virulent strains of agrobacteria. This inhibition was proportional to the number of avirulent or heat-killed cells and only occurred if the treatment was made before, during, or not more than 15 min after the application of virulent agrobacteria. Treatments with bacteria of other species did not have any effect. The LPS of the virulent agrobacteria contain the attachment ligands. Wall, Boivin antigens and purified LPS, from virulent agrobacteria preparations also had an inhibitory effect (Whatley *et al.*, 1976). When the whole cells, of avirulent and virulent agrobacteria, were compared with their LPS there was a correlation in their inhibitory effect. Here, also, there was no inhibition if the treatment was made more than 15 min after the application of the virulent agrobacteria. The ligands appear to be carried by the LPS polysaccharide chains, but the lipid A moiety did not have any inhibitory effect (Lippincott and Lippincott, 1980). The ability of virulent agrobacteria to attach to host sites is coded by Ti plasmid genes and by chromosome genes (Whatley *et al.*, 1978). Avirulent, non-attaching agrobacteria became competent at attachment and tumour induction when Ti plasmid was introduced into them, and their LPS also acquired inhibitory properties. However, when the Ti plasmid was removed from a virulent strain, the cells were still able to attach.

The plant receptors, for the attachment of agrobacteria, are present on the plant cell walls (Lippincott and Lippincott, 1977). Tumour initiation was inhibited by treating the wounds with partially purified wall fragments of the susceptible plant. Suspensions of membranes from the same plant did not have any inhibitory effect. The attachment sites must be relatively specific because the inhibitory effect disappeared when the wall fragments were mixed beforehand with dead, virulent but not avirulent agrobacteria cells. The receptors may be carried in the middle lamella and appear to consist partly, if not completely, of polygalacturonic acid. Polygalacturonic acid was able to inhibit the tumour formation only if it was applied during or not more than 15 min after the application of virulent agrobacteria and inhibition was complete with a dose of 10 mg cm^{-3}. A pectin with 30% methylated polygalacturonic acid residues was about 1000 times less efficient. Other polysaccharides and gums of plant origin and various lectins did not show any inhibitory effect. The degree of methylation of the pectic polymers appears to be critical for the attachment. Wall fragments of monocotyledons (non-hosts), of tumour tissue, and of dicotyledon embryonic tissue, which lack attachment sites, were not able to inhibit tumour initiation (Lippincott and Lippincott, 1978). If, however, these types of tissues were treated with pectin methylesterase they acquired a high inhibitory capacity, which was reversed by pectinmethyltransferase (Lippincott and Lippincott, 1980).

4.1.5. Recognition of bacterial EPS by cell walls

A molecular recognition appears to exist between the EPS of some phytopathogenic bacteria, causing leaf spots, and the mesophyll cell walls of their natural host plants (El-Banoby et al., 1981a,b). When EPS of Pseudomonas phaseolicola, P. lachrymans, P. pisi, and P. glycinea were locally infiltrated in the leaves of their host plants, and then kept in a moist environment, it was observed that the water-soaking persisted only in compatible combinations. In incompatible combinations and the immune swiss-chard, water-soaking disappeared within a few hours. Temporary water-soaking was also obtained when the leaves of all host plants were treated with EPS of P. fluorescens, a saprophytic bacterium. In incompatible combinations of bean and P. phaseolicola, 2–3 h after infiltration, the EPS appeared to be irregularly aggregated, and these aggregates diminished during the following hours and almost disappeared after 12 h. Intercellular fluid, extracted from leaves of a resistant cultivar, caused 17% hydrolysis of EPS in vitro, compared with only 1.5% by fluids from a susceptible cultivar. There was also a significant reduction in the capacity of the EPS incubated with intercellular fluid from a resistant cultivar to induce persistent water-soaking. The enzyme which specifically degrades the EPS seems to be pre-formed and anchored to the walls of resistant cultivars.

4.2. Cell responses following recognition

4.2.1. Xanthomonas vesicatoria/*green pepper*

Green pepper leaves showed differential loss of electrolytes according to whether they had been infiltrated with compatible or incompatible strains of *X. vesicatoria* (Stall and Cook, 1979). In the incompatible combinations the loss of electrolytes increased rapidly within the first 24 h, the period within which the hypersensitive response became macroscopically visible. In the compatible interactions the loss of electrolytes became appreciable only after 36–48 h; the disease symptoms became visible after 4 days, although modifications of the endoplasmic reticulum of the cells could be observed after only 36 h. When both compatible and incompatible bacteria were infiltrated together there was a notable reduction in the loss of electrolytes. It was necessary to inject the compatible cells at least 3 h before the incompatible ones to produce this effect; when the infiltration was anticipated by 6 h the effect was very evident. This effect disappeared when the living compatible bacteria were replaced with heat-killed bacteria of the same strain, but could be observed when the heat-killed compatible bacteria were infiltrated 24 h before the incompatible bacteria. The living cells of the compatible strain are recognized by the plant within 6 h after infiltration when there is no loss of electrolytes or histological modification. The cognons responsible for this are available in the living compatible bacteria, but not in the heat-killed ones. The latter, however, must possess some cognons whose effect is delayed by 24 h. It is possible that there are two different types of cognons, or that the heat treatment makes the same cognons less accessible to the plant cognors. In the first case the rapid effect cognon must be heat sensitive and released by the bacteria in the first hours after infiltration.

The water-soaked state of the green pepper leaves inhibited the HR and made the patterns of electrolyte loss similar to those of the susceptible reaction. The loss of electrolytes in the leaves infiltrated with the incompatible strain increased rapidly within 2–3 h, if the water-soaked state ceased. This indicates that if cellular recognition exists, it must be negative in operation and the bacterial signal, eliciting incompatibility, must be produced within 3 h. When the intercellular spaces of green pepper leaves were infiltrated with incompatible bacteria, suspended in water–agar, the HR was inhibited and the loss of electrolytes proceeded as in the susceptible reaction. Apparently, for the transmission and/or reception of the incompatibility signal it is necessary that the incompatible bacteria are in contact with the plant walls. This interpretation, however, is not unequivocal because the response of the leaves to incompatible bacteria suspended in water-agar, exposed to the air was not reported.

4.2.2. Pseudomonas glycinea/*soybean*

The soybean phytoalexin glyceollin was differentially released by the cotyledons of two soybeans cultivars incubated for 24 h with suspensions of five *P. glycinea* races (Bruegger and Keen, 1979; see also Chapter 13). Each of the two cultivars released greater amounts of glyceollin in the incompatible interactions. A similar response was obtained when the living bacteria were substituted with suspensions of their crude cell walls and with wall materials extracted with SDS. The same materials taken from *P. pisi*, which is incompatible with soybean, also induced in both cultivars large accumulations of glyceollin. The cognon (or elicitor) responsible appears to be carried by components of the bacterial cell walls, but it becomes accessible to the plant cognor only following SDS treatment. The cognon must be a substance with a low molecular weight (<10 kDa), carried by a component of the cell wall with a relatively high molecular weight (>150 kDa); it is freed through pronase or mild base treatment. Bruegger and Keen (1979) suggested that the cognon is a polysaccharide moiety complexed with protein, because of its heat stability (100 °C for 5 min) and its partial sensitivity to periodate treatment. LPS, protein–LPS complexes, and EPS were inactive as elicitors, although more recent information (see Section 2.1.2, Chapter 13) indicates that pure LPS is active in its triethylamine salt form.

4.2.3. Entrapment and immobilization

Cells of *Pseudomonas pisi* infiltrated into the intercellular spaces of tobacco migrated to the mesophyll cell surfaces within 20 min if the leaves were left exposed to the air (Goodman *et al.*, 1976). After 4 h the bacterial cells were completely ensheathed on the plant cell wall by a structure which became increasingly thicker and more complex when additional amounts of vesicles, fibrils, and membrane fragments were integrated into it. In parallel experiments the cells of *P. fluorescens*, a saprophytic bacterium, were entrapped and immobilized as those of *P. pisi*, whereas *P. tabaci*, a compatible bacterium, was not immobilized and only after 6 h occasionally appeared to be entrapped by fibrils. The enveloping materials have been interpreted as being of plant origin, partly detached from the same plant walls and in part released through the walls in connection with cell wall apposition (see Section 4.2.4). This phenomenon, also called encapsulation, has been observed in many other systems (Medeghini Bonatti *et al.*, 1979; Daub and Hagedorn, 1980; Whatley and Sequeira, 1981; Figures 2 and 3). Encapsulation was not generally observed with virulent strains in homologous combinations, with the exception of bean leaves where it was not always observed for incompatible bacteria. The encapsulation response seems to be relatively differential and the compatible bacteria remain more or less free within the host tissues.

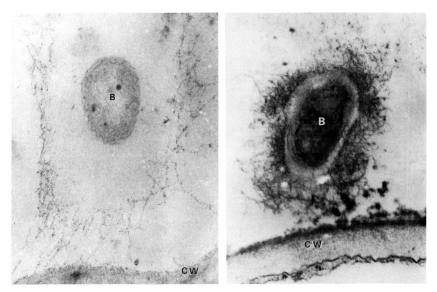

Figure 2. Left: *Pseudomonas syringae* pv. *aptata* cell entrapped by fibrillar material close to the cell wall in tobacco 2 h after infiltration (× 16 000). Right: *Rhizobium trifolii* cell in association with a wall of a clover root hair. Capsular material around the bacterium is in contact with globular electron-dense particles on the plant cell wall surface (× 12 000). B = bacterium; CW = cell wall. (From Medeghini Bonatti *et al.*, 1979, and Dazzo and Hubbell, 1975). Reproduced from the former by permission of Verlagsbuchhandlung Paul Parey and from the latter by permission of the American Society for Microbiology)

The significance of encapsulation as a plant defence response has been seriously questioned. Encapsulation appears to be concerned with the attachment of the bacteria to the plant cell walls rather than with killing or inhibiting them. In fact, in protected tissue, where encapsulation lasts longer, bacteria survive better (Medeghini Bonatti *et al.*, 1979). Hildebrand *et al.*, (1980) concluded that the film entrapping the bacteria was derived from materials dissolved from the plant cell walls during the infiltration of the bacteria which then condense at the air–water interface. However, Whatley and Sequeira (1981) were unable to demonstrate the entrapment of inert materials, such as polystyrene beads or asbestos fibre. When bacterial suspensions are infiltrated, substances of bacterial origin may contribute to the formation of the film. In a bacterial suspension, with a concentration of 10^9 cells cm^{-3}, obtained directly from the washings of the slants with water there were large quantities of EPS and/or LPS (Hildebrand *et al.*, 1980). This can be easily demonstrated if the sterile filtrate is tested with the appropriate antiserum. Similar bacterial substances may form deposits on the cell walls (Graham *et al.*, 1977) or in the intercellular spaces (Medeghini Bonatii *et al.*,

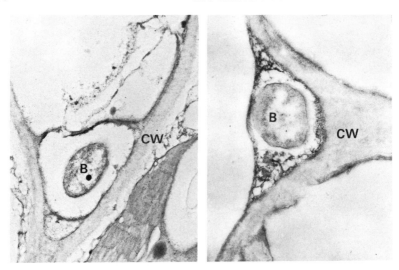

Figure 3. Encapsulation of avirulent *Pseudomonas syringae* pv. *aptata* (left, × 10 500) and virulent *P. syringae* pv. *tabaci* (right, ×15500) in the intercellular spaces of a tobacco leaf 4 and 12 h, respectively, after infiltration. Both the incompatible and the compatible bacteria are segregated, by a closed envelope of fibrillar material. into small niches on the host plant cell walls. B = bacterium; CW = cell wall. [From Medeghini Bonatti *et al.*, 1979, and Mazzucchi *et al.*, 1982. Reproduced from the former by permission of Verlagsbuchhandlung Paul Parey and from the latter by permission of Academic Press Inc. (London) Ltd.]

1979). Hildebrand *et al.*, (1980) suggested that the encapsulation favoured the pathogen, creating a liquid environment around it, ideal for its growth, especially in the early stages of incubation. The nature of the encapsulating material remains unknown. Recently encapsulation was also observed for compatible bacteria (Hildebrand *et al.*, 1980; Mazzucchi *et al.*, 1982). In tobacco leaves a virulent strain of *P. tabaci* appeared to be encapsulated in small niches on the cell walls during a period which fell between the lag phase and the population peak (Figure 3).

In various systems virulent bacteria are entrapped, during the endophytic growth, by fibrillar material in xylem vessel lumena (Horino, 1976; Wallis and Truter, 1978) or in intercellular spaces (Le Normand *et al.*, 1971; Daub and Hagedorn, 1980; Mazzucchi *et al.*, 1982). In the endophytic habitat there is sufficient evidence to indicate that the most likely constituents of the fibrillar network are EPS (Cagle, 1974; Dudman, 1977; Dazzo, 1980; Rudolph, 1980; see Section 4.1.2).

4.2.4. Cell wall appositions

Associated with the attachment and encapsulation phenomena, a plant cell response may be evident. About 2–3 h after infiltration of avirulent *P. pisi* into tobacco mesophyll, on the opposite side of the wall, from where the bacteria were attached, a retraction of the plasmalemma could be observed and an electron-lucent periplasmic area formed in which loose aggregates of microfibrils, mixed together with electron-dense membrane-bound vesicles, accumulated (Politis and Goodman, 1978). After 4 h the wall showed a dome-shape swelling into the cytoplasm. Numerous vesicles were associated both with the swollen part of the wall and with the invaginated plasmalemma. Similar ultrastructure alterations have been described in other plants within a few hours after infiltration of avirulent bacteria (Sequeira *et al.*, 1977; Cason *et al.*, 1977; Roebuck *et al.*, 1978; Medeghini Bonatti *et al.*, 1979; Whatley and Sequeira, 1981); although it is not always easy to distinguish this cellular response from the first signs of HR, there are sound reasons for considering this localized reaction to be a controlled response of the plant cell. In the tobacco leaves it manifests itself long before 6 h, when there is still no sign of hypersensitive cellular decompartmentation (HCD) (Goodman and Plurad, 1971), and lasts longer when the concentration of *P. pisi* is lowered. In other systems, the beginning of the cellular response precedes that of the HCD by 3–14 h. Finally, the response seems to be largely aspecific because it also occurs in tobacco following the attachment of saprophytic bacteria (Goodman *et al.*, 1977) and compatible bacteria (Sequeira *et al.*, 1977). Indeed, wall thickening (papilla formation) appears to be a common response to many infectious agents (see Chapter 11). The cognons and cognors involved in this cellular recognition are practically unknown. The cognons may be carried by the bacterial LPS, because an identical cellular response was obtained in tobacco leaves by treatment with purified LPS (Graham *et al.*, 1977) or protein–LPS complexes of compatible and incompatible bacteria (Mazzucchi and Medeghini Bonatti, unpublished work).

If a rapid localized reaction is accepted as a common response of plant cell to attached bacteria, it is possible to interpret the HR as the effect of a negative cell–cell recognition. During the localized reaction, especially in the initial phases, the plant cell may easily become irreversibly damaged by substances released by living, metabolically active bacteria. In fact, the HR is caused only by bacteria capable of multiplying in the intercellular spaces and the synthesis, by the bacteria, of the causal substances depends on reactions initiated by the host plant (Lyon and Wood, 1976; Sequeira *et al.*, 1977). The so-called HR induction period, calculated as 3–4 h in tobacco and 1–4 h in beans, represents the time necessary for firm attachment of the bacteria to the walls and the beginning of the localized response by the plant cell. The time required for the visible appearance of the damage depends on the particular host–pathogen combination. Saprophytic bacteria, and also certain incompatible pathogenic bacteria, attach themselves to the walls and induce the localized response, but they are not able to multiply. These neutral types of bacteria, do not transmit a signal of incompatibility, and the plant tends to accept and tolerate them. Therefore, they can survive, endophytically, for several days without inducing symptoms. The compatible bacteria do not induce HR because they adopt strategies to avoid the immediate transmission of the incompatibility signal: delaying the attachment (Goodman *et al.*, 1977; Mazzucchi *et al.*, 1982), making the attachment less firm (Goodman *et al.*, 1977; Sequeira and Graham, 1977), inducing a milder or delayed localized response (Sequeira *et al.*, 1977), delaying the release of substances inducing electrolyte loss (Lyon and Wood, 1976; Stall and Cook, 1979), multiplying at a distance from the plant walls within a network of fibrillar material (Hildebrand *et al.*, 1980; Mazzucchi *et al.*, 1982), and inducing fluid water in the intercellular spaces (Goodman, 1972; Huang and Van Dyke, 1978).

5. ACKNOWLEDGEMENTS

I thank Mrs. Baglioni Juliet Macan for her help in translating the manuscript and Mrs. Vaioli Aurora Gamberini for her patience and perfect typing.

REFERENCES

Atkinson, M. M., Huang, J. S., and Van Dyke, C. G. (1981). Adsorption of pseudomonads to tobacco cell walls and its significance to bacterium–host interactions, *Physiol. Plant Pathol.*, **18**, 1–5.

Bateman, D. F. (1978). The dynamic nature of disease, in *Plant Disease. An Advanced Treatise*, Vol. III (Eds. J. G. Horsfall and E. B. Cowling), Academic Press, New York, pp. 53–83.

Bauer, W. D. (1981). Infection of legumes by rhizobia, *Annu. Rev. Plant Physiol.*, **32**, 407–449.

Bayer, M. E. (1981). The fusion sites between outer membrane and cytoplasmic membrane of bacteria: their role in membrane assembly and virus infection, in *Bacterial Outer Membranes* (Ed. M. Inouye), Wiley, Chichester, pp. 167–202.

Beringer, J. E., Brewin, N., Johnson, A. W. B., Schulman, H. M., and Hopwood, D. A. (1979). The *Rhizobium*–legume symbiosis, *Proc. R. Soc. London, Ser. B.*, **204**, 219–233.

Bhuvaneswari, T. V., Turgeon, G., and Bauer, W. D. (1980). Early stages in the infection of soybean (*Glycine max* L. Merr.) by *Rhizobium japonicum*. I. Localization of infectible root cells, *Plant Physiol.*, **66**, 1027–1031.

Bohlool, A. B., and Schmidt, E. L. (1974). Lectins: a possible basis for specificity in the *Rhizobium*–legume root nodule symbiosis, *Science*, **185**, 269–271.

Boyles, J., and Bainton, D. F. (1979). Changing pattern of plasma membrane-associated filaments during the initial phases of polymorphonuclear leukocyte adherence, *J. Cell. Biol.*, **82**, 347–368.

Broughton, W. J. (1978). Control of specificity in legume–*Rhizobium* associations, *J. Appl. Bacteriol.*, **45**, 165–194.

Bruegger, B. B., and Keen, N. T. (1979). Specific elicitors of glyceollin accumulation in the *Pseudomonas glycinea*–soybean host–parasite system, *Physiol. Plant Pathol.*, **15**, 43–51.

Burke, D., Mendonca-Previato, L., and Ballou, C. E. (1980). Cell–cell recognition in yeast: purification of *Hansenula wingei* 21-cell sexual agglutination factor and comparison of the factors from three genera, *Proc. Natl. Acad. Sci. USA*, **77**, 318–322.

Cagle, G. D. (1974). Fine structure and distribution of extracellular polymer surrounding selected aerobic bacteria, *Can. J. Microbiol.*, **21**, 395–408.

Callow, J. A. (1977). Recognition, resistance, and role of plant lectins in host–parasite interactions, *Adv. Bot. Res.*, **4**, 1–49.

Carpita, N., Sabularse, D., Montezinos, D., and Delmer, D. P. (1979). Determination of the pore size of cell walls of living plant cells, *Science*, **205**, 1144–1147.

Cason, E. T., Jr., Richardson, P. E., Brinkerhoff, L. A., and Gholson, R. K. (1977). Histopathology of immune and susceptible cotton cultivars inoculated with *Xanthomonas malvacearum*, *Phytopathology*, **67**, 195–198.

Cason, E. T., Richardson, P. E., Essenberg, M. K., Brinkerhoff, L. A., Johnson, W. M., and Venere, R. J. (1978). Ultrastructural cell wall alterations in immune cotton leaves inoculated with *Xanthomonas malvacearum*, *Phytopathology*, **68**, 1015–1021.

Chen, A. P., and Phillips, D. A. (1976). Attachment of *Rhizobium* to legume roots as the basis for specific interactions, *Physiol. Plant.*, **38**, 83–88.

Clarke, A. E., and Gleason, P. A. (1981). Molecular aspects of recognition and response in the pollen stigma interaction, in *The Phytochemistry of Cell Recognition and Cell Surface Interactions* (Eds F. A. Loewus and C. A. Ryan). Plenum Press, New York, pp. 161–211.

Clarke, A. E., and Knox, R. B. (1978). Cell recognition in flowering plants, *Q. Rev. Biol.*, **53**, 3–28.

Cline, K., and Albersheim, P. (1981). Host–pathogen interactions. XVII. Hydrolysis of biologically active fungal glucans by enzymes isolated from soybean cells, *Plant Physiol.*, **68**, 221–228.

Curtis, A. S. G. (1981). Cell recognition and intercellular adhesion, in *Biochemistry of Cellular Regulation. Vol. IV. The Cell Surface* (Ed. P. Knox), CRC Press, Boca Raton, pp. 151–187.

Daniels, S. L. (1980). Mechanisms involved in sorption of microorganisms to solid surfaces, in *Adsorption of Microorganisms to Surfaces* (Eds G. Bitton and K. C. Marshall), Wiley, Chichester, pp. 8–58.

Daub, M. E., and Hagedorn, D. J. (1980). Growth kinetics and interactions of *Pseudomonas syringae* with susceptible and resistant bean tissues, *Phytopathology*, **70**, 429–436.

Dazzo, F. B. (1980). Adsorption of microorganisms to roots and other plant surfaces, in *Adsorption of Microorganisms to Surfaces* (Eds G. Bitton and K. C. Marshall), Wiley, Chichester, pp. 253–316.

Dazzo, F. B., and Hubbell, D. H. (1975). Cross-reactive antigens and lectins as determinants of symbiotic specificity in *Rhizobium trifolii*–clover association, *Appl. Microbiol.*, **30**, 1017–1033.

Dazzo, F. B., Napoli, C., and Hubbell, H. D. (1976). Adsorption of bacteria to roots as related to host specificity in the *Rhizobium*–clover symbiosis, *Appl. Environ. Microbiol.*, **32**, 166–171.

Drachman, D. B. (1979). Immunopathology of receptors. Introduction, *Fed. Proc. Fed. Am. Soc. Exp. Biol.*, **38**, 2606.

Duckworth, M. (1977). Teichoic acids, in *Surface Carbohydrates of the Prokaryotic Cell* (Ed. I. Sutherland), Academic Press, London, pp. 177–208.

Dudman, W. F. (1977). The role of surface polysaccharide in natural environment, in *Surface Carbohydrates of the Prokaryotic Cell* (Ed. I. Sutherland), Academic Press, London, pp. 357–414.

Edwards, P. R., and Ewing, W. H. (1972). *Identification of Enterobacteriaceae*, Burgess Publishing Company, Minneapolis.

El-Banoby, F. E., Rudolph, K., and Hüttermann, A. (1981a). Biological and physical properties of an extracellular polysaccharide from *Pseudomonas phaseolicola*, *Physiol. Plant Pathol.*, **17**, 291–301.

El-Banoby, F. E., Rudolph, K., and Mendgen, K. (1981b). The fate of extracellular polysaccharide from *Pseudomonas phaseolicola* in leaves and leaf extracts from halo blight susceptible and resistant bean plants (*Phaseolus vulgaris* L.), *Physiol. Plant Pathol.*, **18**, 91–98.

Engström, P., and Hazelbauer, G. L. (1980). Multiple methylation of methyl-accepting chemotaxis proteins during adaptation of *E. coli* to chemical stimuli, *Cell*, **20**, 165–171.

Goodman, R. N. (1972). Electrolyte leakage and membrane damage to relation to bacterial population, pH, and ammonia production in tobacco leaf tissue inoculated with *Pseudomonas pisi, Phytopathology*, **62**, 1327–1331.

Goodman, R. N., and Burkowicz, A. (1970). Ultrastructural changes in apple leaves inoculated with a virulent or an avirulent strain of *Erwinia amylovora, Phytopathol. Z.*, **68**, 258–268.

Goodman, R. N., and Plurad, S. B. (1971). Ultrastructural changes in tobacco undergoing the hypersensitive reaction caused by plant pathogenic bacteria, *Physiol. Plant Pathol.*, **1**, 11–15.

Goodman, R. N., Huang, P. Y., and White, J. A. (1976). Ultrastructual evidence for immobilization of an incompatible bacterium, *Pseudomonas pisi*, in tobacco leaf tissue, *Phytopathology*, **66**, 754–764.

Goodman, R. N., Politis, D. J., and White, J. A. (1977). Ultrastructural evidence of an 'active' immobilization process of incompatible bacteria in tobacco leaf tissue: a resistance reaction, in *Cell Wall Biochemistry* (Eds B. Solheim and J. Raa), Universitetsforlaget, Tromsø, Oslo, pp. 423–438.

Graham, T. L., Sequeira, L., and Huang, T. R. (1977). Bacterial lipopolysaccharides as inducers of disease resistance in tobacco, *Appl. Environ. Microbiol.*, **34**, 424–432.

Haack, A. (1964). Über den Einfluss der Knollchenbakterien auf die Wurzelhaare von Leguminosen und Nichtleguminosen, *Zentralbl. Bakteriol. Parasitenkd. Infektionster., Abt. 2*, **117**, 343–366.

Hamblin, J., and Kent, S. P. (1973). Possible role of phytohaemagglutinin in *Phaseolus vulgaris* L., *Nature (London)*, **245**, 28–30.

Herbert, W. J., and Wilkinson, P. C. (1977). *A Dictionary of Immunology*, Blackwell, Oxford.

Heslop-Harrison, J. (1978). Genetics and physiology of angiosperm incompatibility systems, *Proc. R. Soc. London, Sect. B*, **202**, 73–92.

Hildebrand, D. C., Alosi, M. C., and Schroth, M. N. (1980). Physical entrapment of pseudomonads in bean leaves by films formed at air–water interfaces, *Phytopathology*, **70**, 98–109.

Holl, F. B., and LaRue, T. A. (1976). Genetics of legume plant hosts, in *Proceedings of the 1st International Symposium on Nitrogen Fixation* (Eds W. E. Newton and C. J. Nyman), Washington State University Press, Pullman, WA, pp. 391–399.

Horino, O. (1976). Induction of bacterial leaf blight resistance by incompatible strains of *Xanthomonas oryzae* in rice, in *Biochemistry and Cytology of Plant Parasite Interaction* (Eds K. Tomiyama, J. M. Daly, I. Uritani, H. Oku and S. Ouchi), Kodanska, Tokyo, pp. 43–55.

Huang, J. S., and Van Dyke, C. G. (1978). Interaction of tobacco callus tissue with *Pseudomonas tabaci, P. pisi* and *P. fluorescens, Physiol. Plant Pathol.*, **13**, 65–72.

Hubbell, D. H. (1970). Studies on the root hair 'curling factor' of *Rhizobium, Bot. Gaz.*, **131**, 337–342.

Jann, K., and Jann, B. (1977). Bacterial polysaccharide antigens, in *Surface*

Carbohydrates of the Prokaryotic Cell (Ed. I. Sutherland), Academic Press, London, pp. 247–287.

Kleinman, H. K., Klebe, R. J., and Martin, G. R. (1981). Role of collagenous matrices in the adhesion and growth of cells, *J. Cell Biol.*, **88**, 473–485.

Knox, P. (1981). The adhesion of cells to a solid substratum, in *Biochemistry of Cellular Regulation. Vol. IV. The Cell Surface* (Ed. P. Knox), CRC Press, Boca Raton, pp. 121–149.

Kumarasinghe, R. M. K., and Nutman, P. S. (1977). *Rhizobium*-stimulated callose formation in clover root hairs and its relation to infection, *J. Exp. Bot.*, **28**, 961–976.

Läuchli, A. (1976). Apoplastic transport in tissues, in *Encyclopedia of Plant Physiology. Vol. 2, Part B* (Eds U. Lüttge and M. G. Pitman), Springer Verlag, Berlin, Heildelberg, and New York, pp. 3–34.

Law, I. J., and Strijdom, B. W. (1977). Some observations on plant lectins and *Rhizobium* specificity, *Soil Biol. Biochem.*, **9**, 79–84.

Legocki, R. P., and Verma, D. P. S. (1980). Identification of 'nodule-specific' host proteins (nodulins) involved in the development of *Rhizobium*–legume symbiosis, *Cell*, **20**, 153–163.

Leive, L. L., and Davis, B. D. (1980). Cell envelope; spores, in *Microbiology* (Eds B. D. Davis, R. Dulbecco, H. N. Eisen and H. S. Ginsberg), Harper and Row, Hagerstown, pp. 71–110.

Le Normand, M., Coleno, A., Gourret, J. P., and Fernandes-Arias, H. (1971). Etude de la graisse du haricot et de son agent: *Pseudomonas phaseolicola* (Burkh.) Dowson, *Ann. Phytopathol.*, **3**, 391–400.

Lippincott, J. A., and Lippincott, B. B. (1977). Nature and specificity of the bacterium–host attachment in *Agrobacterium* infection, in *Cell Wall Biochemistry Related to Specificity in Host–Plant Pathogen Interactions* (Eds B. Sölheim and J. Raa), Universitetsforlaget, Oslo, pp. 439–451.

Lippincott, J. A., and Lippincott, B. B. (1978). Cell walls of crown-gall tumors and embryonic plant tissues lack *Agrobacterium* adherence sites, *Science*, **199**, 1075–1077.

Lippincott, J. A., and Lippincott, B. B. (1980). Microbial adherence in plants, in *Bacterial Adherence* (Ed. E. H. Beachey), Chapman and Hall, London, pp. 375–398.

Ljunggren, H. (1969). Mechanism and pattern of *Rhizobium* invasion into leguminous root hairs, *Physiol. Plant*, Suppl. **5**, 1–84.

Lyon, F., and Wood, R. K. S. (1976). The hypersensitive reaction and other responses of bean leaves to bacteria, *Ann. Bot.*, **40**, 479–491.

Marchalonis, J. J. (1980). Molecular interactions and recognition specificity of surface receptor, in *Self/Non-Self Discrimination. Contemporary Topics in Immunobiology*, Vol. 9 (Eds J. J. Marchalonis and N. Cohen), Plenum Press, New York, pp. 255–288.

Marshall, K. C. (1976). *Interfaces in Microbial Ecology*, Harvard University Press, Cambridge, MA.

Marshall, K. C., and Bitton, G. (1980). Microbial adhesion in perspective, in *Adsorption of Microorganisms to Solid Surfaces* (Eds G. Bitton and K. C. Marshall), Wiley, Chichester, pp. 1–5.

Marshall, K. C., Stout, R., and Mitchell, R. (1971). Mechanism of the initial events in the sorption of marine bacteria to surfaces, *J. Gen. Microbiol.*, **68**, 337–348.

Mazzucchi, U., and Pupillo, P. (1976). Prevention of confluent hypersensitive necrosis in tobacco leaves by a bacterial protein–lipopolysaccharide complex, *Physiol. Plant Pathol.*, **9**, 101–112.

Mazzucchi, U., Bazzi, C., and Pupillo, P. (1979). The inhibition of susceptible and hypersensitive reaction by protein–lipopolysaccharide complexes from phytopathogenic pseudomonads. Relationship to polysaccharide antigenic determinants, *Physiol. Plant Pathol.*, **14**, 19–30.

Mazzucchi, U., Bazzi, C., and Medeghini Bonatti, P. (1982). Encapsulation of *Pseudomonas syringae* pv. *tabaci* in relation to its growth in tobacco leaves both pretreated and not pretreated with protein–lipopolysaccharide complexes, *Physiol. Plant Pathol.*, **21**, 105–112.

Medeghini Bonatti, P., Dargeni, R., and Mazzucchi, U. (1979). Ultrastructure of tobacco leaves protected against *Pseudomonas aptata* (Brown et Jamieson) Stevens, *Phytopathol. Z.*, **96**, 302–312.

Moore, R., and Walker, D. B. (1981a). Studies of vegetative compatibility-incompatibility in higher plants. I. A structural study of a compatible autograft in *Sedum telephoides* (*Crassulaceae*), *Am. J. Bot.*, **68**, 820–830.

Moore, R., and Walker, D. B. (1981b). Studies of vegetative compatibility-incompatibility in higher plants. I. A structural study of an incompatible heterograft between *Sedum telephoides (Crassulaceae)* and *Solanum pennellii* (Solanaceae), *Am. J. Bot.*, **68**, 831–842.

Owen, M. J., and Crumpton, J. (1981). The lymphocyte surface and mitogenesis, in *Biochemistry of Cellular Regulation. Vol. IV. The Cell Surface* (Ed. P. Knox), CRC Press, Boca Raton, pp. 21–47.

Ozawa, T., and Yamaguchi, M. (1980). Increase in cellulase activity in cultured soybean cells caused by *Rhizobium japonicum, Plant Cell Physiol.*, **21**, 331–337.

Politis, D. J., and Goodman, R. N. (1978). Localized cell wall appositions: incompatibility response of tobacco leaf cells to *Pseudomonas pisi, Phytopathology*, **68**, 309–316.

Raa, J., Robertsen, B., Sölheim, B., and Tronsmo, A. (1977). Cell surface biochemistry related to specificity of pathogens is and virulence of microorganisms, in *Cell Wall Biochemistry Related to Specificity in Host–Plant Pathogen Interactions* (Eds B. Solheim and J. Raa), Universitetsforlaget, Oslo, pp. 11–30.

Reisert, P. S. (1981). Plant cell surface structure and recognition phenomena with reference to symbioses, *Int. Rev. Cytol.*, Suppl. **12**, 71–112.

Reporter, M., Raveed, D., and Norris, G. (1975). Binding of *Rhizobium japonicum* to cultured soybean root cells: morphological evidence, *Plant Sci. Lett.*, **5**, 73–76.

Robinson, R. A. (1976). *Plant Pathosystems*, Springer-Verlag, Berlin.

Roebuck, P., Sexton, R., and Mansfield, J. M. (1978). Ultrastructural observations on the development of the hypersensitive reaction in leaves of *Phaseolus vulgaris* cv. Red Mexican inoculated with *Pseudomonas phaseolicola* (race 1), *Physiol. Plant Pathol.*, **12**, 151–157.

Roth, I. L. (1977). Physical structure of surface carbohydrates, in *Surface Carbohydrates of the Prokaryotic Cell* (Ed. I. Sutherland), Academic Press, London, pp. 5–26.

Rudolph, K. (1980). Multiplication of bacteria in leaf tissue, *Angew. Bot.*, **54**, 1–9.

Sequeira, L. (1978). Lectins and their role in host–pathogen specificity, *Annu. Rev. Phytopathol.*, **16**, 453–481.

Sequeira, L. (1980). Defenses triggered by the invader: recognition and compatibility phenomena, in *Plant Disease. An Advanced Treatise*, Vol. V (Eds J. G. Horsfall and E. B. Cowling), Academic Press, New York, pp. 179–200.

Sequeira, L. (1981). Induction of host physical responses to bacterial infection: a recognition phenomenon, in *Plant Disease Control* (Eds R. C. Staples and G. H. Toenniessen), Wiley, Chichester, pp. 143–153.

Sequeira, L., and Graham, T. L. (1977). Agglutination of avirulent strains of *Pseudomonas solanacearum* by potato lectin, *Physiol. Plant Pathol.*, **11**, 43–54.

Sequeira, L., Gaard, G., and De Zoeten, G. A. (1977). Interaction of bacteria and host cell walls: its relation to mechanisms of induced resistance, *Physiol. Plant Pathol.*, **10**, 43–50.

Shimshick, E. J., and Herbert, R. R. (1978). Adsorption of rhizobia to cereal roots, *Biochem. Biophys. Res. Commun.*, **84**, 736–742.

Sölheim, B., and Paxton, L. (1981). Recognition in *Rhizobium*–legume systems, in *Plant Disease Control* (Eds R. C. Staples and G. H. Toenniessen), Wiley, Chichester, pp. 71–83.

Sölheim, B., and Raa, J. (1973). Characterization of the substances causing deformation of root hairs of *Trifolium repens* when inoculated with *Rhizobium trifolii*, *J. Gen. Microbiol.*, **77**, 241–247.

Stacey, G., Paau, A. S., and Brill, W. J. (1980). Host recognition in the *Rhizobium*–soybean symbiosis, *Plant Physiol.*, **66**, 609–614.

Stall, R. E., and Cook, A. A. (1979). Evidence that bacterial contact with the plant cell is necessary for the hypersensitive reaction but not the susceptible reaction, *Physiol. Plant Pathol.*, **14**, 77–84.

Sutherland, I. W. (1977). Bacterial exopolysaccharides—their nature and production, in *Surface Carbohydrates of the Prokaryotic Cell* (Ed. I. Sutherland), Academic Press, London, pp. 27–96.

Tsien, H. C., and Schmidt, E. L. (1977). Polarity in the exponential phase of *Rhizobium japonicum* cells, *Can. J. Microbiol.*, **23**, 1274–1284.

Van der Planck, J. E. (1975). *Principles of Plant Infection*, Academic Press, New York,

Van der Planck, J. E. (1978). *Genetics and Molecular Basis of Plant Pathogenesis*, Springer-Verlag, Berlin.

Wallis, F. M., and Truter, S. J. (1978). Histopathology of tomato plants infected with *Pseudomonas solanacearum* with emphasis on ultrastructure, *Physiol. Plant Pathol.*, **13**, 307–317.

Wang, E. A., and Koshland, D. E. (1980). Receptor structure in the bacterial sensing system, *Proc. Natl. Acad. Sci. USA*, **77**, 7157–7161.

Weiss, L. (1970). A biophysical consideration of cell contact phenomena, in *Principles of Adhesion in Biological Systems*, (Ed. R. S. Mainly), Academic Press, London, pp. 1–14.

Whatley, M. H., and Sequeira, L. (1981). Bacterial attachment to plant cell walls, in *The Phytochemistry of Cell Recognition and Cell Surface Interactions* (Eds. A. Loewus and C. A. Ryan), Plenum Press, New York, pp. 213–240.

Whatley, M. H., Margot, J. B., Schell, J., Lippincott, B. B., and Lippincott, J. A. (1978). Plasmid and chromosomal determination of *Agrobacterium* adherence specificity, *J. Gen. Bacteriol.*, **107**, 395–398.

Whatley, M. H., Bodwin, J. S., Lippincott, B. B., and Lippincott, J. A., (1976). Role for *Agrobacterium* cell envelope lipopolysaccharide in infection site attachment, *Infect. Immun.*, **13**, 1080–1083.

Whatley, M. H., Hunter, N., Cantrell, M. A., Hendrick, C., Keegstra, K., and Sequeira, L. (1980). Lipopolysaccharide composition of the wilt pathogen *Pseudomonas solanacearum, Plant Physiol.*, **65**, 557–559.

Yao, P. Y., and Vincent, J. M. (1976). Factors responsible for the curling and branching of clover root hairs by *Rhizobium, Plant Soil*, **45**, 1–16.

Yeoman, M. M., Kilpatrick, D. C., Miedzybrodzka, M. B., and Gould, A. R. (1978). Cellular interactions during graft formation in plants, a recognition phenomenon?, *Symp. Soc. Exp. Biol.*, **32**, 139–159.

Further Reading

Curtis, A. S. G. (1981). Cell recognition and intercellular adhesion, in *Biochemistry of the Cellular Regulation, Vol. IV. The Cell Surface* (Ed. P. Knox), CRC Press, Boca Raton, pp. 151–187.

Heslop-Harrison, J. (1978). *Cellular Recognition Systems in Plants*, Edward Arnold, London.

Knox, R. B., and Clarke, A. E. (1980). Discrimination of self and non-self in plants, in *Self/Non-Self Discrimination. Contemporary Topics in Immunobiology*, Vol. 9 (Eds J. J. Marchalonis and N. Cohen), Plenum Press, New York, pp. 1–36.

Loewus, F. A., and Ryan, A. C. (1981). *The Phytochemistry of Cell Recognition and Cell Surface Interactions. Recent Advances in Phytochemistry*, Vol. 15, Plenum Press, New York.

Reisert, P. S. (1981). Plant cell surface structure and recognition phenomena with reference to symbioses, *Int. Rev. Cytol.*, Suppl. 12, 71–112.

Sequeira, L. (1981). Induction of host physical responses to bacteria infection: a recognition phenomenon, in *Plant Disease Control* (Eds R. C. Staples and G. H. Toenniessen), Wiley, Chichester, pp. 143–153.

Sölheim, B., and Paxton, J. (1981). Recognition in *Rhizobium*–legume systems, in *Plant Disease Control* (Eds R. C. Staples and G. H. Toenniessen), Wiley, Chichester, pp. 71–83.

IV Effects on Host Metabolism

Biochemical Plant Pathology
Edited by J. A. Callow
© 1983 John Wiley & Sons Ltd

15

Bioenergetic and Metabolic Disturbances in Diseased Plants

STEVEN W. HUTCHESON and BOB B. BUCHANAN

Division of Molecular Plant Biology, University of California, Berkeley, CA 94720 USA

1. INTRODUCTION: ENERGY METABOLISM AND ITS REGULATION IN THE PLANT

Plant productivity is the foundation for the existence of life on our planet. Our dependence on plants goes beyond their direct or indirect utilization as a food source: we also use their non-edible products, such as wood and fibre, and our industrial civilization is powered by coal, oil, and natural gas—the fossilized products of plants of millions of years ago. Because of this dependence, we are profoundly affected by diseases of all plants, but especially diseases of plants important to agriculture. Agricultural crops are susceptible to a variety of diseases that severely reduce their yield and quality. It is, therefore, important that we understand the mechanisms through which the causative agents, phytopathogens, alter the cellular processes necessary for plant growth and productivity. If the mechanisms through which phytopathogens alter host processes are known, it may be possible to diminish or abolish the effects of infection through specific manipulation of key components in host tissue by either genetic or chemical means.

A number of cellular processes and constituents necessary for the growth and development of the host plant are affected during the course of infection by pathogens (pathogenesis). It was established early in this century that phytopathogens utilize compounds obtained from the host for their growth and development. In the case of obligate or biotrophic plant parasites such as rusts and powdery mildews, it was observed that placing infected plants in darkness, or removing atmospheric carbon dioxide, arrested the development of the parasite even at advanced stages of pathogenesis (Reed, 1914; Waters, 1928; Trelease and Trelease, 1929; Sempio, 1942). Pathogen development under these conditions became possible only if infected leaves were supplied an external source of carbohydrate. These observations and others have prompted the conclusion that products of either photosynthesis or respiration provide the energy needed for pathogen development.

In addition to obtaining energy-rich compounds from the host, phytopathogens may induce changes in the levels of cellular constituents (i.e. organic acids, amino acids, carbohydrates, and enzymes) in infected or nearby tissues. Concomitant with changes in levels of cellular constituents are alterations in the bioenergetic processes of the host. Numerous studies with a wide variety of etiological agents have shown that, almost universally, infection by phytopathogens elicits a substantial increase in the respiration rate of host tissue. In contrast, the photosynthetic activity of infected tissue generally declines.

Although a great deal is known about the changes in cellular processes that occur during pathogenesis, relatively little is understood at the molecular level about how infectious agents alter specific anabolic (synthetic) and catabolic (degradative) reactions of the host. The purpose of this chapter is to review the pathogen-induced alterations in bioenergetics and related metabolic reactions of the host that result from infection by each of the major classes of phytopathogens—viruses, fungi, and bacteria. Because of the biochemical emphasis and limited scope of this contribution, physiological effects of phytopathogens, such as those which result from alterations in gas exchange and translocation, will not be addressed. These topics were covered by Kosuge (1978) and Talboys (1978).

2. GENERAL ASPECTS OF PHOTOSYNTHESIS

During autotrophic growth, plants obtain their substance and energy through the reactions of photosynthesis—the process by which the electromagnetic energy of sunlight is used for the synthesis of carbohydrates, $[CH_2O]$, and other organic compounds from carbon dioxide and water:

$$CO_2 + 2H_2O \xrightarrow{\text{light}} [CH_2O] + O_2 + H_2O \tag{1}$$

An important by-product of the process is molecular oxygen (derived from water) which escapes into the atmosphere. By liberating oxygen, photosynthesis functions as a gigantic planetary ventilation system which continuously pumps oxygen into the earth's atmosphere to counter the loss of oxygen by respiratory reactions and combustion. The carbohydrates and other compounds formed are either stored or converted into the building blocks necessary for the maintenance and growth of the plant.

The reactions of photosynthesis can be broadly divided into two phases: (a) a light phase in which the electromagnetic energy of sunlight is trapped and converted into biologically useful chemical energy (assimilatory power) and (b) a carbon reduction (or biosynthetic) phase in which the assimilatory power generated by light is used for *de novo* synthesis of carbohydrates and other cellular constituents. This latter phase takes place in a series of enzymatic reactions, some of which, as noted below, are regulated by light.

The conversation of sunlight into chemical energy involves cyclic and non-cyclic photophosphorylation, the two light-dependent processes that jointly provide ATP and reduced ferredoxin (or its product, NADPH) needed for the assimilation of carbon dioxide to carbohydrates and other cellular components (Arnon, 1977). In photosynthetic eucaryotic organisms, both of these processes take place in chloroplasts—the organelles that house the chlorophyll-containing (thylakoid) membrances and the associated enzyme components that catalyze the photosynthetic assimilation of CO_2.

In the reactions of cyclic photophosphorylation, radiant energy is converted into the pyrophosphate bonds of ATP via electron transport. Cyclic photophosphorylation requires no added chemical substrate and is driven entirely by light:

$$ADP + P_i \xrightarrow[\text{ferredoxin}]{\text{light}} ATP \qquad (2)$$

Cyclic photophosphorylation in isolated chloroplasts requires an added catalyst; there is now general agreement that the physiological catalyst of cyclic photophosphorylation is ferredoxin, an iron–sulphur protein which catalyses non-cyclic and cyclic electron transport simultaneously (Arnon, 1977). Cyclic phosphorylation is strongly inhibited by antimycin A, a property that provides a diagnostic tool to study this reaction.

Non-cyclic photophosphorylation differs from its cyclic complement in that the formation of ATP is coupled to an oxidation–reduction reaction in which electrons from water are transferred in the light to ferredoxin (Fd) with a concomitant evolution of molecular oxygen:

$$2H_2O + 4Fd_{ox} + 2ADP + 2P_i \xrightarrow{\text{light}} 4Fd_{red} + 2ATP + O_2 + 4H^+ \qquad (3)$$

The ferredoxin so reduced can be used, in turn, to reduce NADP in a reaction catalysed by a flavoprotein enzyme, ferredoxin–NADP reductase, independently of light:

$$2Fd_{red} + NADP \xrightarrow[\text{reductase}]{\text{Fd–NADP}} NADPH_2 + 2Fd_{ox} \tag{4}$$

The electron transport reactions of non-cyclic photophosphorylation are inhibited specifically by 3-(3.4-dichlorophenyl)-1, 1-dimethylurea (DCMU), a herbicide, and by o-phenanthroline, a metal-chelating agent.

The mechanism by which cyclic and non-cyclic photophosphorylation are generally considered to be effected by the thylakoid membranes of chloroplasts is shown in Figure 1. The heart of the system is a light-driven transport of electrons from two different active chlorophylls (P680, P700) to each of two acceptors (A-II, A-I). In the non-cyclic reactions, electrons from water flow through two photochemical reactions (photosystems II and I) to ferredoxin and then to NADP; in the cyclic reactions, electrons flow through a single photochemical reaction (photosystem I) and cycle in a closed system without net oxidation or reduction. It is to be noted that the light-driven transport of electrons in both cyclic and non-cyclic electron transport is accompanied by a transport of protons that results in the formation of a proton gradient across chloroplast membranes. The discharge of this ion gradient releases energy

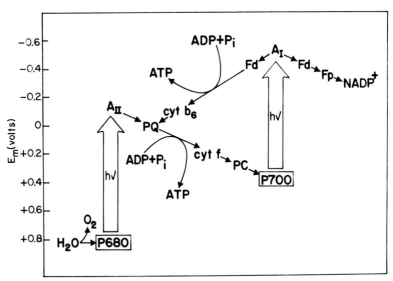

Figure 1. Mechanisms of non-cyclic and cyclic photophosphorylation in chloroplasts. Non-standard abbreviations: P680, reaction centre chlorophyll of photosystem II; A_{II}, primary electron acceptor of photosystem II; PQ, plastoquinone; cyt. f, cytochrome f; PC, plastocyanin; P700, reaction centre chlorophyll of photosystem I; A_I, primary electron acceptor of photosystem I; Fd, ferredoxin; F_p, flavoprotein (ferredoxin–NADP reductase); cyt b_6, cytochrome b_6

that is used for the synthesis of ATP from ADP and P_i (Trebst, 1974; Malkin and Bearden, 1978).

ATP is the substance that is often characterized as the universal energy currency of living cells, be they plants, animals, or bacteria. One of the functions of ATP is to provide the energy needed for the myriad of cellular biosynthetic reactions, including those of photosynthesis. NADPH (or $NADPH_2$) represents the reduced form of nicotinamide adenine dinucleotide phosphate. NADPH is a strong reductant in biosynthetic reactions: its great reducing power enables it to provide electrons (accompanied by protons) for other molecules. In carbon dioxide assimilation, NADPH is the carrier of the electrons and protons (i.e. the hydrogen atoms) required for the conversion of phosphorylated intermediates (i.e. intermediates activated by ATP) to carbohydrates and other organic compounds.

The ATP and NADPH formed in cyclic and non-cyclic photophosphorylation jointly constitute the above-noted assimilatory power needed for the conversion of CO_2 to carbohydrates, which takes place via the photosynthetic carbon cycle (also known as the reductive pentose phosphate cycle or Calvin cycle). According to the cycle, 1 mol of carbon dioxide is converted to the level of carbohydrate with the energy expenditure of 3 mol of adenoside-5'-triphosphate (ATP) and 2 mol of NADPH (Bassham and Calvin, 1957) (Figure 2). Cell components other than carbohydrates are formed by reactions leading from the cycle (Bassham and Calvin, 1957).

The need for ATP and NADPH (or its analogue, NADH) to convert carbon dioxide into carbohydrates is not peculiar to photosynthesis. The same two compounds are required when carbohydrates are formed from carbon

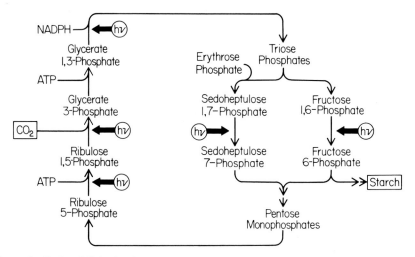

Figure 2. Role of light in the activation of enzymes of the reductive pentose phosphate cycle of CO_2 assimilation in plant photosynthesis

dioxide by non-photosynthetic, autotrophic cells. What distinguishes carbon dioxide assimilation by photosynthetic cells from that by non-photosynthetic cells is the source of energy for the generation of ATP and NADPH. Photosynthetic cells when illuminated generate these two compounds at the expense of the electromagnetic energy of sunlight, whereas non-photosynthetic cells (or photosynthetic cells in the dark) generate them at the expense of some other source of chemical energy via catabolic reactions (i.e. glycolysis, the citric acid cycle, and the oxidative pentose phosphate cycle).

For many years it was assumed that the sole requirement for light in photosynthesis was to provide the reductant (reduced ferredoxin or NADPH) and ATP needed as substrates to drive CO_2 assimilation via the pathway responsible for that process (the reductive pentose phosphate cycle). Results obtained during the past 15 years have caused a revision in that concept (Buchanan, 1980; Buchanan *et al.*, 1979). It is now apparent that, in addition to its substrate function in supplying ATP and $NADPH_2$, light is required catalytically in chloroplasts for the regulation of enzymes of pathways functional in the synthesis as well as in the degradation of energy-rich compounds. The regulatory function of light encompasses both enzyme activation [whereby the reductive pentose phosphate cycle (Figure 2) and related biosynthetic pathways are accelerated] and enzyme inhibition (whereby starch degradation, glycolysis, and the oxidative pentose phosphate pathway

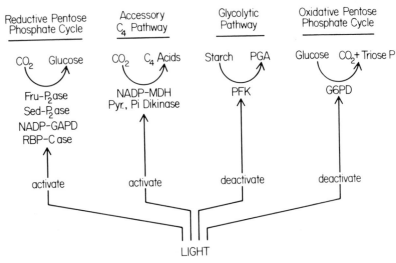

Figure 3. Effect of light on regulatory enzymes of carbon pathways in chloroplasts. Fru-P$_2$ase, fructose 1,6-bisphosphatase; Sed-P$_2$ase, sedoheptulose 1,7-bisphosphatase; NADP–GAPD, NADP–glyceraldehyde 3-phosphate dehydrogenase; RBP-Case, ribulose 1,5-bisphosphate carboxylase/oxygenase; NADP–MDH, NADP–malate dehydrogenase; Pyr., Pi dikinase, pyruvate, P$_i$ dikinase; PFK, phosphofructokinase; G6PD, glucose 6-phosphate dehydrogenase

are impeded). It is because of this dual regulatory function of light that the degradative and synthetic pathways, which share a number of enzymes, can coexist and operate within the confines of the chloroplast (Figure 3).

In fulfilling its regulatory role, light acts through several different mechanisms, which, in each case, are linked to the photosynthetic apparatus (Figure 4). The light absorbed by chlorophyll induces changes in components associated with electron transport and thereby conveys the signal of light to enzymes of metabolic processes that, according to some lines of evidence, need not be restricted to chloroplasts.The light-linked changes used to elicit an alteration of enzyme activity include (i) shifts in the concentration of ions, especially H^+ and Mg^{2+}; (ii) increases in the concentration of metabolites that alter enzyme activity (enzyme effectors), such as ATP and $NADPH_2$; and (iii) changes in the oxidation–reduction state of specific regulatory proteins, such as thioredoxins. These three types of mechanisms appear to act in concert for the regulation of specific chloroplast enzymes during photosynthesis. Each of the changes effected by light must be reversed under conditions prevailing in the dark. It may be important that the enzymes found to undergo activation by light include, among others, those that catalyse reactions that seemingly limit photosynthesis, at least in some plants.

The question might be asked as to why light has a regulatory role in addition to a well documented substrate role in producing the ATP and NADPH (assimilatory power) required for photosynthesis. It appears that the chloroplast uses the changes linked to light to enhance the activity of enzymes of photosynthetic pathways and to suppress the activity of enzymes of degradation. In this manner, the plant is assured that each pathway is fully operative only during the time of greatest need, i.e. the photosynthetic and related biosynthetic enzymes during the day, and the degradative enzymes during the night when the plant must extract its energy from the breakdown of energy-rich compounds that were synthesized during the day. If the plants were unable to link the activity of the indicated enzymes to light, the light energy

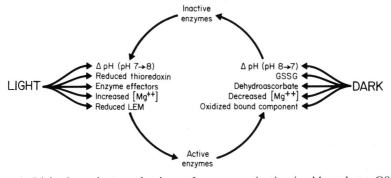

Figure 4. Light-dependent mechanisms of enzyme activation in chloroplasts. GSSG, oxidized glutathione; LEM, light-effect mediator

converted and stored in energy-rich compounds would be spent uselessly in competing degradative reactions.

3. EFFECT OF PHYTOPATHOGENS ON PHOTOSYNTHESIS

Although there is extensive information on the physiological and biochemical changes induced in plants by pathogenic agents, little is known about the effect of disease-causing organisms on photosynthesis at the molecular level. Studies with each of the three major types of etiological agents (viruses, fungi, bacteria) have revealed that the rate of photosynthesis in infected plants may be unchanged, suppressed, or, in some cases, increased. Irrespective of effects on the rate of photosynthesis, infection can alter, in some cases appreciably, the products of photosynthesis. The mechanism by which the alteration is effected is known only in a few instances. We summarize below the effect on photosynthesis of representatives of each of the major types of etiological agents.

3.1. Viruses

After the discovery of viruses by Ivanovski in 1894, and later, independently by Beijerinck, two views emerged to explain the primary effect of virus infection on leaves: (i) the view of Beijerinck and Konig that virus infection is primarily a disease of chloroplasts and manifests itself by chloroplast damage and ultimate disintegration; and (ii) the view of Woods and Ivanovski that the primary effect of virus infection does not involve chloroplasts [see Magyarosy et al., (1973) for references]. According to the second view, virus infection leads to a reduction in the number of chloroplasts but it does not cause major changes in their photosynthetic activity. Chloroplast activity may be altered in the advanced stages of disease, but such changes would result from secondary effects rather than being a primary cause of the diseased conditions.

More recent investigations of a biochemical character have provided evidence that systemic virus infection decreases the rate of photosynthesis by leaves and of photophosphorylation by isolated chloroplasts. In some cases, an initial increase in the rate of photosynthesis has been noted, but in all cases photosynthesis was suppressed in the plants at the later stages of infection. Unfortunately, those investigations provide no evidence as to whether the lower rates of photosynthesis reflect a primary change in chloroplasts caused by the virus or whether the reduction in rates results from a general decline in the metabolic activities of the infected leaf. Furthermore, it is not known whether the increased synthesis of free amino acids and organic acids reported for different types of virus-infected leaves reflects changes in reactions that occur within chloroplasts.

Work from our laboratory has confirmed and extended the conclusions formed at the turn of the century by Woods and Ivanovski (Magyarosy et al.,

1973). Leaves from squash plants (*Cucurbita pepo* L.) systemically infected with squash mosaic virus showed a shift in products from sugars to amino acids (alanine in particular), an increase in cytoplasmic ribosomes, and fewer chloroplasts. There was no difference between the healthy and diseased chloroplasts with respect to most parameters tested: sensitivity of CO_2 assimilation to DCMU and antimycin A; ultracentrifugation profile; products of photosynthetic $^{14}CO_2$ assimilation; activity of the enzymes, ribulose 1, 5-bisphosphate carboxylase/oxygenase and phosphoenolpyruvate carboxylase; chloroplast ultrastructure. Despite the presence of fewer chloroplasts, we found, on a chlorophyll basis, a slight increase in the rate of $^{14}CO_2$ assimilation in infected plants similar to that observed by Zaitlin and Hesketch (1965) and Smith and Neales (1977). It is possible that the inability of plants infected with certain viruses to produce a full complement of chloroplasts is related to a reported reduction in the capacity to synthesize chloroplastic RNAs and proteins (Hirai and Wildman, 1969; Mohamed and Randles, 1972; Randles and Coleman, 1970; Oxelfelt, 1971; Kosuge, 1978).

Results similar to our findings with squash mosaic virus were more recently reported by Platt *et al*., (1979) with virus-infected *Tolmiea menziesii* L. (piggyback plant). They found, in addition, that infected leaves had an increased ratio of chlorophyll *a:b* and a decreased content of the light-harvesting chlorophyll *a+b* protein. They concluded that the infected leaves have a significantly smaller amount of antenna chlorophyll but a similar content of reaction-centre chlorophyll. Relative to their healthy counterparts, the infected leaves showed an enhanced formation of glycine and a diminished formation of sucrose.

In contrast to the evidence summarized above, work with other virus infections has supported the view of Beijerinck and Konig. For example, the research groups of Bové (Goffeau and Bové, 1965) and of Matthews (Bedbrook and Matthews, 1972, 1973) investigated the effect of turnip yellow mosaic virus on photosynthesis. Both groups used *Brassica pekinensis* (Chinese cabbage) in their studies, and it was concluded that the chloroplasts were structurally and biochemically altered by infection: Bové's group reported an increase of both cyclic and non-cyclic photophosphorylation in chloroplasts from infected plants; Matthews' group observed that infection causes a shift in photosynthetic products from sugars to organic acids and amino acids, a shift that was also reported with other viruses by other investigators (Magyarosy *et al*., 1973; Platt *et al*., 1979). Accompanying this shift in products was an increase in the activity of phosphoenolpyruvate carboxylase and aspartate aminotransferase and a decrease in the activity of ribulose 1, 5-bisphosphate carboxylase/oxygenase. It is noteworthy that with turnip yellow mosaic virus, the viral nucleic acid is formed in the vesicles of the chloroplast envelope, and the virus is thus considered to be a parasite of the chloroplast (Matthews, 1973).

It may be concluded from these studies that the disparate views of

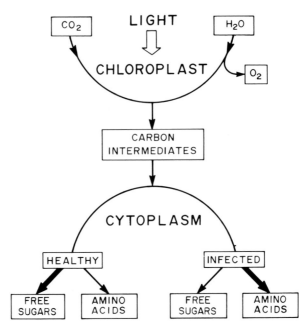

Figure 5. Effect of virus infection on plant photosynthesis (squash mosaic virus/squash plants)

Beijerinck and Konig and of Ivanovski and Woods were both conceptually correct. Thus, depending on the disease, photosynthesis can be altered by viral infection in either of two ways. As exemplified by squash plants (infected with squash mosaic virus) and by piggyback plants (doubly infected with tomato bushy stunt and cucumber virus), virus infection may reduce the number of chloroplasts, but it has little effect on the reactions of photosynthesis or on chloroplast ultrastructure. In these plants, the virus mainly influences the reactions of the leaf cytoplasm (Figure 5). Alternatively, as exemplified by Chinese cabbage plants infected with turnip yellow mosaic virus, virus infection alters the partial reactions of photosynthesis (e.g. photophosphorylation) and thereby impedes plant growth. It appears that the interaction between the host plant and a specific virus determines the course of changes in photosynthesis and in other processes of plants.

3.2. Fungi

A change in the rate of photosynthesis by whole plants or intact leaves has been reported to be induced by fungal infection (Magyarosy *et al.*, 1976; Aust *et al.*, 1977; Rowe and Reid, 1977; Mignucci and Boyer, 1979; Spotts and Ferre, 1979). Infection by fungal pathogens, in most cases, seems to cause a reduction in the rate of photosynthesis. Investigations of the effects of fungal infection in either of two ways. As exemplified by squash plants (infected with laboratory has investigated some of the effects of obligate fungal parasites (particularly rusts and powdery mildew) on the reactions of photosynthesis. During the course of these studies we observed that each of these pathogens effected a preferential inhibition of non-cyclic photophosphorylation (Montalbini and Buchanan, 1974; Magyarosy et al., 1976). With both of these agents, the inhibition stemmed from a diminution of electron flow from water to NADP and, in the one case examined (powdery mildew), infection led to a reduced rate of photosynthetic carbon dioxide assimilation. Chloroplasts isolated from rust and powdery mildew-infected leaves showed a substantial (up to 45%) decrease in the rate of non-cyclic electron transport and the attendant photophosphorylation. Infection had no effect on the coupling of phosphorylation to photosynthetic electron transport (i.e. photophosphorylation) as determined by the ratio of ATP formed to the amount of NADP reduced.

The nature of the parasite-induced block in non-cyclic electron flow was, at first, a mystery. An early idea was that the fungal parasite effects the formation of a toxic compound that is functionally similar to the electron transport inhibitor, DCMU (Montalbini and Buchanan, 1974). It is well known that many pathogenic fungi produce phytotoxins (see Chapter 8; Yoder, 1980) but, nevertheless, this possibility now seems unlikely in view of the fact that extensively washed chloroplast membranes isolated from rust-infected plants showed no relief from parasite-induced inhibition compared to their unwashed counterparts (Montalbini et al., 1981).

An alternative hypothesis would be that parasite infection brings about an alteration in the components of the non-cyclic electron transport chain. Evidence obtained by Magyarosy and Malkin (1978) supports this conclusion. They found that the cytochrome content of the electron transport chain was decreased by about one third in chloroplasts from powdery mildew-infected sugar beet plants. These results suggest that infection by obligate parasites specifically alters the content of certain electron carriers involved in electron transfer between components of the electron transport chain and thereby reduces the rate of non-cyclic electron transport. Significantly, the photosystem I and II reaction centres and the bound iron–sulphur proteins were unaffected by infection.

Besides reducing the rate of non-cyclic electron transport and carbon dioxide assimilation, infection of sugar beet by powdery mildew led to a shift in products from sucrose to amino acids (alanine, glutamate, and aspartate) (Magyarosy et al., 1976). These observations provide an explanation for the reduced capability of sugar beet plants infected with powdery mildew to form sucrose. In this study, there was also a reduction in the activity of enzymes involved in the production of organic acids (malate dehydrogenase, phosphoenol pyruvate carboxylase) (Magyarosy et al., 1976). In separate studies, rust infection decreased the activity of ribulose 1, 5-bisphosphate carboxylase/oxygenase in wheat leaves (Wrigley and Webster, 1966) but had little or no effect on photosynthesis in oat plants (Wynn, 1963).

At present it is not clear how obligate plant parasites such as rusts and powdery mildew alter the amount or the activity of proteins (e.g. chloroplast cytochromes) in host tissue. Protein levels can be altered through changes in either the rate of their synthesis or degradation, or both. There have been studies covering the general changes in host nucleic acids and proteins induced by pathogen infection; however, relatively little is known at the molecular level that can explain the changes observed. These aspects are discussed further in Chapter 18.

Beside altering protein synthesis in host cells, obligate plant parasites can also affect the level of proteins through degradative reactions. During the infection process, fungal mycelia release a myriad of extracellular enzymes. In addition to releasing proteases which could have a direct effect on host proteins (Hislop and Pitt, 1976), invading mycelia also release lipases and compounds, such as tannins, which increase the permeability of host cells (Kosuge and Gilchrist, 1976; Wood, 1977). Protein synthesis would be affected indirectly through the loss of important metabolites needed for protein synthesis. Changes in permeability may also be linked to the changes in chloroplast ultrastructure that occur after infection (Coffee et al., 1972; Heath, 1974; Mlodzianowski and Siwecki, 1975; Magyarosy et al., 1976).

In summary, it seems that the contradictory observations on the rates of photosynthesis and on molecular changes observed in fungus-infected plants, stem basically from the diverse effects of fungal growth during pathogenesis. Depending on the stage of infection, one can observe differential effects on the biochemistry of the affected cells. These seemingly conflicting effects are further complicated by the differential responses that pathogens elicit in susceptible and resistant plants and by the presence of two eucaryotic genomes directing metabolism in infected tissue. Caution, therefore, should be exercized in making generalizations concerning the effect of fungal infection on photosynthesis and related processes with the limited information that is currently available.

3.3. Bacteria

Bacteria are an important group of disease-causing agents that are responsible for extensive crop losses throughout the world. Despite the economic importance of these organisms, little is known about their effects on photosynthesis. In those crops that have been examined, the organisms caused a long-term decline in the rate of photosynthesis (Beckman et al., 1962; Magyarosy and Buchanan, 1975; Crosthwaite and Sheen, 1978) that was accompanied by extensive structural damage to chloroplasts (Braun, 1955; Goodman and Plurad, 1971; Sigee and Epton, 1976).

Saprophytic bacteria, i.e. those which do not elicit a disease response, differ from their pathogenic counterparts in that, when infiltrated into leaves, they cause only an initial decrease in the rate of photosynthesis, and they induce no alterations in chloroplast ultrastructure. Moreover, the photosynthetic products formed in the pathogen- and saprophyte-infiltrated plants are similar to each other and to the healthy controls (Magyarosy and Buchanan, 1975). Although further studies are needed in this area, it appears that the inhibition of photosynthesis and the accompanying ultrastructural changes induced by the pathogens are due, in part, to the effect of toxins that are formed (Mitchell, 1978; Crosthwaite and Sheen, 1978) and released by the invading organisms (see Chapter 8; Yoder, 1980).

4. EFFECTS OF PHYTOPATHOGENS ON RESPIRATORY REACTIONS

Attendant with their effects on photosynthesis, phytopathogens influence other aspects of host metabolism. A common response of plants to infection is an increase in the respiratory activity of infected tissue. In some cases, the increased activity extends beyond the immediate site of infection to include adjacent tissue and other organs of the infected plant. This increase in respiratory activity can be substantial (greater than double), beginning early in pathogenesis and continuing until the affected tissue is necrotic. In other host–pathogen combinations, the increase in respiratory activity is either transitory or non-existent (e.g. certain systemic viral infections). In leaf tissue, high respiratory activity induced by the pathogen would rapidly expend cellular reserves. When the concomitant loss of photosynthetic activity is considered, it is evident this type of host–pathogen interaction would clearly be detrimental to the growth and productivity of the diseased plant.

The present information indicates that the increased respiration of diseased tissue (measured as O_2 uptake) stems from one or more of the following mechanisms: (1) wound respiration due to regenerative and protective activity of tissue physically damaged by the invading pathogen; (2) uncoupling of mitochondrial electron transport from ATP synthesis as a result of the action of toxins released by the pathogen; (3) loss of compartmentation of enzymes and key metabolites because of increased membrane permeability (possibly induced by toxins); (4) enhanced consumption of ATP, NADPH, and other energy-rich compounds through increased biosynthesis; (5) increased levels of enzymes involved in carbohydrate degradation; (6) increased levels of substrates (starch and soluble sugars) that accumulate as a result of blocked translocation; and (7) increased activity of oxidases involved in the biosynthesis of secondary compounds (e.g. polyphenol oxidases and peroxidases).

When the tissues of a plant are disrupted by either mechanical or pathological means, several processes are initiated by the host in attempts to isolate and/or regenerate the traumatized tissue. In wounded storage tissue, mechanical abrasion or cutting leads to the deposition of lignin and suberin in tissue adjacent to the wound [see Kahl (1974) and Uritani and Asahi (1980) for references]. In some cases there is initiation of growth of callus to regenerate the damaged tissue (Uritani and Asahi, 1980). The biosynthetic activity associated with lignin synthesis and tissue regeneration requires energy that may be derived from increased respiratory activity. Accompanying this increased biosynthetic activity is an increase in the number of mitochondria.

A trait of wound respiration that is shared with respiration of diseased tissue is insensitivity to cyanide, a potent inhibitor of cytochrome c oxidase of both plant and animal mitochondria. The cyanide-resistant electron pathway that is characteristic of plant mitochondria seems to branch from the cytochrome c oxidase pathway at a site near coenzyme Q and to utilize an alternative oxidase in the reduction of O_2 to H_2O (Solomos, 1977). Respiration by this alterative pathway results in a lower production of ATP, i.e. only one ATP formed per NADH oxidized instead of three ATPs when the NADH is oxidized via the cytochrome oxidase pathway. Because of the diminished bioenergetic capacity of these cells, higher rates of carbohydrate degradation are required to maintain basic cellular processes.

In addition to effects resulting from physical invasion of tissue, fungal and bacterial pathogens release a number of components that affect the metabolism of the host cells and, in some cases, play a role in the virulence of the pathogen. Such components include digestive enzymes and specific toxins which can alter the properties of host membranes [see Chapter 8 and Yoder (1980) for further information]. One well studied host–pathogen interaction involving toxins is the parasitism of susceptible varieties of corn (i.e. hybrids with Texas male-sterile cytoplasm) by the causal agent of southern corn leaf blight, *Helminthosporum maydis* Race T (HMT). HMT produces a host-selective toxin which has numerous effects on susceptible tissue. This toxin increases the permeability of membranes of susceptible varieties, particularly mitochondrial membranes, and stimulates oxidative reactions in the parent mitochondria. Studies with isolated, susceptible mitochondria indicate that electron transport is uncoupled from ATP synthesis in toxin treated mitochondria (Gregory *et al.*, 1978). Other work suggests that HMT toxin renders mitochondria highly permeable to NAD, a metabolite unable to cross healthy mitochondrial membranes (Gregory *et al.*, 1978). It is evident that the permeability changes induced by HMT toxin would substantially lower the synthetic capabilities of affected cells. Attendant with the loss of ATP synthesis would be a loss of localized pools of reduced pyridine dinucleotides (NADH and NADPH). The altered membrane permeability would allow the uncontrolled oxidation of these key cofactors, depriving the cell of energy needed for biosynthesis. Further work is needed to elucidate the molecular mechanisms whereby compounds such as HMT toxin alter the permeability of cell membranes, as well as to link the responses observed *in vitro* with the symptoms of this disease observed in the field.

Another factor involved in the increased respiration of diseased tissue is the enhanced turnover of ATP and NADPH resulting from the increased biosynthesis of compounds produced by the host in response to wounding or pathogen infection (e.g. lignin, as mentioned above). This increased turnover of ATP and NADPH would in turn stimulate respiratory activity so that a stable pool size of these key metabolites would be maintained. There is evidence that the higher turnover of the energy-rich compounds is associated with an increase in the activity of certain enzymes of glycolysis and/or the oxidative pentose phosphate pathway [see Asahi *et al.,* (1979) and Uritani and Asahi (1980) for references]. In a number of cases there is a disproportionate increase in the activity of enzymes of the oxidative pentose phosphate pathway, possibly due to the reported increased levels of both glucose 6-phosphate dehydrogenase and phosphogluconate dehydrogenase [see Uritani and Ashai (1980) for references]. The increased activity of this pathway appears to be coupled to the increased synthesis of phenylpropanoid compounds that are derived from products of the pathway. Higher plants, unlike animals and bacteria, have the ability to synthesize a wide variety of phenylpropanoid compounds including lignin, flavonoids, isoflavonoids, coumarins, and esters of hydroxycinnamic acids. Certain of the phenylpropanoid derivatives serve as phytoalexins (Chapter 12). Such antimicrobial compounds may be formed as a result of the action of polyphenol oxidase and other oxidases in the host. Numerous investigators, including ourselves, have reported that an increase in oxidase activity (e.g. polyphenol oxidase) is observed during the course of a variety of plant diseases [see Montalbini, *et al.,* (1981) for references]. It seems that this increase may represent an attempt by the host plant to limit infection through the enhanced formation of quinones and quinone-like derivatives that inhibit the growth of etiological agents more effectively than do the parent phenolic compounds. It remains to be seen whether this increase in quinones is related to the increase (doubling) observed in the level of both active and latent forms of polyphenol oxidase in infected plant tissues.

5. CONCLUSION

Our understanding of the molecular events that accompany the establishment and progression of the diseased condition in plants is far from complete. Based on the facts at hand, different types of infectious agents seem not to follow any consistent pattern with respect to biochemical changes induced in the host.

Depending on the agent, contrasting effects can be observed with respect to basic processes, ranging from a decrease in rate or magnitude on the one hand to a stimulation on the other. Thus, it is not yet possible to make generalizations regarding the biochemical changes induced in the host by a group of plant pathogens, and each disease must be considered individually. It is up to the future to determine whether this present situation will be maintained or whether, following the addition of new facts, similar types of diseases can be characterized by a defined set of biochemical parameters.

REFERENCES

Arnon, D. J. (1977). Photosynthesis 1950-75. Changing concepts and perspective in *Photosynthesis. I. Photosynthetic Electron Transport and Photophosphorylation* (Eds A. Tresbst and M. Avron), *Encyclopedia of Plant Physiology*, New Series, Vol. 5, Springer Verlag, Berlin, Heidelberg, and New York, pp. 7–56.

Asahi, T., Kojima, M., and Kosuge, T. (1979). The energetics of parasitism, pathogenism and resistance in plant disease, in *Plant Disease, Vol. IV, How Pathogens Induce Disease* (Eds J. G. Horsfall and E. B. Cowling), Academic Press, New York, pp. 47–74.

Aust, H. J., Domes, W., and Kranz, J. (1977). Influence of CO_2 uptake of barley leaves on incubation period of powdery mildew under different light intensities, *Phytopathology*, **67**, 1469–1472.

Bassham, J. A., and Calvin, M. (1957). *The Path of Carbon in Photosynthesis*, Prentice-Hall, Englewood Cliffs, NJ.

Beckman, C. H., Brun, W. A., and Buddenhagen, I. W. (1962). Water relations in banana plants infected with *Pseudomonas solanacearum. Phytopathology*, **52**, 1144–1148.

Bedbrook, J. R., and Matthews, R. E. F. (1972). Changes in the proportions of the early products of photosynthetic carbon fixation by TYMV infection, *Virology*, **48**, 255–258.

Bedbrook, J. R., and Matthews, R. E. F. (1973). Changes in the flow of early products of photosynthetic carbon fixation associated with the replication of TYMV, *Virology*, **53**, 84–91.

Braun, A. C. (1955). A study on the mode of action of the wildfire toxin, *Phytopathology*, **45**, 659–664.

Buchanan, B. B. (1980). Role of light in the regulation of chloroplast enzymes, *Annu. Rev. Plant Physiol.*, **31**, 341–374.

Buchanan, B. B., Wolosiuk, R. A., and Schürmann, P. (1979). Thioredoxin and enzyme regulation, *Trends Biochem. Sci.*, **4**, 93–96.

Coffee, M. D., Palevitz, B. A., and Allen, P. J. (1972). Ultrastructural changes in rust-infected tissues of flax and sunflower, *Can. J. Bot.*, **50**, 1485–1492.

Crosthwaite, L. M., and Sheen, S. J. (1978). Inhibition of ribulose 1,5-bisphosphate carboxylase by a toxin isolated from *Pseudomonas tabaci, Phytopathology*, **69**, 376–379.

Goffeau, A., and Bové, J. M. (1965). Virus infection and photosynthesis, *Virology*, **27**, 243–252.

Goodman, R. N., and Plurad, S. B. (1971). Ultrastructural changes in tobacco undergoing the hypersensitive reaction caused by plant pathogenic bacteria, *Physiol. Plant Pathol.*, **1**, 11–15.

Gregory, P., Matthews, D. E., York, D. W., Earle, E. D., and Gracen, V. E. (1978). Southern corn leaf blight disease: studies on mitochondrial biochemistry and ultrastructure, *Mycopathologia*, **66**, 105–112.

Heath, M. C. (1974). Chloroplast ultrastructure and ethylene production of senescing and rust-infected cowpea leaves, *Can. J. Bot.*, **52**, 2591–2597.

Hirai, T., and Wildman, S. G. (1969). Effect of TMV multiplication on RNA and protein synthesis in tobacco chloroplasts, *Virology*, **38**, 73–82.

Hislop, E. A., and Pitt, D. (1976). Subcellular organization in host–parasite interactions, in *Physiological Plant Pathology* (Eds R. Heitefuss and P. H. Williams), *Encyclopedia of Plant Physiology*, New Series, Vol. 4, Springer Verlag, Berlin, Heidelberg, and New York, pp. 389–407.

Kahl, G. (1974). Metabolism in plant storage tissue slices, *Bot. Rev.*, **40**, 263–314.

Kosuge, T. (1978). The capture and use of energy by diseased plants, in *Plant Disease, Vol. II, How Plants Suffer from Disease* (Eds J. G. Horsfall and E. B. Cowling), Academic Press, New York, pp. 86–116.

Kosuge, T., and Gilchrist, D. G. (1976). Metabolic regulation in host-parasite interactions, in *Physiological Plant Pathology* (Eds R. Heitefuss and P. H. Williams), *Encyclopedia of Plant Physiology*, New Series, Vol. 4, Springer Verlag, Berlin, Heidelberg, and New York, pp. 679–702.

Magyarosy, A. C., and Buchanan, B. B. (1975). Effect of bacteria on photosynthesis of bean leaves, *Phytopathology*, **65**, 777–780.

Magyarosy, A. C., and Malkin, R. (1978). Effect of powdery mildew infection of sugar beet on the content of electron carriers in chloroplasts, *Physiol. Plant Pathol.*, **13**, 183–188.

Magyarosy, A. C., Buchanan, B. B., and Schürmann, P. (1973). Effect of a systemic virus infection on chloroplast function and structure, *Virology*, **55**, 426–438.

Magyarosy, A. C., Schürmann, P., and Buchanan, B. B. (1976). Effect of powdery mildew infection on photosynthesis by leaves and chloroplasts of sugar beets, *Plant Physiol.*, **57**, 486–489.

Malkin, R., and Bearden, A. J. (1978). Membrane-bound iron–sulphur centers in photosynthetic systems, *Biochim. Biophys. Acta,* **505**, 147–181.

Matthews, R. E. F. (1973). Induction of disease by viruses, *Annu. Rev. Phytopathol.*, **11**, 147–170.

Mignucci, J. S., and Boyer, J. S. (1979). Inhibition of photosynthesis and transpiration in soybean infected by *Microsphaera diffusa, Phytopathology*, **69**, 227–230.

Mitchell, R. E. (1978). Halo blight of beans: toxin production by several *Pseudomonas phaseolicola* isolates, *Physiol. Plant Pathol.*, **13**, 37–49.

Mlodzianowski, F., and Siwecki, R. (1975). Ultrastructural changes in chloroplasts of *Populus tremula* L. leaves affected by the fungus *Melampsora pintorqua*, Braun. Rostr., *Physiol Plant Pathol.*, **6**, 1–3.

Mohamed, N. A., and Randles, J. W. (1972). Effect of tomato spotted wilt virus on ribosomes, ribonucleic acids and Fraction I protein in *Nicotina tabacum* leaves, *Physiol. Plant Pathol.*, **2**, 235–245.

Montalbini, P., and Buchanan, B. B. (1974). Effect of a rust infection on photophosphorylation by isolated chloroplasts, *Physiol. Plant Pathol.*, **4**, 191–196.

Montalbini, P., Buchanan, B. B., and Hutcheson, S. W. (1981). Effect of rust infection on rates of photochemical polyphenol oxidation and latent polyphenol oxidase activity of *Vicia faba* chloroplast membranes, *Physiol. Plant Pathol.*, **18**, 51–57.

Oxelfelt, P. (1971). Development of systemic TMV infection. II. RNA metabolism in systemically infected leaves, *Phytopathol. Z.*, **71**, 247–256.

Platt, S. G., Henriques, F., and Rand, L. (1979). Effect of virus infection on the chlorophyll content, photosynthetic rate and carbon metabolism of *Tolmiea menziesii, Physiol. Plant Pathol.*, **15**, 351–365.

Randles, J. W., and Coleman, D. F. (1970). Loss of ribosomes in *N. glutinosa* L. infected with lettuce necrotic yellows virus, *Virology*, **41**, 459–464.

Reed, G. M. (1914). Influence of light on infection of certain hosts by powdery mildrews, *Science*, **39**, 294–295.

Rowe, J., and Reid, V. (1977). Some aspects of carbon metabolism in the barley–*Helminthosporium teres* complex. I. The effects of infection upon carboxylation *in vivo* and *in vitro, Can. J. Bot.*, **57**, 195–207.

Sempio, C. (1942). Influenzadi alcuni glucidi isomeri sullo suilluppo della ruggine del Fagiolo e di altre malattie fungine, *Riv. Biol.*, **34**, 52–56.

Sigee, D. C., and Epton, H. A. (1976). Ultrastructural changes in resistant and susceptible varieties of *Phaselous vulgaris, Physiol. Plant Pathol.*, **9**, 1–8.

Smith, P. R., and Neales, T. F. (1977). Analysis of the effects of virus infection on the photosynthetic properties of peach leaves, *Aust. J. Plant Physiol.*, **4**, 723–732.

Solomos, T. (1977). Cyanide-resistant respiration in higher plants. *Annu. Rev. Plant Physiol.*, **28**, 279–297.

Spotts, R. A., and Ferre, D. C., (1979). Photosynthesis, transpiration, and water potential of apple leaves infected by *Venturia inaequalis, Phytopathology*, **69**, 717–719.

Talboys, P. W. (1978). Dysfunction of the water system, in *Plant Disease, Vol. III, How Plants Suffer from Disease* (Eds J. G. Horsfall and E. B. Cowling), Academic Press, New York, pp. 141–162.

Tani, T., Yoshikawa, M., and Naito, N. (1973). Template activity of ribonucleic acid extracted from oat leaves infected by *Puccinia coronata, Ann. Phytopathol. Soc. Jpn.*, **39**, 7–13.

Trebst, A. (1974). Energy conservation in photosynthetic electron transport of chloroplasts, *Annu. Rev. Plant Physiol.*, **25**, 423–458.

Trelease, S. F., and Trelease, H. M. (1929). Susceptibility of wheat to mildew as influenced by carbohydrate supply, *Bull. Torrey Bot. Club*, **56**, 65–92.

Uritani, I., and Asahi, T. (1980). Respiration and related metabolic activity in wounded and infected tissue, in *Biochemistry in Plants Vol. 2, Metabolism and Respiration* (Eds P. K. Stumpf and E. E. Conn), Academic Press, New York, pp. 463–487.

Waters, C. W. (1928). The control of teliospore and urediospore formation by experimental methods, *Phytopathology*, **18**, 157–213.

Wood, R. K. S. (1977). Enzymes produced by fungi and bacteria. Their role in pathogenicity, *Ann. Phytopathol.*, **10**, 127–136.

Wrigley, C. W., and Webster, H. L. (1966). The effect of stem rust infection on the soluble proteins of wheat, *Aust. J. Biol. Sci.*, **19**, 895–901.

Wynn, W. K. (1963). Photosynthetic phosphorylation by chloroplasts isolated from rust-affected oats, *Phytopathology*, **53**, 1376–1377.

Yoder, O. C. (1980). Toxins in pathogenesis, *Annu. Rev Phytopathol.*, **18**, 103–129.

Zaitlin, M., and Hesketch, J. D. (1965). The short term effects of infection by tobacco mosaic virus on apparent photosynthesis of tobacco leaves, *Ann. Appl. Biol.*, **55**, 239–263.

Further Reading

Daly, J. M. (1976). The carbon balance of diseased plants: changes in respiration, photosynthesis and translocation, in *Physiological Plant Pathology* (Eds R. Heitefuss and P. H. Williams), *Encyclopedia of Plant Physiology*, New Series, Vol. 4, Springer Verlag, Berlin, Heidelberg, and New York, pp. 450–479.

Kosuge, T. (1978). The capture and use of energy by diseased plants, in *Plant Disease, Vol. III, How Plants Suffer from Disease* (Eds J. G. Horsfall and E. B. Cowling), Academic Press, New York, pp. 86–116.

Kosuge, T., and Gilchrist, D. G. (1976). Metabolic regulation in host–parasite interactions, in *Physiological Plant Pathology* (Eds R. Heitefuss and P. H. Williams), Encyclopedia of Plant Physiology, New Series, Vol. 4, Springer Verlag, Berlin, Heidelber, New York, pp. 679–702.

Merrett, M. J., and Bayley, J. (1969). The respiration of tissues infected by virus, *Bot. Rev.*, **35**, 372–392.

Uritani, I., and Asahi, T. (1980). Respiration and related metabolic activity in wounded and infected tissue, in *Biochemistry of Plants Vol. II, Metabolism and Respiration*, (Eds P. K. Stumpf and E. E. Conn), Academic Press, New York, pp. 463–487.

Biochemical Plant Pathology
Edited by J. A. Callow
© 1983 John Wiley & Sons Ltd

16

The Effects of Disease on the Structure and Activity of Membranes

ANTON NOVACKY

Department of Plant Pathology, University of Missouri, Columbia, MO 65211, USA

'The more central a concept is to active scientific investigation, the more transient our knowledge of it. The biomembrane is a case in point. It is an identifiable, unique piece of the physical and biological world. Much like the old idea of the gene, it is also so rich a concept that it guides investigations and formulations in virtually every field of biology.'

J. S. Beck, 1980.

1. INTRODUCTION

The universal feature of all living organisms is their membrane structure, the plasma membrane which separates the cell interior from the environment outside and organellar membranes which compartmentalize the cell interior. The plasma membrane selectively controls the transport of ions and numerous substrates both passively by virtue of its structural composition (the permeability through the lipid matrix along concentration gradients) and actively by energy-consuming membrane proteins (transport against concentration gradients). This discussion will be limited mostly to the plasma membrane.

Plant cell membranes are composed of approximately equal amounts of lipids and proteins. Lipids include phospholipids (glycerophosphatides) associated with, for example, sterols and glycolipids, forming the basic membrane matrix into which the protein components are integrated. The amphipathic character of phospholipids, i.e. their polar and non-polar moieties, determines their orientation within the membrane. The polar hydrophilic moiety faces the aqueous environment outside the membrane while the hydrophobic non-polar fatty acid chains faces the interior of the membrane. Electron micrographs show the membranes as double-track lines. This image results, perhaps, from fixation changes, but more likely reflects the orientation of the polar region in the lipid matrix. The surface structure of membrane protein molecules must also be amphipathic to be compatible with membrane lipid milieu. The current understanding of the integration of membrane lipids and proteins is best expressed in the fluid-mosaic membrane model of Singer (Singer and Nicholson, 1972). (For a review of previous membrane models from which the fluid mosaic model evolved, see e.g. Kotyk and Janáček, 1977.)

According to Singer's model, proteins may be located deeply within the fluid lipid bilayer (integral proteins), close to the surface (peripheral proteins), or traverse the membrane with a channel in the middle. The overall picture regarding the nature of plant membrane proteins is still incomplete. However, the best known protein component in plant membranes is the K^+-stimulated ATPase (Hodges, 1976) which plays a focal role in active membrane transport. There is much less information concerning the activity of other enzymes within the plasma membrane, e.g. glycosyltransferase (Leonard and Hodges, 1980).

To illustrate the functional characteristic of membrane lipids and proteins, observations with the 'protoplasmic drop,' a phenomenon described first by the Swiss botanist Nägeli more than 120 years ago (see Hope and Walker, 1975), will be briefly discussed. Nägeli noted that the released protoplasmic content from broken cells is quickly surrounded by a new membrane. In recent years such a droplet has served as a model in various membrane studies. In the investigations of Inoue et al. (1973), the protoplasmic content of single cells from the giant alga Nitella was used. The protoplasmic mass released after cell dissection and end wall removal was soon surrounded by a newly formed membrane. This droplet possesed some characteristics of the intact Nitella cell, i.e. a high membrane potential measurable with glass microelectrodes. When Inoue et al. treated the droplet with proteolytic enzymes, the membrane potential declined immediately to a very low value. However, the droplet remained intact. It was only after the application of phospholipase

A, a lipid-hydrolysing enzyme, that the membrane disintegrated and the droplet was destroyed. Hence, these experiments with protoplasmic droplets may be taken as a good demonstration that the skeletal property of membranes are lipids and the functional (active) features of membranes are proteins.

Membrane transport takes place through both the lipid matrix and membrane protein components. The phospholipid matrix is the site of the passive permeation (diffusion) of water and ions and compounds such as amino acids and sugars that follow concentration gradients and are transported with water. The movement against the concentration gradients (an 'uphill' transport) proceeds through the protein component (intrinsic proteins) for which the energy from the metabolism of the cell (ATP) is required (for more details see, e.g. Christensen, 1975).

1.1. Transmembrane potential difference

The result of both the passive and energy-dependent transport processes is an asymmetric distribution of charges across the plasma membrane or the membranes of other cellular organelles. A consequence of this separation is the transmembrane potential difference (PD). Glass microelectrodes are used to measure PD values which denote the electropotential across two membranes: plasma membrane and tonoplast. However, the contribution of tonoplast is minimal (at best a few millivolts). Therefore, PD is considered simply as the potential across the plasma membrane (Jones *et al.*, 1975).

Individual species of higher plants have characteristic values of PD; hence changes in PD can be used as a sensitive indicator of the status of membranes. Two components of PD have been recognized. When the flow of energy to the membrane is blocked by metabolic inhibitors a decline of the potential indicates the abolition of the active, energy-dependent PD component that is commonly attributed as the electrogenic pump (Spanswick, 1981). The remaining value of PD is interpreted as the passive component of the membrane potential resulting from the diffusion along the concentration gradients (Higinbotham *et al.*, 1970). For example, the PD for cotton cotyledons has been assessed as follows: PD = -170 mV, PD under CN^- = -80 mV, therefore the passive PD = -90 mV (Novacky *et al.*, 1976). (For additional details on transmembrane potentials see, e.g. Findlay and Hope, 1976.)

1.2. Electrogenic pump

The K^+ stimulated ATPase demonstrated in cell membranes of many plants (Hodges, 1976) provides the energy-dependent component of PD or the electrogenic pump (Sze and Churchill, 1981). Several studies indicate that the electrogenic pump is the primary energy transducer linking cellular metabolism to membrane transport and that the principal ion transported by this

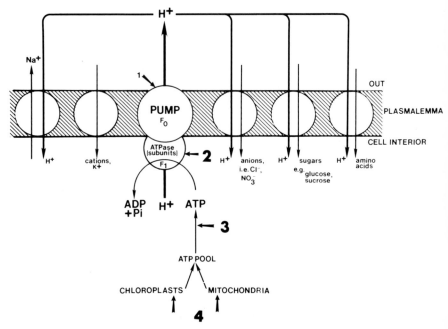

Figure 1. Diagrammatic representation of active (energy-dependent) membrane transport processes in the plasma membrane of higher plants. Currently available evidence suggests that activity of the electrogenic proton pump is central to secondary transport processes which directly or indirectly depend on the membrane H^+ movement. Blank circles in the model are postulated carriers such as H^+ solute co-transport systems. Transport by passive diffusion is not included here. In our simplified scheme we may imagine pathogen interference occurring at points marked with arrows in the diagram. Pathogens may affect the main component of the electrogenic pump (F_0) (1), the ATPase subunits of the pump (F_1) (2), the flow of ATP to the pump (3), or at the ATP generation site, within the mitochondria or chloroplasts (4). Diagram modified according to Marrè (1980)

pump is H^+ (Spanswick, 1981). Thus, the ATPase in the plasma membrane is similar to that of bacteria and mitochondria. As the electrogenic pump removes H^+ from the cell interior a steep chemical potential gradient is generated. This gradient was named the proton-motive force by Mitchell (1967) and is a part of his chemiosmotic hypothesis that recognizes the focal role of this force in active membrane transport. The H^+ pump constitutes the primary active transport mechanism across the membrane. The pH gradient and the negative membrane potential generated by the H^+ pump are the driving forces for the secondary active transport of ions and non-electrolytes such as sugars and amino acids (Poole, 1978) (see Figure 1).

The cell maintains the H^+ pump and regulates the primary and secondary active transport with the ATP either from respiration or photosynthesis. In addition to the role in membrane transport, the electrogenic pump is also believed to play an important role in the cell's pH control 'biophysical pH-stat' (Smith and Raven, 1979).

2. MEMBRANES UNDER DISEASE CONDITIONS

Increased leakage of electrolytes from tissues infected with plant-pathogenic bacteria or fungi (or their toxic metabolites) is generally considered a sign of altered membrane permeability (Wheeler, 1978). Several reviewers (e.g. Hanchey and Wheeler, 1979) caution that the increased efflux of electrolytes and non-electrolytes may result from perforation of the normally continuous plasma membrane of live cells. Such leakage may reflect the loss of viability and not a change in the cell permeability *per se*. Reported dramatic increases in leakage may indicate only the size of the population of dead cells. Electron micrographs of cells in tissues exhibiting electrolyte losses reveal discontinuities in the plasma membrane and often also in the tonoplast (e.g. White *et al.*, 1973). Such cells can no longer be considered viable. Nevertheless, data recording the leakage of electrolytes may be important because they indicate a degree of membrane involvement in pathogenesis.

Alterations in membranes resulting in leakage of electrolytes may be caused by either the activities of a pathogen, e.g. by production of host-specific or non-specific toxins, hydrolytic enzymes, or autocatalytic processes in cells of diseased tissues. The autocatalytic changes may include changes in membrane lipids, e.g. an induction of senescence-like lipid peroxidation or changes induced by stress metabolites. It is well established that various injuries including those induced by pathogenicity elicit a rapid increase in the synthesis of stress metabolites—phytoalexins (Chapter 12). In addition to their possible inhibitory effect on the pathogens, these compounds may have a deleterious effect on plant cells. For example, phaseollin produced a leakage of preloaded Rubidium-86 in bean leaves (Van Etten and Bateman, 1971), pisatin caused the plasma membrane in protoplasts isolated from pea leaves to disappear gradually (Shiraishi *et al.*, 1975) and rishitin rapidly disrupts the viability of cells in potato leaf epidermis (Lyon, 1980).

Leakage may alter quantitatively and qualitatively at different phases of disease development. Leakage may reflect the terminal phase of cell damage when the integrity of the plasma membrane is lost and an uncontrolled efflux accompanies cell death. An example of this is when the cell wall is damaged and the turgor pressure in the cell is no longer counteracted by a rigid wall and the plasma membrane begins to rupture. This happens with diseases in which pathogens produce cell wall-degrading enzymes. The barrier function of the membrane is lost and a rapid efflux of cellular content occurs (Bateman and Basham, 1976).

The fact that the cellular content may leak out of cells in the very early stages of the pathogenesis became evident after the discovery of the host-specific toxin victorin [*Drechslera* (= *Helminthosporium*) *victoriae* toxin] (Scheffer, 1976; Wheeler, 1978). The ability of victorin to reproduce the disease caused by the inciting fungus *D. victoriae* led to intensive studies of the disease processes in general and the role of the plasma membrane in particular. In a cell susceptible to victorin electrolyte leakages detected 3–5 min after contact with the toxin suggests that the plasma membrane may play the crucial role in the disease initiation. Both the fungus and victorin are highly specific. They are injurious only to a limited number of oat cultivars (cv. Victoria and those cultivars derived from it). The specificity of victorin and its toxicity at extremely low concentrations (dilutions of 10^{-7} of a *D. victoriae* culture filtrate) has stimulated great interest in the role that membrane perturbation plays in pathogenesis (Wheeler, 1981).

We may postulate two categories of alterations in membrane function and structure under conditions of disease: (1) changes in membrane lipids, the site of passive permeability, and (2) changes in membrane proteins, the site of energy-dependent membrane transport. These two sites are closely inter-related, i.e. alterations in one membrane component will immediately influence the function of the other. For example, after lipids are removed during purification the enzymatic activity of membrane protein is also lost but regained when the proteins are reconstituted with lipids (Raison, 1980).

In the following discussion we shall examine the evidence for alterations in membranes and discuss experimental data selected to explain permeability changes.

2.1. Interference with membrane lipids

Whereas alterations in membrane lipids that occur during the plant development and ageing have been recognized for some time (Simon, 1974), changes of lipids during pathogenesis have attracted only limited research attention. It is true, however, that permeability to water and to non-electrolytes increase dramatically in several diseases (Thatcher, 1939, 1942) and that non-pathogen wounding (e.g. slicing of storage tissue) leads to an immediate and extensive autocatalytic degradation of membrane lipids (Galliard, 1978). Perhaps it is because studies of lipids during disease development have revealed only small alterations (e.g. Hoppe and Heitefuss, 1974a, b). The experiments of Hoppe and Heitefuss were carried out on the lipids in bean leaves infected with *Uromyces phaseoli* and changes were detected only at the more advanced stages of disease. Nevertheless, in that host–pathogen complex the large amount of unsaturated fatty acids, phosphatidylcholine and phosphatidylethanolamine, is an important finding (Hoppe and Heitefuss, 1975). Such a change in fatty acid saturation may result in alterations in the phase-transition temperature of membrane lipids.

The change in membrane fluidity is related to alterations in permeability during the process of senescence, according to a series of studies by McKersie and Thomson (e.g. 1977, 1978). The phase-transition temperature, the temperature at which the liquid crystalline phase transforms to the gel phase, is increased and as a result of this transition the membrane becomes more permeable and leaky. Portions or some areas of the membrane in the liquid crystalline phase, shift to the gel phase. McKersie and Thompson (1978) have suggested that the presence of the gel phase causes a discontinuity in the membrane bilayer where it interfaces with the adjacent liquid crystalline phase, and at the juncture permeability is increased.

In essence changes found in pathogenesis are often signs of premature or accelarated senescence (Farkas, 1978). We may also assume that in the disease process, alterations of the membrane lipid phase occur that are responsible for increased leakage. Recently, the gel–lipid phase was found in ozone-injured tissue, i.e. ozone injury caused an increase in the phase transition temperature similar to shifts during senescence (Pauls and Thompson, 1980).

The symptoms of ozone injury are characterized by water-soaked lesions which later become necrotic and resemble symptoms accompanying leaf spotting fungal and bacterial diseases. Possibly the ozone damage is a result of the oxidation of unsaturated fatty acids in the membranes (Pauls and Thompson, 1980). Fatty acids liberated during membrane phospholipid degradation are toxic to many cellular processes. Lipoxygenase (previously known as lipoxidase) is an enzyme which specifically catalyses the oxidation of unsaturated fatty acids (Galliard and Chan 1980). Fatty acid hydroperoxides and carbonyl compounds produced by lipoxygenase are potentially destructive to membranes (Galliard, 1978).

A role for lipoxygenase in fungal pathogenesis has recently been established. Lupu et al. (1980) found that the powdery mildew pathogen induced an increased in the activity of lipoxygenase in tabacco. Parallel to this increase, peroxides, the products of enzymatic degradation, accumulated in the infected tissue, which then became leaky. The exogenous application of antioxidants suppressed both lipoxygenase activity and the growth of the mildew pathogen on leaves. Lupu et al. suggested that the endogenous antioxidants may constitute a portion of the plant's defence mechanism against injury.

A clear correlation between lipid peroxidation and the rate of leakage from cells of ageing tobacco tissue was demonstrated by Dhindsa et al. (1981). The accumulation of melondialdehyde, a product of lipid degradation, was closely correlated with the leakage of ^{86}Rb from cells during senescence.

The direct effect of stress metabolites—phytoalexins on the lipid membrane matrix was demonstrated recently. Lyon (1980) incorporated rishitin, a phytoalexin produced by potato, into positively or negatively charged liposomes (prepared from egg lecithin/cholesterol/stearylamine or egg lecithin/cholesterol/phosphatidic acid). The presence of rishitin increased liposome permeability to low molecular weight non-electrolytes, which was related to the change in the transition temperature, i.e. membrane fluidity. Experiments with liposomes were also related to observations of potato leaf epidermal strips incubated with rishitin. An accumulation of Evans Blue, which selectively enters cells with damaged membranes, suggested a rapid effect of rishitin on several cell membranes.

The studies of Thatcher (1939, 1942) (for more details see Wheeler and Hanchey, 1968) provide indirect evidence of alterations in the membrane lipid matrix. Thatcher examined the membrane permeability to non-electrolytes in disease tissue using the plasmometric technique, which is based on microscopic examination of the swelling and contraction of protoplasts as a result of changed external osmoticum of either impermeable or permeable solutes (Stadelmann, 1966). The test cells are first plasmolysed in a hypertonic solution and then deplasmolysed either in water or in a non-electrolyte such as urea. The rate of deplasmolysis indicates the degree of membrane permeability to water or to the non-electrolytes tested. Thatcher's studies showed that during pathogenesis dramatic increases occurred in the transport of water and non-electrolytes. In tobacco infected with *Pseudomonas tabaci* Turner and Novacky (1976) found similar increases in permeability to non-electrolytes with the plasmometric method, whereas the water permeability remained unchanged. However, in tobacco undergoing the bacterial hypersensitive reaction, induced by an incompatible pathogen, *P. pisi*, water permeability decreased several-fold (Turner, 1976; cf. Novacky, 1980).

Some membranologists (Dainty, 1976; Zimmermann, 1977) have criticized the interpretations of the results obtained plasmometrically, arguing that the membrane in the plasmolysed protoplast differs considerably from the membrane in the deplasmolysed state. For example, membrane dehydration may occur during exosmosis, i.e. the efflux of water during plasmolysis. The indirect evidence for the alteration in the lipid membrane matrix is the dramatic change in membrane permeability during disease development and during hypersensitivity. However, the aforementioned uncertainties suggest that values calculated by using the plasmometric method will be re-examined critically with other techniques such as the recently developed pressure probe technique (Zimmermann and Steudle, 1978).

2.2. Interference with membrane proteins

Several studies indicate that many plant pathogens or their toxic metabolites interfere with energy-dependent membrane transport mechanisms governed by membrane proteins. Direct alteration of the ATPase complex or one of its subunits in the membrane may affect the transport activity and related processes. For instance, if the membrane ATPase becomes inhibited, proton pumping may decrease or even cease. If, however, the ATPase is stimulated,

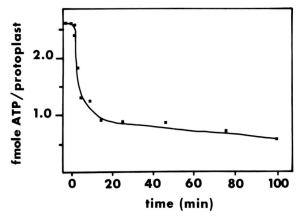

Figure 2. The ATP concentration in suspension of T-cytoplasm maize leaf mesophyll protoplasts incubated in the dark in the presence of HMT toxin (6.2 μg cm^{-3}). Toxin was introduced at zero time. [From Walton *et al.* (1979), *Plant Physiol.*, **63**, 806–810. Reproduced by permission of the American Society of Plant Physiologists]

an increased proton efflux results. An alteration in the ATPase activity may also result from a change in the flow of energy to the pump. In such an instance pathogen-induced change may occur at the site of ATP synthesis in the mitochondria or chloroplasts which decreases the level of cellular ATP. For example, Walton *et al.* (1979) found a 50% decrease in the level of ATP, 4 min after application of HMT toxin to protoplasts of T-cytoplasm corn (Figure 2). Such a drastic decrease in the energy level may also inhibit the plasma membrane ATPase. Another example of inhibition of ATP synthesis is the effect of tentoxin [*Alternaria alternata* (= *A. tenuis*) toxin] on tentoxin-sensitive plants. Tentoxin binds specifically to chloroplast ATPase, thus inhibiting photophosphorylation (Steele *et al.*, 1976; see also Chapter 8). However, even if the level of ATP remains unchanged, this does not guarantee the availability of ATP to the plasma membrane. ATP flow to the membrane may be changed or rerouted to other cellular sites. An example would be pathogen-induced damage where intensive repair processes may require extra ATP energy. This was suggested for cotyledons of cotton (*Gossypium hirsutum*) inoculated with *Xanthomonas malvacearum* (Novacky and Ullrich-Eberius, 1982).

Marrè (1980) listed 17 host-specific and non-specific mycotoxins that are known to interfere with the host plasma membrane. The evidence for such effects, however, is mostly circumstantial (see Section 6, Chapter 8). What is not clearly established is where and when this interference occurs, since pathogens may affect numerous sites within cells with a consequent change in membrane proteins. Several investigators have attempted to establish the minimum time to induce an effect on transport processes with the minimum dose of toxin in question. Unfortunately, until recently, this has been difficult to accomplish because the toxins were only partially purified and effective doses were variable. Some toxins have recently been purified and character- ized and have been used in studies designed to elucidate their membrane effect. Thus far, only a few attempts have been made to demonstrate toxin interference with the plasma membrane *in vitro*.

2.2.1. Inhibition of electrogenic pump

A strong indication that pathogen-induced alterations in transport processes occur across the plasma membrane comes from electrophysiological meas- urements of transmembrane potentials. The best example is with *Drechslera (Helminthosporium) maydis* race T toxin (HMT toxin). A rapid membrane depolarization using a partially purified preparation of HMT toxin in roots of susceptible maize was found to be similar to depolarization with sodium azide. Simultaneously the uptake of K^+, Na^+, Cl^-, and PO_4^{3-} were inhibited (Mertz and Arntzen, 1977, 1978) Figure 3) and the observations were inter- preted as a sign of an inhibition of the electrogenic pump. However, when Kono and Daly (1979) fully purified the HMT toxin and it was used in similar experiments, the membranes of susceptible maize root cells depolarized only

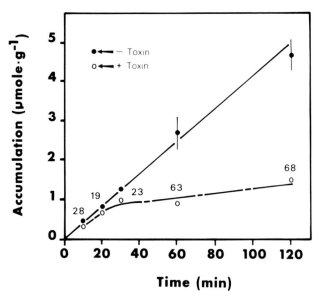

Figure 3. Effect of HMT toxin on potassium accumulation in roots of T-cytoplasm maize. Percentage inhibition of K^+ accumulation occurring during toxin treatment is given by a number near the datum. (●, Untreated; ○, toxin treated). [From Mertz and Arntzen (1977), *Plant Physiol.*, **60**, 363–369. Reproduced by permission of the American Society of Plant Physiologists]

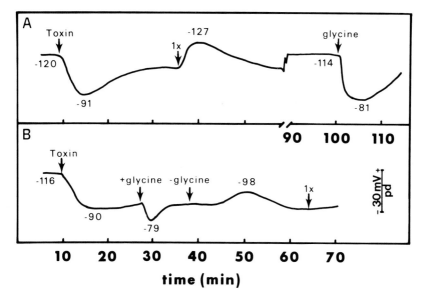

Figure 4. Effect of HMT toxin (1 μg cm^{-3}) on membrane potential of root cells of T-cytoplasm corn. Effects of glycine (50 mM) after toxin treatment (A) or during the toxin treatment (B). (From Novacky and Ball, unpublished work)

transiently (Novacky and Ball, 1982) (Figure 4). Additionally the efficiency of the H$^+$ pump during the exposure to this purified HMT toxin was tested. The normal depolarization and subsequent repolarization by such solutes as glycine is interpreted as H$^+$/solute co-transport followed by an increased activity of the H$^+$ pump (Novacky *et al.*, 1978) (Figure 1). An application of glycine to a cell of susceptible T hybrid corn root previously treated with HMT toxin still caused typical membrane depolarization (Figure 4A). Membranes also depolarized when glycine was applied during the maximum toxin depolarization (Figure 4B). The persistance of the glycine-induced depolarization in the presence of HMT-toxin strongly suggests that the activity of the electrogenic pump is not affected by the exposure to the toxin. The minimum toxin dose required to depolarize root/cell membranes was 0.25 μg cm^{-3} while the effective dose necessary for the inhibition of dark CO_2 fixation and mitochondrial oxidation was in the 5–50 ng cm^{-3} range (Payne *et al.*, 1980). Only longer treatments of root cells with higher toxin concentration (4 μg cm^{-3}) appeared to affect the pump. Membranes never repolarized in the presence of this higher toxin concentration (Frederico *et al.*, 1980).

It was proposed by Payne *et al.* (1980) (see also Daly, 1981) that HMT toxin acts as an ionophore. These compounds form channels (e.g. gramicidin) or function as 'carriers' (e.g. valinomycin) through membranes and facilitate transfer of ions, e.g. K$^+$ and H$^+$, thus collapsing the electrochemical gradients.

The various effects by HMT toxin on cell membranes, including the best known effect on mitochondria of T hybrid maize cells (Miller and Koeppe, 1971) and membrane depolarization (Figure 4A), appears to support this hypothesis. A sudden inflow of protons caused by the protonophoric activity of HMT toxin may change the intracellular H^+ concentration and depolarize the membrane. The cellular 'pH stat' mechanism (Smith and Raven, 1979), sensing this change in its pH, may stimulate the H^+ pump which then partially repolarizes the membrane. Upon removal of the toxin the pump, which is still in an 'activated' state, transiently hyperpolarizes the potential (Figure 4A). During extended treatments toxin may act at other sites in the cell and collapse the pH gradients necessary for generation of ATP in mitochondria and chloroplast membranes and thereby arrest ATP production. This would also effect the electrogenic pump. The attractive hypothesis of Payne *et al.* (1980) postulating ionophoric-like activity of HMT toxin awaits *in vitro* experiments with isolated membranes and ATPases.

If the protonophoric hypothesis concerning HMT toxin activity is correct, then the primary site of action of HMT toxin would more likely be at the lipid matrix rather than at the membrane protein. HMT toxin could penetrate any membrane of the cell, not only the plasma membrane. This would unify data from several studies on HMT toxin that suggest an effect on mitochondria, chloroplasts, and protoplasts (Daly, 1981). The question of specificity of this toxin for membranes of susceptible T hybrid corn cells and not the membranes of normal endosperm maize or membranes of other plant species needs further intensive investigation.

Daly's (1981) ionophore hypothesis concerning changes due to HMT toxin may explain the mode of action of other toxins such as the non-specific toxin

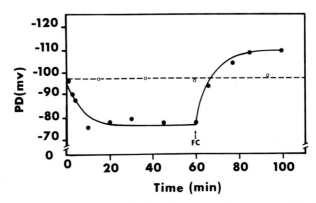

Figure 5. Effects of zearalenone and fusicoccin on membrane potential of corn root cells. Zearalenone was applied at zero time; fusicoccin (FC) (10^{-5} M) was applied at time indicated by arrow. [From Vianello and Macri (1978), *Planta*, **143**, 51–57. Reproduced by permission of Springer-Verlag]

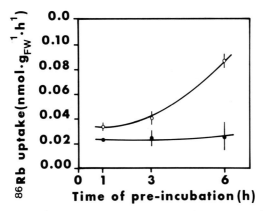

Figure 6. Effect of zearalenone on rubidium-86 uptake in corn roots (○, untreated; ●, zearalenone treated). [From Vianello and Macri (1978), *Planta*, **143**, 51–57. Reproduced by permission of Springer-Verlag]

zearalenone (F-2). This toxin is produced by several *Fusarium* spp. associated with maize and other cereals. Vianello and Macri (1978) presented comprehensive experimental evidence supporting the contention that the primary target site for zearalenone is the ATPase in the plasma membrane. The earliest effect of zearalenone (10 μg cm^{-3}) is a rapid depolarization of the plasma membrane (Figure 5). Another experiment designed to demonstrate alteration of plasma membrane by F-2 is an increased ion leakage detected 30 min after toxin application; the inhibition of Rb$^+$ uptake (denoting the cation species) was detected at 60 min (Figure 6) and a change in H$^+$ extrusion was detected after 90 min (Figure 7). Vianello and Macri (1978) also tested the

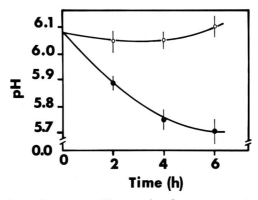

Figure 7. Effect of zearalenone on H$^+$ extrusion from corn roots measured as pH change in the bathing solution (●, control; ○, zearalenone treated). [From Vianello and Macri (1978), *Planta*, **143**, 51–57. Reproduced by permission from Springer-Verlag]

effect of zearalenone on the isolated ATPase and found 35% inhibition. Fusicoccin (see Section 2.2.2) stimulated the electrogenic pump even though treated with zearalenone (Figure 5). This may be interpreted as evidence that zearalenone does not inhibit the main proton channel through the ATPase complex or possibly that fusicoccin can bypass the inhibition.

2.2.2. *Stimulation of electrogenic pump*

Membrane transport processes may be stimulated in some instances by the attack of pathogens. The most reasonable explanation for the enhanced uptake of 3-*O*-methylglucose (Figure 8), L-amino acids, and 2-aminobutyrate in squash hypocotyls infected with *Hypomyces solani* (Hancock, 1975) is an activation of the electrogenic pump which stimulates the H^+/solute co-transport system of the infected tissue. No fungal metabolites responsible for this stimulation have been identified so far. Hancock and Huisman (1981) suggested that the activation of uptake could be a defence machanism of the plant to withdraw sugars and amino acids from the intercellular space, thus making the interior of the plant a less acceptable environment for pathogens.

There are few cases where the increased membrane transport is related to toxic metabolites of pathogens. The host-specific toxin produced by *Dreschlera* (= *Helminthosporium) carbonum* (HC toxin), the non-specific toxins (cotylenins) produced by *Cladosporium* sp. and the non-specific toxin (fusicoccin) produced by *Fusicoccum amygdali* all caused a significant increase in solute uptake (Marrè, 1980).

Fusicoccin is the best known mycotoxin of this group because its growth-promoting characteristics are similar to those of auxins (Marrè, 1979). Fusicoccin stimulation of the plasma membrane ATPase causes activation of H^+ efflux, resulting in hyperpolarization of the membrane potential (Marrè *et*

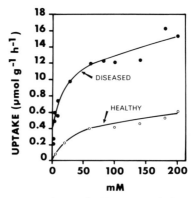

Figure 8. Uptake of 3-*o*-methylglucose in *Hypomyces*-infected squash hypocotyls (○, healthy; ●, diseased). [From Hancock (1979), *Plant Physiol.*, **44**, 1267–1272. Reproduced by permission of the American Society of Plant Physiologists]

al., 1974). This stimulates those transport processes which depend on the membrane potential such as the K⁺ uptake (see Figure 1). An increased H⁺ efflux then accounts for the increase of H⁺ co-transport with sugars and amino acids (Böcher *et al.*, 1980).

In vivo observation of stimulation by fusicoccin is complemented by *in vitro* experiments with the binding of [³H]fusicoccin to isolated fractions of the plasma membrane (Dohrmann *et al.*, 1977). These experiments strongly indicate that the site of fusicoccin activity is the plasma membrane multiunit ATPase complex. Treatments which decrease the intracellular ATP level and thus inactivate the electrogenic pump often severely retard fusicoccin stimulation (Marré, 1980). However, under long-day growing conditions which deplete the level of ATP in *Lemna gibba* and consequently stall the electrogenic pump, the addition of fusicoccin can still activate the pump resulting in membrane hyperpolarization (Böcher *et al.*, 1980) (Figure 9). This may mean that fusicoccin mobilizes ATP from previously unavailable sources.

The binding experiments of Tognoli *et al.* (1979) and Stout and Cleland (1980) indicate however, that the site of fusicoccin is not identical with the membrane ATPase complex: the fusicoccin–membrane protein complex can be separated electrophoretically from the main ATPase system. The site of action of fusicoccin is at best only at the periphery of the ATPase system. This

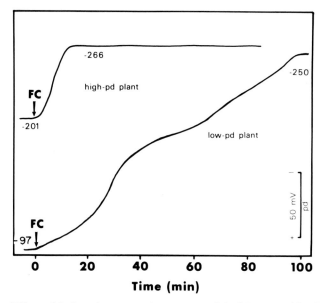

Figure 9. Effect of fusicoccin on membrane potential of *Lemna gibba* G1 plants with low (−97 mV) or high (−201 mV) membrane potentials (pd). [From Böcher *et al.* (1980), *Plant Sci. Lett.*, **18**, 215–220. Reproduced by permission of Elsevier/North Holland Scientific Publishers Ltd.]

notion is also supported by Chastain *et al.* (1981), who found that fusicoccin stimulation of ATPase takes place even if a specific inhibitor of ATPase appears to block or inactivate the major proton passage through ATPase.

As previously indicated by Stout and Cleland (1980), the natural function of the fusicoccin-binding protein is still unknown. *Fusicoccum amygdali* attacks only a very limited number of hosts, but the metabolic product fusicoccin is active in all higher plants tested so far (Marrè, 1979). This may indicate that fusicoccin mimics some still undiscovered growth regulator. Perhaps *F. amygdali* and its product fusicoccin is a unique exception among plant pathogens in producing specific membrane-affecting molecules. We know more about fusicoccin simply because fusicoccin has attracted the attention of plant pathologists and physiologists more than any other metabolite produced by plant pathogens. However, it may be a more or less typical case of disease development in which the pathogen directly affects the membrane function of host cells.

3. CONCLUSION

This chapter has outlined many host–pathogen associations where disease development is manifested by changes in the function of cell membranes. In addition to plant pathogens that produce host-specific toxins many studies have been concerned with the bacterially induced hypersensitive reaction. In this situation as well as with host-specific toxins the alterations of the plasma membrane were postulated and documented by many investigators (for more details, see e.g. Sequeira, 1976; Goodman, 1972). The experimental results and ultrastructural observations suggest changes in cell membranes. However, there are many unanswered questions, the plasma membrane has not been uniquivocally demonstrated to be the target site. Many of these questions will need much additional research before they can be answered.

New insight into this fascinating field of the role of membranes in disease initiation no doubt will be brought about by the introduction of innovative techniques such as the *in vitro* study of ATPase in microsomal vesicles (Sze and Churchill, 1981) and the use of the radiolabelled mycotoxins as well as synthetic toxins (Suzuki *et al.,* 1981).

REFERENCES

Bateman, D. F., and Basham, H. G. (1976). Degradation of plant cell walls and membranes by microbial enzymes, in *Encyclopedia of Plant Physiology*, New Series, Vol. 4 (Eds R. Heitefuss and P. H. Williams), Springer Verlag, Berlin, Heidelberg, and New York, pp. 316–355.
Beck, J. S. (1980). *Biomembrane Fundamentals in Relation to Human Biology*, McGraw-Hill, New York.
Böcher, M., Fisher, E., Ullrich-Eberius, C., and Novacky, A. (1980). Effect of

fusicoccin on the membrane potential, on the uptake of glucose and glycine, and on the ATP level in *Lemna gibba* G1., *Plant Sci. Lett.*, **18**, 215–220.

Chastain, C. J., LaFayette, P. R., and Hanson, J. B. (1981). Action of protein synthesis inhibitors in blocking electrogenic H⁺ efflux from corn roots, *Plant Physiol.*, **67**, 832–835.

Christensen, H. N. (1975). *Biological Transport*, W. A. Benjamin, Reading, MA.

Dainty, J. (1976). Water relations in plant cells, in *Encyclopedia of Plant Physiology*, New Series, Vol. 2A (Eds U. Lüttge and M. G. Pitman), Springer Verlag, Berlin, Heidelberg, and New York, pp. 12–35.

Daly, J. M. (1981). Mechanism of action, in *Toxins in Plant Disease* (Ed. R. D. Durbin), Academic Press, New York, pp. 331–394.

Dhindsa, R. S., Plumb-Dhindsa, P., and Thorpe, T. A. (1981). Leaf senescence: correlated with increased levels of membrane permeability and lipid peroxidations, and decreased levels of superoxides dismutase and catalase, *J. Exp. Bot.*, **32**, 93–101.

Dohrmann, U., Hertel, R., Pesci, P., Cocucci, S. M., Marrè, E., Randazzo, G., and Ballio, A. (1977). Localization of *in vitro* binding of the fungal toxin fusicoccin to plasma-membrane-rich fractions from corn coleoptiles, *Plant Sci. Lett.*, **9**, 291–299.

Farkas, G. (1978). Senescence and plant disease, in *Plant Disease, An Advanced Treatise*, Vol. 3 (Eds J. G. Horsfall and E. B. Cowling), Academic Press, New York, pp. 391–412.

Findlay, G. P., and Hope, A. B. (1976). Electrical properties of plant cells: methods and findings, in *Encyclopedia of Plant Physiolocy*, New Series, Vol. 2A (Eds U. Lüttge and M. G. Pitman) Springer Verlag, Berlin, Heidelberg, and New York, pp. 53–92.

Frederico, R., Scalia, S., Ballio, A., Cocucci, M., Ballarin-Denti, A., and Marrè, E. (1980). Inhibition of fusicoccin-inicied electrogenic proton extrusion in susceptible maize by *Helminthosporium maydis* race T toxin, *Plant Sci. Lett.*, **17**, 129–134.

Galliard, T. (1978). Lipolytic and lipoxygenase enzymes in plants and their action in wounded tissues, in *Biochemistry of Wounded Plant Tissues* (Ed. G. Kahl), Walter de Gruyter, Berlin, and New York, pp. 155–201.

Galliard, T., and Chan, H. W.-S. (1980). Lipoxygenases, in *Biochemistry of Plants*, Vol. 4 (Ed. P. K. Stumpf), Academic press, New York, pp. 131–161.

Goodman, R. N. (1972). Electolyte leakage and membrane damage in relation to bacterial population, pH, and ammonia production in tobacco leaf tissue inoculated with *Pseudomonas pisi, Phytopathology*, **62**, 1327–1331.

Hanchey, P., and Wheeler, H. (1979). The role of host cell membranes, in *Recognition and Specificity, in Plant Host–Pathogen Interactions* (Eds J. M. Daly and I. Uritani), University Park Press, Baltimore, pp. 193–210.

Hancock, J. G. (1975). Influence of *Hypomyces* infection and excision on amino acid and 3-*O*-methylglucose uptake by squash hypocotyls, *Physiol. Plant Pathol.*, **6**, 65–73.

Hancock, J. G., and Huisman, O. C. (1981). Nutrient movement in host–pathogen systems, *Annu. Rev. Phytopathol.*, **19**, 309–333.

Higinbotham, N., Graves, J. S., and Davis, R. F. (1970). Evidence for an electrogenic ion transport pump in cells of higher plants, *J. Membrane Biol.*, **3**, 210–222.

Hodges, T. K. (1976). ATPases associated with membranes of plant cells, in *Encyclopedia of Plant Physiology*, New Series, Vol. 2A (Eds U. Lüttge and M. G. Pitman) Springer Verlag, Berlin, Heidelberg, and New York, pp. 260–283.

Hope, A. B., and Walker, N. A. (1975). *The Physiology of Giant Algal Cells*, Cambridge University Press, Cambridge.

Hoppe, H. H., and Heitefuss, R. (1974a). Permeability and membrane lipid metabolism of *Phaseolus vulgaris* infected with *Uromyces phaseoli*. II. Changes in lipid concentration and ^{32}P incorporation into phospholipids, *Physiol. Plant Pathol.*, **4**, 11–24.

Hoppe, H. H., and Heitefuss, R. (1974b). Permeability and membrane lipid metabolism of *Phaseolus vulgaris* infected with *Uromyces phaseoli*. III. Changes in relative concentration of lipid-bound fatty acids and phospholipase activity, *Physiol. Plant Pathol.*, **4**, 25–36.

Hoppe, H. H., and Heitefuss, R. (1975). Permeability and membrane lipid metabolism of *Phaseolus vulgaris* infected with *Uromyces phaseoli*. IV. Phospholipids and phospholipid fatty acids in healthy and rust-infected bean leaves resistant and susceptible to *Uromyces phaseoli, Physiol. Plant Pathol.*, **5**, 263–271.

Inoue, I., Tetsuo, U., and Kobatake, Y. (1973). Structure of excitable membranes formed on the surface of protoplasmic drops isolated from *Nitella*. I. Conformation of surface membrane determined from the reactive index and from enzyme actions, *Biochim. Biophys. Acta*, **298**, 653–663.

Jones, M. G. K., Novacky, A., and Dropkin, V. H. (1975). Transmembrane potentials of parenchyma cells and nematode-induced transfer cells, *Protoplasma*, **85**, 15–37.

Kono, Y., and Daly, J. M. (1979). Characterization of the host-specific pathotoxin produced by *Helminthosporium maydis*, race T, affecting corn with Texas male sterile cytoplasm, *Bioorg. Chem.*, **8**, 391–397.

Kotyk, A., and Janáček, K. (1979). *Membrane Transport*, Academia, Prague.

Leonard, R. T., and Hodges, T. K. (1980). The plasma membrane, in *Biochemistry of Plants*, Vol. 1 (Ed. N. E. Tolbert), Academic Press, New York, pp. 163–182.

Lupu, R., Grossman, S., and Cohen, Y. (1980). The involvement of lipoxygenase and antioxidants in pathogenesis of powdery mildew on tobacco plants, *Physiol. Plant Pathol.*, **16**, 241–248.

Lyon, G. D. (1980). Evidence that the toxic effect of rishitin may be due to membrane damage, *J. Exp. Bot.*, **31**, 957–966.

Marrè, E. (1979). Fusicoccin: a tool in plant physiology, *Annu. Rev. Plant Physiol.*, **30**, 273–288.

Marrè, E. (1980). Mechanism of action of phytotoxins affecting plasmalemma function, *Prog. Phytochem.*, **6**, 253–284.

Marrè, E., Lado, P., Ferroni, A., and Ballarin-Denti, A. (1974). Transmembrane potential increase induced by auxin, benzyladenine and fusicoccin. Correlation with proton extrusion and cell enlargement, *Plant Sci. Lett.*, **2**, 257–265.

McKersie, B. C., and Thompson, J. E. (1977). Lipid crystalization in senescent membranes from cotyledons, *Plant Physiol.*, **59**, 805–807.

McKersie, B. D., and Thompson, J. E. (1978). Phase behavior of chloroplast and microsomal membranes during leaf senescence, *Plant Physiol.*, **61**, 639–643.

Mertz, S. M., and Arntzen, C. J. (1977). Selective inhibition of K^+, Na^+, Cl^- and PO_4^{3-} uptake in *Zea mays* L. by *Bipolaris (Helminthosporium) maydis* race T pathotoxin, *Plant Physiol.*, **60**, 363–369.

Mertz, S. M., and Arntzen, C. J. (1978). Depolarization of the electogenic transmembrane electropotential of *Zea mays* L. by *Bipolaris (Helminthosporium) maydis* race T toxin, azide, cyanide, dodecylsuccinic acid, or cold temperature, *Plant Physiol.*, **62**, 781–783.

Miller, R. J., and Koeppe, D. E. (1971). Southern corn leaf blight: susceptible and resistant mitochondria, *Science*, **173**, 67–69.

Mitchell, P. (1967). Translocation through natural membranes, *Adv. Enzymol.*, **29**, 33–87.

Novacky, A. (1980). Disease-related alteration in membrane function, in *Plant Membrane Transport: Current Conceptual Issues* (Eds R. M. Spanswick, W. J. Lucas and J. Dainty, Elsevier/North Holland Biomedical Press, Amsterdam, pp. 369–378.

Novacky, A., and Ball, E. (1983). Membrane potential in susceptible maize cells treated with *Helminthosporium maydis* race T toxin, *Plant Sci. Let.*, Submitted for publication.

Novacky, A., and Ullrich-Eberius, C. I. (1982). Relationship between membrane potential and ATP level in *Xanthomonas malvacearum* infected cotton cotyledons, *Physiol. Plant Pathol.*, **21**, 237–249.

Novacky, A., Karr, L. A., and Van Sambeek, J. W. (1976). Using electrophysiology to study plant disease development, *BioScience*, **26**, 499–504.

Novacky, A. Ullrich-Eberius, C. I., and Lüttge, U. (1978). Membrane potential changes during transport of hexoses in *Lemna gibba* G1., *Planta*, **138**, 263–270.

Payne, G. Kono, Y., and Daly, J. M. (1980). A comparison of purified host-specific toxin from *Helminthosporium maydis*, race T, and its acetate derivative on oxidation by mitochondria from susceptible and resistant plants, *Plant Physiol.*, **65**, 785–791.

Pauls, K. P., and Thompson, J. E. (1980). *In vitro* stimulation of senescence-related membrane damage by ozone-induced lipid peroxidation, *Nature (London)*, **283**, 504–506.

Poole, R. J. (1978). Energy coupling for membrane transport, *Annu. Rev. Plant Physiol.*, **29**, 437–460.

Raison, J. K. (1980). Membrane lipids: structure and function, in *Biochemistry of Plants*, Vol 4. (Ed. P. K. Stumf), Academic Press, New York, pp. 57–83.

Scheffer, R. P. (1976). Host-specific toxins in relation to pathogenesis and disease resistance, in *Encyclopedia of Plant Physiology*, New Series, Vol. 4 (Eds R. Heitefuss and P. H. Williams), Springer Verlag, Berlin, Heidelberg, and New York, pp. 247–269.

Sequeira, L. (1976). Induction and suppression of the hypersensitive reaction caused by phytopathogenic bacteria: specific and non-specific components, in *Specificity in Plant Diseases* Eds R. K. S. Wood and A. Graniti), Plenum Press, New York, pp. 289–306.

Shiraishi, T., Oku, H., Isono, M., and Ouchi, S. (1975). The injurious effect of pisatin on the plasma membrane of pea, *Plant Cell Physiol.*, **16**, 939–942.

Simon, E. W. (1974). Phospholipids and plant membrane permeability, *New Phytol.*, **73**, 377–420.

Singer, S. J., and Nicolson, G. L. (1972). The fluid mosaic model of the structure of cell membranes, *Science*, **175**, 720–731.

Smith, F. A., and Raven, J. A. (1979). Intracellular pH and its regulation, *Annu. Rev. Plant Physiol.*, **30**, 289–311.

Spanswick, R. M. (1981). Electrogenic ion pumps, *Annu. Rev. Plant Physiol.*, **32**, 267–289.

Stadelmann, E. J. (1966). Evaluation of turgidity, plasmolysis and deplasmolysis of plant cells, in *Methods in Cell Physiology*, Vol. 2 (Ed. D. M. Prescott), Academic Press, New York, pp. 143–216.

Steele, J. A., Uchytil, T. F., Durbin, R. D., Bhatnaga, P., and Rich, D. H. (1976). Chloroplast coupling factor I: a species-specific receptor for tentoxin, *Proc. Natl. Acad. Sci. USA*, **73**, 2245–2248.

Stout, R. G., and Cleland, R. E. (1980). Partial characterization of fusicoccin binding to receptor sites on oat root membranes, *Plant Physiol.*, **66**, 353–359.

Suzuki, Y., Tegtmeier, K., Daly, J. M., and Knoche, H. W. (1981). Comparison of *Helminthosporium maydis* race T toxin to synthetic compounds for specificity and toxicity, *Phytopathology*, **71**, 907 (Abstr).

Sze, H., and Churchill, K. A. (1981). $Mg^{2+}/KC1$-ATPase of plant plasma membranes is an electrogenic pump, *Proc. Natl. Acad. Sci. USA*, **78**, 5578–5582.

Thatcher, F. S. (1939). Osmotic and permeability relations in the nutrition of fungus parasites, *Am. J. Bot.*, **26**, 449–458.

Thatcher, F. S. (1942). Futher studies of osmotic and permeability relation in parasitism, *Can. J. Res. Sect. C.,* **20,** 283–311.

Tognoli, L., Beffagna, N., Pesci, P., and Marrè, E. (1979). On the relationship between ATPase and fusicoccin binding capacity of crude and partially purified microsomal preparations from maize coleoptile, *Plant Sci. Lett.,* **16,** 1–14.

Turner, J. G. (1976). The non-electrolyte permeability of tobacco cell membranes in tissues inoculated with *Pseudomonas pisi* and *Pseudomonas tabaci, PhD* Thesis, University of Missouri, Columbia.

Turner, J. G., and Novacky, A. (1976). Effect of *Pseudomonas tabaci* and *P. pisi* on permeability of tobacco protoplasts to nonelectrolytes, *Proc. Am. Phytopathol. Soc.,* **3,** 260–261.

Van Etten, H. D., and Bateman, D. F. (1971). Studies on the mode of action of the phytoalexin phaseolin, *Phytopathology,* **61,** 1363–1372.

Vianello, A., and Macri, F. (1978). Inhibition of plant cell membrane transport phenomena induced by zearalenone (F-2), *Planta,* **143,** 51–57.

Walton, J. D., Earle, E. D., Yoder, O. C., and Spanswick, R. M. (1979). Reduction of adenosine triphosphate levels in susceptible maize mesophyll protoplasts by *Helminthosporium maydis* race T toxin, *Plant Physiol.,* **63,** 806–810.

Wheeler, H. (1978). Disease alterations in permeability and membranes, in *Plant Disease, An Advanced Treatise,* Vol. 3 (Eds J. G. Horsfall and E. B. Cowling), Academic Press, New York, pp. 327–347.

Wheeler, H. (1981). Role in pathogenesis, in *Toxins in Plant Disease* (Ed. R. D. Durbin), Academic Press, New York, pp. 477–494.

Wheeler, H., and Hanchey, P. (1968). Permeability phenomena in plant disease, *Annu. Rev. Phytopathol.,* **6,** 331–350.

White, J. A., Calvert, O. H., and Brown, M. F. (1973). Ultrastructural changes in corn leaves after inoculation with *Helminthosporium maydis,* race T, *Phytopathology,* **63,** 296–300.

Zimmermann, U. (1977). Cell turgor pressure regulation and turgor pressure-mediated transport processes, in *Integration of Activity in Higher Plants* (Ed. D. H. Jennings), *Soc. Exp. Biol. Symp.,* No. 31, pp. 117–154.

Zimmermann, U., and Steudle, E. (1978). Physical aspects of water relations of plant cells, *Adv. Bot. Res.,* **6,** 45–117.

Further Reading

Christensen, H. N. (1975). *Biological Transport,* W. A. Benjamin, Reading, Ma.

Hope, A. B., and Walker, N. A. (1975). *The Physiology of Giant Algal Cells.* Cambridge University Press, Cambridge.

Kotyk, A., and Janáček, K. (1979). *Membrane Transport,* Academia, Prague.

Lüttage, U., and Higinbotham, N. (1979). *Transport in Plants,* Springer Verlag, New York, Heidelberg and Berlin.

Lüttge, U., and Pittman, M. G. (Eds) (1976). Transport in plants. II. Part A, Cells: Part B, tissues and organs, in *Encyclopedia of Plant Physiology,* New Series (Eds A. Pirson and H. H. Zimmermann), Springer Verlag, Berlin, Heidelberg, and New York, pp.

Nobel, P. (1974). *Introduction to Biophysical Plant Physiology,* Freeman, San Francisco.

Biochemical Plant Pathology
Edited by J. A. Callow
© 1983 John Wiley & Sons Ltd

17

Phenylpropanoid Metabolism and its Regulation in Disease

M. LEGRAND

Laboratoire de Virologie, Institut de Biologie Moléculaire et Cellulaire, 15 rue Descartes, 67000 Strasbourg, France

1. INTRODUCTION

General phenylpropanoid metabolism leads to the formation of activated cinnamic acids which are the precursors of numerous naturally occurring compounds, namely flavonoids, lignin, and various bound forms of cinnamic acids. The activated cinnamic acids are derived biosynthetically from phenylalanine and all except one of the enzymes involved in this pathway are

367

now known. Phenylpropanoid metabolism is affected by a wide variety of stimuli, for example light, wounding, chemicals, and microbial or viral infection. The activation of the pathway in diseased plants has often been implicated as a mechanism of resistance of the host against the pathogen. In recent years considerable progress has been made in the understanding of regulation of enzymatic activities involved in the pathway.

The objectives of this chapter are (a) to describe the biosynthetic pathway of activated cinnamic acids and their role as intermediates in plant metabolism, (b) to focus on recent progress in enzymatic steps of the pathway, (c) to discuss some experimental techniques which are available for regulation studies, and (d) to discuss more deeply some selected papers dealing with regulation mechanisms of the pathway in diseased plants.

2. PHENYLPROPANOID METABOLISM

2.1. The general phenylpropanoid pathway

The general phenylpropanoid pathway can be defined as the metabolic sequence leading to the synthesis of activated cinnamic acids which are the precursors for a wide variety of compounds. Here we shall describe briefly the sequence of reactions involved in this pathway, but further details can be found in more specialized reviews (Amrhein and Zenk, 1977; Grisebach, 1977; Gross, 1977, 1978; Towers and Chi-Kit Wat, 1979).

The sequence of enzymatic reactions leading from phenylalanine to activated cinnamic acids is presented in Figure 1. With the exception of the still hypothetical ferulic acid 5-hydroxylase, all of the enzymes involved in the pathway have been characterized. The first reaction of the pathway is the deamination of phenylalanine to *trans*-cinnamic acid, which is catalysed by phenylalanine ammonia-lyase (PAL). Cinnamic acid is then converted to *p*-coumaric acid by *trans*-cinnamic acid 4-monooxygenase, a membrane-bound enzyme which is associated with cytochrome P-450 and requires NADPH as electron donor. Preparations from many plants and fungi have been shown to catalyse directly the formation of *p*-coumaric acid from tyrosine (Camm and Towers, 1973). *p*-Coumaric acid is subsequently hydroxylated by *p*-coumaric acid 3-monooxygenase to yield caffeic acid. Caffeic acid is then methylated in the 3-position to ferulic acid by *o*-diphenol *O*-methyltransferases (OMTs), which require *S*-adenosylmethionine as methyl donor. These enzymes have been studied in detail in tobacco leaves (Legrand *et al.*, 1976, 1978; Collendavelloo *et al.*, 1981), in tulip anthers (Sütfeld and Wiermann, 1978), and in cell suspension cultures of soybean (Poulton *et al.*, 1976) and tobacco (Tsang and Ibrahim, 1979). It appears that OMTs of plants differ greatly in their number, their physico-chemical properties, and their specificity towards the numerous *o*-diphenolic substrates. The same enzymatic system catalyses the methylation of 5-hydroxyferulic acid to

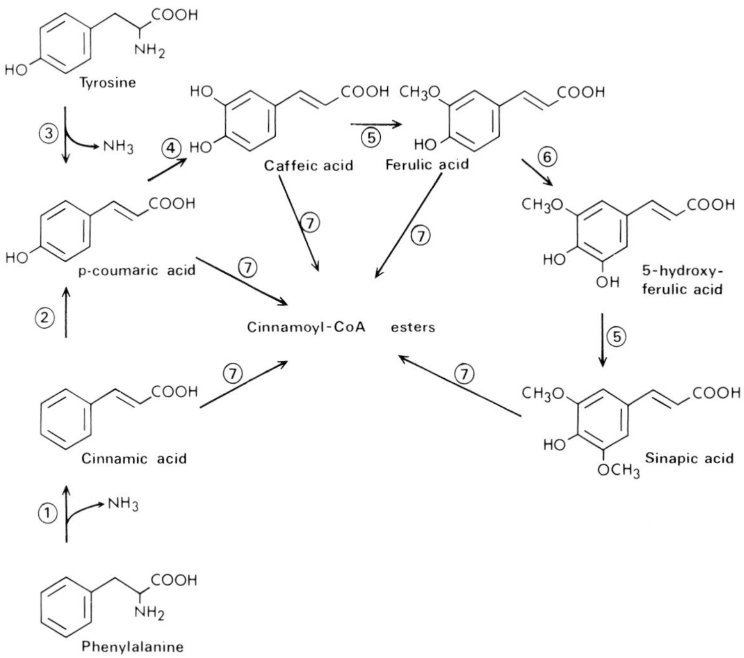

Figure 1. The general phenylpropanoid pathway. 1, Phenylalanine ammonia-lyase; 2, *trans*-cinnamic acid-4-monooxygenase; 3, tyrosine ammonia-lyase; 4, *p*-coumaric acid-3-monooxygenase; 5, *o*-diphenol-*O*-methyltransferases; 6, ferulic acid-5-hydroxylase (hypothetical); 7, cinnamic acids : CoA ligase

sinapic acid (reaction in Figure 1). After reduction to the corresponding alcohols (see Figure 2), ferulic and sinapic acids are incorporated into lignins. Higuchi *et al.* (1977) proposed that the relative affinities of OMTs for caffeic and 5-hydroxyferulic acids represent one of the possible biochemical explanations for the known differences between the lignin from gymnosperms and angiosperms. The last step of general phenylpropanoid metabolism, the formation of cinnamoyl-CoA esters, is mediated by cinnamic acids: CoA ligases. It is an endergonic reaction requiring ATP which undergoes cleavage to AMP and pyrophosphate. In a number of plants ligases with different affinities for the different hydroxycinnamic acids of the pathway have been detected. In a few cases isoenzymes have been characterized (for literature see Gross, 1977).

2.2. Cinnamoyl-CoA esters as central intermediates

Cinnamoyl-CoA esters are the precursors of cinnamyl alcohols, flavonoids, and numerous esters and amides. Figure 2 shows the biosynthetic routes of

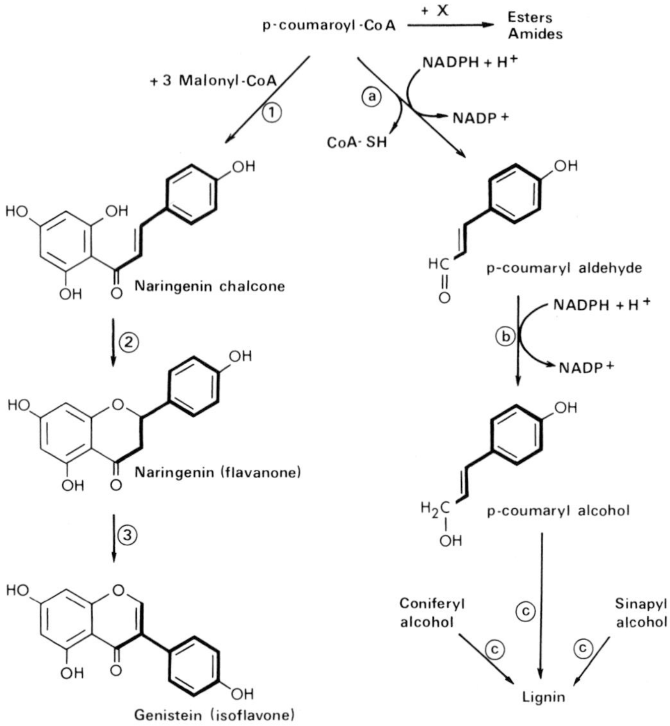

Figure 2. The biosynthetic routes of *p*-coumaroyl-CoA ester. 1, Chalcone synthase; 2, chalcone–flavanone isomerase; 3, this reaction has not been demonstrated. (a) Cinnamoyl CoA : NADP$^+$ oxidoreductase; (b) cinnamyl alcohol dehydrogenase; (c) peroxidase. The carbon skeleton coming from *p*-coumaric acid is outlined with thick bonds.

p-coumaroyl-CoA which have been demonstrated by enzymatic studies *in vitro* (except for 3).

   p-Coumaroyl-CoA can enter the pathway of flavonoids by condensation with three molecules of malonyl-CoA. The product of the reaction was thought to be the flavanone naringenin but has been shown recently to be the chalcone (Heller and Hahlbrock, 1980; Sütfeld and Wiermann, 1981). Naringenin chalcone is then converted by an isomerase to the flavanone naringenin. The flavanone is a branch point between the anthocyanidins and the isoflavonoids, a subclass of flavonoids of great interest in phytopathology. Isoflavonoids have been demonstrated to behave as phytoalexins in many Leguminosae. Unfortunately, no enzymes are known which can catalyse the synthesis of 4′, 7-dihydroxyflavanone (the precursor of numerous isoflavonoids with 4′, 7-substitution) and the rearrangement of flavanones to isoflavones. A metabolic grid leading to the different isoflavonoids has been postulated from radiolabelling experiments (Grisebach and Ebel, 1978; VanEtten and Pueppke, 1976) but the enzymes involved in these reactions are unknown. Recently the prenylation of 3, 6α, 9-trihydroxypterocarpan has been obtained with cell-free extracts from soybean cotyledons (Zähringer *et al.*, 1979). The prenylated products are the precursors of glyceollin isomers, the phytoalexins of soybean. The prenyltransferase was detected only in preparations from elicitor treated cotyledons and not from wounded cotyledons, indicating that the enzyme activity must be induced by elicitor application.

Another possible fate for the *p*-coumaroyl-CoA ester is reduction to *p*-coumaryl alcohol, which is one of the three building units of lignin (Figure 2). Two enzymes are involved in this process: (1) cinnamoyl-CoA: $NADP^+$ oxidoreductase, producing the corresponding aldehyde from the thioester, and (2) cinnamyl alcohol dehydrogenase, reducing the aldehyde to alcohol (Gross *et al.*, 1973; Rhodes and Wooltorton, 1975; Gross, 1977; Grisebach, 1977). The same enzyme activities catalyse the reduction of feruloyl-CoA and sinapoyl-CoA esters to coniferyl and sinapyl alcohols, respectively. The final step of lignin synthesis is the polymerization of the three cinnamyl alcohols mediated by the peroxidase–H_2O_2 system. Cell wall-bound peroxidases are likely to be involved in both the required H_2O_2 production and in the oxidation of the alcohols to free radicals which polymerize into lignin. (for more details and literature see Gross, 1978).

Finally, thioesters of cinnamic acids may lead to the synthesis of conjugates like esters of quinic or shikimic acids. The transferases catalysing these reactions have recently been found in cell-free extracts (Ulbrich and Zenk, 1979, 1980).

3. REGULATION OF PHENYLPROPANOID SYNTHESIS DURING DISEASE

3.1. Induced increase of phenylpropanoid production

There are numerous reports on the accumulation of various phenylpropanoid derivatives in plants as a response to infection. Recent reviews have dealt with the increased synthesis of phenolics (Friend, 1981), lignins (Friend, 1976, 1981; Vance, *et al.*, 1980; see also Chapter 11), and isoflavonoid phytoalexins (VanEtten and Pueppke, 1976; Grisebach and Ebel, 1978; Friend, 1981) in diseased plants. Here we shall discuss some selected papers in which measurements of enzyme activities involved in the pathway have been reported.

3.1.1. *Lignin*

Infection of *Raphanus sativus* by *Peronospora parasitica* increased the amount of lignin (Asada *et al.*, 1975). Chemical analysis demonstrated that the lignin extracted from infected material contained fewer methoxy groups than that from healthy material. This situation may arise from a lower synthesis of sinapyl alcohol compared with that of *p*-coumaryl and coniferyl alcohols as a result of infection. If this assumption is right, the regulating enzymes could be the *O*-methyltransferases, CoA ligases, CoA ester reductases, or alcohol dehydrogenases. Another biochemical explanation for the observed differences between lignin from healthy and infected materials would be a lower incorporation rate of sinapyl alcohol catalysed by peroxidases. In fact, 16 isoperoxidases have been isolated from roots of Japanese radish (Asada *et al.*, 1975). They have very different affinities for each of the three alcohols and variable concentrations of isoenzymes may account for the relative increase in guaiacyl lignin in diseased tissues. The lignified tissues showed an increase in PAL activity which could be blocked by metabolic inhibitors. Therefore the authors concluded that there was *de novo* synthesis of the enzyme induced by infection. This experimental approach will be discussed below.

Friend and collaborators studied the response of potato tubers to infection by *Phytophthora infestans*. They established that in the case of an incompatible host–parasiste interaction there was a relatively rapid deposition of lignin-like material compared with that in susceptible combination (Friend *et al.*, 1973; Friend, 1976). This lignin-like material yielded *p*-hydroxybenzaldehyde, vanillin, and syringaldehyde on alkaline nitrobenzene oxidation but did not stain with any common lignin stains. On alkaline hydrolysis it yielded *p*-coumaric and ferulic acids. Friend suggested that this material consists of these hydroxycinnamic acids esterified to a carbohydrate component of the cell wall. In the case of incompatible reactions increases of PAL and OMT activities have been shown to occur only in the area where the fungus was found (Friend and Thornton, 1974; Friend, 1976). Hence these increases in enzyme activities seem to be associated with the deposition of lignin-like material and with resistance of potato to *P. infestans*.

Lignified papilla formation has been proposed as a mechanism of resistance of *Phalaris arundinacea* to *Helminthosporium avenae* (Vance and Sherwood, 1976). Within 18 h after inoculation, infected leaves had a higher lignin content than uninoculated controls. Higher activities of PAL, hydroxycinnamic acid: CoA ligases, and peroxidases were detected in resistant tissues. In cycloheximide-treated leaves the increased synthesis of lignin and the enhanced activity of the enzymes were inhibited and the tissues became susceptible. The authors suggested that lignin biosynthesis at the site of attempted fungal penetration may play an important role in the resistance of reed canarygrass.

Localized lignin formation in the leaf epidermal cell wall has also been implicated as the mechanism of resistance of wheat to non-pathogens (Ride, 1975). The deposition of lignin was preceded by increases in PAL and OMT activities (Maule and Ride, 1976). The synthesis and properties of lignin in relation to fungal infection are considered in more detail in Chapter 11.

Phenylpropanoid metabolism is strikingly stimulated in virus-infected plants and increased lignification has been proposed as one of the processes that reduce spread of the virus from cell to cell (Wu, 1973; Conti *et al.*, 1974; Faulkner and Kimmins, 1975; Kimmins and Wuddah, 1977). Hypersensitive tobacco leaves infected by tobacco mosaic virus (TMV) show sharp increases in activities of PAL, CAH, CoA ligases, OMTs, and peroxidases (Legrand *et al.*, 1976). These stimulated activities are localized in the layers of cells surrounding the necrotic lesion and the enzymatic stimulus is spread radially in advance of necrosis. The zones of stimulation can be revealed by their fluorescence under UV radiation and their areas were taken as a measure of the number of stimulated cells. The amplitude of enzyme stimulation was evaluated in tobacco leaves infected by different strains of TMV inducing lesions of variable size. The ratio of the amplitude of total stimulation over the fluorescent areas was calculated and the values reflected the intensity of enzymatic stimulation at the level of one stimulated cell. These calculations showed that the intensity of stimulation per cell was correlated with the efficiency of the localizing mechanism that limits virus spread to the vicinity of lesions. O-Methyltransferase activity has been intensively studied in healthy and TMV-infected tobacco leaves (Legrand *et al.*, 1978; Collendavelloo *et al.*, 1981). Three enzymes, the activities of which are increased after infection, have been separated. The three enzymes have different affinities for various o-diphenolic substrates. One of them is specialized in the methylation of phenylpropanoid-type substrates (caffeic and 5-hydroxyferulic acids) whereas the other two accept a wide variety of substrates. The manner in which phenylpropanoid units are incorporated into lignin has already been described. Indeed, an increased synthesis of normal lignin has been demonstrated in infected tobacco leaves (Massala, 1981). The methylation of other substrates of OMTs (such as catechol or homocatechol) leads also to phenoxymethoxy products which are excellent candidates for oxidation by peroxidases into free radicals. These radicals exist in several resonance forms and can polymerize readily. Work is now in progress to investigate whether such lignin-like polymers are implicated in the mechanism of localization of the virus in hypersensitive tobacco.

The role of increased synthesis of phenylpropanoids in resistance of tobacco to TMV has been recently investigated by using aminooxy compounds (Massala *et al.*, 1980; Massala, 1981). These compounds have been shown to be competitive inhibitors of PAL *in vitro* and *in vivo* (Amrhein *et al.*, 1976). Tobacco leaves were supplied with aminooxyacetate (AOA) or α-aminooxy-β-phenylpropionic acid (AOPP). This treatment increased the size of the lesions and weakened the mechanism of localization of the virus, as shown by the presence of virus at the edges of the large lesions of treated leaves. It was shown that aminooxy compounds had inhibited the activity of PAL *in vivo* and had reduced the flux of phenylpropanoid derivatives without suppressing the increased synthesis of PAL and OMT induced by infection. This point is crucial because it demonstrates that the stimulus itself and protein synthesis in general are not affected by the treatment, in contrast with what is observed when metabolic inhibitors are supplied. Inhibitors of protein or nucleic acid synthesis such as actinomycin D and cycloheximide have been extensively used in phytopathology in order to investigate the role of a given metabolic pathway in the resistance of plants to pathogens. A parallel decrease in the activity of the studied pathway and in the level of resistance of the host upon treatment by metabolic inhibitors has often been taken as a proof of the involvement of this pathway in the resistance mechanism. This approach can be criticized because of its lack of specificity; any primary mechanism responsible for resistance and requiring protein synthesis would also be inhibited after treatment by metabolic inhibitors. In contrast, the feeding of aminooxy compounds specifically decreased the flow towards phenylpropanoids by inhibiting PAL activity *in vivo*. The data obtained with PAL inhibitors strongly support the argument for an active role of the enhanced production of phenylpropanoid derivatives in hypersensitive resistance to viral infection.

3.1.2. *Isoflavonoids*

There are numerous reports in the literature on the accumulation of isoflavonoids in infected plants (for references see VanEtten and Pueppke, 1976; Friend, 1981), but few of these papers deal with the regulation of isoflavonoid biosynthesis. Progress has recently been made in studies on the interaction between soybean and *Phytophthora megasperma* var. *sojae* (Pms) (see also Chapters 2 and 13). Increases in PAL, chalcone–flavanone isomerase, and peroxidase activities have been detected in resistant or susceptible cultivars of soybean after wounding or inoculation of hypocotyls by Pms (Partridge and Keen, 1977). The authors concluded that the increases in enzymatic activities were due to mechanical damage of the plant tissue. Treatment of cell suspension cultures by a glucan elicitor isolated from Pms cell walls stimulated PAL activity and glyceollin accumulation (Ebel *et al.*, 1976; see also Chapters 2 and 13). Pms elicitor induced incorporation of mevalonate into glyceollin and increased the activities of PAL and chalcone synthase in soybean cotyledons (Zähringer *et al.*, 1978). In contrast with the results obtained with hypocotyls (Partridge and Keen, 1977), no changes in enzyme activities were found in wounded control cotyledons. Incubation of the elicitor-treated cotyledons with an aminooxy compound reduced the level of glyceollin (Grisebach and Ebel, 1978). This demonstrates that *in vivo* PAL activitiy controls, at least partially, glyceollin production in soybean cotyledons.

Whether PAL activity controls phytoalexin production in plant–parasite combinations has been debated because mechanical damage itself can enhance PAL activity (Camm and Towers, 1973). In several host–parasite systems increases in the activity of PAL have been associated with phytoalexin production (Ebel *et al.*, 1976; Grisebach and Ebel, 1978; Gustine *et al.*, 1978; Zähringer *et al.*, 1978; Dixon and Lamb, 1979), but work on other host–parasite models is needed in order to establish whether this mechanism of isoflavonoid production is general.

Concerning chalcone synthase, its activity has been shown to be stimulated by elicitor treatment in soybean cotyledons (Zähringer *et al.*, 1978) but not in parsley cells (Hahlbrock *et al.*, 1981). Unfortunately, no data are available on the effect of elicitor on cell cultures from soybean in which glyceollin accumulates (Ebel *et al.*, 1976). Moreover, the channelling towards one isoflavonoid or another must be controlled through an enzyme involved in a later reaction of the pathway. Since very little is known of the enzymology of this part of the pathway, it is impossible at present to have a general view on the regulation of isoflavonoid synthesis.

Recently, Grisebach's group has succeeded in introducing an isopentenyl substituent into the isoflavone genistein (see Figure 2) and 2'-hydroxygenistein using cell-free extracts from *Lupinus albus* (Schröder *et al.*, 1979). In further work the same group characterized the transferase which catalyses the prenylation of 3,6α,9-trihydroxypterocarpan in cell-free preparations from soybean cotyledons. The transferase activity is induced by Pms elicitor treatment and has not been detected in wounded cotyledons (Zähringer *et al.*, 1979). The prenylated products are presumed to be the immediate biosynthetic precursors of glyceollin isomers. The rates of synthesis of 3,6α,9-trihydroxypterocarpan, the substrate of the transferase, have been determined (Moesta and Grisebach, 1980). The rates increased after induction with both biotic (glucan) and abiotic (Hg^{2+}) elicitors and this is in agreement with the role of trihydroxypterocarpan as an intermediate in glyceollin biosynthesis and with an increased rate of synthesis of glyceollin after treatment by elicitors. Somewhat different conclusions on the effects of biotic and abiotic elicitors have been reached by Yoshikawa, and are discussed in Section 3.4, Chapter 13.

At localized sites of infection of soybean by *P. megasperma sojae*, Yoshikawa *et al.* (1978) observed higher levels of glyceollin in incompatible than in compatible combinations. Pulse-labelling and chase experiments were used to study tentatively the rates of glyceollin biosynthesis and degradation (Yoshikawa *et al.*, 1979; Moesta and Grisebach, 1981). No differences in labelling between compatible and incompatible interactions were detected after pulsing with [^{14}C]phenylalanine and it was suggested that the differences in glyceollin accumulation were due to differences in the rates of glyceollin degradation (Yoshikawa *et al.*, 1979; see also Section 3.4, Chapter 13). However, one criticism of this interpretation is that in these experiments, radioactivity in glyceollin could not be chased after incubation for a few hours with non-labelled phenylalanine and it even increased in resistant hypocotyls. This is a typical situation encountered in a system which is not in a steady state: not only the pool of glyceollin but also the pools of intermediates are likely to be increased by infection. These pools of variable size can dilute or trap the radioactivity to a different extent in the different materials, which hampers the interpretation of the data. Finally, it is always difficult to draw conclusions from negative chase experiments. These problems relating to interpretation of pulse-chase experiments are illustrated by the fact that Moesta and Grisebach (1981), who used $^{14}CO_2$ as precursor and performed chase experiments with air, came to different conclusions. In fact, it should be recognized that pulse-labelling and chase experiments provide unequivocal information only if several experimental conditions are fulfilled. First, a labelled precursor metabolically close to the substance under study should be fed to overcome the problems arising from varying pools of intermediates. Secondly, on feeding the radioactive precursor, the radioactivity must enter the studied pathway rapidly and specifically in order to allow a very short pulse of labelling. If this is not the case radioactivity could be trapped in some unknown compounds and enter the studied pathway after the end of the pulse, distorting the results of the chase. The third condition is that the radioactivity should be efficiently chased from the studied substance into the product(s) of the catabolism by feeding the unlabelled precursor. These conditions were not fulfilled in the studies of glyceollin accumulation in infected hypocotyls of soybean. In conclusion, further research is required in this controversial area before it can be accepted that phytoalexin accumulation is controlled at the level of degradation.

3.2. Regulation of some enzyme activities involved in the pathway

An increase in enzyme activity may arise from three major mechanisms: (1) an increased rate of *de novo* synthesis of the enzyme, (2) a decreased rate of its degradation, or (3) an activation of an inactive form of the enzyme. There are few examples where the mechanisms controlling the increase of enzyme activities following infection have been investigated and, with one exception, all of them deal with PAL. There are five techniques available for testing the nature of an increase in enzyme activity. Among these, four have been routinely used to study the regulation of enzyme activity under various stimuli: (1) the use of metabolic inhibitors, (2) radiolabelling, (3) immunological methods, and (4) density labelling. With the development of plant molecular biology, there is now another possible approach: the isolation and *in vitro* translation of mRNAs.

3.2.1. *Metabolic inhibitors*

Metabolic inhibitors have been extensively used in phytopathology. A few examples have already been cited in which inhibitors have been supplied in order to investigate the relationship between the increased production of some metabolites after infection and the establishment of resistance. Metabolic inhibitors have also been used for the study of enzyme regulation. Indeed, *de novo* synthesis has often been claimed to be the mechanism underlying an increase of enzyme activity when there is no increase in activity on application of inhibitors. In fact, as already emphasized in this chapter, the lack of specificity of this approach hampers the interpretation of the data. It can be said that this type of experimentation cannot give reliable information on the mechanisms of regulation unless the treatment by inhibitors is effective on protein or nucleic acid synthesis without affecting the changes in enzyme activity. That would demonstrate that increases in enzymatic activity are independent of protein or nucleic acids synthesis.

3.2.2. *Radiolabelling*

Radiolabelling requires the purification of the enzyme to a single homogeneous protein and the measurement of its specific radioactivity. It should be pointed out that an increase in total radioactivity incorporated into the protein does not prove unequivocally an increased rate of synthesis of the enzyme induced by the stimulus, since this situation might equally well arise from a decreased rate of enzyme degradation. Hence a comparative study of the specific radioactivity of enzymatic protein in stimulated and unstimulated materials must be made. An additional problem is that the supply of radioactive precursor may be disturbed by infection. Such a situation has been encountered in the case of TMV-infected tobacco leaves bearing local lesions (Duchesne *et al.*, 1977). Transport of radiolabelled amino acids through the petioles is strongly affected by lesion formation and consequently the labelling of the amino acid pool was changed in infected leaves. This problem has been overcome by floating the infected tissue on isotonic radioactive solution after removal of the lower epidermis. In fact, radiolabelling appears to be a particularly suitable method with cell suspension cultures and has been used very successfully in studies of regulation of enzyme levels by light (for references see Hahlbrock and Grisebach, 1979). The incorporation rates of [^{35}S]methionine into the proteins of cell suspension cultures of parsley are so rapid that the authors were able to show incorporation of radioactivity into PAL protein after only 20 min of labelling (Betz *et al.*, 1978). PAL was isolated from crude protein extracts by immunoprecipitation and its radioactivity was measured after purification of the subunit on SDS-polyacrylamide gels. The values of radioactivity were used to calculate the rates of synthesis and degradation of the enzyme under various conditions of irradiation. The same techniques were used to follow the changes in the rate of PAL synthesis *in vivo* upon treatment of parsley cell with Pms elicitor (Hahlbrock *et al.*, 1981).

3.2.3. Immunological techniques

Immunological techniques require the preparation of pure protein in sufficient quantity to raise an antiserum in an animal. Details on basic techniques in immunology and their application in studying plant enzymes can be found in a review (by Daussant *et al.*, 1977). Immunological techniques are of great value in characterizing an enzymatic protein as catalytically active or inactive. Coupled with the radiolabelling method, quantitative immunoprecipitation has been used to demonstrate an increased *de novo* synthesis of PAL after mechanical damage of root tissue of sweet potato (Tanaka and Uritani, 1977) and in elicitor-treated cells (Hahlbrock *et al.*, 1981).

3.2.4. Density labelling

In past years, density labelling has yielded much information about enzyme regulation in plants (for a review see Johnson, 1977). This method consists in comparing the labelling of the stimulated and unstimulated enzymes by a heavy isotope which can be ^{15}N, ^{18}O, or deuterium. Deuterium is the most widely used heavy isotope and is usually supplied as deuterated water. The labelled enzyme is separated from the unlabelled one by isopycnic centrifugation. Density labelling does *not* require extensive purification of the enzyme which is detected on the gradient by its activity. The position of the peak of enzyme activity gives the mean buoyant density of the enzyme. The density shift of the labelled enzyme from the native unlabelled enzyme is a direct measure of the specific labelling of the protein. The profiles expected if different control mechanisms are operating have been discussed in detail (Lamb and Rubery, 1976; Johnson, 1977). Only one point ought to be stressed here: an increase in the rate of *de novo* synthesis of the enzyme is the sole mechanism which can account for a higher density shift of the stimulated enzyme from infected material compared with that of the unstimulated enzyme from healthy material. In fact, this situation was observed in the two cases where density labelling was used to study the modulation of PAL activity by infection. In hypersensitive tobacco leaves infected by TMV the mechanism underlying the increase in activity of PAL has been demonstrated to be the enhanced rate of *de novo* synthesis of the enzyme (Duchesne *et al.*, 1977). *De novo* synthesis of PAL has also been reported to occur in french bean cells treated by an elicitor preparation obtained from cell wall of *Colletotrichum lindemuthianum* (Dixon and Lamb, 1979). In this material PAL activity is dependent on elicitor concentration and at high concentrations of elicitor the level of enzymic activity is under the dual control of enzyme production and enzyme degradation (Lawton et al., 1980).

Density labelling has been used in the author's laboratory to investigate the regulation of OMT activity in tobacco. The regulation of this activity is particularly interesting since, as already mentioned, OMT activity arises from three different enzymes in tobacco leaves. Moreover, the activities of all three are increased by infection but not to the same extent. The three enzymes from healthy and infected leaves were separated by chromatography on DEAE-cellulose and the density shifts estimated for each. The data demonstrated that increased rates of de novo synthesis were responsible for the increases in activity of all three enzymes.

A condition for obtaining meaningful data is that infection does not alter the proportion of density label in the amino acid pools from which the enzyme is synthesized. This has been verified in the case of hypersensitive tobacco by comparing the labelling of acid phosphatase extracted from healthy and infected leaves. Acid phosphatase activity is not changed after infection and its density shift reflects the labelling of the pool of amino acids. The labelling of the pool of amino acids can also be monitored either by measuring the labelling of proteins synthesized in bacteria supplied with the amino acids extracted from the experimental plant material (Johnson and Smith, 1977) or by measuring the maximum labelling of the studied enzyme, provided that the experimental conditions allow the separation of the pre-existing unlabelled enzyme from the newly synthesized labelled one (Lawton et al., 1980).

3.2.5. mRNA isolation and translation in vitro

These techniques have been developed very recently and have been used for regulation studies in plants in only a few cases. It is now possible to isolate polyribosomal RNAs, and fractionate and translate them in a cell-free protein-synthesizing system. These cell-free systems are usually derived from wheat germ or rabbit reticulocyte lysates and use labelled amino acids as substrates. This has been achieved in Hahlbrock's laboratory with the mRNAs of PAL, 4-coumaroyl-CoA synthase, chalcone synthase, and UDP apiose synthase (an enzyme of flavonoid glycoside pathway) from plant cell suspension cultures (Hahlbrock and Grisebach, 1979; Hahlbrock, 1981). The translation products were precipitated by specific antibodies and the newly synthesized protein subunits were characterized on SDS-polyacrylamide gels. The radioactivity incorporated into these subunits was taken as a measure of the activity of translatable mRNA present in the preparations. By this method it has been shown that the light-induced changes in enzyme activities were due to increased activity of translatable mRNAs in irradiated cells.

The activity of PAL-mRNA has also been measured in suspension cultured parsley cells treated with glucan elicitor from *P. megasperma* var. *sojae* (Hahlbrock *et al.*, 1981). The authors showed that the changes in PAL activity were caused by corresponding changes in the mRNA activity for this enzyme. This is an elegant and unquestionable way of demonstrating that the treatment by the elicitor induced an increased rate of synthesis of the enzyme. Although these techniques are rather difficult to handle they appear to be very promising and must provide useful tools for studying enzyme regulation in diseased plants in the future.

4. CONCLUSION

The inducibility of phenylpropanoid metabolism following various stimuli offers a fertile model for investigating the regulation of secondary plant metabolism. An obvious prerequisite for the study of the regulation of an enzymatic step is the characterization of the enzyme involved in the reaction. This has been achieved for most of the reactions leading to lignin synthesis but, in contrast, many gaps remain in our knowledge of the enzymology of the isoflavonoid pathway. Although changes in the activities of some enzymes involved in phenylpropanoid metabolism have been measured in diseased plants, PAL and OMTs are the sole enzymes for which the regulatory mechanism governing their activity has been elucidated. At present new techniques offer themselves for investigating the regulation of plant metabolism. Investment of time and effort is now necessary to apply these biochemical methods to the study of intact plant–pathogen interactions. Besides the techniques already described in this paper, it is now possible to clone cDNA made from purified plant mRNAs. The use of cloned cDNA as a hybridization probe allows the measurement of the amount of a particular mRNA present in plant preparations. In this way the relative rates of transcription of corresponding genes can be compared in healthy and infected tissues. The use of such methods in the future should be an important step towards the understanding of the mechanism of plant disease resistance at the molecular level.

5. ACKNOWLEDGEMENT

I thank Dr. B. Fritig for helpful discussions and for reviewing the manuscript.

REFERENCES

Amrhein, N., and Zenk, M. H. (1977). Metabolism of phenylpropanoid compounds, *Physiol. Veg.*, **15**, 251–260.
Amrhein, N., Gödeke, K. H., and Kefeli, V. I. (1976). The estimation of relative intracellular phenylalanine ammonia-lyase (PAL) activities and the modulation *in vivo* and *in vitro* by competitive inhibitors, *Ber. Dtsch. Bot. Ges.*, **89**, 247–259.

Asada, Y., Ohguchi, T., and Matsumoto, I. (1975). Lignin formation in fungus-infected plants, *Rev. Plant Protec. Res.*, **8**, 104–113.

Betz, B., Schäfer, E., and Hahlbrock, K. (1978). Light induced phenylalanine ammonia lyase in cell-suspension cultures of *Petroselinum hortense*. Quantitative comparison of rates of synthesis and degradation, *Arch. Biochem. Biophys.*, **190**, 126–135.

Camm, E. L., and Towers, G. H. N. (1973). Review article. Phenylalanine ammonia lyase, *Phytochemistry*, **12**, 961–973.

Collendavelloo, J., Legrand, M., Geoffroy, P., Barthelemy, J., and Fritig, B. (1981). Purification and properties of the three *o*-diphenol-*O*-methyltransferase of tobacco leaves, *Phytochemistry*, **20**, 611–616.

Conti, G. G., Vegetti, G., Favali, M. A., and Bassi, M. (1974). Phenylalanine ammonia-lyase and polyphenoloxidase activities correlated with necrogenesis in cauliflower mosaic virus infection, *Acta Phytopathol. Hung.*, **9**, 185–193.

Daussant, J., Laurière, C., Carfantan, N., and Skakoun, A. (1977). Immunochemical approaches to questions concerning enzyme regulation in plants, in *Regulation of Enzyme Synthesis and Activity in Higher Plants* (Ed. H. Smith), Academic Press, London, pp. 197–223.

Dixon, R. A., and Lamb, C. J. (1979). Stimulation of *de novo* synthesis of *L*-phenylalanine ammonia-lyase in relation to phytoalexin accumulation in *Colletotrichum lindemuthianum* elicitor-treated cell suspension culture of french bean (*Phaseolus vulgaris*), *Biochim. Biophys. Acta*, **586**, 453–463.

Duchesne, M., Fritig, B., and Hirth, L. (1977). Phenylalanine ammonia-lyase in tobacco mosaic virus-infected hypersensitive tobacco: Density labelling evidence of *de novo* synthesis, *Biochim. Biophys. Acta*, **485**, 465–481.

Ebel, J., Ayers, A. R., and Albersheim, P. (1976). Response of suspension-cultured soybean cells to the elicitor isolated from *Phytophthora megasperma* var *sojae*, a fungal pathogen of soybeans, *Plant Physiol.*, **57**, 775–779.

Faulkner, G., and Kimmins, W. C. (1975). Staining reactions of the tissue bordering lesions induced by wounding, tobacco mosaic virus, and tobacco necrosis virus in bean, *Phytopathology*, **65**, 1396–1400.

Friend, J. (1976). Lignification in infected tissue, in *Biochemical Aspects of Plant–Parasite Relationships* (Eds J. Friend and D. R. Threlfall), Academic Press, London, pp. 291–303.

Friend, J. (1981). Plant phenolics, lignification and plant disease, *Prog. Phytochem.*, **7**, 197–261.

Friend, J., and Thornton, J. D. (1974). Caffeic acid-*O*-methyltransferase, phenolase and peroxidase in potato tuber tissue inoculated with *Phytophthora infestans*, *Phytopathol. Z.*, **81**, 56–64.

Friend, J., Reynolds, S. B., and Aveyard, M. A. (1973). Phenylalanine ammonia lyase, chlorogenic acid and lignin in potato tuber tissue inoculated with *Phytophthora infestans*, *Physiol. Plant Pathol.*, **3**, 495–507.

Grisebach, H. (1977). Biochemistry of lignification, *Naturwissenschaften*, **64**, 619–625.

Grisebach, H., and Ebel, J. (1978). Phytoalexins, chemical defense substances of higher plants?, *Angew. Chem. Int. Ed. Engl.*, **17**, 635–647.

Gross, G. (1977). Biosynthesis of lignin and related monomers, *Recent Adv. Phytochem.*, **11**, 141–184.

Gross, G. G. (1978). Recent advances in the chemistry and biochemistry of lignin, *Recent Adv. Phytochem.*, **12**, 177–220.

Gross, G. G., Stöckigt, J., Mansell, R. L., and Zenk, M. H. (1973). Three novel enzymes involved in the reduction of ferulic acid to coniferyl alcohol in higher plants: ferulate: CoA ligase, feruloyl-CoA reductase and coniferyl alchol oxidoreductase, *FEBS Lett.*, **31**, 283–286.

Gustine, D. L., Sherwood, R. T., and Vance, C. P. (1978). Regulation of phytoalexin synthesis in jackbean callus cultures. Stimulation of phenylalanine ammonia-lyase and O-methyltransferase, *Plant Physiol.*, **61**, 226–230.

Hahlbrock, K. (1981). Messenger RNA induction as a prerequisite for flavonoid biosynthesis in parsley, *Hoppe-Seyler's Z. Physiol. Chem.*, **362**, 8–9 (Abstr.).

Hahlbrock, K., and Grisebach, H. (1979). Enzymatic controls in the biosynthesis of lignin and flavonoids, *Annu. Rev. Plant Physiol.*, **30**, 105–130.

Hahlbrock, K., Lamb, C. J., Purwin, C., Ebel, J., Fautz, E., and Schäfer, E. (1981). Rapid response of suspension-cultured parsley cells to the elicitor from *Phytophthora megasperma* var *sojae*. Induction of the enzymes of general phenylpropanoid metabolism, *Plant Physiol.*, **67**, 768–773.

Heller, W., and Hahlbrock, K. (1980). Highly purified 'Flavanone synthase' from parsley catalyses the formation of naringenin chalcone, *Arch. Biochem. Biophys.*, **200**, 617–619.

Higuchi, T., Shimada, M., Nakatsubo, F., and Tanahashi, M. (1977). Differences in biosynthesis of guaiacyl and syringyl lignins in woods, *Wood Sci. Technol.*, **11**, 153–167.

Johnson, C. B. (1977). The use of density labelling techniques in investigation into the control of enzyme levels, in *Regulation of Enzyme Synthesis and Activity in Higher Plants* (Ed. H. Smith), Academic Press, London, pp. 225–243.

Johnson, C. B., and Smith, H. (1977). Phytochrome control of amino acid synthesis in cotyledons of *Sinapis alba, Phytochemistry*, **17**, 667–670.

Kimmins, W. C., and Wuddah, D. (1977). Hypersensitive resistance: determination of lignin in leaves with a localized virus infection, *Phytopathology*, **67**, 1012–1016.

Lamb, C. J., and Rubery, P. H. (1976). Interpretation of the rate of density labelling of enzymes with ^2H_2O. Possible implications for the mode of action of phytochrome, *Biochim. Biophys. Acta*, **421**, 308–318.

Lawton, M. A., Dixon, R. A., and Lamb, C. J. (1980). Elicitor modulation of the turnover of L-phenylalnine ammonia-lyase in french bean cell suspension cultures, *Biochim. Biophys. Acta*, **633**, 162–175.

Legrand, M., Fritig, B., and Hirth, L. (1976). Enzymes of the phenylpropanoid pathway and the necrotic reaction of hypersensitive tobacco to tobacco mosaic virus, *Phytochemistry*, **15**, 1353–1359.

Legrand, M., Fritig, B., and Hirth, L. (1978). O-Diphenol-O-methyltransferases of healthy and tobacco-mosaic-virus-infected hypersensitive tobacco, *Planta*, **144**, 101–108.

Massala, R. (1981). Utilisation d'inhibiteurs compétitifs de la phénylalanine ammoniac lyase (PAL) en pathologie végétale: application á l'étude du rôle des phénylpropanoides dans la résistance hypersensible aux virus, *Thèse de Doctorat d'Etat*, Strasbourg.

Massala, R., Legrand, M., and Fritig, B. (1980). Effect of α-aminooxyacetate, a competitive inhibitor of phenylalanine ammonia-lyase, on the hypersensitive resistance of tobacco to tobacco mosaic virus, *Physiol. Plant Pathol.*, **16**, 213–226.

Maule, A. J., and Ride, J. P. (1976). Ammonia-lyase and O-methyltransferase activities related to lignification in wheat leaves infected with *Botrytis, Phytochemistry*, **15**, 1661–1664.

Moesta, P., and Grisebach, H. (1980). Effects of biotic and abiotic elicitors on phytoalexin metabolism in soybean, *Nature (London)*, **286**, 710–711.

Moesta, P., and Grisebach, H. (1981). Investigation of the mechanism of glyceollin accumulation in soybean infected by *Phytophthora megasperma* f. sp. *glycinea*, *Arch. Biochem. Biophys.*, **212**, 462–467.

Partridge, J. E., and Keen, N. T. (1977). Soybean phytoalexins: rates of synthesis are not regulated by activation of initial enzymes in flavonoid biosynthesis, *Phytopathology*, **67**, 50–55.

Poulton, J., Grisebach, H., Ebel, J., Schaller-Hekeler, B., and Hahlbrock, K. (1976). Two distinct S-adenosyl-L-methionine: 3,4-dihydric phenol 3-O-methyltransferases of phenylpropanoid metabolism in soybean cell suspension cultures, *Arch. Biochem. Biophys.*, **173**, 301–305.

Ride, J. P. (1975). Lignification in wounded wheat leaves in response to fungi and its possible role in resistance, *Physiol. Plant Pathol.*, **5**, 125–134.

Rhodes, M. J. C., and Wooltorton, L. S. C. (1975). Enzymes involved in the reduction of ferulic acid to coniferyl alcohol during the aging of disks of swede root tissue, *Phytochemisty*, **14**, 1235–1240.

Schröder, G., Zähringer, U., Heller, W., Ebel, J., and Grisebach, H. (1979). Biosynthesis of antifungal isoflavonoids in *Lupinus albus*. Enzymatic prenylation of genistein and 2'-hydroxygenistein, *Arch. Biochem. Biophys.*, **194**, 635–636.

Sütfeld, R., and Wiermann, R. (1978). The occurrence of two distinct SAM: 3,4-dihydric phenol 3-O-methyltransferases in tulip anthers, *Biochem. Physiol. Pflanz.*, **172**, 111–123.

Sütfield, R., and Wiermann, R. (1981). Purification of chalcone synthase from tulip anthers and comparison with the synthase from *Cosmos* petals, *Z. Naturforsch., Teil C*, **36**, 30–34.

Tanaka, Y., and Uritani, I. (1977). Synthesis and turnover of phenylalanine ammonia-lyase in root tissue of sweet potato injured by cutting, *Eur. J. Biochem.*, **73**, 255–260.

Towers, G. H. N., and Chi-Kit Wat (1979). Phenylpropanoid metabolism, *Planta Med.* **37**, 97–114.

Tsang, Y. F., and Ibrahim, R. K. (1979). Two forms of O-methyltransferase in tobacco cell suspension culture, *Phytochemistry*, **18**, 1131–1136.

Ulbrich, B., and Zenk, M. H. (1979). Partial purification and properties of hydroxycinnamoyl-CoA: quinate hydroxycinnamoyl transferase from higher plants, *Phytochemistry*, **18**, 929–933.

Ulbrich, B., and Zenk, M. H. (1980). Partial purification and properties of p-hydroxycinnamoyl-CoA: shikimate-p-hydroxycinnamoyl transferase from higher plants, *Phytochemistry*, **19**, 1625–1629.

Vance, C. P., and Sherwood, R. T. (1976). Regulation of lignin formation in reed canarygrass in relation to disease resistance, *Plant Physiol.*, **57**, 915–919.

Vance, C. P., Kirk, T. K., and Sherwood, R. T. (1980). Lignification as a mechanism of disease resistance, *Annu. Rev. Phytopathol.*, **18**, 259–298.

Van Etten, H. D., and Pueppke, S. G. (1976). Isoflavonoid phytoalexins, in *Biochemical Aspects of Plant–Parasite Relationships* (Eds J. Friend and D. R. Threlfall), Academic Press, London, pp. 239–289.

Wu, J. H. (1973). Wound-healing as a factor in limiting the size of lesions in *Nicotiana glutinosa* leaves infected by the very mild strain of tobacco mosaic virus (TMV-VM), *Virology*, **51**, 474–484.

Yoshikawa, M., Yamauchi, K., and Masago, H. (1978). Glyceollin: its role in restricting fungal growth in resistant soybean hypocotyls infected with *Phytophthora megasperma* var. *sojae, Physiol. Plant Pathol.*, **12**, 73–82.

Yoshikawa, M., Yamauchi, K., and Masago, H. (1979). Biosynthesis and biodegradation of glyceollin by soybean hypocotyls infected with *Phytophthora megasperma* var. *sojae, Physiol. Plant Pathol.*, **14**, 157–169.

Zähringer, U., Ebel, J., and Grisebach, H. (1978). Induction of phytoalexin synthesis in soybean. Elicitor-induced increase in enzyme activities of flavonoid biosynthesis and

incorporation of mevalonate into glyceollin, *Arch. Biochem. Biophys.*, **188**, 450–455.
Zähringer, U., Ebel, J., Mulheim, L. J., Lyne, R. L., and Grisebach, H. (1979). Induction of phytoalexin synthesis in soybean. Dimethylallylpyrophosphate: trihydroxypterocarpan dimethylallyl transferase from elicitor-induced cotyledons, *FEBS Lett.*, **101**, 90–92.

Further Reading

Friend, J. (1981). Plant phenolics, lignification and plant disease, *Prog. Phytochem.*, **7**, 197–261.
Gross, G. (1977). Biosynthesis of lignin and related monomers, *Recent Adv. Phytochem.*, **11**, 141–184.
Hahlbrock, K., and Grisebach, H. (1979). Enzymatic controls in the biosynthesis of lignin and flavonoids, *Annu. Rev. Plant Physiol.*, **30**, 105–130.
VanEtten, H. D., and Pueppke, S. G. (1976). Isoflavonoid phytoalexins, in *Biochemical Aspects of Plant–Parasite Relationships* (Eds J. Friend and D. R. Threlfall), Academic Press, London, pp. 239–289.

Biochemical Plant Pathology
Edited by J. A. Callow
© 1983 John Wiley & Sons Ltd

18

Transcription and Translation in the Diseased Plant

L. C. van Loon

Department of Plant Physiology, Agricultural University, 6703 BD Wageningen The Netherlands

and J. A. Callow

Department of Plant Biology, University of Birmingham, Birmingham, UK

1. INTRODUCTION

Pathogens cause a great variety of symptoms, varying from hardly visible local discolourations to severe perturbations of plant growth and development. The same pathogen may induce very different types of symptoms on different plant species and often even on different varieties or cultivars of a single species. Such differences in reaction depend on genetic variation within the host plant population and form the basis for selection and breeding for resistance. Genetically determined symptom expression is subject to the same regulatory control mechanisms as normal plant growth and development. However, owing to the presence of the pathogen, such genetic information is expressed in an untimely or uncoordinated way, thus giving rise to the particular symptoms of disease.

Conversely, the same plant species may react very differently to infection with different pathogens or even races or strains of a single pathogen. The symptoms of disease that are caused by infection of a susceptible plant are thus characteristic of the specific combination of host and pathogen, and reflect the interaction of the genetic constitution of both. Since genetic information is ultimately expressed in nucleic acids and proteins via the processes of transcription and translation, specific alterations in nucleic acid and protein metabolism are fundamental to pathogenesis. With fungi and bacteria, the metabolism of the pathogen gives rise to macromolecules and metabolites that may be transferred to host cells, and are involved in recognition and the initiation of host reactions or function as toxins causing disorganization, destruction, and decay of host cell structure and metabolism. Viruses do not possess a metabolism of their own and thus depend on the cellular machinery of the host plant, interfering directly with host cell transcription and translation. The amount of genetic information in viruses is so limited that it is questionable whether specific 'symptom-inducing' functions are present. Rather, host symptoms appear to arise as a result of the interference of the virus with normal patterns of metabolism and development. Viruses thus offer the opportunity of studying the reaction of the host in the absence of extensive metabolism of the pathogen, and to relate changes in nucleic acid and protein metabolism directly to the development of symptoms of disease.

The development of many fungal parasites on their hosts is controlled by a gene-for-gene relationship involving host genes for resistance and parasite genes for virulence. Most current models of the molecular events involved in resistance favour an initial recognition event at the host–parasite interface between complementary products of host genes for resistance and parasite genes for avirulence, the resulting product or 'dimer' then acting as a trigger, via secondary messengers, to initiate changes in gene expression and enzyme synthesis leading to the development of appropriate 'barriers' to limit fungal growth (Callow, 1983; see Chapter 13). Ellingboe (1982) has argued on genetic grounds that the initial act of recognition at the host–parasite interface may itself constitute the mechanism limiting pathogen development in some as yet undefined way. This would appear to relegate changes in gene expression during resistance to some secondary effect, the consequence but not the cause of resistance. Notwithstanding this evidence, which is incomplete, it is clear from work carried out mainly with biotrophic pathogens that in both compatible and incompatible interactions there are changes in host gene expression.

2. INTERACTION OF VIRUSES WITH THEIR HOSTS

2.1. The viral strategy

During the infection process the protein coat of an attached virus particle remains behind at the cell surface. The freed nucleic acid enters and finds its way to the subcellular site where multiplication takes place. Some viruses are found exclusively in the nucleus, others in chloroplasts; the majority, however, are in the cytoplasm, sometimes associated with specific membranous structures.

In most plant viruses the nucleic acid consists of single-stranded (ss) RNA that functions both as the genetic material for virus replication and as the messenger (m)RNA for translation of virus-specific proteins. In several groups of plant viruses the genetic information is divided between two or more nucleoprotein particles each containing a characteristic nucleic acid component with one or more specific functions. A small number of known plant viruses posess double-stranded (ds) RNA or ss- or ds-DNA. In the latter two cases, DNA is transcribed into RNA which serves exclusively a messenger function.

2.2. Viral replication

During the replication of the ss-RNA viruses two ds-structures arise, the 'replicative form', consisting of the original (+) strand paired with a newly synthesized complementary (−) strand, and the 'replicative intermediate', in which nascent (+) strands are continuously displaced from the (−) strand by newly initiated (+) strands. Ds-RNA and RNA-dependent RNA polymerases that catalyse the formation of such ds-structures also occur in non-infected plants (Ikegami and Fraenkel-Conrat, 1979). The role of these enzymes is unknown and, although they are greatly stimulated upon infection, it is not clear how far they play a role in viral RNA replication. So far, plant virus-specific RNA polymerases (replicases) have not been isolated to a sufficient extent either to synthesize several rounds of viral (+) strands or to identify their polypeptide composition.

Efficient and synchronous infection is required in order to investigate the course of RNA replication. This is usually accomplished by inoculation of isolated plant protoplasts. When virus-infected tissues are homogenized, particulate fractions containing cytoplasmic membranes yield a bound replicase activity which is associated with template viral RNA. Such enzymes show little or no response to exogenous RNA. However, from the same membranes, a template-free, soluble, unbound enzyme can be extracted that requires RNA for activity and does not display template specificity. Only in the case of brome mosaic virus (BMV) has the solubilized enzyme been shown to prefer homologous viral RNA as template (Hardy et al., 1979).

In tobacco infected with tobacco mosaic virus (TMV), soluble and bound replicases show parallel changes in activity with time after inoculation (Duda et al., 1973) and behave identically with the RNA-dependent RNA polymerase present in uninfected tissues in all properties tested (Romaine and Zaitlin, 1978). Also, a membrane-bound RNA-dependent RNA polymerase from cowpea mosaic virus (CPMV)-infected cowpea leaves appears to be identical with the RNA polymerase from non-infected leaves (Dorssers et al., 1982). These observations suggest that host enzymes are involved in the formation of the complementary strand to the infecting viral RNA. That virus-coded proteins also play a role during viral RNA replication is evident from the multiplication of multipartite genome viruses. For instance, the genetic information of CPMV is divided between two separately encapsidated RNAs: the larger B (bottom) component and the smaller M (middle) component. The third, top component does not contain RNA. B-RNA is replicated in cowpea protoplasts in the absence of M-RNA, but M-RNA is completely dependent on B-RNA, suggesting that an essential part of CPMV replicase is encoded by B-RNA (Goldbach et al., 1980). In addition, mutants of both TMV and alfalfa mosaic virus (AMV) have been described in which replication is temperature-sensitive, but the in vitro activity of the membrane-bound replicase is not (Dawson and White, 1979; Linthorst et al., 1980).

The group of plant rhabdoviruses contains ss-RNA of the (−) type. The virus particles contain a constitutive RNA-dependent RNA polymerase which transcribes the (−) RNA into (+) strands that function as the mRNA. When freed from their enveloping membrane, these viruses are able to replicate in vitro without added enzyme, although a stimulation in the rates of both transcription and replication is observed when extracts from infected plants are added (Flore and Peters, in preparation). The ds-RNA virus, wound tumour virus, likewise carries its own RNA transcriptase activity. The smallest known type of pathogen, the viroids, consist of naked circular ss-DNA, and do not have the capacity to code for a viroid-specific replicase. These small RNAs are copied by the host DNA-dependent RNA polymerase II, which is normally responsible for the synthesis of mRNA (Rackwitz et al., 1981).

Host functions are necessary for the multiplication of RNA viruses, as shown by experiments with the inhibitor of DNA-dependent RNA synthesis, actinomycin D (AMD). It might be expected that viral RNA synthesis, which does not involve a DNA intermediary, is not influenced by such inhibitors. However, after inoculation of tobacco varieties that react to TMV with systemic mosaic symptoms, AMD decreasingly inhibits virus multiplication during the first 24 h. Cycloheximide and 2-thiouracil (inhibitors of protein synthesis on cytoplasmic ribosomes and virus replication, respectively) are effective for a much longer time (Dawson and Schlegel, 1976).

A similar sensitivity to AMD was observed by Rottier *et al.* (1979) after inoculation of cowpea protoplasts with CPMV. However, while AMD completely inhibited virus replication, synthesis of virus-specific proteins still proceeded. Therefore, AMD does not interfere with the messenger function of the viral RNA, but with its replication. Because other viruses have also proved sensitive to AMD, provided that the inhibitor is applied immediately after inoculation, transcription of the host genome appears to be required during the period of initiation of rapid virus multiplication. Whether this host activity is necessary for the formation of a functional virus–replicase complex, or for the creation of a suitable environment for replication of the virus in the cell, is unknown at present. It would seem conceivable that this contribution of the host determines how far a plant can function as a host for the virus. However, host plant specificity is not usually expressed at the level of virus multiplication: whereas CPMV does not infect tobacco and, conversely, cowpea is not infected by the common strain of TMV, both viruses multiply to similar extents in isolated protoplasts of both cowpea and tobacco (Huber *et al.*, 1977; Huber, 1979). Reactions comparable to the symptoms that develop when each of the two viruses infects its own host do not occur under these conditions. Apparently, virus multiplication occurs in a manner analogous to the fundamental processes regulating nucleic acid metabolism in the plant cell. Only rarely is resistance maintained at the cellular level (Beier *et al.*, 1977; Maule *et al.*, 1980). In principle, therefore, each plant cell appears to be able to replicate every plant virus.

2.3. Viral transcription and translation

The complete nucleotide sequence (8024 nucleotides) is known for the ds-DNA virus, cauliflower mosaic virus (CaMV) (Franck *et al.*, 1980). CaMV DNA encompasses six possible cistrons, one of which codes for the viral coat protein and another for a protein of about 66 000 daltons (P66) that is part of viral inclusion bodies. CaMV RNA transcript mapping experiments have identified both a large full-genome-length transcript and a 19 S RNA from the same strand. The 3'-end of the 19 S transcript is polyadenylated and encodes the synthesis of P66 (Odell *et al.*, 1981; Covey and Hull, 1981). No other virus-coded proteins have been identified so far.

Except for the rhabdoviruses, mentioned above, the genome of the ss-RNA viruses functions directly as the mRNA. Eukaryotic messengers are generally monocistronic and RNA viruses of plants have evolved to adapt to this condition: their information for a number of different gene products is expressed by messengers with only one initiation site for protein synthesis. This is achieved by post-translational cleavage of a full-length transcript, the generation of smaller, subgenomic, messengers, a multipartite genome consisting of two or more different RNA species, or a combination of these strategies.

Like other eukaryotic mRNAs, many plant viral RNAs are capped at the 5'-end with 7-methylguanosine, joined by the 5'-hydroxyl group of its ribose through a triphosphate bridge and an inverted 5'–5'-linkage to the second nucleotide. Others have a small protein (VPg) covalently bound to the 5'-end. Some viral RNAs, like many mRNAs, have 3'-terminal poly(A). However, many plant viral RNAs have the property of accepting amino acids at their 3'-ends (Van Vloten-Doting and Neeleman, 1980). For instance, BMV specifically accepts tyrosine. BMV RNA binds to elongation factor EF 1, but does not form a ternary complex with GTP. When tyrosine is accepted, it is not transferred into nascent protein. If the 3'-terminus of BMV RNA is chemically modified so as to no longer accept tyrosine, the RNA is still fully active as a message *in vitro*. Thus, the tRNA-like region of BMV RNA does not function as a true tRNA and it is unlikely that it plays a role in translation (Hall, 1979).

Translational products of viral RNA have been identified both *in vivo* by comparison of the products of protein synthesis in host plants or protoplasts upon infection, and *in vitro* by translation of viral nucleic acids in cell-free protein-synthesizing systems. By comparing the changes in the proteins synthesized upon infection of protoplasts from different host plants (Huber, 1979) or in both plant tissue and a cell-free protein-synthesizing system (Bruening *et al.*, 1976), together with their peptide maps (Rezelman *et al.*, 1980), newly induced virus-specific proteins are unequivocally distinguished from host proteins stimulated by infection. Ambiguities in the products observed in different cell-free systems have recently been attributed to partial readthrough due to the presence of suppressor tRNAs in several natural tRNA populations (Bienz and Kubli, 1981).

It has been shown for the comoviruses only, of which CPMV is the type member, that the complete viral RNAs are translated. Both the B-RNA, comprising 2.0×10^6 daltons and the M-RNA, of 1.4×10^6 daltons, give rise to large precursor polypeptides, which are processed after transcription by a protease encoded on the B-RNA (Pelham, 1979; Franssen *et al.*, 1982). Whereas M-RNA directs the synthesis of two overlapping polypeptides as the result of two initiation sites for protein synthesis (Franssen *et al.*, 1982), the B-RNA-encoded primary translation product is apparently cleaved in different ways (Rezelman *et al.*, 1980; Goldbach *et al.*, 1982) (Figure 1A). This

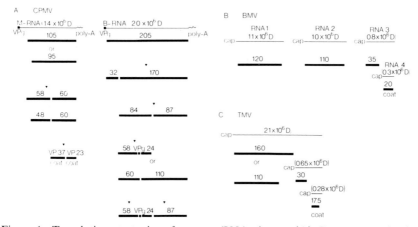

Figure 1. Translation strategies of some ss-RNA viruses. (A) Cowpea mosaic virus; (B) brome mosaic virus; (C) tobacco mosaic virus. The viral RNAs are indicated by thin lines, the translational products by solid bars. Numbers refer to the molecular weights of the polypeptides ×10⁻³. (Adapted from Pelham, 1979; Rezelman *et al.*, 1980; Franssen *et al.*, 1982; Goldbach *et al.*, 1982; Kaesberg, 1976; and Beachy *et al.*, 1976)

allows several different proteins to be specified by a minimal amount of RNA.

This situation appears unique in that other RNA viruses, with coding capacities of up to 3×10^6–4×10^6 dalton RNA, give rise to no more than three or four protein products. *In vitro*, BMV, which systemically infects wheat and for which the wheat embryo system is thus homologous, is translated into four proteins. Like other bromoviruses and AMV, BMV contains four RNAs, with molecular weights of 1.1×10^6 (RNA 1), 1.0×10^6 (RNA 2), 0.8×10^6 (RNA 3) and 0.3×10^6 (RNA 4), respectively, packaged into three types of particles. RNA 1 and 2 each code for a large polypeptide corresponding to almost the entire coding capacity, whereas RNA 3 has two cistrons, one 'open' and giving rise to a polypeptide of about 35 000 daltons, and one 'closed' and identical with RNA 4, which codes for the coat protein (Kaesberg, 1976) (Figure 1B). Only RNAs 1, 2, and 3 are necessary for infection. During infection, RNA 4 is generated, possibly through cleavage or specific partial transcription of RNA 3. RNA 4 has a regulatory function in that it is able to switch off the translation of the larger RNAs when replication has increased the number of RNA molecules to a certain critical proportion. The resulting abundant synthesis of coat protein thus ensures encapsidation of all RNAs to take place (Zagorski, 1978).

In most monopartite genome viruses, such as TMV, monocistronic messengers are generated during infection. Full-length TMV RNA (2.1×10^6 daltons) is translated exclusively into two large polypeptides with overlapping sequences and comprising roughly two thirds of the total coding capacity. However, in infected tissues, also a small (0.28×10^6 daltons) and an intermediate size (0.65×10^6 daltons) virus-specific RNA are found, both of which contain the 3'-terminus and code for the coat protein and for a 30 000 dalton polypeptide, respectively. The intermediate RNA thus contains the code for the 30 000 dalton polypeptide in an 'open' cistron and that for the coat protein in a 'closed' cistron, analogous to the situation with BMV RNA 3 (Beachy *et al.*, 1976; Bruening *et al.*, 1976) (Figure 1C).

Except for the coat protein, which is by far the predominant product *in vivo*, the nature and significance of these viral translation products are obscure.The largest polypeptides are synthesized relatively early in the infection process (Sakai and Takebe, 1974; Paterson and Knight, 1975) and may be involved in RNA replication. Similarly, AMV RNAs 1 plus 2 can replicate in the absence of RNA 3, indicating that information encoded by these RNAs is involved in replication (Nassuth *et al.*, 1981). Also, the smaller non-coat protein appears at the same time as replicase activity and has been suggested as a possible replicase or subunit of replicase (Hariharasubramaian *et al.*, 1973). It is probable that polyvirus RNAs, such as potato virus Y, like CaMV, code for the protein of the intracellular inclusion bodies characteristic of that group of viruses.

These assigned functions do not appear to be compatible with those of specific disease-inducing proteins. Although the capsid proteins of different strains of a virus usually differ in amino acid sequence, it is unlikely that such differences are responsible for the apparent variation in symptoms. Thus, the common strain of TMV, which causes extensive mosaic and malformations of the young developing leaves, has a coat protein identical with that of the 'masked' strain, which does not induce visible symptoms on tobacco. With multipartite viruses in which the code for coat protein is present on the RNA of one of the particles, the information for symptom type is usually not connected with that for coat protein when particles from different strains are recombined. The conclusion that the coat protein is not responsible for symptom type is supported by the fact that viroids do not have a protein coat; one or more polypeptide products that could be encoded by the viroid RNA have not been found (Sänger, 1980). In potato spindle tuber viroid, containing only 359 nucleotides, strains exist that induce symptoms varying from very mild to very severe. These differences are determined by a small number (2–10) of nucleotide substitutions (Dickson *et al.*, 1979) (Figure 2).

Thus, minor changes in RNA sequence can have profound effects on symptom expression. This is also apparent when virus particles or RNA are subjected to mutagenic treatment *in vitro*. During treatment of TMV RNA with nitrous acid, deamination of only one nucleotide base transforms the property of inducing systemic mosaic symptoms on *Nicotiana sylvestris* to one of induc-

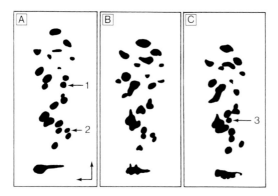

Figure 2. Autoradiograms of two-dimensional oligonucleotide patterns obtained by digestion with RNase T1 of a ^{125}I-labelled severe (A), intermediate (B), and mild (C) strain of potato spindle tuber viroid. Mild strains are characterized by the presence of spot no. 3 and the absence of spots nos. 1 and 2. Both spot nos. 1 and 2 are found in severe strains. Intermediate strains contain only spot no. 2. (After Dickson *et al.*, 1979. Reproduced by permission of Macmillan Journals Ltd.)

ing local necrotic lesions (Mundry and Gierer, 1958). In spite of many attempts, the reverse mutation from local lesions to systemic mosaic has never been observed. Therefore, the transition from systemic mosaic symptoms to local lesions has the character of a deletion. Not only from TMV, but also from AMV, cucumber mosaic virus (CMV), and CPMV one can easily obtain spontaneous or artificial mutants that have lost the ability to spread systemically within the host. These observations indicate that the viral genome contains a specific function that enables the virus to escape localization in necrotic lesions, to multiply abundantly, and to spread systemically throughout the entire plant.

Mundry and Gierer (1958) found that mutagenic treatment of TMV yielded about eight times as many viable mutants as the percentage transition from systemic mosaic symptoms to local lesions. The genetic information concerned comprised about one third of the length of the TMV RNA. These mutants could be distinguished in up to 13 classes on the basis of symptom type. These observations seem difficult to reconcile with the small number of proteins encoded by the viral RNA. A similar diversity of changes in symptom expression is observed after mutagenesis of CPMV (De Jager, 1976). Recombination of CPMV RNAs from different strains and mutants showed that both the M- and the B-RNA contain functions specifying systemic spread of the virus and the induction of chlorosis, necrosis, and local lesion morphology.

3. HOST TRANSCRIPTION IN RESPONSE TO VIRUSES AND
BIOTROPHIC FUNGI

3.1. Host rRNA transcription

3.1.1. Fungal infections

A number of studies have claimed to demonstrate striking effects of fungal infection on host rRNA synthesis and have been the subject of earlier reviews (Callow, 1976; Chakravorty and Shaw, 1977). Unfortunately, much of this work is unconvincing and here a critical assessment of the current state of knowledge will be attempted in the light of some of the major technical difficulties.

(a) The number of host cells responding to infection, especially in the early stages of the interaction, may be relatively few and therefore the chances of detecting small, but perhaps critical changes in host transcription may be low.

(b) Changes following infection may be superimposed on, and are often not distinguished from, those normally occurring during leaf senescence. This has led to ambiguity and contradiction. Thus, Tani *et al.* (1973) clearly stated that there was no net increase in total rRNA in rust-infected oat leaves. The total rRNA content of a susceptible, infected leaf was up to 1.74 times that of healthy controls because the rate of decrease during senescence of uninfected leaves was apparently greater. However, Chakravorty and Shaw (1977) have apparently interpreted the data of Tani *et al.* to imply net synthesis of rRNA.

A detailed comparison of the effects of mildew infection on rRNA metabolism in cucumber leaves was made by Callow (1973). In higher plant leaves, depending on age, about 80% of total RNA is in the form of rRNA, and of this about 70–50% is found in the 80S cytoplasmic ribosomes (principally as 1.3×10^6 and 0.7×10^6 daltons mol. wt.), the remainder being found in the 70S chloroplastic ribosomes (1.1×10^6 and 0.6×10^6 daltons) (Figure 3). During the course of leaf expansion in young cucumber leaves infected by powdery mildew (*Erysiphe cichoracearum*), the normal accumulation of total RNA was retarded. Fractionation of the RNAs by gel electrophoresis revealed that this was primarily due to reduced accumulation of chloroplastic rRNAs (chl. rRNA, Figure 3). Pulse-labelling with ^{32}Pi showed a parallel and specific reduction of incorporation into newly synthesized chl. rRNAs. Further, in infections of older leaves, the more rapid loss of chl. rRNA compared with cyt. rRNA associated with senescence was accentuated. Thus, the major effect of mildew infection on rRNAs appears to be retardation of chl. rRNA synthesis and accelerated senescence. In fact, the more rapid loss of chl. rRNAs appears to be a consistent feature of infection by biotrophs, being also reported for oat leaves infected by *Puccinia coronata avenae* (Tani *et al.*, 1973) and maize leaves infected by the smut *Ustilago maydis* (Callow, 1976).

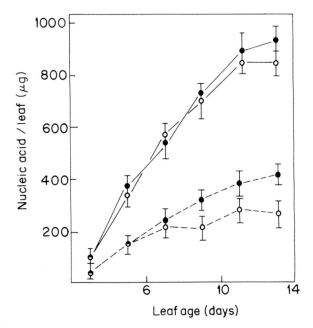

Figure 3. Cytoplasmic and chloroplastic rRNAs in healthy and mildew-infected cucumber leaves (leaves infected when 3 days old and still expanding). (●——●) cytoplasmic rRNA (i.e. $1.3 \times 10^6 + 0.7 \times 10^6$), healthy; (○——○) cytoplasmic rRNA, infected; (●- - - -●) chloroplastic rRNA (i.e. $1.1 \times 10^6 + 0.6 \times 10^6$), healthy; (○- - - -○) chloroplastic rRNA, infected. 95% confidence limits shown. (After Callow, 1973; reproduced by permission of Academic Press)

Bennet and Scott (1971a) also reported a 50% decrease in barley chloroplast ribosomes 5 days after mildew infection.

(c) Interpretation of the results of experiments using labelled precursors to follow rates of synthesis is complicated by the fact that the precursor pool sizes supporting rRNA synthesis may alter on infection. Particularly unsatisfactory is the use of ^{32}Pi. Since biotrophic fungi appear to sequester host orthophosphate in their hyphae as polyphosphate (Bennet and Scott, 1971b), the intracellular orthophosphate pools in infected tissues may be considerably smaller. Thus, demonstrations of increased ^{32}Pi incorporation into RNAs of infected tissues (Wolf, 1968; Chakravorty and Shaw, 1971; Tani et al., 1973) must be treated with great caution.

(d) Workers frequently present data on a relative rather than absolute basis, rendering detailed assessment difficult and promoting erroneous interpretation [see (b) above]. Thus, Tani et al. (1973) expressed their data as ratios of the level of radioactivity of RNAs in infected tissues relative to those of uninoculated controls, or per unit of DNA (which presumably includes some fungal DNA), whilst Wolf (1968) presented data only in terms of specific activity.

(e) Many of the earlier investigations were conducted prior to the development of routine techniques to permit separate assessment of changes in the quite distinct chloroplastic and cytoplasmic ribosome systems. Thus studies reporting changes in total RNA (Quick and Shaw, 1964), high molecular weight RNA (Wolf, 1968) and total rRNA (Chakravorty and Shaw, 1971) are relatively less informative than those in which defined chloroplastic and cytoplasmic rRNAs or ribosomes are separated (Bennet and Scott, 1971a; Callow, 1973, 1976; Tani et al., 1973).

(f) The overriding difficulty, however, is the problem of separately assessing the contributions of host and parasite to the observed changes. Whilst the use of gel electrophoresis permits a separate and unequivocal assessment of changes in chl. rRNAs, the cyt. rRNAs of host and fungus and the complete ribosomal particles are indistinguishable in terms of molecular size or sedimentation. Thus, in the absence of any other information, the demonstration of a doubling of the 80S ribosome content in barley leaves 9 days after powdery mildew infection (Bennet and Scott, 1971a) is probably due to the contribution of the developing fungus, since despite the fact that superficial mycelium and spores were brushed from the leaves before extraction, in a heavy mildew infection the contribution from the remaining haustoria in the host epidermis must be substantial. Similarly, the maintenance of cyt. rRNA levels in rust-infected oats (Tani et al., 1973) in the face of declining levels in senescing controls almost certainly reflects increased fungal rRNA synthesis (Tani and Yamamoto, 1979).

On the other hand, changes in RNA synthesis have been reported at times which are sufficiently early to preclude extensive pathogen development. Thus, Chakravorty and Shaw (1971), notwithstanding the complications discussed in (c) above, demonstrated a two- to threefold increase in the rate of incorporation of ^{32}Pi into total rRNA within 48 h of inoculation of flax with a compatible race of flax rust, and in the absence of net changes in the amount of RNA. Furthermore, the nucleotide composition of the newly synthesized rRNA was similar to that of uninfected flax rRNA, and different from that of germinating rust uredospores. It was therefore suggested that this early rRNA synthesis was of host origin and circumstantial support for this conclusion was provided by an earlier cytophotometric and autoradiographic study showing increased host nucleolar RNA content 2 days after infection (Bhattacharya et al., 1965; Bhattacharya and Shaw, 1967). Large increases in cyt. rRNA synthesis in maize smut galls are also probably of host origin since the base composition of the cyt. rRNAs more closely reflected that of uninfected maize, and the changes were associated with large (up to eight-fold) increases in host nucleolar mass (Figure 4; Callow, 1975, 1976).

In conclusion, then, the balance of evidence suggests that the major effect

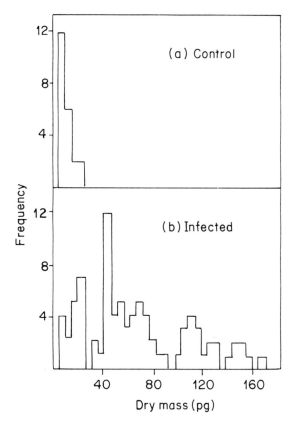

Figure 4. Dry masses of nuclei from 2nd leaf sheaths of 14-day-old maize seedlings infected with sporidia of *Ustilago maydis* after 7 days. Dry masses measured by laser interferometry. (After Callow, 1975; reproduced by permission of Blackwell Scientific Publications)

of infection of leaf tissue by biotrophic pathogens is an acceleration of the normal senescence-associated decline in chl. rRNA and 70S ribosomes. In contrast to other reviewers, this author concludes that there is no strong evidence to support a general conclusion that infection stimulates the synthesis of host rRNA, most of the increases reported probably being due to fungal rRNAs. This is not to deny the possibility of such changes in the host, but simply reflects the equivocal nature of most data obtained so far.

3.1.2. *Viral infections*

Many symptoms of virus infection resemble disturbances in hormone-regulated processes of growth and development, such as stunting, wilting, and malformations. Often, the growth of developing leaves is arrested (Bos, 1978). In typical mosaic symptoms, light green to yellow leaf areas alternate with dark green ones. The light areas show many features of senescing leaves; the dark ones, on the contrary, are physiologically young. Already before the appearance of the light green–dark green mosaic symptoms on the young, developing leaves of tobacco, the common strain of TMV inhibits the synthesis of chl. rRNA, without having an effect on cyt. rRNA (Figure 5). The *flavum* strain, which induces severe yellow–green mosaic symptoms, inhibits the synthesis of chl. rRNA more strongly than the common strain and causes an earlier degradation of these RNAs. Moreover, TMV *flavum* also inhibits the synthesis of cyt. rRNA (Fraser, 1969). Decreases in leaf chl. rRNA synthesis and, under certain conditions, also cyt. rRNA occur in several host–virus combinations with symptoms of systemic mottling, mosaic, or yellowing. Such effects are characteristically similar to early (light green) and later (yellow) stages of leaf senescence (Woolhouse, 1974).

In spite of the more severe symptoms, TMV *flavum* multiplies far less than the common strain (cf. Figure 5). Moreover, both strains do not multiply in chloroplasts, but on cytoplasmic membranes. Regulation of natural leaf senescence likewise is a cytoplasmic process. This suggests that TMV interferes with the regulation of normal leaf development by inhibiting chloroplast functioning and inducing premature senescence.

3.2. Host mRNA transcription

Possibly of greater relevance to the crucial events involved in the acceptance or rejection of a potential parasite are changes in patterns of gene transcription and translation leading to the synthesis of new protein and enzyme products in the host controlling the regulation of host metabolic pathways.

There is substantial circumstantial evidence to support such changes in relation to both compatible and incompatible interactions (Chakravorty and Shaw, 1977). For example, Flynn *et al.* (1976) demonstrated that host RNA polymerase II was stimulated in activity and showed altered catalytic properties (e.g. template preference) 2–4 days after infection of wheat by compatible races of wheat stem rust. The template activity of host chromatin was also increased on infection, with new size classes of RNA being synthesized. Other workers have used inhibitors of RNA and protein synthesis to identify changes in transcription and translation in the control of resistance responses (see Chapters 11 and 17). However, recent developments in technique now permit the direct isolation and *in vitro* translation of specific mRNAs and their quantitative and qualitative assessment (see also Chapter 17).

Pure *et al.* (1979) isolated polysomes from healthy wheat leaves and those infected with a compatible race of wheat stem rust 3 days after inoculation.

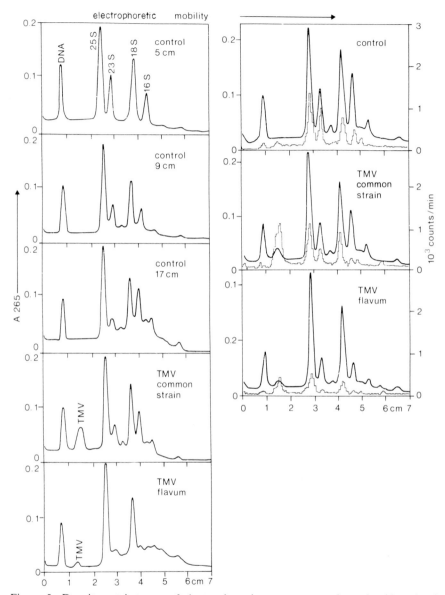

Figure 5. Densitometric traces of electrophoretic patterns on polyacrylamide gels of nucleic acids from healthy tobacco leaves of different lengths (control) and 4 days after inoculation of a 9 cm long leaf with either the common or the *flavum* strain of tobacco mosaic virus (left). Infection with the *flavum* strain leads to the loss of 23S and 16S chl. rRNA. The incorporation patterns (right) indicate the incorporation of ^{32}P 2 days after inoculation. Both strains inhibit the incorporation into chl. rRNA; in addition, the *flavum* strain also inhibits the synthesis of 25S and 18S cyt. rRNA. (From Fraser, 1969. Reproduced by permission of Springer-Verlag)

Translation of the polysomal mRNA in a wheat embryo cell-free system generated a range of polypeptides and specific differences in the size classes of polypeptide were detected when translation products from infected and control extracts were compared on polyacrylamide gels. Controls with the chain initiation inhibitor aurinetricarboxylic acid showed that there was no re-initiation of new polypeptide chains in the translation system. Hence, the size classes of newly synthesized polypeptides probably reflected the size classes of the polysomal mRNA molecules on extraction and thus differences in mRNA transcription. It was suggested that the specific differences detected were unlikely to be due to the presence of fungal polysomes in the extracts of infected plants since, at 3 days, the amount of fungal mycelium was held to be minimal (sporulating pustules did not appear until 12–14 days after inoculation), and changes in polysomal mRNA appeared to be independent of inoculum density. Neither of these lines of evidence can be considered to be totally conclusive however.

Tani and Yamamoto (1978, 1979) demonstrated a specific stimulation of incorporation of ^{3}H-uridine into poly(A)-rich RNA (mRNA) 12 h after inoculation of oat leaves with an incompatible rust race. Cordycepin, an inhibitor of polyadenylation, blocked mRNA synthesis and stimulated growth of infection hyphae and haustorial development, suggesting a causal relationship between the specific mRNA synthesis after 12 h and the expression of resistance. No stimulation of incorporation into mRNA was detected in a compatible combination. Whilst this may suggest that compatibility does not depend on specific transcription, it is doubtful whether this technique would have been sensitive enough to reveal minor changes in mRNA transcription of a small number of genes.

Incompatibility in the interaction between *Phytophthora megasperma* f.sp. *glycinea* (syn. f.sp. *sojae*) and soybean also appears to involve rapid quantitative and qualitative changes in gene transcription (see also Chapters 2, 13, and 17). Within 4 h of inoculation with an incompatible, but not a compatible, race, poly(A)-rich mRNA was synthesized several times more rapidly than in uninoculated controls (Yoshikawa *et al.*, 1977). Translation of mRNAs in a cell-free system revealed specific qualitative differences in the patterns of polypeptide synthesized (Yoshikawa, 1980). Inhibitors of transcription (AMD) and translation (blasticidin S) both reduced the level of resistance and the accumulation of the phytoalexin, glyceollin. The *de novo* synthesis of specific proteins in this system has yet to be demonstrated. However, Hahlbrock *et al.* (1981) used anti-PAL antiserum to precipitate products of an *in vitro* protein-synthesizing system to show that elicitor-induced increases in PAL activity in cultured parsley cells were correlated with increases in specific PAL mRNA.

Only in the case of tobacco systemically infected with TMV has the effect of the virus on host mRNA been specifically investigated. TMV infection did not cause any alteration in the concentration of host polyadenylated mRNA. Moreover, these mRNAs from healthy and virus-infected plants had similar size distribution and poly(A) tails of similar lengths (Fraser and Gerwitz, 1980). In contrast, Coutts and Wagih (1981) observed an enhancement of poly(A) mRNA synthesis up to 24 h after inoculation in half-leaves of cowpea adjacent to halves infected with tobacco necrosis virus (TNV). Since local lesion formation during a hypersensitive reaction is associated with an effective virus-localizing mechanism, such synthetic activity might also comprise transcription of specific resistance genes. However, whereas hypersensitivity towards TMV is governed by the dominant N gene in tobacco, no corresponding mRNA has so far been found and inhibitors of DNA-dependent RNA synthesis such as AMD at most only weaken the localising mechanism without preventing the resistance reaction. Furthermore, resistance against viruses is genetically determined as often by recessive as by dominant genes, whereas the resistance reaction may be accompanied by similarly increased activity under both circumstances.

4. HOST TRANSLATION

4.1. Polyribosomes

In systemic virus infections and infections by biotrophic fungi the reduced amount of, in particular, chloroplast ribosomes diminishes the capacity of the host to synthesize chloroplast proteins.

Dyer and Scott (1972) reported a remarkable and specific decrease in chloroplast polyribosomes within 24 h of mildew infection of a compatible barley cultivar. This could result from the reduction in chl. rRNA and 70S ribosome synthesis or their accelerated breakdown (Section 3.1), but could also be caused by reduced rates of mRNA synthesis, reduced availability of tRNAs or substrates for protein synthesis, or the activity of RNases (see Section 4.2), or may be an artefact of extraction. The last explanation is less likely since Simpson et al. (1979) have demonstrated reduced in vivo populations of total polysomes in mildewed barley (see also Section 4.2). The net effect, however, must be a reduced synthesis of proteins encoded within the chloroplast genome. Since these include proteins involved in both the carboxylative and phosphorylative reactions of photosynthesis (Prebble, 1981), these effects on RNA synthesis could form the basis of the often reported reduction in photosynthetic rate in mildewed leaves (Chapter 15). For example, Callow (unpublished observations) has shown that 10 days after infection of cucumber leaves by mildew, the leaf contains very little fraction I protein (ribulose bisphosphate oxygenase/carboxylase).

Cytoplasmic protein synthesis may also be affected. Decreases in the amount of both chloroplast and cytoplasmic ribosomes correlate with symptom severity after infection of tobacco with different strains of CMV (Roberts and Wood, 1981). However, upon infection of older tobacco leaves with TMV, the normal loss of cytoplasmic ribosomes during senescence is retarded (Fraser, 1973). The rate of host protein synthesis is reduced by up to 75% during virus multiplication but then recovers. Since the rates of transcription or turnover of host mRNA were not altered by infection, this reduction is controlled at the translational level, possibly by competition between the mRNAs for virus and host proteins (Fraser and Gerwitz, 1980). However, the accelerated senescence of virus-infected leaves will itself lead to changes within the population of mRNAs and, consequently, affect host protein synthesis similarly to normal leaf ageing. Loss of ribosomes following the appearance of symptoms of lettuce necrotic yellows virus in *Nicotiana glutinosa* was correlated also with decreases in the length of polysomes and reduced protein synthesis (Randles and Coleman, 1972), phenomena occurring naturally during leaf senescence.

Host protein synthesis has frequently been reported to be inhibited in hypersensitive host–virus combinations. However, ribosomes are more numerous in zones around local necrotic lesions, reflecting an enormous synthetic activity. Upon administration of labelled amino acids, these precursors are not evenly distributed and reach only low specific activity in the zone around the developing lesions (De Laat and Van Loon, 1981). As protein content is usually increased upon infection of resistant plants, protein synthesis appears to be increased rather than decreased.

4.2. Enzyme changes

Pathogenesis and symptom development are accompanied by extensive metabolic alterations, reflected by changes in the activity of many enzymes. These alterations are to a large extent connected with the type and severity of the developing symptoms and are essentially similar after infection with fungi, bacteria, or viruses. For two fairly extreme symptom types due to a single virus, a systemic mosaic disease such as tobacco mosaic induced by TMV on *n* gene-containing tobacco varieties and a hypersensitive reaction in which necrotic local lesions develop on *N* tobaccos, a compilation of those data that have been well established is shown in Table 1. In mosaic-diseased leaves not only the effects on the synthesis of rRNAs, discussed above, but also several enzyme alterations appear similar to those occurring in non-infected plants after leaves have fully expanded and enter the phase of senescence, or when leaves are detached and become, therefore, subject to accelerated ageing. Thus, the changes in enzyme activities also support the idea that when mosaic symptoms are induced, part of the developmental programme normally expressed only during leaf senescence is already turned on in young, developing leaves.

Table 1. Comparison of changes in enzyme activities[a]

Enzyme	Systemic mosaic[b]	Natural senescence	Artificial ageing	Wounding or injury	Hypersensitive necrosis[b]
Ribulosebisphosphate carboxylase	−	−	−	0	0
Phosphoenolpyruvate carboxylase	−	−			+
Glycolate oxidase		−			0
Respiratory activity		−	+	+	+
Glucose-6-phosphate dehydrogenase	0	0	++	++	++
6-Phosphogluconate dehydrogenase	0	0	++	++	++
Aromatic biosynthesis					+
Phenylalanine ammonia-lyase	−	−	+	+	++
Cinnamic acid 4-hydroxylase					++
Caffeic acid O-methyltransferase					+
Polyphenoloxidase	+	−	++	++	+
Peroxidase	++	++	++	++	++
Catalase	−	+			
Ribonuclease	+	+	++	++	++
Acid phosphatase	0	0	++	++	+
Protease	−	−	+	+	+
1-Aminocyclopropane-1-carboxylic acid synthase	−	−	−	+	+

[a]+, Increase; −, decrease; 0, no change.
[b]virus-induced.

In contrast, most of the changes occurring during a hypersensitive reaction are similar to those occurring in non-infected plants as a result of artificial ageing, stress or injury, and differ from alterations during natural senescence. With increased symptom severity, from mottling, chlorosis, mosaic, and yellowing up to ring-spotting, enzyme changes in different virus–host plant combinations approach more and more a pattern resembling artificial ageing and wounding or injury, until upon systemic necrosis the changes are virtually identical with those characteristic of a hypersensitive reaction.

The nature of the changes in these enzymes has been explained in only a few cases. Density-labelling with D_2O has shown that increases in the key enzymes of phenylpropanoid biosynthesis, PAL and caffeic acid O-methyltransferase, result from *de novo* synthesis during hypersensitivity (see Chapter 17). Induction of phytoalexin biosynthesis in french bean cells treated with a fungal elicitor also involves *de novo* synthesis of PAL (Dixon and Lamb, 1979).

RNase undergoes substantial changes in activity during viral, bacterial, and fungal infection (Chakravorty and Shaw, 1977). In a number of compatible rust infections this increase occurs in two well defined phases, early and late, with only the early phase in incompatible interactions. These quantitative changes in activity are accompanied by changes in enzyme properties including substrate preference and in flax and wheat at least part of the change is due to novel RNase molecules distinct in properties from those of either non-infected host or parasite, although it is not clear whether these are the products of *de novo* enzyme synthesis or modifications of existing forms of enzyme. It has been suggested that the significance of these new RNases lies in their potential for post-transcriptional processing of new precursor RNA molecules synthesized following infection.

More precise information on this potential role has been obtained for mildew infections. Chakravorty and Scott (1979) isolated two RNase fractions from mildewed barley leaves. Qualitative changes in the properties of a minor, microsomal RNase fraction were detected as early as 24–48 h after inoculation but increased specific activities were detected only at 48 h. Changes in the specificity of the major, soluble leaf RNase for synthetic substrates were not detected until later stages of infection. Both enzymes were believed to be of host origin and showed different properties and substrate specificities. In experiments with mixed chloroplastic and cytoplasmic polysomes, polysomal mRNAs were hydrolysed *in vitro* by the microsomal enzyme only (Simpson *et al.*, 1979). The enzymes from healthy and infected leaves were equally effective and appeared to hydrolyse mRNAs at identical cleavage sites and it is not entirely clear, therefore, what the precise significance of the altered substrate specificity for synthetic polynucleotides following infection is. Simpson *et al.* (1979) suggested that the significance of the changes in microsomal RNase in infected leaves lies in the earlier demonstration (see Section 4.1) that within 24 h of infection of barley by a susceptible mildew race there is a loss of host polyribosomes and they used the peptidyl-puromycin transferase reaction to show that this breakdown in polysomes occurs *in vivo*. However, the earlier demonstration of Dyer and Scott (1972) showed that polysome breakdown in compatible interactions was specific for chloroplastic polysomes, whereas the minor microsomal RNase does not appear to be specific for polysome mRNA hydrolysis *in vitro*. At present therefore, it is difficult to assess the significance of changes in RNase in relation to the important events in host–pathogen interactions.

4.3. Pathogenesis-related proteins

In several virus–host plant combinations the development of symptoms may be accompanied by the appearance of one or more new host-specific protein components. Such proteins have been found in several tobacco species after infection with a number of different viruses inducing symptoms varying from a very faint green mottling to systemic necrosis (Van Loon and Van Kammen, 1970; Kassanis *et al.*, 1974; Tas and Peters, 1977), in cucumber infected with tomato spotted wilt virus, CMV, and TNV (Tas and Peters, 1977; Andebrhan *et al.*, 1980), in cowpea infected with TNV (Coutts, 1978), in several test plants infected with the citrus exocortis viroid (Conejero *et al.*, 1979), in *Gomphrena globosa* infected with any of three necrotizing viruses (Redolfi *et al.*, 1982), and in beans infected with tomato bushy stunt virus, AMV, or TNV (Redolfi and Cantisani, 1981). A comparable protein has been described to be greatly stimulated, if not newly induced, in tomato leaves infected with the fungus *Cladosporium fulvum* (De Wit and Bakker, 1980).

These proteins are particularly evident when hypersensitive necrosis occurs. When a total soluble protein extract from hypersensitively reacting tobacco leaves is subjected to electrophoresis in polyacrylamide gels, four new proteins, designated 1a, b, c, and 2, are evident (cf. Figure 6). These proteins have been named 'pathogenesis-related proteins' (PRs). The same proteins are induced in tobacco after infection with fungi or bacteria that induce necrosis (Gianinazzi *et al.*, 1980). However, they do not appear as a result of senescence or artificial ageing, mechanical or chemical wounding, or water, drought, or salt stress. In locally infected plants the proteins are not confined to the affected leaves, but substantial quantities may be found in healthy plant parts. Since the proteins are also induced in virus–host plant combinations in which mild symptoms develop without necrosis, these obser-vations suggest that their appearance is more than just a result of senescence, ageing, stress, or injury.

PRs are selectively extracted at low pH, where they remain soluble, whereas the majority of the leaf proteins are not. In hypersensitively reacting tobacco leaves, additional PRs, provisionally designated N-S, are thus

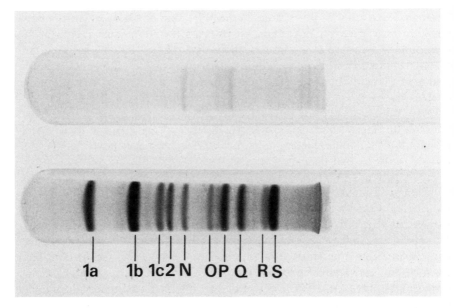

Figure 6. Electrophoretic patterns on polyacrylamide gels of the pH 3-soluble, pathogenesis-related proteins-containing fraction from Samsun NN tobacco leaves, 17 days after inoculation with water (upper) or tobacco mosaic virus (lower)

revealed (Figure 6). The proteins are related in that 1a, b, and c, and possibly P and Q, are charge isomers of about 15 000 daltons. Likewise, 2 and N are charge isomers. PR 1a and b are very similar in amino acid composition and peculiar in that they contain about 10% aromatic amino acids (Antoniw *et al.*, 1980). All ten PRs are resistant to the endogenous tobacco proteases and to trypsin and chymotrypsin, but it appears unlikely that they are just stable end-products of proteolysis. They are generated neither by repeated freezing and thawing of non-infected leaves, nor by treatment of protein extracts from non-infected leaves with proteases or detergents. Furthermore, when infected leaves are extracted in the presence of inhibitors of protease activity, the pattern remains unaffected.

Since protease activity is strongly increased in hypersensitively reacting tobacco, the PRs appear to be extremely well adapted to function in an environment subject to accelerated proteolysis such as occurs in ageing and dying cells. However, at least 60% of the PRs is present outside the cellular plasma membrane, even in non-infected plant parts. They can be washed from the intercellular spaces under conditions where no cytoplasmic enzymes are released and they are present in only very small amounts in protoplasts from such leaves. They are not isozymic forms of a number of the enzymes known to increase after infection and, thus far, their function is unknown.

There is indirect evidence, however, that PRs may be involved in limiting virus multiplication or spread and, in particular, are responsible for the state of acquired resistance in tobacco, because when these proteins are present, the multiplication and spread of TMV are invariably reduced. However, after primary infection, PRs become apparent only when symptoms are already expressed and the localizing reaction has been initiated. In cucumber and cowpea, PRs are not present in leaves distant from the necrotic tissue, although those leaves are at least partially resistant to challenge inoculation (Wagih and Coutts, 1981). Similarly, the presence of the stimulated protein in tomato leaflets opposite to inoculated ones did not prevent fungal growth upon challenge (De Wit and Bakker, 1980).

PRs may accumulate to up to 10% of the total soluble protein fraction in tobacco. Owing to the uneven distribution of amino acids taken up by hypersensitively reacting leaves (Duchesne et al., 1977; De Laat and Van Loon, 1981), labelling of PRs as evidence of de novo synthesis has not been demonstrated in virus-infected leaves, but their appearance can be inhibited by AMD. In contrast, in tomato leaves inoculated with an avirulent race of Cladosporium fulvum, $^{35}SO_4^{2-}$ was preferentially incorporated at the site of migration of the stimulated host protein on gels, suggesting that this protein is indeed synthesized de novo after infection (De Wit and Bakker, 1980).

It was shown recently that PRs are induced naturally in healthy tobacco plants when these are in the phase of monocarpic senescence when they flower abundantly and then die (Fraser, 1981), or when healthy tobacco or cucumber leaves are subjected to mannitol-induced osmotic stress during which the cells become plasmolysed (Wagih and Coutts, 1981). They are also found in tobacco callus, whether habituated or grown on a hormone-containing nutrient medium (Antoniw et al., 1981). Antoniw et al. (1981) attribute the occurrence of PRs in callus to the presence in the medium of the cytokinin benzyladenine and the auxins 2,4-D and IAA, because these growth regulators by themselves, when injected into leaves at high concentrations, induce PRs. Tobacco PRs have also been induced artificially by application of polyacrylic acid (Gianinazzi and Kassanis 1974), ethephon, from which ethylene is released in plant tissues (Van Loon, 1977), the immediate natural precursor of ethylene, ACC, and benzoic acid and its derivatives salicylic acid and aspirin (White, 1979). All of these chemicals induce resistance against further infection.

Ethylene not only induces the appearance of PRs in non-infected tobacco plants, but also increases the activity of those enzymes which are characteristically stimulated during a hypersensitive reaction (Van Loon, 1982). Stimulation of ethylene synthesis is itself a typical response to stress or injury, and from hypersensitively reacting leaves large quantities of ethylene evolve with a peak at the time when symptoms become apparent. This stimulation is initiated by a large increase in ACC-synthase activity (cf. Table 1). These observations suggest that ethylene is the physiological inducer of PRs in tobacco.

4.4. Host-mediated antiviral activity

Extracts from several plants prevent the development of symptoms when applied at the same time as or shortly before virus inoculation. Although most inhibitors may interfere with virus entry into the leaf, pokeweed antiviral peptide, purified from the leaves of *Phytolacca americana,* inhibits protein synthesis on eukaryotic ribosomes by interfering with the elongation factor EF 2-mediated translocation step (Irvin *et al.,* 1980). Similar inhibitory proteins from carnation leaves may act in the same way (Stirpe *et al.,* 1981). However, these proteins do not inhibit protein synthesis in systems employing homologous ribosomes from pokeweed and carnation, respectively, and therefore interfere only with mechanical transmission of viruses from these hosts.

As a result of a primary infection with some viruses, fungi, or bacteria, several plant species acquire a non-specific resistance against further virus infection, expressed throughout the entire plant (Ross, 1966; Loebenstein, 1972). Such systemic acquired resistance is assumed to depend on the movement of antiviral factors, produced in infected cells, to other regions of a plant, conferring resistance to yet uninfected cells and thereby limiting virus spread. In tobacco cultivars that react hypersensitively to TMV, the development of induced resistance is blocked by AMD (Loebenstein, 1972), suggesting transcription to be required. A superficially similar resistance can be induced artificially by treating plants with yeast RNA, synthetic ds-RNA, polyanions (Loebenstein, 1972), polyacrylic acid (Gianinazzi and Kassanis, 1974), ethephon (Van Loon, 1977), aspirin (White, 1979), or an inhibitor extracted from *Boerhaavia diffusa* roots. The last inhibitor causes different plants to produce a highly active antiviral agent (AVA) that is neither host- nor virus-specific and prevents infection at both treated and untreated plant parts. Moreover, production of AVA is inhibited when AMD is applied soon after inhibitor treatment (Verma and Awasthi, 1980). A putative antiviral factor (AVF) from *N. glutinosa* has been characterized by Sela (1981) as a phosphoglycoprotein. AVF is released early after infection by hypersensitively reacting tobacco plants, apparently as a result of phosphorylation of an activating enzyme, and, when mixed with TMV, interferes with infection of *Datura stramonium.* It has been suggested that minimal amounts of AVF suffice to inhibit virus translation and replication, but so far conclusive evidence has not been presented.

5. CONCLUSION

Adaptation to pathological conditions of biotic or abiotic origin may involve major shifts in metabolism, concurrent with specific alterations at the transcriptional and translational levels. The similarities in the types of symptoms induced by infections with fungi, bacteria, and viruses indicate that most of the underlying metabolic changes are reactions of the host plant. Patterns emerge showing similarities to senescence, ageing, stress, and injury, suggesting that we may learn more about pathogenesis when we understand more fully how senescence and ageing are regulated and by what mechanisms stress and injury affect plant metabolism. Almost all changes described so far become evident only after symptoms have become apparent and do not indicate how pathogenesis is initiated. However, the ability of the host to react in the particular way it does is constitutively present in a 'cryptic' form, and is evoked by the triggering action of the infecting pathogen. The genetic information underlying symptom expression may never be expressed during normal plant development or, if it is, upon infection it is expressed in an aberrant fashion, resulting in abnormal patterns of growth and development.

REFERENCES

Andebrhan, T., Coutts, R. H. A., Wagih, E. E., and Wood, R. K. S. (1980). Induced resistance and changes in the soluble protein fraction of cucumber leaves locally infected with *Colletotrichum lagenarium* or tobacco necrosis virus, *Phytopathol. Z.*, **98**, 47–52.

Antoniw, J. F., Ritter, C. E., Pierpoint, W. S., and Van Loon, L. C. (1980). Comparison of three pathogenesis-related proteins from plants of two cultivars of tobacco infected with TMV, *J. Gen Virol.*, **47**, 79–87.

Antoniw, J. F., Kueh, J., Walkey, D. G. A., and White, R. F. (1981). The presence of pathogenesis-related proteins in callus of Xanthi-nc tobacco, *Phytopathol. Z.*, **101**, 179–184.

Beachy, R. N., Zaitlin, M., Bruening, G., and Israel, H. W. (1976). A genetic map for the cowpea strain of TMV, *Virology*, **73**, 498–507.

Beier, H., Siler, D. J., Russell, M. L., and Bruening, G. (1977). Survey of susceptibility to cowpea mosaic virus among protoplasts and intact plants from *Vigna sinensis* lines, *Phytopathology*, **67**, 917–921.

Bennet, J., and Scott, K. J. (1971a). Ribosome metabolism in mildew-infected barley leaves, *FEBS Lett.*, **16**, 93–95.

Bennet, J., and Scott, K. J. (1971b). Inorganic polyphosphates in the wheat stem rust fungus and in rust-infected wheat leaves, *Physiol. Plant Pathol.*, **1**, 185–198.

Bhattacharya, P. K., and Shaw, M. (1967). The physiology of host–parasite relations. XVIII. Distribution of tritium-labelled cytidine, uridine and leucine in wheat leaves infected with the stem rust fungus, *Can. J. Bot.*, **45**, 555–563.

Bhattacharya, P. K., Naylor, J. M., and Shaw, M. (1965). Nucleic acid and protein changes in wheat leaf nuclei during rust infection, *Science*, **150**, 1605–1607.

Bienz, M., and Kubli, E. (1981). Wild-type $tRNA_G^{Tyr}$ reads the TMV RNA stop codon. but Q base-modified $tRNA_Q^{Tyr}$ does not, *Nature (London)*, **294**, 188–190.

Bos, L. (1978). *Symptoms of Virus Disease in Plants*, 3rd Ed., PUDOC, Wageningen.

Bruening, G., Beachy, R. N., Scalla, R., and Zaitlin, M. (1976). *In vitro* and *in vivo* translation of the ribonucleic acids of a cowpea strain of tobacco mosaic virus, *Virology*, **71**, 498–517.

Callow, J. A. (1973). Ribosomal RNA metabolism in cucumber leaves infected by *Erysiphe cichoracearum*, *Physiol. Plant Pathol.*, **3**, 249–257.

Callow, J. A. (1975). Endopolyploidy in maize smut neoplasms induced by the maize smut fungus, *Ustilago maydis, New Phytol.*, **75**, 253–257.

Callow, J. A. (1976). Nucleic acid metabolism in biotrophic infections, in *Biochemical Aspects of Plant–Parasite Relationships*, (Eds J. Friend and D.R. Threlfall), Academic Press, London, pp. 305–330.

Callow, J. A. (1983). Recognition in higher plant–fungal pathogen interactions, in *Cellular Interactions in Plants* (Eds H. Linskens, and J. H. Heslop-Harrison), *Encyclopaedia of Plant Physiology*, New Series, Springer Verlag, Berlin, Heidelberg, and New York, in press.

Chakravorty, A. K., and Scott, K. J. (1979). Changes in two barley leaf ribonuclease fractions during infection by the powdery mildew fungus, *Physiol. Plant Pathol.*, **14**, 85–97.

Chakravorty, A. K., and Shaw, M. (1971). Changes in the transcription pattern of flax cotyledons after inoculation with flax rust, *Biochem. J.*, **123**, 551–557.

Chakravorty, A. K., and Shaw, M. (1977). The role of RNA in host–parasite specificity, *Annu. Rev. Phytopathol.*, **15**, 135–151.

Conejero, V., Picazo, I., and Segado, P. (1979). Citrus exocortis viroid (CEV): protein alterations in different hosts following viroid infection, *Virology*, **97**, 454–456.

Coutts, R. H. A. (1978). Alterations in the soluble protein patterns of tobacco and cowpea leaves following inoculation with tobacco necrosis virus, *Plant Sci. Lett.*, **12**, 189–197.

Coutts, R. H. A., and Wagih, E. E. (1981). Alterations in RNA and protein metabolism in uninoculated half-leaves of cowpea adjacent to tobacco necrosis virus infected halves, *Plant Sci. Lett.*, **21**, 51–59.

Covey, S. N., and Hull, R. (1981). Transcription of cauliflower mosaic virus DNA. Detection of transcripts, properties, and location of the gene encoding the virus inclusion body protein, *Virology*, **111**, 463–474.

Dawson, W. O., and Schlegel, D. E. (1976). The sequence of inhibition of tobacco mosaic virus synthesis by actinomycin D, 2-thiouracil, and cycloheximide in a synchronous infection, *Phytopathology*, **66**, 177–181.

Dawson, W. O., and White, J. L. (1979). A temperature-sensitive mutant of tobacco mosaic virus deficient in synthesis of single-stranded RNA, *Virology*, **93**, 104–110.

De Jager, C. P. (1976). Genetic analysis of cowpea mosaic virus mutants by supplementation and reassortment tests, *Virology*, **70**, 151–163.

De Laat, A. M. M., and Van Loon, L. C. (1981). Complications in interpreting precursor/product relationships by labeling experiments. Methionine as the precursor of ethylene in tobacco leaves, *Z. Pflanzenphysiol.*, **103**, 199–205.

De Wit, P. J. G. M., and Bakker, J. (1980). Differential changes in soluble tomato leaf proteins after inoculation with virulent and avirulent races of *Cladosporium fulvum* (syn. *Fulvia fulva*), *Physiol. Plant Pathol.*, **17**, 121–130.

Dickson, E., Robertson, H. D., Niblett, C. L., Horst, R. K., and Zaitlin, M. (1979). Minor differences between nucleotide sequences of mild and severe strains of potato spindle tuber viroid, *Nature (London)*, **277**, 60–62.

Dixon, R. A., and Lamb, C. J. (1979). Stimulation of *de novo* synthesis of L-phenylalanineammonia-lyase in relation to phytoalexin. accumulation in *Colletotrichum lindemuthianum* elicitor-treated cell suspension-cultures of french bean (*Phaseolus vulagris*), *Biochim. Biophys. Acta*, **586**, 453–463.

Dorssers, L., Zabel, P., Van der Meer, J., and Van Kammen, A. (1982). Purification of a host-encoded RNA-dependent RNA polymerase from cowpea mosaic virus infected cowpea leaves, *Virology*, **116**, 236–249.

Duchesne, M., Fritig, B., and Hirth, L. (1977). Phenylalanine ammonia-lyase in tobacco mosaic virus-infected hypersensitive tobacco. Density-labelling evidence of *de novo* synthesis, *Biochim. Biophys. Acta*, **485**, 465–481.

Duda, C. T., Zaitlin, M., and Siegel, A. (1973). *In vitro* synthesis of double-stranded RNA by an enzyme system isolated from tobacco leaves, *Biochim. Biophys. Acta*, **319**, 62–71.

Dyer, T. A., and Scott, K. J. (1972). Decrease in chloroplast content of barley leaves infected by powdery mildew, *Nature (London)*, **236**, 237–238.

Ellingboe, A. H. (1982). Genetical aspects of active defense, in *Active Defense Mechanisms in Plants* (Ed. R. K. S. Wood), Plenum Press, New York and London, pp. 179–192.

Flynn, J. G., Chakravorty, A. K., and Scott, K. J. (1976). Changes in the transcription pattern of wheat leaves during early stages of infection by *Puccinia graminis tritici, Proc. Aust. Biochem. Soc.*, **9**, 44 (Abstr.).

Franck, A., Guilley, H., Jonard, G., Richards, K., and Hirth, L. (1980). Nucleotide sequence of cauliflower mosaic virus DNA, *Cell*, **21**, 285–294.

Franssen, H., Goldbach, R., Broekhuijsen, M., Moerman, M., and Van Kammen, A. (1982). Expression of middle-component RNA of cowpea mosaic virus: *in vitro* generation of a precursor to both capsid proteins by a bottom-component RNA-encoded protease from infected cells, *J. Virol.*, **41**, 8–17.

Fraser, R. S. S. (1969). Effects of two TMV strains on the synthesis and stability of chloroplast ribosomal RNA in tobacco leaves, *Mol. Gen. Genet.*, **106**, 73–79.

Fraser, R. S. S. (1973). The synthesis of tobacco mosaic virus RNA and ribosomal RNA in tobacco leaves, *J. Gen. Virol.*, **18**, 267–279.

Fraser, R. S. S. (1981). Evidence for the occurrence of the "pathogenesis-related" proteins in leaves of healthy tobacco plants during flowering, *Physiol. Plant Pathol.*, **19**, 69–76.

Fraser, R. S. S., and Gerwitz, A. (1980). Tobacco mosaic virus infection does not alter the polyadenylated messenger RNA content of tobacco leaves, *J. Gen. Virol.*, **46**, 139–148.

Gianinazzi, S., and Kassanis, B. (1974). Virus resistance induced in plants by polyacrylic acid, *J. Gen. Virol.*, **23**, 1–9.

Gianinazzi, S., Ahl, P., Cornu, A., and Scalla, R. (1980). First report of host b-protein appearance in response to a fungal infection in tobacco, *Physiol. Plant Pathol.*, **16**, 337–342.

Goldbach, R., Rezelman, G., Zabel, P., and Van Kammen, A. (1980). Independent replication and expression of B-component RNA of cowpea mosaic virus, *Nature (London)*, **286**, 297–300.

Goldbach, R., Rezelman, G., Zabel, P., and Van Kammen, A. (1982). Expression of the bottom-component RNA of cowpea mosaic virus: evidence that the 60-kilodalton VPg precursor is cleaved into single VPg and a 5-kilodalton polypeptide, *J. Virol.*, **42**, 630–635.

Hahlbrock, K., Lamb, C. J., Purwin, C., Ebel, J., Fautz, G., and Schaffer, E. (1981). Rapid response of suspension cultured parsley cells to the elicitor from *Phytophthora megasperma* var. *sojae, Plant Physiol.*, **67**, 768–773.

Hall, T. C. (1979). Transfer RNA-like structures in viral genomes, *Int. Rev. Cytol.*, **60**, 1–26.

Hardy, S. F., German, T. L., Loesch-Fries, L. S., and Hall, T. C. (1979). Highly active template-specific RNA-dependent RNA polymerase from barley leaves infected with brome mosaic virus, *Proc. Natl. Acad. Sci. USA,* **76**, 4956–4960.

Hariharasubramanian, V., Hadidi, A., Singer, B., and Fraenkel-Conrat, H. (1973).

Possible identification of a protein in brome mosaic virus infected barley as a component of viral RNA polymerase, *Virology*, **54**, 190–198.

Huber, R. (1979). Proteins synthesized in tobacco mosaic virus infected protoplasts, *Meded. Landbouwhogesch. Wageningen*, 79–15.

Huber, R., Rezelman, G., Hibi, T., and Van Kammen, A. (1977). Cowpea mosaic virus infection of protoplasts from Samsun tobacco leaves, *J. Gen. Virol.*, **34**, 315–323.

Ikegami, M., and Fraenkel-Conrat, H. (1979). Characterization of double-stranded ribonucleic acid in tobacco leaves, *Proc. Natl. Acad. Sci. USA*, **76**, 3637–3640.

Irvin, J. D., Kelly, T., and Robertus, J. D. (1980). Purification and properties of a second antiviral protein from *Phytolacca americana* which inactivates eukaryotic ribosomes, *Arch. Biochem. Biophys.*, **200**, 418–425.

Kaesberg, P. (1976). Translation and structure of the RNAs of brome mosaic virus, in *Animal Virology* (Eds D. Baltimore, A. S. Huang, and C. F. Fox), Academic Press, New York, pp. 555–565.

Kassanis, B., Gianinazzi, S., and White, R. F. (1974). A possible explanation of the resistance of virus-infected plants to second infection, *J. Gen. Virol.*, **23**, 11–16.

Linthorst, H. J. M., Bol. J. F., and Jaspars, E. M. J. (1980). Lack of temperature sensitivity of RNA-dependent RNA polymerases isolated from tobacco plants infected with *ts* mutants of alfalfa mosaic virus, *J. Gen. Virol.*, **46**, 511–515.

Loebenstein, G. (1972). Localization and induced resistance in virus-infected plants, *Annu. Rev. Phytopathol.*, **10**, 177–206.

Maule, A. J., Boulton, M. I., and Wood, K. R. (1980). Resistance of cucumber protoplasts to cucumber mosaic virus: a comparative study, *J. Gen. Virol.*, **51**, 271–279.

Mundry, K. W., and Gierer, A. (1958). Die Erzeugung von Mutationen des Tabakmosaikvirus durch chemische Behandlung seiner Nucleinsäure *in Vitro, Z. Vererbungsl.*, **89**, 614–630.

Nassuth, A., Alblas, F., and Bol, J. F. (1981). Localization of genetic information involved in the replication of alfalfa mosaic virus, *J. Gen. Virol.*, **53**, 207–214.

Odell, J. T., Dudley, R. K., and Howell, S. H. (1981). Structure of the 19S RNA transcript encoded by the cauliflower mosaic virus genome, *Virology*, **111**, 377–385.

Paterson, R., and Knight, C. A. (1975). Protein synthesis in tobacco protoplasts infected with tobacco mosaic virus, *Virology*, **64**, 10–22.

Pelham, H. R. B. (1979). Synthesis and proteolytic processing of cowpea mosaic virus proteins in reticulocyte lysates, *Virology*, **96**, 463–477.

Prebble, J. N. (1981). *Mitochondria, Chloroplasts and Bacterial Membranes*, Longmans, London and New York.

Pure, G. A., Chakravorty, A. K., and Scott, K. J. (1979). Cell-free translation of polysomal mRNA isolated from healthy and rust-infected wheat leaves, *Physiol. Plant Pathol.*, **15**, 201–205.

Quick, W. A., and Shaw, M. (1964). The physiology of host–parasite relations. XIV. The effect of rust infection on the nucleic acid content of wheat leaves, *Can. J. Bot.*, **42**, 1531–1540.

Rackwitz, H.-R., Rohde, W., and Sänger, H. L. (1981). DNA-dependent RNA polymerase II of plant origin transcribes viroid RNA into full-length copies, *Nature (London)*, **291**, 297–301.

Randles, J. W., and Coleman, D. (1972). Changes in polysomes in *Nicotiana glutinosa* L. leaves infected with lettuce necrotic yellows virus, *Physiol. Plant Pathol.*, **2**, 247–258.

Redolfi, P., and Cantisani, A. (1981). Protein changes and hypersensitve reaction in virus infected bean leaves, *Abstr. Fifth Int. Congr. Virology, Strasbourg*, No. P26/06, p. 264.

Redolfi, P., Vecchiati, M., and Gianinazzi, S. (1982). Changes in the soluble leaf protein constitution of *Gomphrena globosa* during the hypersensitive reaction to different viruses, *Phytopathol. Z.*, **103**, 48–54.

Rezelman, G., Goldbach, R., and Van Kammen, A. (1980). Expression of bottom component RNA of cowpea mosaic virus in cowpea protoplasts, *J. Virol.*, **36**, 366–373.

Roberts, P. L., and Wood, K. R. (1981). Decrease in ribosome levels in tobacco infected with a chlorotic strain of cucumber mosaic virus, *Physiol. Plant Pathol.*, **19**, 99–111.

Romaine, C. P., and Zaitlin, M. (1978). RNA-dependent RNA polymerase in uninfected and tobacco mosaic virus-infected tobacco leaves: viral-induced stimulation of a host polymerase activity, *Virology*, **86**, 241–253.

Ross, A. F. (1966). Systemic effects of local lesion formation, in *Viruses of Plants* (Eds A. B. R. Beemster and J. Dijkstra), North-Holland, Amsterdam, pp. 127–150.

Rottier, P. J. M., Rezelman, G., and Van Kammen, A. (1979). The inhibiton of cowpea mosaic virus replication by actinomycin D, *Virology*, **92**, 299–309.

Sakai, F., and Takebe, I. (1974). Protein synthesis in tobacco mesophyll protoplasts induced by tobacco mosaic virus infection, *Virology*, **62**, 426–433.

Sänger, H. L. (1980). Viroids: biology, structure and possible functions, in *Genome Organization and Expression in Plants* (Ed. C. J. Leaver), Plenum Press, New York, pp. 553–601.

Sela, I. (1981). Interferon-like substance from virus-infected plants, *Perspect. Virol.*, **11**, 129–139.

Simpson, R. S., Chakavorty, A. K., and Scott, K. J. (1979). Selective hydrolysis of barley leaf polysomal messenger RNAs during the early stages of powdery mildew infection, *Physiol. Plant Pathol.*, **14**, 245–258.

Stirpe, F., Williams, D. G., Onyon, L. J., Legg, R. F., and Stevens, W. A. (1981). Dianthins, ribosome-damaging proteins with anti-viral properties from *Dianthus caryophyllus* L. (carnation), *Biochem. J.*, **195**, 399–405.

Tani, T., and Yamamoto, H. (1978). Nucleic acid and protein synthesis in association with the resistance of oat leaves to crown rust, *Physiol. Plant Pathol.*, **12**, 113–121.

Tani, T., and Yamamoto, H. (1979). RNA and protein synthesis and enzyme changes during infection, in *Recognition and Specificity in Plant Host–Parasite Interactions* (Eds J. M. Daly and I. Uritani), Japan Scientific Societies Press, Tokyo, pp. 273–286.

Tani, T., Yoshikawa, M., and Naito, N. (1973). Effect of rust infection of oat leaves on cytoplasmic and chloroplast ribosomal ribonucleic acids, *Phytopathology*, **63**, 491–494.

Tas, P. W. L., and Peters, D. (1977). The occurrence of a soluble protein (E_1) in cucumber cotyledons infected with plant viruses, *Neth. J. Plant Pathol.*, **83**, 5–12.

Van Loon, L. C. (1977). Induction by 2-chloroethylphosphonic acid of viral-like lesions, associated proteins and systemic resistance in tobacco, *Virology*, **80**, 417–420.

Van Loon, L. C. (1982). Regulation of changes in proteins and enzymes associated with active defence against virus infection, in *Active Defense Mechanisms in Plants* (Ed. R. K. S. Wood), Plenum Press, New York and London, pp. 247–273.

Van Loon, L. C., and Van Kammen, A. (1970). Polyacrylamide disc electrophoresis of the soluble leaf proteins from *Nicotiana tabacum* L. var. 'Samsun' and 'Samsun NN'. II. Changes in protein constitution after infection with tobacco mosaic virus, *Virology*, **40**, 199–211.

Van Vloten-Doting, L., and Neeleman, L. (1980). Translation of plant virus RNAs, in *Genome Organization and Expression in Plants* (Ed. C. J. Leaver), Plenum Press, New York, pp. 511–527.

Verma, H. N., and Awasthi, L. P. (1980). Occurrence of a highly active antiviral agent in plants treated with *Boerhaavia diffusa* inhibitor, *Can. J. Bot.*, **58**, 2141–2144.

Wagih, E. E., and Coutts, R. H. A. (1981). Similarities in the soluble protein profiles of leaf tissue following either a hypersensitive reaction to virus infection or plasmolysis, *Plant Sci. Lett.*, **21**, 61–69.

White, R. F. (1979). Acetylsalicylic acid (aspirin) induces resistance to tobacco mosaic virus in tobacco, *Virology*, **99**, 410–412.

Wolf, G. (1968). On the incorporation of ^{32}P into various nucleic acid fractions of rust-infected primary leaves of wheat, *Neth. J. Plant Pathol.*, **74**, (Suppl.), 19–23.

Woolhouse, H. W. (1974). Longevity and senescence in plants, *Sci. Prog.*, **61**, 123–147.

Yoshikawa, M. (1980). Resistance mechanisms underlying the soybean–*Phytophthora megasperma* f. sp. *glycinea* infection, *Plant Prot.*, **34**, 248–256.

Yoshikawa, M., Masago, H., and Keen, N. T. (1977). Activated synthesis of poly(A)-containing messenger RNA in soybean hypocotyls infected with *Phytophthora megasperma* var. *sojae, Physiol. Plant Pathol.*, **10**, 125–138.

Zagorski, W. (1978). Translational regulation of expression of the brome-mosaic-virus RNA genome *in vitro, Eur. J. Biochem.*, **86**, 465–472.

Further Reading

Fraenkel-Conrat, H. and Wagner, R. R. (Eds) (1977). *Comprehensive Virology, Vol. 11, Regulation and Genetics. Plant Viruses*, Plenum Press, New York.

Atabekov, J. G., and Morozov, S. Y. (1979). Translation of plant virus messenger RNAs, *Adv. Virus Res.*, **25**, 1–91.

Hall, T. C., and Davies, J. W. (Eds) (1979), *Nucleic Acids in Plants*, Vol. 2, CRC Press, Boca Raton, FL.

Matthews, R. E. F. (1981), *Plant Virology*, 2nd ed., Academic Press, New York.

Goodman, R. N., Király, Z., and Wood, K. R. (1983), *The Biochemistry and Physiology of Infectious Plant Disease*, 2nd ed., Van Nostrand, Princeton, NJ.

Biochemical Plant Pathology
Edited by J. A. Callow
© 1983 John Wiley & Sons Ltd

19

Hormones and Metabolic Regulation in Disease

ERICH F. ELSTNER

Institut für Botanik und Mikrobiologie, Biochemisches Labor, Technische Universität, Arcisstrasse 21, 8000 München 2, West Germany

1. INTRODUCTION

There are numerous reports of changes in the concentrations of growth regulators (plant hormones) during the development of host–parasite relationships and/or the expression of disease symptoms. However, most tests reported in the past were based on unspecific bioassays which did not take into account (a) possible 'stress' situations (handling, wounding, flooding, drought) and thus artifactually introduced changes accompanying the test, (b) the extremely complex relationships and regulatory steps existing between the different growth regulators, and (c) the production of compounds by many pathogens in addition to phytohormones that act or function similarly to natural, endogenous host growth regulators. Certain toxins (e.g. helminthosporal, victorin, fusicoccin) or growth factors (e.g. γ-aminobutyric acid) produced by the pathogen may interfere with the mechanisms of symptom induction (Sequeira, 1973; Daly and Knoche, 1976; Peters and Lippincott, 1976). Natural or synthetic steroids may, similar to indoleacetic acid (IAA) and its oxidized metabolites (Grambow and Tücks 1979), influence the growth and development of the pathogen and thus also the host, as shown in the case of the sexual reproduction of *Phytophtora cactorum* (Nes *et al.,* 1980). Even when plants are protected against diseases, for example by spraying with certain insecticides, the primary cause of the disease (i.e. the insects) may be removed but development-determining factors can be secondarily introduced. Thus, metabolites of the insecticide carbofuran have been shown to act as inhibitors of IAA oxidation, with kinetics that differ from other well known naturally occurring inhibitors (Lee *et al.,* 1980).

All growth regulators, independent of their provenance or structure, may be divided into those compounds which induce 'long-term' effects (probably via nucleic acid metabolism) and those resulting in 'short-term' effects mediated mainly via membrane alterations or direct influence on existing enzymic systems. In both cases, the activities and concentrations of certain enzymes may be changed by activation of non-active or latent enzymes, by *de novo* synthesis of enzymes, or by the production of activators, inhibitors, or degradative processes. In most cases, it is extremely difficult or even impossible to assess accurately the validity of the assumption that metabolic changes, hormonal fluctuations, disease symptoms, and development of the host–parasite relationship are causally and not just incidentally connected. In 1973, Sequeira stated that there were no adequate chemical techniques for the quantitative determination of *all* phytohormones. Today we possess very

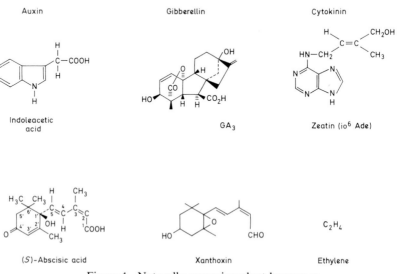

Figure 1. Naturally occurring plant hormones

sensitive methods for the simultaneous determination of 'standard' phyto-hormones (Figure 1) (Weiler, 1981), so that we can look forward in the near future to the resolution of many as yet unresolved problems.

In this chapter the involvement of the different phytohormones in the development of certain host–parasite relationships and corresponding disease symptoms will be outlined.

2. AUXINS

Auxins (Figure 1) are derivatives of the amino acid tryptophan and are involved in the natural regulation of plant growth, e.g. elongation of develop-ing plant tissues. The changes in auxin concentration after infection may thus cause abnormal growth responses in the host. Little information is available, however, on how these changes in auxin level are brought about; there are several possibilities:

(a) increased or decreased synthesis by host and/or pathogen;
(b) increased or decreased oxidative degradation by host and/or pathogen;
(c) compartmentalization, binding, or their reverse;
(d) conjugation (chemical modification) or deconjugation; or
(e) changes in the transport forms or mechanisms (Heitefuss and Williams, 1976; Moore 1979; MacMillan, 1980; Skoog, 1980).

In the case of auxin-mediated growth responses, two separate components may operate consecutively, viz. a fast cell elongation reaction mediated by IAA via cell wall acidification ('first response'; Cleland and Rayle, 1978) and a slower, second component, also mediated by IAA, but working via gene activation ('second response'). Since in many cases auxins and cytokinins have been shown to act as regulators of ethylene formation on the one hand, while cytokinins may also be involved as regulators of the second response of cell elongation on the other, it is obvious that an extremely complicated situation has to be envisaged after triggering or suppressing the hormone recognition 'sensors' of the host.

Changes in auxin levels have been reported in numerous fungal, bacterial, and viral diseases. Some well known plant diseases will be briefly surveyed in this context.

Crown galls are plant tumours induced by *Agrobacterium tumefaciens* (see also Chapter 5). *A. tumefaciens* in liquid culture produces IAA-like growth substances and IAA in turn is necessary for the growth of crown gall tissue, which contains about ten times the amount of various types of auxins, e.g. IAA, indoleacetonitrile (IAN), and other indole derivatives, found in normal callus tissue. The above-mentioned estimations have been criticized, how-ever, since artefacts due to overestimations inherent in 'crude assays' based on fresh weight (different tissues do not have the same density) and different extraction techniques, cannot be excluded (see Sequeira, 1973).

Two elegant approaches to the problem concerning the role of auxins in gall formation have been reported. In one study, mutants of the olive-and oleander-knot pathogen *Pseudomonas savastanoi* which are resistant to α-methyltryptophane (MT), an inhibitor of anthranilate synthase (an enzyme involved in IAA biosynthesis), were shown to possess an altered pathway for IAA biosynthesis (Schmidt and Kosuge, 1979). Isolates which were able to accumulate IAA caused gall formation while MT-sensitive isolates that could not synthesize and thus accumulate IAA in the presence of MT did not induce gall production. In another report, a plasmid-containing strain of *A. tumefaciens* (C-58) was found to produce 5–10 times more IAA in the presence of tyrosine (a precursor in the biosynthesis of IAA) than the plasmid-less strain (NT1). This high IAA production was restored after the reintroduction of the plasmid into NT1 (Liu and Kado, 1979), showing that both oncogenicity and IAA production are transferred by the same plasmid.

The clubroot disease of certain Cruciferae is characterized by root deformations induced by the pathogen *Plasmodiophora brassicae*. There are indications that the pathogen might interfere with the compartmentalization in the host of the indole derivative glucobrassicin and the enzyme glucosinolase. On hydrolysis by this enzyme, glucobrassicin yields the auxin-active 3-indoleacetonitrile (IAN) and eventually IAA, which in turn are thought to be responsible for the observed morphological changes after infection (Butcher *et al.*, 1974). Root nodules in legumes are induced by different strains of the genus *Rhizobium*. The large amounts of auxins measured in the nodules are probably of host origin, synthesized as a result of an alteration of the host's metabolism by the symbiont.

2.1. Auxin metabolism and disease development

In addition to conjugation, the enzymic oxidative degradation of IAA by peroxidase (POD) constitutes an important regulatory mechanism in the control of auxin concentration. This reaction is very complex and involves activators (Mn^{2+}, monophenols) and inhibitors (o-diphenols), in addition to requiring IAA as the electron donor. Changes in the above effectors directly introduced by the parasite often cannot be differentiated from the side effects which accompany the infection process (e.g. mechanical effects) and which would also yield similar results, for example by enhancing polyphenol oxidase (PO) activity and thus secondarily increasing the concentrations of diphenols via monophenol oxidation. The priority of effects presents an obvious problem in this case and therefore needs extremely careful consideration.

The products of IAA oxidation are mainly 3-methyleneoxindole and indolealdehyde (Moore, 1979). IAA oxidation has been suggested as being responsible for the outcome of several host–parasite interactions and connected symptoms. Disease resistance, for example, in certain cases has been found to be correlated with IAA decarboxylation. In wheat leaves infected by *Puccinia graminis*, the rate of $^{14}CO_2$ release from ^{14}C-labelled IAA was slower than that from the leaves of healthy plants. Near-isogenic lines of wheat which only differed by one temperature-sensitive gene (Sr6 locus) were resistant at *ca.* 20 °C and susceptible at 25 °C. After exhibiting approximately the same rates of IAA decarboxylation in both susceptible and resistant lines during the initial stages of infection, a stimulation of decarboxylation in resistant tissue after 2–3 days was observed. Enzyme studies showed, however, that apparently no direct correlation existed between disease resistance and POD activity. Since POD as well as PO has been shown to be strongly increased under various different stress situations, the question again arises as to whether the POD changes often observed in relation to certain developments in the host–parasite interaction might be a consequence of secondary effects, rather than an integral factor in the chain of events ultimately responsible for either resistance or susceptibility (Daly and Knoche, 1976).

Hence it is possible that IAA decarboxylation occurs merely as a consequence of the changed metabolism. One function of the increased POD activity (possibly resulting from the induction of isoenzymes) might be seen in lignin formation (Solheim and Raa, 1977) where the true electron donor for the POD is NAD(P)H instead of IAA. In the case of experimental exogenous application of IAA, the latter might be used as a substrate instead of NAD(P)H (and thus be decarboxylated). The true function of POD might thus be seen in H_2O_2 production in the cell wall (where a large amount of POD is located in a bound form) for the purposes of lignification (Gross *et al.*, 1977; Mäder *et al.*, 1980) and possibly, in a similar manner to polymorphonuclear leucocyte in animals, as the driving force of an extremely potent defence system against invaders (Roos, 1977). This function, however, would be independent of the function(s) of IAA or its metabolic products.

The mechanism of oxidase regulation is also very contradictory and speculative. The results of Mn^{2+} and monophenol-stimulated and *o*-diphenol-inhibited activity *in vitro* are complicated by (a) non-specific binding on to different macromolecules of the vast amounts of the different phenolics found in the plant cell, (b) by specific (?) auxin 'protectors' (Haard, 1978), and (c) by the possible induction of POD and PO by other metabolic changes ('stress') as well as by other phytohormones (see Section 6). The phenolic regulators, in turn, may be produced via ethylene induction as reported for tobacco plants infected by *Peronospora tabacina* where severe stunting was accompanied by the accumulation of scopoletin, *o*- and *p*-coumaric acids (stimulators of auxin oxidation), and ethylene formation. Ethylene treatment induced similar changes in phenolics (Reuveni and Cohen, 1978).

After infestation by nematodes, increased levels of chlorogenic acid (an inhibitor of auxin oxidation) have been measured, which were suggested to be responsible for the decreased auxin oxidation and thus increased tissue growth and gall formation (Knypl *et al.,* 1975). Since infestation by nematodes causes mechanical wounding and would result in ethylene formation, the above finding might also represent an ethylene effect.

Transfer RNA has also been reported to inhibit IAA decarboxylation, this inhibition being reversed by H_2O_2 (Lee, 1980). Since H_2O_2 is a product of peroxidase-catalysed NAD(P)H oxidation (Gross *et al.,* 1977) possibly involved in lignin formation ('wound healing'), another pair of as yet poorly understood counteracting regulators of IAA oxidation might exist. Peroxidase, on the other hand, was increased in both susceptible sr6 and resistant Sr6 lines of wheat after treatment with ethylene at 20 °C. The susceptibility of the sr6 lines to wheat stem rust remained unchanged, but the resistant Sr6 line became susceptible (see Section 6 and Daly and Knoche, 1976). Similarly, an increase of POD in tobacco has been shown not to be directly involved in 'induced' resistance (Nadolny and Sequeira, 1980), whereas a possible role was proposed in cotton resistance against bacterial blight (Venere, 1980) and in reed canary grass (possibly via lignification) (Vance *et al.,* 1976). The relationship between production of inhibitors of IAA oxidation (*o*-diphenols) from monophenolic precursors, concomitant production of toxic *o*-quinones, and finally lignification of the invaded and/or wounded tissue (together with associated changes in ethylene formation and action) deserves increased appreciation.

2.2. Metabolic regulation and IAA

As in the case of several other host–parasite systems, following infection of sunflower with *Puccinia carthami,* the hypocotyl shows increased elongation concomitant with enhanced respiration, reduced regulation of the glycolytic pathway, and augmented pentose phosphate-pathway activities. These changes are correlated with increased IAA concentrations and biosynthesis of polyols (conversion of hexoses into pentitols), which represent the reserves of the developing uredospores (Daly and Knoche, 1976). The involvement of IAA in an overall defence system, including the activation of metabolic processes leading finally to the synthesis of a physical barrier ('long-term defence') was recently speculated upon (Beckman, 1976). It was suggested that after a recognition process, an increased release of phenolics following membrane changes and IAA synthesis was initiated and regulated. IAA acting via enhancement of a proton pump (Cleland and Rayle, 1978) and the pentose phosphate pathway is a possible initiator of the production of precursors for cell wall material and other mechanical barriers synthesized in the 'plasticized' cell wall. These complex processes, which include overlapping, contrasting, and interfering systems, may be further complicated. For example, it has been observed that the products of the oxidative degradation of IAA, 3, 3'-indolylmethane (BIM), and 3-methyleneoxindole stimulate mycelial growth and transition from germ tube to mycelial growth in *Puccinia graminis* (Grambow and Tücks, 1979). This effect was antagonized by IAA and indole-3-aldehyde. The authors hypothesized that the steady-state concentrations of IAA and its various metabolites may control the development of this rust fungus.

3. CYTOKININS

Cytokinins (Figure 1) are N^6-substituted derivatives of adenine. In the healthy plant, they seem to be synthesized mainly in the roots, transported in the xylem, and further metabolized in different plant organs, bound to certain receptors, or incorporated into a specificity-determining position of tRNA. Cytokinins activate several enzymes, probably via *de novo* synthesis, and induce the metabolism and transportation of the various important components of cell nutrition concomitant with a general stimulation of metabolic activities and cell division. In contrast to the growth factors abscisic acid and ethylene, which often induce symptoms of senescence, the cytokinins might be called 'juvenility factors' in several cases (Moore, 1979; Skoog, 1980).

Differentiated cells around the sites of infection may undergo division accompanied by increased protein synthesis and respiration. These areas with new metabolic activities become 'metabolic sinks' and thus exhibit a reduced expression of senescence symptoms. Areas affected by fungal infection (e.g. mildews) often retain chlorophyll longer than uninfected zones. Aqueous extracts from spores of several fungi have been shown to contain compounds which induce the formation of such 'green islands.' One might thus assume that the synthesis of cytokinins (or cytokinin-like factors) can occur in both the pathogen and the host (Sequeira, 1973; Heitefuss and Williams, 1976; Moore, 1979; Skoog, 1980). For example, in the case of maize meristems infected with *Ustilago maydis,* an increase in the degree of infection was correlated with enhanced cytokinin activity. The polar cytokinins zeatin and zeatin riboside were present in both healthy and infected plants, whereas polar growth promotors were produced by the fungus and accumulated only in the infected plants (Mills and Van Staden, 1978). In peach leaves infected with *Taphrina deformans,* a new cytokinin and increased IAA levels have been reported in addition to the cytokinins present in the healthy leaves (Sziráki *et al.,* 1975). It has further been demonstrated that metabolites are accumulated in infected leaves exposed to $^{14}CO_2$ and that radioactive glucose moves towards the sites of exogenous kinetin application (in addition to transport over greater distances). In the 'fasciation' disease of peas caused by *Corynebacterium fasciens,* the characteristic loss of apical dominance and lateral bud development can also be induced by cytokinin application, substituting for the infection. In the case of the 'false broomrape' disease, a root disorder of tobacco, or the 'clubroot' disease of crucifers, increased cytokinin/ auxin ratios or cytokinin and auxin levels, respectively are likely to be at least partially responsible for the observed symptoms (Heitefuss and Williams, 1976)

A role for cytokinins was postulated in acquired resistance of tobacco hypersensitive to tobacco mosaic virus (TMV), since increases in the cytokinin content in systemically resistant lines were correlated with the suppression of local lesion development induced by TMV after exogenous application of kinetin (N^6-furfuryladenine, a synthetic cytokinin). The number of virus particles was not changed after kinetin treatment, however (Balazs *et al.*, 1977), whereas the application of IAA seemed to inhibit TMV multiplication. Kinetin apparently decreased lesion size and lesion enlargement, i.e. only the symptoms of this disease, while IAA seemed to be involved in the stimulation of factors responsible for virus localization (Van Loon, 1979). A similar effect was brought about by ethephon (a synthetic ethylene releaser) and it has been speculated that an 'interferon-like antiviral factor' (AVF) is ultimately responsible for the suppression of TMV multiplication (Sela, 1981; see also Chapter 18).

The therapeutic action of kinetin against viral diseases, on the other hand, seems to depend on its concentration during treatment; high concentrations ($1.0–10$ mg dm^{-3}) have been reported to lower the ring-spot virus levels in *Nicotiana glutinosa* (Tavantzis *et al.*, 1979). In a similar manner, *ca.* 0.2 mg dm^{-3} of kinetin suppressed the hypersensitive reaction in *N. tabacum* against race 0 of *Phytophthora parasitica* var. *nicotianae*, so that heavy colonization could occur, whereas increased concentrations of IAA reversed the kinetin effects. The balance of the phytohormones thus seems to represent one modifying principle in the host response, regulating the balance between disease susceptibility and resistance (Haberlach *et al.*, 1978).

Plant tumour and gall formation (generally disorganized growth), might also involve induction by cytokinins (or cytokinin-like factors) released from *Agrobacterium tumefaciens, Rhizobium japonicum,* or *Plasmodiophora brassicae* (Heitefuss and Williams, 1976; MacMillan, 1980), possibly in a cooperative manner with increased auxin levels.

The mechanism of the induction of cell division by cytokinins is unknown, although an increase in endogenous cAMP levels via inhibition of cAMP phosphodiesterase by cytokinins has been suggested. cAMP as a second messenger has also been proposed as a mediator for other plant growth regulators (for references see Heitefuss and Williams, 1976), but these reports have to be considered with reservation, since cAMP determinations in plant tissues are extremely problematic (Amrhein, 1977).

With regard to the mechanism of retardation of senescence symptoms (e.g.

pigment degradation), a comparison with the cytokinin-like activities of certain iron chelators (Atkin and Neilands, 1972) or phenylureas (Lee *et al.,* 1981) allows the speculation that cytokinins activate an enzymic defence against the deleterious effects of certain 'active oxygen' species produced during unsaturated fatty acid oxidation (for example after mechanical wounding of plant tissue; Elstner and Konze, 1976). These enzymes could include superoxide dismutase (E.C. 1.15.1.1), catalase (E.C.1.11.1.6), and peroxidase(s) (E.C. 1.11.1.7) which are involved in the defence against deleterious oxygen species such as the superoxide anion ($O_2^{\cdot-}$) and hydrogen peroxide (H_2O_2), which in turn are thought to be responsible for the initiation of membrane, pigment, and DNA destruction and thus the onset of senescence (Elstner, 1982).

4. GIBBERELLINS

Gibberellins (Figure 1) are polycyclic terpenoids (with a basic *ent*-gibberellane structure) containing 19 or 20 carbon atoms. About 80 different structures have so far been described, including biochemically and physiologically active and inactive metabolites (Heitefuss and Williams, 1976). In addition to the induction and activation of various enzymes and metabolic pathways, such as the mobilization of compartmentalized storage polymers during germination, a role in cell elongation has been suggested distinct to that proposed for auxins (Skoog, 1980).

A great number of microorganisms have been shown to produce gibberellins (Heitefuss and Williams, 1976), but the role of these compounds in plant disease is far from clear, even in the 'foolish seedling' disease of rice, caused by *Gibberella fujikuroi* where gibberellins were first detected. Certain fungi, such as *Puccinia punctiformis,* cause morphological changes in the host (as does *G. fujikuroi* in rice seedlings), which can be mimicked by GA_3 treatment. IAA application does not induce similar effects. It is important to note that transformations of certain gibberellins (GA_1 and GA_2 to GA_3) were found to depend on both plant age and the development of the disease. In some cases, stunting after viral or fungal infection has been correlated with decreased levels of certain gibberellins (e.g. GA_3), where the levels of abscisic acid and IAA remained unchanged. Application of exogenous gibberellins increased the relative infectivity of citrus exocortis viroid (CEV) in *Gynura aurantiaca* plants (Rodriquez *et al.*, 1978). In this context both the influence and the changes of other growth-regulating effectors, such as the phytochrome system, are of great importance. The formation of active gibberellins from inactive precursors (Metzger and Zeevart, 1980) or release from membrane binding (Cooke *et al.,* 1975) under the influence of light have been reported.

5. ABSCISIC ACID AND XANTHOXIN

Growth inhibitors are widely distributed in both plants and microorganisms and belong to a large variety of chemically different classes of compounds (Moore, 1979; Skoog, 1980). Abscisic acid (ABA) and xanthoxin (Figure 1) are the best known and most often described growth inhibitors. Both are terpenoids synthesized from mevalonate. ABA might be derived either directly from farnesyl pyrophosphate or by oxidative cleavage of carotenoids (e.g. violaxanthin) via the intermediate xanthoxin. As with other growth regulators, ABA seems to be metabolized via conjugation (for example glycosylation; MacMillan, 1980). Since abscisic acid as well as xanthoxin seems to be synthesized and stored in the plastids, changes in plastid membrane permeability would regulate the distribution and thus the availability of these compounds for certain biological functions, for example in the guard cells of stomata where ABA induces closure.

Stunting of plants is often observed after infection and is accompanied in some cases (see Section 4) by an increased concentration of growth inhibitors. In the case of tobacco infected with *Pseudomonas solanacearum,* apparently the host and not the parasite produces ABA (Sequeira, 1973). In some cases, increased ABA levels are accompanied by changes in IAA (e.g. decarboxylation) and increased ethylene production. Wilt-diseased plants are especially subjected to increases in ABA and ethylene, probably as a result of increased water losses and hence desiccation. In turn, abscisic acid formation, stomatal closure, and eventual defoliation might be one response by the host in trying to regulate water loss (Audus, 1972).

Antagonism between gibberellins and ABA was suggested as being responsible for reduced growth in cucumber hypocotyls infected with cucumber mosaic virus (CMV) (Aharoni *et al.,* 1977). After infection decreased gibberellin-like activities and increased ABA levels were measured. ABA does not only seem to counteract gibberellins in causing growth retardation, but also acts against cytokinins and auxins by increasing the number of virus-induced local lesions and allowing or enhancing virus multiplication in leaves of tobacco infected by tobacco mosaic virus (TMV) (Balazs *et al.,* 1973), suggesting that the production by the host of an 'antiviral factor' (Sela, 1981) might be suppressed by abscisic acid.

6. ETHYLENE

The gaseous phytohormone ethylene (Figure 1) induces various responses in higher plants, such as germination, root formation, fruit ripening, flowering, growth inhibition, senescence (pigment degration), and defoliation, depending on the individual plant species and the age of the plant. In addition to these functions during the normal life cycle of plants, ethylene is formed after mechanical stress or after infection (Abeles, 1973). The term 'stress' includes wounding, temperature changes, water losses, flooding, radiation and light extremes, changes in gravity, and application of chemicals or biochemicals (Hislop *et al.*, 1973; Heitefuss and Williams, 1976; Lieberman, 1979; Paradies and Elstner, 1980). The function of ethylene evolved after these various influences is not clear, although increased symptoms of senescence such as abscission might serve a biological purpose (see Section 5). Furthermore, ethylene might be involved in wound healing reactions such as lignification and suberization (Kahl, 1978).

6.1. Ethylene and induced resistance against parasites

Resistance against parasites might be induced by the following chain of events (Friend and Threlfall, 1976; Horsfall and Cowling, 1980). After a recognition process due to parasite (elicitor ?)–host interactions, a signal may be produced which could be ethylene or fortified by ethylene. This 'signal reaction' could initiate further reactions or changes, possibly via other receptors (for example in membranes), eventually leading to increases in enzyme activities. The activities of these enzymes migh then produce disadvantageous conditions for the parasite through the formation of phytoalexins, lytic or oxidative environments, or barriers (Figure 2). During the expression of these activities, the host has also to suffer from losses of parts of the tissue due to destruction of cell organelles and finally whole cells by necrosis. These visible symptoms are often referred to as the hypersensitive response. Induced resistance after ethylene treatment has been reported in some cases, for example after treatment of tomato plants with ethylene a certain degree of resistance against *Verticillium albo-atrum* was observed (Pegg, 1976) (Figure 2). The circumstantial evidence for such a role of ethylene as a signal mediating the recognition mechanism and host defence responses will now be examined.

After infection of *Vigna sinensis* with CMV, an increase in the free radical species of fatty acids, malondialdehyde (i.e. thiobarbituric acid-sensitive material, an indicator of unsaturated fatty acid peroxidation), a decrease in unsaturated fatty acids, and an increase of ethylene evolution were measured prior to macroscopically detectable necrotic lesions. The increased conductance due to ion leakage through membranes of the inoculated tissue shortly before local lesion formation was suppressed by free radical scavengers. Fatty acid peroxidation and hence production of alkyl, alkoxy, and peroxy radicals together with ion release from the infected tissue were taken as indication that lipid peroxidation was one early metabolic event during symptom expression (Kato, 1976) Similarly, increase in conductivity, changes in lipid concentration, and enhanced phospholipase activity have been determined as indicators of the loss of membrane integrity (Hoppe and Heitefuss, 1974) as well as ethylene formation (Montalbini and Elstner, 1977) after the penetration of stomata of *Phaseolus vulgaris* by *Uromyces phaseoli*. In addition,

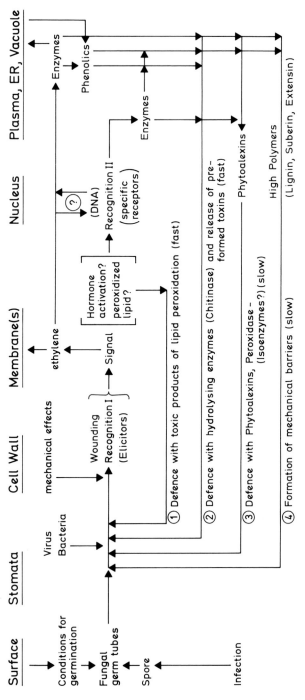

Figure 2. Biochemical events during host–parasite interaction

different enzymes considered to be involved in the overall defence system and which are enhanced following wounding, infection, or treatment with elicitors of phytoalexins, are also enhanced in parallel with, or shortly after, increased ethylene formation or after ethylene treatment. These include β-1,3-glucanases, chitinase, phenylalanine ammonia lyase (PAL), cinnamic acid-4-hydroxylase, peroxidase (POD), and polyphenoloxidase(s) (PO), as well as cell wall materials such as the hydroxyproline-rich glycoprotein extensin (Abeles, 1973; Esquerré-Tugaye and Toppan, 1977; Horsfall and Cowling, 1980).

As in the case of the other growth regulators, however, contradictory results make it very difficult to draw any definite conclusions between resistance and the ethylene-induced effects and/or observations. A more straightforward approach to the question of the role of ethylene in plant resistance has been adopted by examining the effects of ethylene on phytoalexin production. It is today widely accepted that phytoalexins play an important role in the overall defence system of plants against parasites (see Chapter 12) and in some cases there are positive correlations between ethylene formation (or ethylene treatment), phytoalexin production, and enhanced resistance (Paradies and Elstner, 1980).

On the other hand, recent work in the author's laboratory has shown that there is apparently no direct connection between ethylene evolution and phytoalexin accumulation. In the case of *Phaseolus vulgaris* inoculated with *Uromyces phaseoli*, two distinct periods of ethylene evolution have been observed one accompanying the penetration of the fungal hyphae through the stomata and the other occurring during the necrotic reaction of hypersensitive varieties. Hypocotyl segments of hypersensitive varieties showed an increased evolution of ethylene when incubated with cell wall preparations (elicitors) from *U. phaseoli* compared with a susceptible variety (Paradies *et al.*, 1979). It was speculated that ethylene, as already proposed for other host–parasite relationships, might be the signal for the induction of resistance. However, treatment of leaves with ethylene or 1-aminocyclopropane-1-carboxylic acid [ACC, the direct precursor of ethylene (Adams and Yang, 1979)] did not increase the resistance of the susceptible variety (Paradies and Elstner, 1980). In soybean cotyledons, treatment with an elicitor of phytoalexin synthesis isolated from cell walls of *Phytophthora megasperma* var. *sojae* induced ethylene formation. After a delay of *ca.* 1 h, PAL activity was strongly increased and after *ca.* 5 h synthesis of the phytoalexin glyceollin (Chapter 2) was induced. By simultaneous treatment with elicitor and aminoethoxyvinylglycine (AVG, an inhibitor of ethylene formation from methionine) and/or the ethylene precursor ACC, it was clearly established that the induction processes of both ethylene formation and phytoalexin production could be uncoupled from one another.

Thus ethylene does not appear to function as a signal or messenger between elicitor recognition and the induction of phytoalexins (Paradies *et al.*, 1980).

The resistance-inducing effects of ethylene may therefore be coincidental. To consider another example, the increased resistance against anthracnose in ethylene-treated mandarin oranges might be based on the facts that (a) green fruits are susceptible to the disease whereas (ethylene) ripened mature fruits are not and (b) the pathogen *Colletotrichum gloeosporioides* requires ethylene for the formation of penetration hyphae. Since mature fruits evolve much less ethylene than fruits during ripening and also exhibit decreased penetrability owing to lignification of the flavedo layer, the observed resistance might well be indirect.

As far as virus infections are concerned, systemic diseases without necrotic symptoms do not seem to be accompanied by a characteristic evolution of ethylene during a certain stage of disease development. However, in viral infections which do develop necrotic symptoms, characteristic periods of ethylene formation are observed which have been suggested as being responsible for the development of chlorosis (Kato, 1976; Marco and Levy, 1979).

Finally, ethylene formation can be used in certain cases as an additional indicator in the search for fungicides (Walther *et al.*, 1981) and in plant breeding programmes directed towards finding resistant lines, by taking advantage of the ethylene outburst during the penetration of fungal gérm tubes (Montalbini and Elstner, 1977) or during the expression of necrotic symptoms (Koch *et al.*, 1980).

REFERENCES

Abeles, F. B. (1973). *Ethylene in Plant biology*, Academic Press, New York.

Adams, D. O., and Yang, S. F. (1979). Ethylene biosynthesis: identification of 1-aminocyclopropane-1-carboxylic acid, *Proc. Natl. Acad. Sci. USA*, **76**, 170–174.

Aharoni, N., Marco, S., and Levy, P. (1977). Involvement of gibberellins and abscisic acid in the suppression of hypocotyl elongation in CMV-infected cucumbers, *Physiol. Plant Pathol.*, **11**, 189–194.

Amrhein, N. (1977). The current status of cyclic AMP in higher plants, *Annu. Rev. Plant Physiol.*, **28**, 123–132.

Atkin, C. L., and Neilands, J. B. (1972). Leaf infections: siderochromes (natural polyhydroxamates) mimic the 'green island' effect, *Science*, **176**, 300–302.

Audus, L. J. (1972). *Plant Growth Substances*, Leonard Hill, London.

Balazs, E., Gaborjanyi, R., and Kiraly, Z. (1973). Leaf senescence and increased virus suspectibility in tobacco: the effect of abscisic acid, *Physiol. Plant Pathol.*, **3**, 341–346.

Balazs, E., Siraki, J., and Kiraly, Z. (1977). The role of cytokinins in the systemic acquired resistance of tobacco hypersensitive to tobacco mosaic virus, *Physiol. Plant Pathol.*, **11**, 29–37.

Beckman, C. H. (1976). Defenses triggered by the invader: physical defenses, in *Plant Disease. An Advanced Treatise.* Vol. V. (Eds J. G. Horsefall and E. B. Cowling), Academic Press, New York, pp. 225–245.

Butcher, D. N., El-Tigani, S., and Ingram, D. S. (1974). The role of indole glucosinolates in the club root disease of the Cruciferae, *Physiol. Plant Pathol.*, **4**, 127–140.

Cleland, R. E., and Rayle, D. L. (1978). Auxin, H$^+$-excretion and cell elongation, *Bot. Mag. Tokyo*, 1 (special issue), 125–139.

Cooke, R. J., Saunders, P. F., and Vendrick, R. E. (1975). Red light induced production of gibberellin-like substances in homogenates of etiolated wheat leaves and in suspensions of intact etioplasts, *Planta*, **124**, 319–325.

Daly, J. M., and Knoche, H. W. (1976). Hormonal involvement in metabolism of host-parasite interactions, in *Biochemical Aspects of Plant–Parasite Relationships* (Eds J. Friend and D. R. Threlfall), Academic Press, New York, pp. 117–133.

Elstner, E. F. (1982). Oxygen activation and oxygen toxicity, *Annu. Rev. Plant Physiol.*, **33**, 73–96.

Elstner, E. F., and Konze, J. R. (1976). Effect of point freezing on ethylene and ethane production by sugar beet leaf discs, *Nature (London)*, **263**, 351.

Esquerré-Tugayé, and Toppan, A. (1977). Hydroxyproline-rich glycoproteins of the plant cell wall as an inhibitory environment for the growth of a pathogen in its host, in *Cell Wall Biochemistry Related to Specificity in Host–Plant Pathogen Interactions* (Eds B. Solheim and J. Raa), Universitetsforlaget, Tromso, Oslo, and Bergen, pp. 000–000.

Friend, J., and Threlfall, D. R. (Eds) (1976). *Biochemical Aspects of Plant–Parasite Relationships*, Academic Press, New York.

Grambow, H. J., and Tücks, M. T. (1979). The differential effects of indole-3-acetic acid and its metabolites on the development of *Puccinia graminia* f. sp. *tritici in vitro*, *Can. J. Bot.*, **57**, 1765–1768.

Gross, G. G., Janse, C., and Elstner, E. F. (1977). Involvement of malate, monophenols, and the superoxide radical in hydrogen peroxide formation by isolated cell walls from horseradish (*Armoracia lapathifolia* Gilib.), *Planta*, **136**, 271–276.

Haard, N. F. (1978). Isolation and partial characterization of auxin protectors from *Synchytrium endobioticum*-incited tumors in potato, *Physiol. Plant Pathol.*, **13**, 223–232.

Haberlach, G. T., Budde, A. D., Sequeira, L., and Helgeson, J. P. (1978). Modification of disease resistance of tobacco callus tissues by cytokinins, *Plant Physiol.*, **62**, 522–525.

Heitefuss, R., and Williams, P. H. (Eds) (1976). *Physiological Plant Pathology, Encyclopedia of Plant Physiology*, Vol. 4, Springer Verlag, Berlin, Heidelberg, and New York.

Hislop, E. C., Hoad, G. V., and Archer, S. A. (1973).in *Fungal Pathogenicity and the Plant's Response* (Eds R. J. W. Byrde and C. V. Cutting), Academic Press, New York, pp. 87–117.

Hoppe, H. H., and Heitefuss, R. (1974). Permeability and membrane lipid metabolism of *Phaseolus vulgaris* infected with *Uromyces phaseoli*, *Physiol. Plant Pathol.*, **4**, 4–36.

Horsfall, J. G., and Cowling, E. B. (Eds) (1980). *Plant Disease. An Advanced Treatise. Vol. V, How Plants Defend Themselves*, Academic Press, New York.

Kahl, G. (Ed.) (1978). *Biochemistry of Wounded Plant Tissues*, Walter de Gruyter.

Kato, S. (1977). Ethylene production during lipid peroxidation in cowpea leaves infected with cucumber mosaic virus, *Ann. Phytopathol. Soc. Jpn.*, **43**, 587–589.

Koch, F., Baur, M., Burba, M., and Elstner, E. F. (1980). Ethylene formation by *Beta vulgaris* leaves during systemic (beet mosaic virus and beet mild yellowing virus, BMV + BMYV) or necrotic (*Cercospora beticola* Sacc.) diseases, *Phytopathol. Z.* **98**, 40–46.

Knypl, J. S., Chylinska, K. M., and Brzeski, M. W. (1975). Increased level of chlorogenic acid and inhibitors of indole-3-acetic acid oxidase in roots of carrot infested with the northern root-knot nematode, *Physiol. Plant Pathol.*, **6**, 51–64.

Lee, E. H., Bennet, J. H., and Heggestad, H. E. (1981). Retardation of senescence in red clover leaf discs by a new antiozonant, *N*-[2-(2-oxo-1-imidazolidinyl) ethyl]-*N'*-phenylurea, *Plant Physiol.*, **67**, 347–350.

Lee, T. T. (1980) Transfer RNA-peroxidase interaction, *Plant Physiol.*, **66**, 1012–1014.

Lee, T. T., Starratt, A. N., Jevnikar, J. J., and Stoessl, A. (1980). New phenolic inhibitors of the peroxidase-catalyzed oxidation of indole-3-acetic acid, *Phytochemistry*, **19**, 2277–2280.

Lieberman, M. (1979). Biosynthesis and action of ethylene, *Annu. Rev. Plant Physiol.*, **30**, 533–591.

Liu, S. F., and Kado, C. I. (1979). Indoleacetic acid production: a plasmid function of *Agrobacterium tumefaciens* C 58, *Biochem. Biophys. Res. Commun.*, **90**, 171–178.

MacMillan, J. (Ed.) (1980). *Hormonal Regulation of Development*, I, *Encyclopedia of Plant Physiology*, Vol. 6, Springer Verlag, Berlin, Heidelberg, and New York.

Mäder, M., Ungemach, J., and Schloss, P. (1980). The role of peroxidase isoenzyme groups of *Nicotiana tabacum* in hydrogen peroxide formation, *Planta*, **147**, 467–470.

Marco, S., and Levy, P. (1979). Involvement of ethylene in the development of cucumber mosaic virus-induced chlorotic lesions in cucumber cotyledons, *Physiol. Plant Pathol.*, **14**, 235–244.

Metzger, J. D., and Zeevart, J. A. D. (1980). Effect of photoperiod on the levels of endogenous gibberellins in spinach as measured by combined gas chromatopraphy-selected ion current monitoring, *Plant Physiol.*, **66**, 844–846.

Mills, L. J., and Van Staden, J. (1978). Extraction of cytokinins from maize, smut tumors of maize and *Ustilago maydis* cultures, *Physiol. Plant Pathol.*, **13**, 73–80.

Montalbini, P., and Elstner, E. F. (1977). Ethylene evolution by rust-infected, detached bean (*Phaseolus vulgaris* L.) leaves susceptible and hypersensitive to *Uromyces phaseoli* (Pers.) Wint., *Planta*, **135**, 301–306.

Moore, T. C. (1979). *Biochemistry and Physiology of Plant Hormones*, Springer Verlag, Berlin, Heidelberg, and New York.

Nadolny, L., and Sequeira, L. (1980). Increases in peroxidase activities are not directly involved in induced resistance in tobacco, *Physiol. Plant Pathol.*, **16**, 1–8.

Nes, W. R., Patterson, G. W., and Bean, G. A. (1980). Effect of steric and nuclear changes in steroids and triterpenoids on sexual reproduction in *Phytophthora cactorum*, *Plant Physiol.*, **66**, 1008–1011.

Paradies, I., and Elstner, E. F. (1980). Host–parasite interactions: experiments to induce ethylene production in higher plants and on the role of ethylene to produce modulations of disease symptoms and to initiate defence mechanisms, *Ber. Dtsch. Bot. Ges.*, **93**, 635–657.

Paradies, I., Hümme, H., Hoppe, H. H., Heitefuss, R., and Elstner, E. F. (1979). Induction of ethylene formation in bean (*Phaseolus vulgaris*) hypocotyl segments by preparations isolated from germ tube cell walls of *Uromyces phaseoli*, *Planta*, **146**, 193–197.

Paradies, I., Konze, J. R., Elstner, E. F., and Paxton, J. (1980). Ethylene: indicator but not inducer of phytoalexin synthesis in soybean, *Plant Physiol.*, **66**, 1106–1109.

Pegg, G. F. (1976). The response of ethylene-treated tomato plants to infection by *Verticillium albo-atrum*, *Physiol. Plant Pathol.*, **9**, 215–226.

Peters, K. E., and Lippincott, J. A. (1976). Formation of a growth factor (GABA) by *Agrobacterium* inoculated leaves and the effect of GABA on tumor initiation and growth, *Physiol. Plant Pahtol.*, **9**, 331–338.

Reuveni, M., and Cohen, Y. (1978). Growth retardation and changes in phenolic compounds, with special reference to scopoletin, in mildewed and ethylene-treated tobacco plants, *Physiol. Plant Pathol.*, **12**, 179–189.

Rodriquez, J. C., Garcia-Martinez, J. L., and Flores, R. (1978). The relationship between plant growth substance content and infection of *Gynura aurantiace* DC by citrus exocortis viroid, *Physiol. Plant Pathol.*, **13**, 355–363.

Roos, D. (1977). Oxidative killing of microorganisms by phagocytic cells, *Trends Biochem. Sci.*, **2**, 61–64.

Schmidt, M., and Kosuge, T. (1979). The role of indole-3-acetic acid accumulation by alpha methyl tryptophan-resistant mutants of *Pseudomonas savastanoi* in gall formation on oleanders, *Physiol. Plant Pathol.*, **13**, 203–214.

Sela, J. (1981). Antiviral factors form virus-infected plants, *Trends Biochem. Sci.*, **6**, 31–33.

Sequeira, L. (1973). Hormone metabolism in diseased plants, *Annu. Rev. Plant Physiol.*, **24**, 353–380.

Skoog, F. (Ed.) (1980). *Plant Growth Substances*, Springer Verlag, Berlin, Heidelberg, and New York.

Solheim, B., and Raa, J. (Eds) (1977). *Cell Wall Biochemistry Related to Specificity in Host–Plant Pathogen Interactions*, Universitetsforlaget, Tromsö, Oslo, and Bergen.

Sziraki, J., Balázs, E., and Király, Z. (1975). Increased levels of cytokinin and indoleacetic acid in peach leaves infected with *Taphrina deformans, Physiol. Plant Pathol.*, **5**, 45–50.

Tavantzis, S. M., Smith, S. H., and Witham, F. H. (1979). The influence of kinetin on tobacco ringspot virus infectivity and the effect of virus infection on the cytokinin activity in intact leaves of *Nicotiana glutionosa* L., *Physiol. Plant Pathol.*, **14**, 227–233.

Vance, C. P., Anderson, J. O., and Sherwood, R. T. (1976). Soluble and cell wall peroxidase in reed canary-grass in relation to disease resistance and localized lignin formation, *Plant Physiol.*, **57**, 920–922.

Van Loon, L. C. (1979). Effect of auxin on the localization of tobacco mosaic virus in hypersensitively reacting tobacco, *Physiol. Plant Pathol.*, **14**, 213–226.

Venere, R. S. (1980). Role of peroxidase in cotton resistant to bacterial blight, *Plant Sci. Lett.*, **20**, 47–56.

Walther, H. F., Hoffmann, G. M., and Elstner, E. F. (1981). Ethylene formation by germinating, *Drechslera graminea*-infected barley (*Hordeum sativum*) grains: a simple test for fungicides, *Planta*, **151**, 251–255.

Weiler, E. M. (1981). Dynamics of endogenous growth regulators during the growth cycle of a hormone-autotrophic plant cell culture, *Naturwissenschaften*, **67**, 377.

Further Reading

Heitefuss, R., and Williams, P. H. (Eds) (1976). *Physiological Plant Pathology, Encyclopedia of Plant Physiolology*, New Series, Vol. 4, Springer Verlag, Berlin, Heidelberg and New York, Section 5.

Moore, T. C. (1979). *Biochemistry and Physiology of Plant Hormones*, Springer Verlag, Berlin, Heidelberg, and New York.

MacMillan, J. (Ed.) (1980). *Hormonal Regulation of Development, I, Encyclopedia of Plant Physiology*, Vol. 6, Springer Verlag, Berlin, Heidelberg, and New York.

V. Epilogue

Biochemical Plant Pathology
Edited by J. A. Callow
© 1983 John Wiley & Sons Ltd

20

Biochemical Plant Pathology and Plant Disease Control

K. J. BRENT

Long Ashton Research Station, University of Bristol, Long Ashton, Bristol BS18 9AF UK

1. INTRODUCTION

The past century has seen enormous advances in the technology of plant protection. Man has learnt to recognize plant diseases and their causative organisms, and to assess the risk of damage to crops. He has found out how to avoid or suppress disease development, by adjusting cultural practices, breeding resistant cultivars, or applying chemicals. Thus the modern farmer has available many effective methods for controlling diseases, and good advice on how and when to use them.

This progress has been achieved mainly through the combined efforts of agronomists, plant breeders, plant pathologists, and organic chemists, who have conducted large, empirical 'screening' programmes in the laboratory and field to select, from many candidates, new cultivars or chemicals for introduction into agriculture. Has biochemical plant pathology made any contribution, and to what extent are biochemical studies likely to aid future progress in plant disease control? In this chapter we shall discuss these questions first in the context of chemical and biological methods of disease control and of their modes of action, and then in relation to different areas of biochemical plant pathology.

2. DISEASE CONTROL BY CHEMICAL APPLICATIONS

2.1. Modes of action and discovery

About 90 chemicals are now used to control fungal and bacterial diseases of plants. World sales of fungicides in 1978 were estimated to be US$1500m (Anon., 1979); sales figures for bactericides have not been published, but are undoubtedly much smaller (probably less than US$100m). These chemicals have a wide variety of structures (see, for example, Woodcock, 1977; Bent, 1979; Worthing, 1981). They have been found by trial and error, either by 'random' screening of diverse chemicals, or by testing analogues of derivatives of compounds with known activity.

On average, about 30 000 compounds are screened to produce one successful new product. A development cost of US$16–25m is now typical for a totally new product, which must therefore command a large potential market. This in turn means that compounds which act against many different plant pathogens, and which act on diseases of major crops, are particularly sought after.

Some fungicides work mainly as surface protectants: these include the older fungicides, such as sulphur, cuprous oxide, copper oxychloride, organomercurials, dithiocarbamates (e.g. maneb, thiram), phthalimides (e.g. captan, captafol), and dinitrophenols, and also some newer materials, such as chlorothalonil and the dicarboximides (e.g. iprodione, vinclozolin). Most of these compounds are known to be general enzyme inhibitors (Kaars Sijpesteijn, 1970; Lyr, 1977). They are highly toxic to fungal spores and hyphae, in which they readily accumulate at high concentrations ($>100~\mu g~g^{-1}$ fresh wt). They are relatively safe to plants and animals because uptake is much lower (Byrde and Richmond, 1976; Richmond, 1977). Some of them have a broad spectrum of action, but others affect only particular groups of fungi. For example, dinocap (a dinitrophenol) works only on powdery mildew, whereas dithiocarbamates do not control this disease although they are active against many others. The reason for this selectivity between fungal species is not clear, but differences in uptake may be responsible.

Most fungicides introduced since 1965 are 'systemic.' Examples are benzimidazoles (e.g. benomyl, carbendazim), 2-aminopyrimidines (e.g. ethirimol), pyrimidine carbinols (e.g. fenarimol), triazoles (e.g. triadimefon), and acylalanines (e.g. metalaxyl). They penetrate plant surfaces and move long distances in the plant.They are translocated mainly in the transpiration stream, towards shoot or leaf tips. Downward movement from treated leaves or from shoots into roots is much less common; at present the best example of a fungicide which does this is fosetyl-Al (aluminium tris-O-ethylphosphonate). The systemic fungicides are taken up by fungal and plant cells to similar, relatively low concentrations (generally $<5\ \mu g\ g^{-1}$ fresh wt) and act at specific biochemical sites in the target organisms rather than as general enzyme toxicants (Table 1). Some have a narrow spectrum of action; for example, 2-aminopyrimidines act only on powdery mildews, oxathiins act mainly on Basidiomycetes, and RH 124 (4-n-butyl-1,2,4-triazole), which is extremely selective, apparently affects only one species of rust, *Puccinia recondita*.

A good deal of knowledge is now accumulating about the biochemical mode of action of systemic fungicides (Kaars Sijpesteijn, 1977, 1982). Most of them are directly toxic to the fungi they control and suppress fungal growth *in vitro*. However, a few do not affect fungi *in vitro*, and their action appears to be mediated by the presence of host tissue. Examples are shown in Table 1. In some cases disease control may depend on a mixture of direct fungicidal action and defensive response by the host, and these are discussed later.

How useful is this increasing knowledge of mode of action? It did not aid the invention of any of the compounds listed; all of the modes of action mentioned above were discovered well after the compounds concerned were known to be good fungicides, and none of them had a correctly predicted mode of action. The compounds or their progenitors were initially made and tested for a variety of interesting but fortuitous reasons. For example ethirimol arose from a group of pyrimidines originally made as candidate insecticides analogous to diazinon, pyrimidine carbinols were originally made as plant growth regulators, and metalaxyl came from a group of compounds synthesized as wild oat herbicides. However, despite this past dependence on serendipity, most crop protection scientists believe that an understanding of biochemical and physiological modes of action has a considerable practical value, in addition to its intrinsic worth as 'basic' scientific knowledge, and some aspects of this are discussed below.

Although most major plant diseases can now be controlled reasonably well by fungicides, new chemicals are still needed. All existing fungicides have important limitations. These are hard to summarize, because they differ between the fungicides and between crop diseases. In many situations fungicides still have to be applied repeatedly; the use of ten or more fungicide sprays per season is not uncommon in some crops. Therefore, a more persistent biological effect is needed. This may depend on increased stability towards weathering or metabolic breakdown, but more often it requires a compound to move by systemic transport or by surface redistribution from the treated parts to the younger, disease-prone parts of the plant as they develop. Compounds sel-

Table 1. Mode of action of some agricultural fungicides and development of resistance

Fungicide or fungicide group	Approx. 1st date major use	Capacity for systemic action	Probable site of action	Approx. spectrum of useful activity	Development of resistant populations in crops
Sulphur	1800	–	Multi-site (enzyme inhibition)	Powdery mildews, some others	–
Copper compounds	1885	–	Multi-site (enzyme inhibition)	Broad	–
Organomercurials	1925	–	Multi-site (enzyme inhibition)	Broad	+ (v. limited)
Dithiocarbamates	1940	–	Multi-site (enzyme inhibition)	Broad (not powdery mildews)	–
Phthalimides	1955	–	Multi-site (enzyme inhibition)	Broad (not powdery mildews)	–
Dinitrophenols	1955	–	Oxidative phosphorylation	Powdery mildews (and mites)	–
Organotins	1955	–	Oxidative phosphorylation	Broad	+ (v. limited)
Aromatic hydrocarbons	1940	–	?	*Botrytis*, some others	+
Dicarboximides	1975	–	?	*Botrytis* and related fungi	+ (if selection pressure heavy

Benzimidazoles	1970	+	Mitosis (spindle microtubules)	Broad (not Oomycetes)	+ (widespread)
2-Aminopyrimidines	1970	+	Purine metabolism	Powdery mildews only	+
Phosphorothiolates	1965	+	Phospholipid synthesis?	Rice blast	+ (v. limited)
Oxathiins	1970	+	Succinic dehydrogenase	Basidiomycetes	+ (v. limited)
Kasugamycin	1965	+	Protein synthesis	Rice blast	+ (v. limited)
Pyrimidine carbinols	1978	+	Sterol synthesis (C-14 demethylation)	Broad (not Oomycetes)	−
Triazoles	1976	+	Sterol synthesis (C-14 demethylation)	Broad (not Oomycetes)	+ (partial)
Triforine	1972	+	Sterol synthesis (C-14 demethylation)	Broad (not Oomycetes)	−
Morpholines	1970	+	Sterol synthesis (Δ^{14} reduction)	Broad (not Oomycetes)	−
Acylalanines	1978	+	RNA synthesis?	Oomycetes only	+ (widespread)
Fosetyl-Al	1978	+	Host metabolism?	Some Oomycetes	−

dom have sufficient breadth of action to cope with all the different pathogens which are likely to damage a crop. Some lack the ability to penetrate into plants and eradicate disease that may already be present. Certain diseases are either poorly controlled or are not affected at all by chemicals. These include all the diseases caused by bacteria and viruses (except for vector control by insecticides), and many of the soil-borne diseases such as vascular wilts and root rots.

The causes of these difficulties are diverse, but efforts to overcome them are limited by a lack of knowledge of the biochemistry of selective action, and of the chemical and biological factors that govern the uptake and movement of compounds in plants. Hence continued work on biochemical mode of action of fungicides, including detailed studies of physico-chemical interactions with target sites at the molecular level, is vital, as is also research on the mechanisms of pesticide redistribution in and on plants after application.

Possibly an even more pressing reason for discovering new compounds is to cope with problems of resistance. Current strategies are to minimize exposure of the pathogen to any single group of fungicides, to use fungicides together with other, 'non-chemical' measures, and to apply mixed fungicide formulations or mixed programmes, based on compounds with different modes of action. Although existing fungicides exert a wider range of different biochemical effects than do insecticides, there is still a need for a greater diversity of biochemical actions. This applies especially to those diseases which depend on a dangerously small range of fungicides for their control; a good current example is eye-spot of cereals (*Pseudocercosporella herpotrichoides*), against which only the benzimidazole fungicides are very effective.

It is not possible yet to 'tailor-make' or 'custom-build' molecules with all of the properties needed for a desired fungicide or bactericide. These properties are themselves hard to specify precisely, and may conflict with each other. Say, for example, that we want a compound which will control powdery mildew, downy mildew and *Botrytis* in grape vine, when applied as a protectant or as a curative treatment at monthly intervals. We need a material which will

(a) Inhibit the growth of the three fungi at various stages of colonisation in grape-vine tissue;
(b) not damage grape-vine tissues;
(c) penetrate leaves or fruit and reach the existing infections;
(d) move into regions of new growth produced during the monthly interval;
(e) persist in effective amounts for at least 1 month;
(f) not persist in the environment, and not affect wine fermentation by yeasts;
(g) be feasible to manufacture economically, stable in storage, and patentable;
(h) have a durable action to which organisms will not adapt.

Each of these features itself involves a complex set of processes; for example, toxicity to the target organisms depends on penetration to the biochemical site of action, perhaps through surface polysaccharide, cell wall, plasmalemma, and cytoplasm, and then on selective binding to a specific receptor site.

We cannot, at present, work out a combination of structural features that will specify all of these daunting requirements, and indeed such a combination may not always exist. However, our slowly increasing knowledge of how fungicides work and how they move in plants is already beginning to offer some guidance. With regard to the three heterogeneous disease targets quoted in the above example, biochemical studies indicate that triazoles or pyrimidine carbinols, which are highly effective on powdery mildew, will not suffice because they are inhibitors of sterol biosynthesis and the downy mildew is known not to synthesize sterols. Extension of their action to the downy mildew would require incorporation of an extra toxophore, for example a group inhibiting Oomycete wall biosynthesis, without altering too much the overall physical parameters of the molecule which determine penetration and permit binding to the sterol-oxidizing target enzymes. To obtain movement into and through grape tissue would depend more on intuition than on science at present, although we know that partition coefficient is a major factor (Hartley and Graham-Bryce, 1980; Briggs, 1981).

The more we can comprehend mechanisms of fungicidal action, the more secure and useful will our guidelines become. Modern computer techniques will be needed, not only to store and retrieve data but also to work out structural combinations needed to satisfy a particular set of biological properties. Equally we need to know better why the available chemicals are selective; modes of non-action can be as important as modes of action. In our example above, whereas we understand why a fungicide of the triazole group will not work against the downy mildew and that minor modifications would not affect this, we have no idea why current triazoles will not affect *Botrytis*, which does produce its own sterols. It may be an uptake problem, or result from metabolic inactivation, both of which might be overcome by relatively simple molecular modifications. Biochemical parameters are increasingly entering the thinking of fungicide chemists, and enriching the study of structure–activity relationships which formerly rested almost entirely on structural variations based on a knowledge of classical organic chemistry. Biochemical knowledge can lead to alternative lines of chemical invention, even if the hypotheses on which they are based turn out to be wrong, as is often the case at present!

2.2. Acquired resistance

Fungicide biochemistry has acquired an added practical interest in the last decade because of the appearance of fungicide-resistant forms. This has marred the performance of several groups of systemic fungicides in many crops (Georgopoulos, 1977; Dekker, 1981). Examples are listed in Table 1. Some earlier examples of resistance concerned a few surface fungicides with non-specific modes of action, e.g. hexachlorobenzene, biphenyl, organomercurials, and dodine. However, these cases arose after some years of use and then only in certain areas and with a few diseases. Resistance has evolved more quickly and widely to some of the systemic fungicides; this has caused failures of disease control and has led to the withdrawal from some uses of the fungicides concerned.

Mechanisms of resistance to fungicides are not clearly understood. Unlike insecticide resistance, which is caused predominantly by oxidative metabolism of the insecticide, fungicide resistance appears to result from a variety of causes, but mainly from changes in the biochemical sites of action. Thus comparative studies of fungicidal action in sensitive and resistant forms can help to unravel general mechanisms of action, particularly mechanisms of selectivity between different fungal genera or species.

If a strain of a fungus arises that resists the action of a particular fungicide, it is often found to be resistant simultaneously to related fungicides; for example, benomyl-resistant fungi generally also resist carbendazim, thiophanate-methyl and thiabendazole. All of these fungicides share the same biochemical site of action. If this site becomes insensitive to one fungicide it is likely, although certainly not inevitable, that it will become insensitive to related fungicides. Fungi with related mechanisms of action are often recognized by their similar structures. However, this is not always so: fenarimol, triadimefon, and triforine (Figure 1), for example, are not obviously related and were discovered independently. However, they all appear to act by inhibiting ergosterol synthesis at one stage (C-14 demethylation) and fungal mutants show clear cross-resistance between them (De Waard and Fuchs, 1982). When a fungicide is developed, it is particularly important to know whether its mode of action is highly specific and thus confers a greater risk of resistance, and whether its action is similar to that of other fungicides with which cross-resistance may arise.

As mentioned earlier, an increasingly used strategy of avoidance of resistance is to sell a fungicide mixture; for example, a mixed formulation of

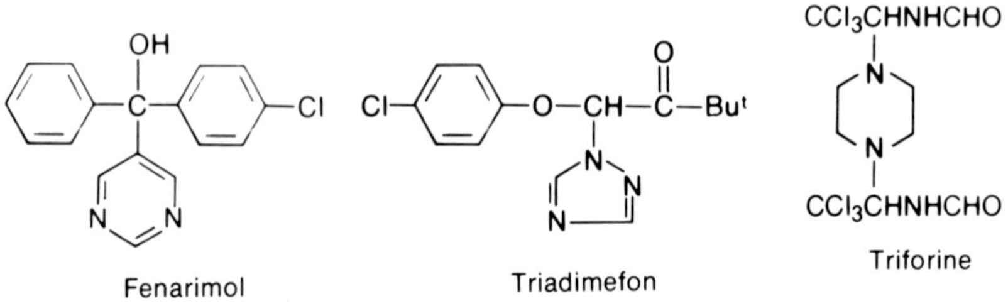

Fenarimol   Triadimefon   Triforine

Figure 1. Chemically dissimilar inhibitors of ergosterol synthesis

metalaxyl and mancozeb is sold in some countries for potato blight control. Another strategy is to tank-mix two or three fungicides, or to use them in sequence in a mixed spray programme. Mixed formulations generally contain a specific and a non-specific fungicide (usually an old, reliable surface fungicide, such as mancozeb). Mixed programmes tend to use fungicides with different specific modes of action. In either case we must know the biochemical mode of action of each fungicide, in order to choose companion compounds correctly.

Mechanisms of action and of resistance in the morpholine fungicides tridemorph and fenpropimorph have a special interest at present. These are now known to inhibit ergosterol biosynthesis in sensitive fungi (Kerkenaar et al., 1979; Kato et al., 1980). Thus strains of fungi resistant to other ergosterol inhibitors, such as the triazoles (e.g. triadimefon, propiconazole) or the pyrimidine carbinols (e.g. fenarimol), might be expected to resist fenpropimorph also. Such cross-resistance was reported for laboratory mutants of Ustilago maydis (Barug and Kerkenaar, 1979). However, field samples of barley powdery mildew that are partially resistant to the triazoles have remained sensitive to fenpropimorph (Holloman, personal communication). Also, the morpholines and triazoles appear to act on different stages in the biosynthetic pathway to ergosterol, these being Δ^{14} reduction and 14-C demethylation, respectively (Table 1). Possibly laboratory mutants have a common mechanism of resistance, which in the case of fenarimol and Aspergillus nidulans appears to depend on active efflux of the fungicide (De Waard and Van Nistelrooy, 1980; De Waard and Fuchs, 1982), whereas field resistance may result from alterations at the site of action. If so, the different site of action of the morpholines compared with the other ergosterol inhibitors may be very important in practice. At present there is much concern about possible resistance problems amongst the sterol-inhibitor fungicides, in view of their very extensive and repeated use in European cereal crops, and signs of shifts towards insensitivity were observed in recent surveys (Fletcher and Wolfe, 1981). If it is confirmed that morpholines have a different site of action and do not exert the same selection pressures in the field, they could very usefully be applied in mixtures and alternating programmes with other sterol inhibitors.

2.3. Novel target sites

Studies of the biochemistry of fungal growth and morphogenesis may reveal metabolic processes which occur in fungi, but not in higher plants or animals. Such systems would have obvious attraction as a target site for candidate fungicides. Well known examples are chitin biosynthesis and lysine biosynthesis. Since chitin is a major wall compound in many fungi and also in insects but not in plants or mammals, a specific inhibitor of some step or steps in its synthesis would have considerable value in crop protection. Since acetyl glucosamine is also a wall compound in bacteria, amino-sugar derivatives might even act on this group. Certain fungicides (e.g. polyoxin) and insecticides (e.g. diflubenzuron) are in fact known to act by interfering with chitin biosynthesis; however, their actions are restricted to particular groups of fungi or insects.

Lysine biosynthesis in fungi is peculiar in that it proceeds via an α-amino adipic acid pathway rather than via diaminopimelic acid as in higher plants. This difference would seem to offer scope for specific inhibition, although it has not yet been exploited successfully.

Specific inhibition of sporulation processes would be valuable in stopping epidemics; however, neither specific inhibitors nor specific biochemical processes involved in sporulation have yet been identified.

3. INDIRECTLY ACTING CHEMICALS

Most fungicides are directly fungitoxic, and the relative potency and selectivity of their effects *in vitro* tend to match their performance *in vivo*. Some fungicides, whilst not fungitoxic *per se*, are converted to fungitoxicants within the plant. Another approach to disease control is to use chemicals that work indirectly either by altering the pathogenicity of the fungus or by altering biochemical processes in the host plant. These may have little or no action *in vitro*, but can stop plant infections. Examples have been reviewed by Langcake (1981). One much cited case is WL28325 (2,2-dichloro-3,3-dimethyl-cyclopropanecarboxylic acid). This compound protects rice plants against rice blast disease (caused by *Pyricularia oryzae*), and yet has little action *in vitro*. Treated plants respond to infection by a hypersensitive reaction and by production of phytoalexin-like compounds (momilactones) (Cartwright *et al.*, 1980), and it has been suggested that WL28325 in some way makes an existing response of the host plant more rapid and potent. This chemical specifically controls rice blast; presumably it affects a particular regulatory mechanism in the rice plant–*Pyricularia oryzae* combination.

Fosetyl-Al (aluminium tris-*O*-ethylphosphonate) is used increasingly to control certain downy mildews and some diseases caused by *Phytophthora* spp. (potato late blight is a notable exception). Usually, it can move out of sprayed leaves and down to the roots. It does not affect fungal growth *in vitro*, and it stimulates defence reactions and phytoalexin synthesis in infected grape-vine and tomato plants (Bompeix *et al.*, 1980). Again, the effects on diseases are highly selective, and further work on its mode of action is needed. It does affect sporulation *in vitro*, and enhances chemical inhibition of mycelial growth *in vitro* in the presence of butanol (Buchenauer and Sikora, 1981), so that actions both on the fungus and on the host plant may well be involved.

A complication in the study of indirect action is that many chemicals, for example dithiocarbamate and mercury fungicides and surfactants used as formulation additives, can elicit phytoalexin synthesis (Reilly and Klarman, 1972; Hargreaves, 1981). Another difficulty is that a partial effect of a fungicide, slowing down or modifying the growth of an invading fungus rather than stopping it, may permit the defence mechanism of the host to work better or even stimulate it by releasing more fungal elicitor. Metalaxyl is another systemic fungicide acting on Oomycetes. It inhibits fungal growth *in vivo* only after formation of haustoria and the appearance of necrotic lesions similar to those which form in resistant crop varieties (Crute, 1978; Staub *et al.*, 1978). Metalaxyl is highly active *in vitro* against growth and sporulation of Oomycetes; however, it does not affect spore germination. Probably the initial growth of the pathogen *in vivo* is enough to stimulate a hypersensitive response and to elicit phytoalexin release as observed by Ward *et al.* (1980). However, the rapid evolution of resistant forms observed with this fungicide in Holland, Eire, and elsewhere (Davidse *et al.*, 1981; Cooke, 1981), and the correlation of resistance tests *in vitro* with such failures of field performance strongly support the view that fungitoxicity is the primary mode of action.

Many other chemicals and compounds with 'non-fungicidal' effects on plant diseases were discussed by Langcake (1981). These include effects of plant growth regulators (e.g. indoleacetic acid, morphactins, chlormequat, ethephon), especially against root diseases and vascular wilts, effects of several rice blast fungicides on melanin synthesis in the fungus associated with reduced pathogenicity (Woloshuk *et al.*, 1981), effects of inhibitors of the activity or biosynthesis of fungal pectolytic and enzymes, e.g. rufianic acid (Grossmann, 1968) and monosaccharides (Patil and Dimond, 1968) or of cutinases, e.g. diisopropyl fluorophosphate (Maiti and Kolattukudy, 1979; see also Chapter 6). Although there are few chemical successes yet, this approach does appear to offer rational, predictive opportunities for chemical synthesis. Since most indirect screening programmes are now based on plant diseases rather than on *in vitro* tests, indirectly acting compounds are likely to be detected as well as direct fungicides, provided they act fairly quickly.

4. BIOLOGICAL CONTROL AGENTS

There are very few examples of the commercial application of biological agents to control plant diseases. Very little biochemical research has been associated with these uses, and only two cases will be mentioned briefly.

The soil fungus *Trichoderma viride* can be applied to plants to suppress the growth of certain pathogenic fungi. One of the few practical uses of *T. viride* to be adopted so far is the control of silver leaf disease of plum trees caused by *Chondrostereum purpureum*. If pellets containing mycelium are inserted into holes drilled in the trunk, they protect against infection for many months;

application of spore suspensions to pruning wounds is also effective (Corke, 1980). The mechanism of protection is not known. Several antibiotics which act on *C. purpureum* are produced by *T. viride in vitro,* but they are detectable only in extremely small amounts in the wood, and not at all in the fruit of treated trees (Burden *et al.,* 1982). Other possible mechanisms are elicitation of phytoalexins or release of antagonistic enzymes. Further research to clarify this question is warranted.

A harmless, mild strain of tomato mosaic virus has been inoculated into young tomato plants to 'vaccinate' them against subsequent attacks by damaging, severe strains (Rast, 1972). Again the mechanism is unknown. There have been numerous reports of the experimental induction by various means of resistance in plants to virus infection, and evidence of an association between acquired resistance and the synthesis of new proteins in the host has been reported by several workers (see Matthews, 1981).

5. PLANT BREEDING

The discovery and development of disease-resistant cultivars have contributed enormously to plant disease control. However, the role of biochemistry in plant breeding has been virtually nil, and the classical procedures of crossing and selection continue to provide the new cultivars. The future may be different. If the recent advances in genetic manipulation in microorganisms can be extended to higher plants, and many workers in this field believe that this will become feasible within the next 5 years, then the identification and transfer of disease resistance genes could permit the production of novel resistant cultivars. According to the particular disease, these may provide resistance where none was available before, or make existing resistance more effective or more durable. The plant pathogen *Agrobacterium tumefasciens* is a promising vehicle for transferring into plants genes from microorganisms or from other plants. The presence of these genes in a potential donor plant, or in a recipient, may be detected and monitored by compatibility tests with relevant pathogens; however, a more direct and precise way to identify genes would be to determine their biochemical modes of action. No clear understanding yet appears to exist of the nature or the function of the products of resistant genes bred into plant cultivars and clearly this must be a major aim for the biochemical plant pathologist.

6. MECHANISMS OF PATHOGENESIS

The more we know about how pathogens invade and damage their hosts, the more informed and rational can be our approach to devising countermeasures, and the more we shall see opportunities for the design of chemical control agents or for making genetic changes. It is impossible to separate completely attack and defence mechanisms, since they interrelate. For convenience, however, we shall consider in this section biochemical activities which centre upon the pathogen. In some cases appressoria, infection pegs, or hyphae are not produced fully in resistant varieties. Close physical association between spores or germ tubes and plant cuticular waxes has been noted; if germinating spores are removed, 'footprints' of flattened plant wax are left behind. Since compatibility is often expressed very early in the infection process, 'recognition' and 'antagonism' must arise in this close initial contact and presumably involve interacting surface polymers of the pathogen and host. A knowledge of the subtle macromolecular matchings between pathogen and plant surfaces could permit chemical or genetic disruption. Roles postulated for plant lectins include the determination of highly specific initial plant–pathogen responses, and further information on this topic is required (Etzler, 1981).

Subsequent invasion may involve enzymic attack of the cuticle (see Chapter 10) and of cell wall pectins and cellulose. Inhibitors of these enzymes or of their biosynthesis could control diseases. Some experimental examples have been mentioned above. Endogenous inhibitors in plants have been detected (Anderson and Albersheim, 1972; Fielding, 1981)—these appear to be proteins, although their active sites may be much smaller. It seems worth noting also that certain Actinomycetes produce oligosaccharides which are potent and selective inhibitors of mammalian α-glucosidases (e.g. α-amylase); an example is acarbose (BAYg5421), which consists of a cyclitol unit, an amino-sugar, and two D-glucose units (Müller et al., 1980), and corresponding β-inhibitors would be of much interest.

The way in which the protoplasts in plant tissues are killed by pathogens is still not clearly understood, although it is known that pectolytic enzymes can disrupt protoplasts in situ very quickly (Hislop et al., 1979; Byrde, 1982). Treatments which inhibit this process of protoplasmic disruption may have potential in disease control. Another action of carbohydrases, where enhancement may be advantageous, is the release of fragments from the walls of the pathogen or the plant which may act as elicitors of phytoalexins and may even exert other 'hormonal' activities (Albersheim et al., 1981). Detailed studies of the formation and action of oligosaccharides in healthy and diseased plants could be very rewarding.

Inhibition of the action or formation of phytotoxins of lower molecular weight produced by pathogens has long been regarded as a possible method of disease control. A few experimental cases have been reported. One example concerns helminthosporoide, an α-galactoside phytotoxin of *Helminthosporium sacchari*, which causes eye-spot symptoms in sugar cane; this action and also its binding to a receptor protein in the plant membranes can be reversed by other α-galactosides such as raffinose (Strobel, 1973). Also, the screening of plant cells or protoplasts in vitro against toxins might provide an alternative method of selecting novel, resistant genotypes. Obvious toxin production occurs only in a minority of plant diseases, but these do include the vascular wilts which are so difficult to control directly.

Some diseases clearly affect crop growth and yields by generally damaging plant tissues, and others are known to produce specific toxins. However, as Hirst (1975) has pointed out, effects of plant disease on the normal physiology and metabolism of the host plant are generally overlooked, and comparative studies of metabolic disturbance by pathogens in plants and plant tissues should be encouraged.

7. MECHANISMS OF PLANT DEFENCE

Crop protection scientists are attracted by 'non-host' resistance because it is complete, durable, and broad-spectrum. Thus the potato plant resists attack by most plant pathogens; it is not, and probably never has been and never will be, invaded by the fungi that cause apple scab or wheat eyespot. The understanding and simulation of non-host resistance seem an urgent and important target for study (Heath, 1981). By analogy with our current views on the durability of action of synthetic fungicides, a diverse mixture of mechanisms may be at work.

Although phytoalexin treatment can give some control of disease (Ward *et al.*, 1975), initial hopes that they could be applied to plants effectively as broad-spectrum fungicides have faded. They tend not to perform well as sprays or root drenches in comparison with existing fungicides (Rathmell and Smith, 1980). They are less fungitoxic than conventional fungicides, they are not systemic, and they are readily metabolized. They are also difficult chemicals to make in quantity, and some of them are toxic to mammals. Simpler chemical analogues of phytoalexins could possibly be useful, if they retain the fungitoxic part of the structure. However, since they work well *in situ* against many pathogens, it is the systems which control their rate of production and the amounts and sites of accumulation that deserve most attention. Hypersensitivity or local lesion response, which is a very desirable phenomenon for crop protection, does appear to depend (at least for fungi) on the prompt formation of phytoalexins.

Exogenous elicitors, whether abiotic or produced by potential pathogens, appear to act by killing or damaging plant cells and are thus unlikely *per se* to be applicable for plant disease control. However, endogenous (or constitutive) elicitors produced by the plant (Bailey, 1982) are not known to damage cells and may govern phytoalexin production more directly. Their identification and the clarification of their action (for example, whether they act by derepression of biosynthesis or by releasing bound phytoalexins) are urgent research targets.

It is important to know why phytoalexin systems do not prevent all plant diseases. Sometimes insufficient phytoalexins appear, and sometimes they appear but fail to control the pathogen. We need to distinguish between these situations, since in the first case chemical treatment or plant breeding, or even modifications in crop husbandry, could alter the plant response and promote effective resistance. In the second case, presumably the organism is phytoalexin-resistant, and it is hard to see how this situation could be improved. However, a better knowledge of the mode of action of phytoalexins might reveal opportunities for synergistic enhancement of action.

Other forms of plant defence require biochemical investigation. The most exciting prospect must be the elucidation of the mechanisms of differential responses between host cultivars and pathogen races, for example in the cereal rusts and powdery mildews. The early reactions, involving deposition of 'reaction material' or lignification, require detailed exploration, especially the triggering mechanisms, and their biochemical links to the major resistance genes should be traced. It is notable that phenylthiourea, which protects cucumber plants against scab, markedly enhances lignification, possibly by increasing the availability of phenolic precursors (Kaars Sijpesteijn, 1969).

Induced defence systems which operate over distances in the plant are attractive research targets, since potent systemic chemicals are no doubt involved and these could provide novel leads for synthesis of chemical control agents. Such systems have been known for some years (e.g. Kuč, 1975) and it is perhaps surprising that the chemical basis has yet to be identified. Systemic resistance against viruses (e.g. Ross, 1966) is of special interest because of the great difficulties of finding selective chemical inhibitors of virus development.

Space does not permit discussion of many other important phenomena such as adult plant resistance, organ specificity (e.g. root *versus* shoot responses), and nutritional interactions (why does nitrogen enhance powdery mildews but inhibit rusts?). However, the author believes that advances in all these areas of biochemical plant pathology, as in those discussed above, will yield rich, long-term rewards both in improved chemical control and in achieving better disease resistance or even immunity. In this field, biological effort throughout the world is substantial and communication between workers seems good. Now the biochemical and chemical components of research are becoming increasingly important. Indeed, it is perhaps advances in the area of macromolecular specificity ('recognition'), embracing a knowledge of coding and of critical sites within interacting glycoproteins and polysaccharides, that may yield the most valuable results in the longer term.

REFERENCES

Albersheim, P., Darvill, A. G., McNeil, M., Valent, B. S., Hahn, M. G., Lyon, G., Sharp, J. K., Derjardins, A. E., Snellman, M. W., Ross, L. M., Borre, K. R., Aman, P., and Franzen, L. (1981). Structure and function of complex carbohydrates in regulating plant-microbe interactions, *Pure Appl. Chem.*, **53**, 79–88.

Anderson, A. J., and Albersheim, P. (1972). Host–pathogen interactions. V. Comparison of the abilities of proteins isolated from three varietes of *Phaseolus vulgaris* to inhibit the endopolygalacturonases secreted by three races of *Colletotrichum lindemuthianum, Physiol. Plant Pathol.*, **2**, 339–346.

Anon. (1979). A look at world pesticide markets, *Farm Chem.*, **142**(9), 61–68.

Bailey, J. A. (1982). Mechanisms of phytoalexin accumulation, in *Phytoalexins* (Eds J. A. Bailey, and J. W. Mansfield), Blackie, Glasgow, pp. 289–318.

Barug, D., and Kerkenaar, A. (1979). Cross-resistance of UV-induced mutants of *Ustilago maydis* to various fungicides which interfere with ergosterol biosynthesis, *Med. Fac. Landbouwwet. Rijksuniv. Gent*, **41**, 421–427.

Bent, K. J. (1979). Fungicides in perspective: 1979, *Endeavour (New Series)*, **3**, 7–14.

Bompeix, G., Ravise, A., Raynall, G., Fettuche, F., and Durand, M. C. (1980). Stimulation des réactions de défense et de synthèse de phytoalexines chez la vigne et la tomate, sous l'influence du tris-*O*-ethyl phosphonate d'aluminium, in *Abstracts of 18ème Colloque de la Société Française de Phytopathologie, Toulouse, April 1980.*

Briggs, G. S. (1981). Relationships between chemical structure and the behaviour and fate of pesticides, *Proc. 1981 Br. Crop Prot. Conf.*, 701–710.

Buchenauer, H., and Sikora, I. (1981). Wirkung von Aluminiumfosethyl (Aliette) auf *Phytophthora* cactorum, *Meded. Fac. Landbouwwet. Rijksuniv. Gent*, **46**, 889–896.

Burden, R. S., Kemp, M. S., and Woodgate-Jones, P. (1982). Mechanisms of biological control with *Trichoderma viride*, *Rep. Long Ashton Res. Stn. 1980.*

Byrde, R. J. W. (1982). Fungal 'pectinases': from ribosome to plant cell wall, *Trans. Br. Mycol. Soc.*, in press.

Byrde, R. J. W., and Richmond, D. V. (1976). A review of the selectivity of fungicides in agriculture and horticulture, *Pestic. Sci.*, **7**, 372–378.

Cartwright, D., Lankcake, P., and Ride, J. P. (1980). Phytoalexin production in rice and its enhancement by a dichlorocyclopropane fungicide, *Physiol. Plant Pathol.*, **17**, 259–267.

Cooke, L. R. (1981). Resistance to metalaxyl in *Phytophthora infestans* in Northern Ireland, *Proc. 1981 Br. Crop Prot. Conf.*, 641–649.

Corke, A. T. K. (1980). Biological control of tree diseases, *Rep. Long Ashton Res. Stn. 1979*, 190–198.

Crute, I. R. (1978). Studies on new systemic fungicides active against *Bremia lactucae*, in *Abstracts of 3rd International Congress of Plant Pathology, Munich, August 1980.*

Davidse, L. C., Looyen, D., Turkensteen, L. J., and Van der Wall, D. (1981). Occurrence of metalaxyl resistant strains of *Phytophthora infestans* in Dutch potato fields, *Neth. J. Plant Pathol.*, **87** (Suppl. 1), 91–103.

Dekker, J. (1981). Impact of fungicide resistance on disease control, *Proc. 1981 Br. Crop Prot. Conf.*, 857–863.

De Waard, M. A., and Van Nistelrooy, J. G. M. (1980). An energy-dependent efflux mechanism for fenarimol in a wild-type strain and fenarimol-resistant mutants of *Aspergillus nidulans, Pestic. Biochem. Physiol.*, **11**, 255–266.

De Warrd, M. A., and Fuchs, A. (1982). Resistance to ergosterol biosynthesis inhibitors, in *Fungicide Resistance in Crop Protection* (Eds J. Dekker, and S. G. Georgopoulos), PUDOC, Wageningen, pp. 86–98.

Etzler, M. E. (1981). Are lectins involved in plant-fungus interactions?, *Phytopathology*, **71**, 744–746.

Fielding, A. H. (1981). Natural inhibitors of fungal polygalacturonases in infected fruits, *J. Gen. Microbiol.*, **123**, 337–381.

Fletcher, J. T., and Wolfe, M. S. (1981). Insensitivity of *Erysiphe graminis* f. sp. *hordei* to triadimefon, triadimenol and other fungicides, *Proc. 1981 Br. Crop Prot. Conf.*, 633–640.

Georgopoulos, S. G. (1977). Development of fungal resistance to fungicides, in *Antifungal Compounds* (Eds H. D. Sisler and M. R. Siegel), Vol. II, Marcel Dekker, New York, pp. 439–495.

Grossman, F. (1968). Studies on the therapeutic effect of pectolytic enzyme inhibitor, *Neth. J. Plant Pathol.*, **74**, (Suppl. 1), 91–103.

Hargreaves, J. A. (1981). Accumulation of phytoalexins in cotyledons of French bean (*Phaseolus vulgaris* L.) following treatment with Triton (*t*-octylphenol polyethoxyethanol) surfactants, *New Phytol.*, **87**, 733–741.

Hartley, G. S., and Graham-Bryce, I. G. (1980), Physical Principles of Plant Behaviour, Vol. 2, Academic Press, London, pp. 544–657.

Heath, M. C. (1981). Non-host resistance, in *Plant Disease Control: Resistance and Susceptibility* (Eds R. C. Staples and G. H. Toenniessen), Wiley, New York, pp. 201–217.

Hirst, J. M. (1975). The role of plant pathology, *Proc 8th Br. Insect. Fung. Conf.*, **3**, 721–729.

Hislop, E. C., Keon, J. P. R., and Fielding, A. H. (1979). Effects of pectin lyase from *Monilinia fructigena* on viability, ultrastructure and localisation of acid phosphatase of cultured apple cells, *Physiol. Plant Pathol.*, **14**, 371–381.

Kaars Sijpesteijn, A. (1969). Mode of action of phenylthiourea, a therapeutic agent for cucumber scab, *J. Sci. Food Agric.*, **20**, 403–405.

Kaars Sijpesteijn, A. (1970). Biochemical modes of action of agricultural fungicides, *World Rev. Pest Control*, **9**, 85–92.

Kaars Sijpesteijn, A. (1977). Effects on fungal pathogens, in *Systemic Fungicides* (Ed. R. W. Marsh) Longmans, London, pp. 131–159.

Kaars Sijpesteijn, A. (1982). Mechanism of action of fungicides, in *Fungicide Resistance in Crop Protection* (Eds J. Dekker and S. G. Georgopoulos), PUDOC, Wageningen, in press.

Kato, T., Shoami, M., and Kawase, Y. (1980). Comparison of tridemorph with buthiobate in antifungal mode of action, *J. Pestic Sci.*, **5**, 69–79.

Kerkenaar, A., Barug, D., and Kaars Sijpesteijn, A. (1979). On the antifungal mode of action of tridemorph, *Pestic. Biochem. Physiol.*, **12**, 195–204.

Kuĉ, J., Shockley, G., and Kearney, K. (1975). Protection of cucumber against *Colletotrichum lagenarium* by *Colletotrichum lagenarium*, *Physiol, Plant Pathol.*, **7**, 195–199.

Langcake, P. (1981). Alternative chemical agents for controlling plant disease, *Phil. Trans. R. Soc. London, Ser. B*, **295**, 83–101.

Lyr, H. (1977). Mechanism of action of fungicides, in *Plant Disease—An Advanced Treatise. Vol. I. How Disease is Managed*, Academic Press, New York, pp. 239–261.

Maiti, I. B., and Kolattukudy, P. E. (1979). Prevention of fungal infection by specific inhibition of cutinase, *Science*, **205**, 507–508.

Matthews, R. E. F. (1981). *Plant Virology*, 2nd ed., Academic Press, New York, pp. 410–413.

Müller, L., Junge, B., Frommer, M., Schmidt, D., and Truscheit, E. (1980). Acarbose (BAYg5421) and homologous selective α-glucosidase inhibitors from Actinoplanacae, in *Enzyme Inhibitors* (Ed. U. Brodbeck), Verlag, Chemie, Weinheim, pp. 123–162.

Patil, S. S., and Dimond, A. E. (1968). Repressions of polygalacturonase synthesis in *Fusarium oxysporun* f. sp. *lysopersici* by sugars, and its effect on symptom reduction in infected tomato plants, *Phytopathology*, **58**, 676–682.

Rast, A. T. B. (1972). MII-16, an artificial symptomless mutant of tobacco mosaic virus for seedling inoculation of tomato crops, *Neth. J. Plant Pathol.*, **78**, 110–112.

Rathmell, W. G., and Smith, D. A. (1980). Lack of activity of selected isoflavonoid phytoalexins as protectant fungicides, *Pestic. Sci.*, **11**, 568–572.

Reilly, J. J., and Klarman, W. L. (1972). The soybean phytoalexin, hydroxyphaseollin, induced by fungicides, *Phytopathology*, **62**, 1113–1115.

Richmond, D. V. (1977). Permeation and migration of fungicides in fungal cells, in *Antifungal Compounds* (Eds M. R. Siegel and H. D. Sisler), Marcel Dekker, New York, pp. 251–276.

Ross, A. F. (1966). Systemic effects of local lesion formation, in *Viruses of Plants* (Eds A. B. R. Beemster and J. Dijkstra), North Holland, Amsterdam, pp. 127–150.

Staub, T., Dahmen, H., and Schwinn, F. (1978). Effects of Ridomil on the development of *Plasmopara viticola* and *Phytophthora infestans* on their host plants, *Z. Pflanzenkr. Pflanzenschutz*, **87**, 83–91.

Strobel, G. A. (1973). The helminthosporide-binding protein of sugar-cane, *J. Biol. Chem.*, **248**, 1321–1328.

Ward, E. W. B., Lazarovits, G., Stoessel, P., Barrie, S. D. and Unwin, C. H. (1980). Glyceollin production associated with control of Phytophthora-rot of soybeans by the systemic fungicide metalaxyl, *Phytopathology*, **70**, 738–740.

Ward, E. W. B., Unwin, C. H., and Stoessel, A. (1975). Experimental control of late blight of tomatoes with capsidiol, the phytoalexin from peppers, *Phytopathology*, **65**, 168–169.

Woloshuk, C. P., Wolkow, P. M., and Sisler, M. D. (1981). The effect of three fungicides, specific for the control of rice blast disease, on the growth and melanin biosynthesis by *Pyricularia oryzae* Cav., *Pestic. Sci.*, **12**, 86–90.

Woodcock, D. (1977). Structure–activity relationships, in *Systemic Fungicides* (Ed. R. W. Marsh), Longmans, London, pp. 32–84.

Worthing, C. (1981). *The Pesticide Manual*, 6th ed., British Crop Protection Council, London.

Further Reading

Bailey, J. A., and Mansfield, J. W. (1982). *Phytoalexins*, Blackie, Glasgow.

Corbett, J. R. (1979). Resistance and selectivity: biochemical basis and practical problems, *Proc. 1979 Br. Crop Prot. Conf.*, 717–730.

Dekker, J., and Georgopoulos, S. G. (1982). *Fungicide Resistance in Crop Protection*, PUDOC, Wageningen.

Marsh, R. W. (1977). *Systemic Fungicides*, 2nd ed., Longmans, London.

Misato, T., Kakiki, K., and Hori, M. (1979). Chitin as a target for pesticide action: progress and prospects, in *Advances in Pesticide Science (Zurich, 1978)*, Part 3 (Ed. H. Geissbuhler), pp. 458–464.

Shephard, M. C. (1981). Factors which influence the biological performance of pesticides, *Proc. 1981 Br. Crop Prot. Conf.*, 711–721.

Siegel, M. R., and Sisler, H. D. (1977). *Antifungal Compounds. Vol. 2, Interactions in Biological and Ecological Systems*, Marcel Dekker, New York.

Staples, R. C., and Toenniessen, G. H. (1981). *Plant Disease Control: Resistance and Susceptibility*, Wiley, New York.

White, R. F., and Antoniw, J. F. (1981). Current reasearch and future prospects for direct control of virus diseases, *Proc. 1981 Br. Crop Prot. Conf.*, 759–768.

Subject Index

Author Index

469